AF594928

# Special Issue in Honor of Professor James D. McChesney on the Occasion of His 80th Birthday

# Special Issue in Honor of Professor James D. McChesney on the Occasion of His 80th Birthday

Editors

**Muhammad Ilias**
**Dhammika Nanayakkara**

MDPI • Basel • Beijing • Wuhan • Barcelona • Belgrade • Manchester • Tokyo • Cluj • Tianjin

*Editors*
Muhammad Ilias
University of Mississippi
USA
Dhammika Nanayakkara
University of Mississippi
USA

*Editorial Office*
MDPI
St. Alban-Anlage 66
4052 Basel, Switzerland

This is a reprint of articles from the Special Issue published online in the open access journal *Molecules* (ISSN 1420-3049) (available at: https://www.mdpi.com/journal/molecules/special_issues/James_D_McChesney).

For citation purposes, cite each article independently as indicated on the article page online and as indicated below:

LastName, A.A.; LastName, B.B.; LastName, C.C. Article Title. *Journal Name* **Year**, *Volume Number*, Page Range.

**ISBN 978-3-0365-2712-3 (Hbk)**
**ISBN 978-3-0365-2713-0 (PDF)**

# Contents

**About the Editors** . . . . . . . . . . . . . . . . . . . . . . . . . . . . . . . . . . . . . . . . . . **vii**

**Larry A. Walker and N. P. Dhammika Nanayakkara**
Special Issue: "James D. McChesney, Vision, Passion and Leadership in the Development of Plant-Derived Natural Products"
Reprinted from: *Molecules* **2021**, *26*, 7415, doi:10.3390/molecules26247415 . . . . . . . . . . . . . . **1**

**Solomon Tesfaye, Kaleab Asres, Ermias Lulekal, Yonatan Alebachew, Eyael Tewelde, Mallika Kumarihamy and Ilias Muhammad**
Ethiopian Medicinal Plants Traditionally Used for the Treatment of Cancer, Part 2: A Review on Cytotoxic, Antiproliferative, and Antitumor Phytochemicals, and Future Perspective
Reprinted from: *Molecules* **2020**, *25*, 4032, doi:10.3390/molecules25174032 . . . . . . . . . . . . . **7**

**Rita M. Moraes, Antonio Luiz Cerdeira and Miriam V. Lourenço**
Using Micropropagation to Develop Medicinal Plants into Crops
Reprinted from: *Molecules* **2021**, *26*, 1752, doi:10.3390/molecules26061752 . . . . . . . . . . . . . . **35**

**Wilmer H. Perera and James D. McChesney**
Approaches toward the Separation, Modification, Identification and Scale up Purification of Tetracyclic Diterpene Glycosides from *Stevia rebaudiana* (Bertoni) Bertoni
Reprinted from: *Molecules* **2021**, *26*, 1915, doi:10.3390/molecules26071915 . . . . . . . . . . . . . . **49**

**Abbas Ali, Nurhayat Tabanca, Betul Demirci, Vijayasankar Raman, Jane M. Budel, K. Hüsnü Can Baser and Ikhlas A. Khan**
Insecticidal and Biting Deterrent Activities of *Magnolia grandiflora* Essential Oils and Selected Pure Compounds against *Aedes aegypti*
Reprinted from: *Molecules* **2020**, *25*, 1359, doi:10.3390/molecules25061359 . . . . . . . . . . . . . **79**

**Mei Wang, Ping Yu, Amar G. Chittiboyina, Dilu Chen, Jianping Zhao, Bharathi Avula, Yan-Hong Wang and Ikhlas A. Khan**
Characterization, Quantification and Quality Assessment of Avocado (*Persea americana* Mill.) Oils
Reprinted from: *Molecules* **2020**, *25*, 1453, doi:10.3390/molecules25061453 . . . . . . . . . . . . . **93**

**Loanda Aparecida Cabral Rudnik, Paulo Vitor Farago, Jane Manfron Budel, Amanda Lyra, Fernanda Malaquias Barboza, Traudi Klein, Carla Cristine Kanunfre, Jessica Mendes Nadal, Matheus Coelho Bandéca, Vijayasankar Raman, Andressa Novatski, Alessandro Dourado Loguércio and Sandra Maria Warumby Zanin**
Co-Loaded Curcumin and Methotrexate Nanocapsules Enhance Cytotoxicity against Non-Small-Cell Lung Cancer Cells
Reprinted from: *Molecules* **2020**, *25*, 1913, doi:10.3390/molecules25081913 . . . . . . . . . . . . . **111**

**Gari Vidal Ccana-Ccapatinta, Jennyfer Andrea Aldana Mejía, Matheus Hikaru Tanimoto, Milton Groppo, Jean Carlos Andrade Sarmento de Carvalho and Jairo Kenupp Bastos**
*Dalbergia ecastaphyllum* (L.) Taub. and *Symphonia globulifera* L.f.: The Botanical Sources of Isoflavonoids and Benzophenones in Brazilian Red Propolis
Reprinted from: *Molecules* **2020**, *25*, 2060, doi:10.3390/molecules25092060 . . . . . . . . . . . . . . **131**

**Sarayut Radapong, Satyajit D. Sarker and Kenneth J. Ritchie**
Oxyresveratrol Possesses DNA Damaging Activity
Reprinted from: *Molecules* **2020**, *25*, 2577, doi:10.3390/molecules25112577 . . . . . . . . . . . . . **139**

**Ya-Ming Xu, A. Elizabeth Arnold, Jana M. U'Ren, Li-Jiang Xuan, Wen-Qiong Wang and A. A. Leslie Gunatilaka**
Teratopyrones A–C, Dimeric Naphtho-$\gamma$-Pyrones and Other Metabolites from *Teratosphaeria* sp. AK1128, a Fungal Endophyte of *Equisetum arvense*
Reprinted from: *Molecules* **2020**, *25*, 5058, doi:10.3390/molecules25215058 . . . . . . . . . . . . . . **147**

**Narayan D. Chaurasiya, Jacob Midiwo, Pankaj Pandey, Regina N. Bwire, Robert J. Doerksen, Ilias Muhammad and Babu L. Tekwani**
Selective Interactions of *O*-Methylated Flavonoid Natural Products with Human Monoamine Oxidase-A and -B
Reprinted from: *Molecules* **2020**, *25*, 5358, doi:10.3390/molecules25225358 . . . . . . . . . . . . . . **157**

**Chunmei Zhai, Jianping Zhao, Amar G. Chittiboyina, Yonghai Meng, Mei Wang and Ikhlas A. Khan**
Newly Generated Atractylon Derivatives in Processed Rhizomes of *Atractylodes macrocephala* Koidz
Reprinted from: *Molecules* **2020**, *25*, 5904, doi:10.3390/molecules25245904 . . . . . . . . . . . . . . **171**

**Ilias Muhammad, Melissa R. Jacob, Mohamed A. Ibrahim, Vijayasankar Raman, Mallika Kumarihamy, Mei Wang, Taha Al-Adhami, Charlotte Hind, Melanie Clifford, Bethany Martin, Jianping Zhao, J. Mark Sutton and Khondaker Miraz Rahman**
Antimicrobial Constituents from *Machaerium* Pers.: Inhibitory Activities and Synergism of Machaeriols and Machaeridiols against Methicillin-Resistant *Staphylococcus aureus*, Vancomycin-Resistant *Enterococcus faecium*, and Permeabilized Gram-Negative Pathogens
Reprinted from: *Molecules* **2020**, *25*, 6000, doi:10.3390/molecules25246000 . . . . . . . . . . . . . **183**

**Mallika Kumarihamy, Luiz H. Rosa, Natascha Techen, Daneel Ferreira, Edward M. Croom Jr., Stephen O. Duke, Babu L. Tekwani, Shabana Khan and N. P. Dhammika Nanayakkara**
Antimalarials and Phytotoxins from *Botryosphaeria dothidea* Identified from a Seed of Diseased *Torreya taxifolia*
Reprinted from: *Molecules* **2021**, *26*, 59, doi:10.3390/molecules26010059 . . . . . . . . . . . . . . . **199**

**Wilmer H. Perera, Siddanagouda R. Shivanagoudra, Jose L. Pérez, Da Mi Kim, Yuxiang Sun, Guddadarangavvanahally K. Jayaprakasha and Bhimanagouda S. Patil**
Anti-Inflammatory, Antidiabetic Properties and In Silico Modeling of Cucurbitane-Type Triterpene Glycosides from Fruits of an Indian Cultivar of *Momordica charantia* L.
Reprinted from: *Molecules* **2021**, *26*, 1038, doi:10.3390/molecules26041038 . . . . . . . . . . . . . . **211**

**Lauren Girard, Kithsiri Herath, Hernando Escobar, Renate Reimschuessel, Olgica Ceric and Hiranthi Jayasuriya**
Development of UHPLC/Q-TOF Analysis Method to Screen Glycerin for Direct Detection of Process Contaminants 3-Monochloropropane-1,2-diol Esters (3-MCPDEs) and Glycidyl Esters (GEs)
Reprinted from: *Molecules* **2021**, *26*, 2449, doi:10.3390/molecules26092449 . . . . . . . . . . . . . . **229**

**Jingjing Li, Dongge Xu, Lingling Wang, Mengyu Zhang, Guohai Zhang, Erguang Li and Susu He**
Glycyrrhizic Acid Inhibits SARS-CoV-2 Infection by Blocking Spike Protein-Mediated Cell Attachment
Reprinted from: *Molecules* **2021**, *26*, 6090, doi:10.3390/molecules26206090 . . . . . . . . . . . . . . **239**

# About the Editors

**Muhammad Ilias** (research professor, National Center for Natural Products Research, School of Pharmacy, University of Mississippi) obtained a B. Pharm. (Hons.) and M. Pharm. in 1976 and 1978, respectively, from the Department of Pharmacy, University of Dhaka, Bangladesh, and a Ph.D. in phytochemistry in 1985 from the Department of Pharmaceutical Chemistry, University of Strathclyde, Glasgow, Scotland, United Kingdom. He worked at the Department of Pharmacy, University of Dhaka, Bangladesh as a lecturer and then as an assistant professor from 1979 to 1987. After post-doctoral training (1987–1989) through the Department of Pharmacognosy, School of Pharmacy, University of Mississippi, he worked as a researcher from 1989 to 1998 at the Medicinal, Aromatic and Poisonous Plant Research Center, College of Pharmacy, King Saud University, Riyadh, Saudi Arabia. Then, in 1998, he joined as a research assistant professor at the National Center for Natural Products Research (NCNPR), School of Pharmacy, University of Mississippi, and currently serves there as a research professor. Dr. Ilias's primary research area is focused on the isolation and structure elucidation of bioactive compounds from plants with a special emphasis on anticancer, anti-infective (i.e., antimicrobial, antimalarial, and antileishmanial), neuroprotective, and psychoactive agents. He has developed various projects and grants in these areas for the last 23 years, funded by the National Institute of Health (NCCIH and NIGMS), Department of Defense, Medicine for Malaria Venture, and Chromodex Inc., where he acted as a PI or a CO-PI, and also served as an investigator of NCNPR's USDA-ARS Center grant for more than 15 years. In the chemistry and bioactivity of anti-infective natural products, he is collaborating with the University of Nairobi, Kenya, and the Muhimbili University Health and Allied Sciences, Der-Es-Salam, Tanzania, exploring traditional plants from East and South East Africa. He is currently involved in a collaborative research project with the AMR Research Group, Institute of Pharmaceutical Sciences, School of Cancer & Pharmaceutical Sciences, King's College London, to pursue natural product drug discovery for antibiotic-resistant bacteria. He is also pursuing the development of a library for natural by-products for neurodegenerative diseases, notably for Alzheimer's disease. For anticancer drug discovery at the NCNPR, he focuses on screening natural products against a panel of luciferase reporter genes that assess the activity of many cancer-related signaling pathways, which determines potential anticancer lead compounds. He has also collaborated with pediatric brain tumor research at CRUK, the University of Cambridge, on natural product anticancer drug discovery with Professor Richard Gilbertson's group. In addition, Dr. Ilias is also involved in the large-scale isolation of bioactive/marker compounds from plants and developing their validated isolation methodology. He has developed new environmentally friendly chromatographic technology and devices, named "Chromatorotor" (US20140224740A1) and "Spin Chromatography System" (UM # 8390), for the application of centrifugal preparative chromatography using various types of binder-free sorbents, especially reversed-phase (C18) silica gel. He has published his research outcomes extensively in peer-reviewed scientific articles, review papers, and book chapters and currently serves as an editorial board member regarding molecule and phytochemical analysis.

**Dhammika Nanayakkara** (research professor) earned a B.Sc. in chemistry in 1975 and a Ph.D. in natural products chemistry in 1981 from the University of Peradeniya in Sri Lanka. After post-doctoral training at the Department of Pharmacognosy in the School of Pharmacy at the University of Illinois Chicago, he joined the Department of Pharmacognosy in the School of Pharmacy at the University of Mississippi as a post-doctoral fellow in 1986. He then went on to join the Research Institute of Pharmaceutical Sciences (RIPS) at the same university as a research assistant professor. He moved to the National Center for Natural Products Research (NCNPR) after it was established as a part of the RIPS in 1995, where he currently holds a research professor position. At the RIPS and NCNPR, he has been engaged in researching and developing natural-products-based anti-infective drugs and agrochemicals. This involves the isolation and identification of biologically active compounds from natural sources by bioassay-guided fractionation methods, as well as optimization of lead compounds by chemical modifications. His work has resulted in the identification of several new anti-infective compounds from natural sources. He has also been the PI or co-PI in several research projects regarding the development of 8-aminoquinolines, an important class of anti-infective drugs with promising utility in the treatment of malaria, Pneumocystis pneumonia, and leishmaniasis. These projects have been funded by the World Health Organization, the National Institutes of Health, Medicines for Malaria Venture, the Bill & Melinda Gates Foundation, and the Department of Defense. In addition to leading these projects, he also served as the synthetic organic chemist in the selection, design, and synthesis of all the 8-aminoquinoline analogs and labeled compounds required for biological testing, metabolic studies, and analytical work. This work led to the discovery of an 8-aminoquinoline analog, NPC1161B, with potent activity against Pneumocystis pneumonia, leishmaniasis, and all stages of malaria infection. He has also collaborated with scientists in the Natural Products Utilization Research Unit (NPURU) of the USDA-ARS on several projects to develop natural-product-based agrochemicals for pest management. One of these collaborative projects led to the development of a natural-product-based algaecide for controlling noxious bacteria in freshwater aquaculture. He has co-authored more than 125 research papers and two international patents.

Editorial

# Special Issue: "James D. McChesney, Vision, Passion and Leadership in the Development of Plant-Derived Natural Products"

**Larry A. Walker * and N. P. Dhammika Nanayakkara**

National Center for Natural Products Research, The University of Mississippi, University, MS 38677, USA; dhammika@olemiss.edu
* Correspondence: lwalker@olemiss.edu

**Citation:** Walker, L.A.; Nanayakkara, N.P.D. Special Issue: "James D. McChesney, Vision, Passion and Leadership in the Development of Plant-Derived Natural Products". *Molecules* **2021**, *26*, 7415. https://doi.org/10.3390/molecules26247415

Received: 29 November 2021
Accepted: 3 December 2021
Published: 7 December 2021



It is a distinct pleasure for me to offer something in recognition of and tribute to Dr. James Dewey McChesney (Figure 1) in this special issue of Molecules. I have had the pleasure of knowing Jim for more than 40 years, ever since I arrived as a young faculty member at the University of Mississippi School of Pharmacy. We worked closely together until 1996, but even since that time, our collaborative relationship and close friendship have grown, and I have benefitted immensely from his mentoring, his scientific acumen, his vast experience, and his thought leadership in the natural products field.

**Figure 1.** Dr. James D. McChesney.

Jim was born near Hatfield, a small town in Missouri, in 1939 and completed a BSc in chemical technology from Iowa State University in 1961. In 1964, he received an MA in botany from the University of Indiana, and, in the following year, he received a Ph.D. in organic chemistry from the same institution. Soon after, he joined the Department of Botany and later the School of Pharmacy at the University of Kansas and became a Professor of Botany and Medicinal Chemistry. In 1978, he joined the Department of Pharmacognosy in the School of Pharmacy at The University of Mississippi as chair and in 1986 he became the Director of the Research Institute of Pharmaceutical Science. In 1996, he joined NaPro Biotherapeutics in Boulder, Colorado as the Vice President of Development. He became the chief scientific officer of Tapestry Pharmaceuticals and then ChromaDex of Boulder, Colorado in 2003. He moved back to his farm home near Etta, Mississippi in 2009 and founded his own companies (Ironstone Separations, Arbor Therapeutics, and Cloaked Therapeutics) in order to pursue his interest in the commercial development of natural products.

Dr. McChesney has done outstanding work throughout his long career in the development of plant-derived secondary metabolites as drug candidates, agrochemicals and food additives, as evidenced by his extensive publication record. Along the way, he has pioneered a number of methods for chromatographic separation, characterization, and analysis.

But beyond the discovery and analysis of phytochemicals, Jim has pioneered many impressive advances in the synthesis of important natural products, intermediates, analogs, derivatives, and metabolites. His accomplishments here feature a substantial body of work on the semi-synthesis of taxanes and their intermediates, which are the subject of numerous patents late in his career. These efforts led to the discovery and development of the third generation taxane analog TPI-287. This tubulin-binding and microtubule-stabilizing agent can also cross the blood-brain barrier, and has been advanced in a number of clinical trials for the treatment of several types of cancer and neurodegenerative diseases. In addition to his contributions to the preclinical and clinical development of taxanes, for more than 40 years he substantially advanced our understanding of the 8-aminoquinoline antimalarials, of which primaquine is the prototype. The 8-aminoquinolines are a critically important antimalarial drug class because of their unique activity on the dormant liver stages of the parasite. His efforts on 8-aminoquinoline chemistry, synthesis and metabolism led to important new insights; these efforts resulted in the development of NPC1161B, a drug candidate which is not only active against all stages of the life cycle of plasmodium in humans, but also highly effective in the treatment of leishmaniasis and *Pneumocystis* pneumonia. This compound is currently at the early clinical development stage.

But these are by no means the only examples. He developed an array of synthetic approaches, including a number of stereoselective routes, for artemisinins antimalarial drugs, cytokinins, hydrophenanthrenes, and several other natural-product or natural-product-inspired classes.

Jim has also contributed immensely to overcome the inherent challenges in the development of nature-based products as potential drugs, agrochemicals and food additives. His work in this area focused on searching for plant varieties with a high content of active compounds—particularly in plant parts such as leaves, in order to harvest them in a sustainable manner—as well as micropropagating these strains in order to rapidly mass produce plants with uniform chemical profiles. His research on galanthamine and podophyllotoxin exemplify these efforts. He also led efforts to improve the production and processing of glycosides from *Stevia* for commercial markets.

Dr. McChesney was chair of the Department of Pharmacognosy at Ole Miss until 1987, leading a small, but highly collaborative and productive group of natural products researchers. Early in his career, he had developed ties with scientists at several universities in Brazil, a relationship which continued throughout his career, notably including a Fulbright Fellowship in 1985 in Fortaleza. During this time and over the next decade as Director of the Research Institute of Pharmaceutical Sciences (RIPS), he was nurturing an ambitious vision for a national effort in the discovery and development of natural products.

This was one of Jim's signal career accomplishments: the conception and implementation of the National Center for Natural Products Research (NCNPR). Working with USDA scientists, university administrators and legislators, he devised a plan for this center, which was envisioned as a partnership between the state, the USDA, and the industry. In the early 1990s, the project began to take shape in earnest under Jim's guidance. When the doors opened in 1995, it was truly a testament to his scientific leadership and vision, and the recognition of his acumen in government, political, and scientific circles.

Jim is known and respected as a hardnosed scientist. His mind is keenly tuned to absorbing and processing and thinking critically about research problems. The breadth and depth of his understanding always illuminate and challenge other researchers, and may sometimes even intimidate. On the other hand, those who come to know Jim find in him a true friend, a tender-hearted man, a gentleman in his own rugged way, a citizen of global awareness and allegiance with compassion for all humanity.

Jim McChesney has been an eminent scholar and a tireless workman. He still is, even into his ninth decade. This special issue includes several articles that represent a small sampling of his research legacy in the form of contributions from his students, colleagues and co-workers through whom he continues in leading the unraveling of the wonders of the natural world. The issue focuses on the areas of natural-product drug discovery and dietary supplements with a special emphasis on the isolation of biologically active constituents, the determination of their mechanisms of action and structure activity relationships, and the development of analytical methods. Thirteen research papers and three reviews appear in this issue.

In the first article, Li et al. [1] reported that the antiviral natural product, glycyrrhizic acid, prevents SARS-CoV-2 infection by complexing with the S-protein, including one mode of binding at the angiotensin-converting enzyme-receptor site, and thereby blocking viral entry into the host cells.

The development of a UHPLC/Q-TOF analytical method to detect process contaminants 3-monochloro-propane-1,2-diol esters and glycidyl esters in glycerin was reported by Girad et al. [2]. Glycerin is widely used as a food additive and these contaminants have been identified as potential toxins, with the kidneys as their main target.

Perera et al. [3] described the anti-inflammatory and antidiabetic properties of one new and six known curcubitane glycosides from *Momordica charantia* fruits, including in silico docking studies of these compounds with $\alpha$-amylase and $\beta$-glucosidase.

Kumarihamy et al. [4] described the bioassay-guided isolation and identification of secondary metabolites with antimalarial and phytotoxic activity from the endophytic fungus *Botryosphearia dothidea*.

Muhammad et al. [5] reported the identification of several antimicrobial constituents from *Machaerium* Pers. and their inhibitory activities and synergism against methicillin-resistant *Staphylococcus aureus*, vancomycin-resistant *Enterococcus faecium*, and permeabilized Gram-negative pathogens.

Zai et al. [6] described the isolation and identification of thermally induced products formed in the traditional way of thermally processing of rhizomes of the Chinese medicinal plant *Atractylodes macrocephala*. These compounds are reported not to be cytotoxic towards mammalian cancer and noncancer cell lines and are devoid of any activity against pathogenic micro-organisms.

Chaurasiya et al. [7] reported the inhibition and kinetics of human monoamine oxidase A- and B-six by O-methylated flavonoids that were isolated from five different plants. The analysis of enzyme-inhibition kinetics was used to determine the mechanism of inhibition and molecular docking studies were used to suggest the possible binding site.

Xu et al. [8] described the bioassay-guided isolation and identification of dimeric naphtho-pyrones and other metabolites from the endophytic fungus *Teratosphaeria* sp. AK1128 and their cytotoxic activity.

Radapong et al. [9] demonstrated that oxyresveratrol, the major constituent in the Thai medicinal plant *Artocapus lakoocha*, possesses pro-oxidant activity facilitated by the presence of copper, as well as significant DNA-damaging activity.

Ccana-Ccapatinta et al. [10] reported the identification of *Dalbergia ecastaphyllum* and *Symphonia globulifera* as the botanical sources of polyprenylated benzophenones in Brazilian red propolis by carrying out phytochemical and chromatographic analyses of plants that were collected by field surveys.

Rudnik et al. [11] reported that the cytotoxic activity of curcumin against non-small-cell lung cancer cells could be enhanced by formulating it as polymeric nanoparticles with or without methotrexate.

Wang et al. [12] reported the characterization and quantification of oils from different parts of avocado fruits and used these results to assess the quality of avocado oils on the market as well as to suggest approaches to identify adulterated products.

Ali et al. [13] reported the insecticidal and biting deterrent activities of *Magnolia grandiflora* essential oils and selected pure compounds against the mosquito *Aedes aegypti*.

Three review papers also appeared in this issue. Perera and McChesney [14] reviewed the approaches to the separation, modification, identification, and scale-up purification of tetra-cyclic diterpene glycosides from *Stevia rebaudiana*. These glycosides are used as sweetening agents in the USA, Europe, and several other countries.

The difficulty in procuring plant material in sufficient quantities is a major impediment for the development of natural-product-based pharmaceuticals and herbal medicines. Micropropagation techniques can be used to overcome this problem. Moraes et al. [15] reviewed the application of micropropagation techniques to mass produce medicinal plants and other important crops. They cited examples of how micropropagation techniques have successfully been used to generate phenotype lines that elicit the desired biological activity and to mass produce high-yielding chemotypes in order to isolate biologically active compounds on a commercial scale.

Tesfaye et al. [16] reviewed the anticancer compounds isolated from Ethiopian medicinal plants that are traditionally used in the treatment of cancer. This review included 119 papers appeared in the scientific literature between 1968 and 2020, and the structures of anticancer compounds that were isolated from 27 medicinal plants belonging to 18 different families.

**Author Contributions:** L.A.W. conceived and wrote initial draft and edited final version. N.P.D.N. modified initial draft, organized and provided narrative on cited publications. All authors have read and agreed to the published version of the manuscript.

**Funding:** This research received no external funding.

**Institutional Review Board Statement:** Not applicable.

**Informed Consent Statement:** Not applicable.

**Data Availability Statement:** Not applicable.

**Conflicts of Interest:** The authors declare no conflict of interest.

## References

1. Li, J.; Xu, D.; Wang, L.; Zhang, M.; Zhang, G.; Li, E.; He, S. Glycyrrhizic Acid Inhibits SARS-CoV-2 Infection by Blocking Spike Protein-Mediated Cell Attachment. *Molecules* **2021**, *26*, 6090. [CrossRef] [PubMed]
2. Girard, L.; Herath, K.; Escobar, H.; Reimschuessel, R.; Ceric, O.; Jayasuriya, H. Development of UHPLC/Q-TOF Analysis Method to Screen Glycerin for Direct Detection of Process Contaminants 3-Monochloropropane-1,2-diol Esters (3-MCPDEs) and Glycidyl Esters (GEs). *Molecules* **2021**, *26*, 2449. [CrossRef] [PubMed]
3. Perera, W.H.; Shivanagoudra, S.R.; Pérez, J.L.; Kim, D.M.; Sun, Y.; Jayaprakasha, G.K.; Patil, B.S. Anti-Inflammatory, Antidiabetic Properties and In Silico Modeling of Cucurbitane-Type Triterpene Glycosides from Fruits of an Indian Cultivar of *Momordica charantia* L. *Molecules* **2021**, *26*, 1038. [CrossRef] [PubMed]
4. Kumarihamy, M.; Rosa, L.H.; Techen, N.; Ferreira, D.; Croom, E.M., Jr.; Duke, S.O.; Tekwani, B.L.; Khan, S.; Nanayakkara, N.P.D. Antimalarials and Phytotoxins from *Botryosphaeria dothidea* Identified from a Seed of Diseased *Torreya taxifolia*. *Molecules* **2021**, *26*, 59. [CrossRef] [PubMed]
5. Muhammad, I.; Jacob, M.R.; Ibrahim, M.A.; Raman, V.; Kumarihamy, M.; Wang, M.; Al-Adhami, T.; Hind, C.; Clifford, M.; Martin, B.; et al. Antimicrobial Constituents from *Machaerium* Pers.: Inhibitory Activities and Synergism of Machaeriols and Machaeridiols against Methicillin-Resistant *Staphylococcus aureus*, Vancomycin-Resistant *Enterococcus faecium*, and Permeabilized Gram-Negative Pathogens. *Molecules* **2020**, *25*, 6000. [CrossRef] [PubMed]
6. Zhai, C.; Zhao, J.; Chittiboyina, A.G.; Meng, Y.; Wang, M.; Khan, I.A. Newly Generated Atractylon Derivatives in Processed Rhizomes of *Atractylodes macrocephala* Koidz. *Molecules* **2020**, *25*, 5904. [CrossRef] [PubMed]
7. Chaurasiya, N.D.; Midiwo, J.; Pandey, P.; Bwire, R.N.; Doerksen, R.J.; Muhammad, I.; Tekwani, B.L. Selective Interactions of *O*-Methylated Flavonoid Natural Products with Human Monoamine Oxidase-A and -B. *Molecules* **2020**, *25*, 5358. [CrossRef] [PubMed]

8. Xu, Y.-M.; Arnold, A.E.; U′Ren, J.M.; Xuan, L.-J.; Wang, W.-Q.; Gunatilaka, A.A.L. Teratopyrones A–C, Dimeric Naphtho-γ-Pyrones and Other Metabolites from *Teratosphaeria* sp. AK1128, a Fungal Endophyte of *Equisetum arvense*. *Molecules* **2020**, *25*, 5058. [CrossRef] [PubMed]
9. Radapong, S.; Sarker, S.D.; Ritchie, K.J. Oxyresveratrol Possesses DNA Damaging Activity. *Molecules* **2020**, *25*, 2577. [CrossRef] [PubMed]
10. Ccana-Ccapatinta, G.V.; Mejía, J.A.A.; Tanimoto, M.H.; Groppo, M.; Carvalho, J.C.A.S.d.; Bastos, J.K. *Dalbergia ecastaphyllum* (L.) Taub. and *Symphonia globulifera* L.f.: The Botanical Sources of Isoflavonoids and Benzophenones in Brazilian Red Propolis. *Molecules* **2020**, *25*, 2060. [CrossRef] [PubMed]
11. Rudnik, L.A.C.; Farago, P.V.; Manfron Budel, J.; Lyra, A.; Barboza, F.M.; Klein, T.; Kanunfre, C.C.; Nadal, J.M.; Bandéca, M.C.; Raman, V.; et al. Co-Loaded Curcumin and Methotrexate Nanocapsules Enhance Cytotoxicity against Non-Small-Cell Lung Cancer Cells. *Molecules* **2020**, *25*, 1913. [CrossRef] [PubMed]
12. Wang, M.; Yu, P.; Chittiboyina, A.G.; Chen, D.; Zhao, J.; Avula, B.; Wang, Y.-H.; Khan, I.A. Characterization, Quantification and Quality Assessment of Avocado (*Persea americana* Mill.) Oils. *Molecules* **2020**, *25*, 1453. [CrossRef] [PubMed]
13. Ali, A.; Tabanca, N.; Demirci, B.; Raman, V.; Budel, J.M.; Baser, K.H.C.; Khan, I.A. Insecticidal and Biting Deterrent Activities of *Magnolia grandiflora* Essential Oils and Selected Pure Compounds against *Aedes aegypti*. *Molecules* **2020**, *25*, 1359. [CrossRef] [PubMed]
14. Perera, W.H.; McChesney, J.D. Approaches toward the Separation, Modification, Identification and Scale up Purification of Tetracyclic Diterpene Glycosides from *Stevia rebaudiana* (Bertoni) Bertoni. *Molecules* **2021**, *26*, 1915. [CrossRef] [PubMed]
15. Moraes, R.M.; Cerdeira, A.L.; Lourenço, M.V. Using Micropropagation to Develop Medicinal Plants into Crops. *Molecules* **2021**, *26*, 1752. [CrossRef] [PubMed]
16. Tesfaye, S.; Asres, K.; Lulekal, E.; Alebachew, Y.; Tewelde, E.; Kumarihamy, M.; Muhammad, I. Ethiopian Medicinal Plants Traditionally Used for the Treatment of Cancer, Part 2: A Review on Cytotoxic, Antiproliferative, and Antitumor Phytochemicals, and Future Perspective. *Molecules* **2020**, *25*, 4032. [CrossRef] [PubMed]

Review

# Ethiopian Medicinal Plants Traditionally Used for the Treatment of Cancer, Part 2: A Review on Cytotoxic, Antiproliferative, and Antitumor Phytochemicals, and Future Perspective

**Solomon Tesfaye [1,*], Kaleab Asres [1], Ermias Lulekal [2], Yonatan Alebachew [1], Eyael Tewelde [1], Mallika Kumarihamy [3] and Ilias Muhammad [3,*]**

1 School of Pharmacy, College of Health Sciences, Addis Ababa University, Churchill Street, 1176 Addis Ababa, Ethiopia; kaleab.asres@aau.edu.et (K.A.); yonalebachew@gmail.com (Y.A.); eyaeltd@gmail.com (E.T.)

2 Department of Plant Biology and Biodiversity Management, College of Natural and Computational Sciences, The National Herbarium, Addis Ababa University, 34731 Addis Ababa, Ethiopia; ermias.lulekalm@aau.edu.et

3 National Center for Natural Products Research, Research Institute of Pharmaceutical Sciences, School of Pharmacy, University of Mississippi, University, MS 38677, USA; mkumarih@olemiss.edu

* Correspondence: soltesfa2010@gmail.com (S.T.); milias@olemiss.edu (I.M.); Tel.: +251-930-518-816 (S.T.); +1-662-915-1051 (I.M.)

Academic Editors: Clementina Manera and Derek J. McPhee
Received: 5 August 2020; Accepted: 2 September 2020; Published: 3 September 2020

**Abstract:** This review provides an overview on the active phytochemical constituents of medicinal plants that are traditionally used to manage cancer in Ethiopia. A total of 119 articles published between 1968 and 2020 have been reviewed, using scientific search engines such as ScienceDirect, PubMed, and Google Scholar. Twenty-seven medicinal plant species that belong to eighteen families are documented along with their botanical sources, potential active constituents, and in vitro and in vivo activities against various cancer cells. The review is compiled and discusses the potential anticancer, antiproliferative, and cytotoxic agents based on the types of secondary metabolites, such as terpenoids, phenolic compounds, alkaloids, steroids, and lignans. Among the anticancer secondary metabolites reported in this review, only few have been isolated from plants that are originated and collected in Ethiopia, and the majority of compounds are reported from plants belonging to different areas of the world. Thus, based on the available bioactivity reports, extensive and more elaborate ethnopharmacology-based bioassay-guided studies have to be conducted on selected traditionally claimed Ethiopian anticancer plants, which inherited from a unique and diverse landscape, with the aim of opening a way forward to conduct anticancer drug discovery program.

**Keywords:** medicinal plants; cancer; Ethiopia; phytochemistry

## 1. Introduction

Cancer is a major global health challenge that affects millions of people annually across the world. Recent estimates showed about 18.1 million new cases of cancer and 9.6 million cancer-related deaths worldwide [1]. Moreover, due to population growth, aging, and increased prevalence of key risk factors, this figure is expected to rise in the coming years. According to the same report, different from other parts of the world, cancer death (7.3%) is higher than cancer incidence (5.2%) in Africa. This is mainly attributed to lack of adequate health care facilities as well as professionals, lack of early cancer detection system, and poor access to chemotherapeutic treatments. Due to these and other factors, including socio-economic conditions, the majority of the population of Africa has relied on traditionally used medicinal herbs and/or plants as a monotherapy or in combination with clinically approved anticancer drugs.

Medicinal plants have been a rich source of clinically effective anticancer agents for the past few decades. Over 60% of the currently used anticancer drugs are either directly derived from plants or inspired by their novel phytochemicals [2] and/or unique ligands as secondary metabolites. In spite of such success, the importance of medicinal plants as a source of leads for anticancer drug discovery was marginalized in comparison with other advanced approaches. This could be due to issues associated with intellectual property rights and securing not enough amounts of plant material which results in the slowness of working with natural products [3]. However, despite these drawbacks, medicinal plant-based drug discovery and development has made a comeback to find potent and affordable natural products with a new mechanism of action and better toxicological profile due to structural diversity of natural product small molecules (NPSM). For instance, among small molecules approved for cancer treatment between 1940 and 2014, 49% are derived and/or originated from natural products [4].

Ethiopia inherited a unique array of fascinating flora from its diverse landscape. Due to the geographical location and diversity, which favors the existence of different habitat and vegetation zones, Ethiopia is home to a variety of plant species. The Ethiopian flora is estimated to contain 6027 species of higher plants of which more than 10% are estimated to be endemic [5]. Different authors have compiled ethnobotanical and ethnopharmacological profiles and reviews of Ethiopian traditionally used medicinal plants [6,7]. However, published reports regarding isolated bioactive compounds of traditionally used Ethiopean medicinal plants, especially those with cytotoxic properties are scant. However, investigations conducted on plants with cytotoxic properties out side Ethiopia, include the study on *Catha edulis* Forsk [8,9], *Artemisia annua* L., *Rumex abyssinicus* Jacq. [9]., *Carissa spinarum* L., *Dodonaea angustifolia* L.f., *Jasminum abyssinicum* Hochst. ex DC., *Rumex nepalensis* Spreng., *Rubus steudneri* Schweinf. and *Verbascum sinaiticum* Benth. [10], *Viola abyssinica* Steud. ex Oliv. [11], *Xanthium strumarium* L. [12], *Senna singueana* (Del). Lock [13], *Glinus lotoides* L. [14], *Kniphofia foliosa* Hochst [15], *Sideroxylon oxyacanthum* Baill., *Clematis simensis* Fresen, and *Dovyalis abyssinica* (A. Rich) Warburg [16]. Thus, for further evaluation, identification, or modification of anticancer leads, thorough review of the chemistry and pharmacology of medicinal plants from relatively uncovered traditional medical systems is crucial. Therefore, in continuation of our previous mini-review [17], in which we documented both ethnobotanical and ethnopharmacological evidence of Ethiopian anticancer plants involving mostly the cytotoxic and antioxidant activities of crude extracts, here, in this review, we comprehensively document the cytotoxic and antiproliferative constituents from anticancer plants those traditionally used in Ethiopia. The secondary metabolites reported from each medicinal plant species are categorized based on the class of natural products they belong to.

## 2. Traditional Uses of Selected Plants

A total of 27 anticancer traditional medicinal plants that belong to 18 botanical families and 27 genera are identified in this review. The botanical families Euphorbiaceae and Cucurbitaceae were the most dominant, represented with 15% and 11% of the selected plant species, respectively (Figure 1). All of the reviewed plants have direct traditional uses for treating either ailments with cancer-like symptoms (determined by traditional practitioner) or for laboratory-confirmed cancer cases. Besides treating cancer, the plants selected in this review are also cited for their various traditional uses, including for the treatment of eczema, leprosy, rheumatism, gout, ringworm, diabetes, respiratory complaints, warts, hemorrhoid, syphilis, and skin diseases (Table 1). The output calls for the need for further phytochemical and pharmacological investigation giving priority to those plants which have been cited most for their use to treat cancer.

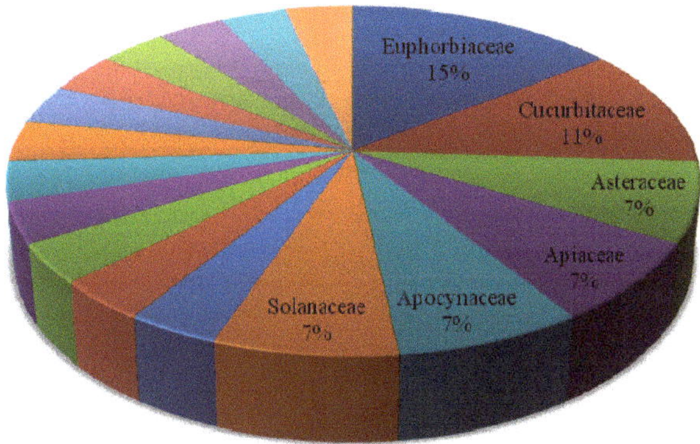


**Figure 1.** Major plant families (in %) of reviewed plants species vegetation zone of Ethiopia [18] (the unmarked blocks are other species).

**Table 1.** General traditional use of selected Ethiopian medicinal plants.

| Botanical Name (Family) | Illnesses/Symptoms Claimed to Be Treated Traditionally |
|---|---|
| *Bersama abyssinica* Fresen. (Melianthaceae) | Antispasmodic [19]; tumor [20] |
| *Carissa spinarum* (Apocynaceae) | Skin cancer [21] |
| *Catharanthus roseus* (L.) G. Don (Apocynaceae) | Cancer, liver infection, Wound, rheumatism [22] |
| *Centella asiatica* (L.) Urb. (Apiaceae) | Genital infection [23]; gastritis, evil eye, swelling [24]; Throat cancer [21] |
| *Croton macrostachyus* Hochst. Ex Delile (Euphorbiaceae) | Stomach ache, typhoid, worm expulsion, wounds, malaria [25]; wounds, malaria and gonorrhea [26]; tumor [27]; skin cancer, wound, ring worm [28]; cancer [29] |
| *Cucumis prophetarum* (Cucurbitaceae) | Skin cancer, cough, stomach-ache, diarrhoea [30]; wound, swollen body part [7] |
| *Ekebergia capensis* Sparrm. (Meliaceae) | Weight loss in children, stabbing pain, bovine tuberculosis [29]; cancer [6] |
| *Euphorbia tirucalli* L. (Euphorbiaceae) | Tumors [27]; wart, wounds [31] |
| *Ferula communis* L. (Apiaceae) | Gonorrhea [32]; Lung cancer [33] |
| *Gloriosa superba* (Colchicaceae) | Snake bite, impotence, stomach-ache [34]; tumors [35] |
| *Jatropha curcas* L. (Euphorbiaceae) | Abdominal pain [36]; rabies [25]; tumor [27,37] |
| *Juncus effusus* L. (Juncaceae) | Wound, stomach ache, bleeding after delivery, muscle cramps, tumors [27] |
| *Kniphofia foliosa* Hochst (Asphodelaceae) | Cervical cancer [21] |
| *Lagenaria siceraria* (Molina) Standl. (Cucurbitaceae) | Diarrhea, vomiting [38]; gonorrhea [39]; wound [25]; cough, cancer [28] |
| *Linum usitatissimum* (Linaceae) | Gastritis [40,41] |
| *Maytenus senegalensis* (Celastraceae) | Stomach-ache [42]; snake bite, tonsillitis, diarrhoea [43]; tumors [20] |
| *Olea europaea* subsp. Cuspidate (Wall. ex. G. Don) Cif. (Oleaceae) | Stomach problems, malaria, dysentery [44]; Eye disease [45]; wound [46]; brain tumor [47] |
| *Plumbago zeylanica* L. (Plumbaginaceae) | Cancer [26]; external body swelling, internal cancer, bone cancer [7]; cancer, cough, snake bite, swelling [31] |
| *Podocarpus falcatus* (Podocarpaceae) | Cancer [34]; amoeba, gastritis [6]; rabies [48] |
| *Premna schimperi* Engl. (Verbenaceae) | Antiseptic [49]; cancer [35] |
| *Prunus africana* (Hook.f.) Kalkman (Rosaceae) | Breast cancer [21]; benign prostatic hyperplasia, prostate gland hypertrophy [26] |
| *Ricinus communis* L. (Euphorbiaceae) | Rabies [48]; dysentery [50]; stomach ache [34,51]; Liver disease [52]; tooth ache [31]; breast cancer [28] |
| *Solanum nigrum* (Solanaceae) | Painful and expanding swelling on finger [7]; cancer [27] |
| *Vernonia amygdalina* Delile (Asteraceae) | Tonsillitis [34]; cancer [6] |
| *Vernonia hymenolepis* A. Rich. (Asteraceae) | Tumor [6,40,41] |
| *Withania somnifera* (Solanaceae) | Snake bite [53]; chest pain [54]; cancer [27] |
| *Zehneria scabra* (L.F. Sond) (Cucurbitaceae) | Fever, head ache [55]; tumor [56]; eye disease, wart [45] |

## 3. Phytochemistry of Ethiopian Anticancer Plants

The present review reports secondary metabolites isolated from 27 plants that are traditionally used to treat different types of cancer in Ethiopia. Phytochemical investigations of traditionally used Ethiopian anticancer plants have led to the isolation of compounds that belong to different classes of natural products [10,57]. In this review, we have not included plants those displayed compounds with very low cytotoxic/antiproliferative activity (i.e., $IC_{50}$ (Concentration that inhibited cell proliferation by 50%)/$ED_{50}$ (Effective dose for 50% of the population) > 50 µg/mL or > 100 µM, in most cases, except few where compounds tested against a panel of cell lines) or plants from which no anticancer compounds were isolated/reported. This review compiled and discussed the potential anticancer/antiproliferative agents based on the types of secondary metabolites, such as terpenoids, phenolic compounds, alkaloids, steroids, and lignans.

### 3.1. *Terpenoids*

Terpenoids are classified according to the number of their isoprene unit as hemi-, mono-, di-, tri-, tetra-, and polyterpenes [58]. Various studies reported that the anticancer activity of terpenoids is due to the inhibition of inflammation, cancer cell proliferation, angiogenesis and metastasis, and induction of programmed cell death [59]. Triterpenoids are one important class of terpenoids, which contain isopentenyl pyrophosphate oligomers [60]. They are biosynthesized by plants through cyclization of 30-carbon intermediate squalene and include various structural subclasses [61]. Several triterpenoids have been shown to have anticancer activity.

Among the different types of triterpenoids, pentacyclic triterpnoids display the most potent anti-inflammatory and anticancer activity [62]. Addo et al. [63] reported the isolation of two new nagilactones along with seven known from the root of *Podocarpus falcatus* (Thunb.) collected from Berga forest, Addis Alem, central Ethiopia. *P. falcatus*is traditionally used to treat jaundice, gastritis, and amoeba [6]. Among the isolated compounds 16-hydroxynagilactone F (**1**), 2*β*,16-dihydroxynagilactone F (**2**), 7*β*-hydroxymacrophyllic acid, nagilactone D (**3**), 15-hydroxynagilactone (**4**), and nagilactone I (**5**) (Figure 2) showed potent antiproliferative activity against HT-29 cell line ($IC_{50}$ < 10 µM) (Table 2). *Premna schimperi*, another traditionally used Ethiopian plant, also showed cytotoxic activity against L929, RAW264.7, and SK.N.SH with $IC_{50}$ values of 11 ± 2.3, 10 ± 2.3, and 1.5 ± 0.3 µg/mL, respectively [57]. The methanolic extract of another commonly used Ethiopian plant, *Croton macrostachyus*, was also shown to possess cytotoxic activity against HTC116 cell line [64]. A diterpenoid compound methyl 2-(furan-3-yl)-6*α*,10*β*-dimethy-l4-oxo-2,4,4*α*,5,6,6*α*,10*α*,10*β*-octahydro-1*H*-benzo[f]isochromene-7-carboxylate) (**6**), demonstrated a moderate cytotoxic activity ($IC_{50}$ = 50 µg/mL). The compound was shown to trigger caspase mediated apoptotic cell death. 3*β*-Hydroxylup-20(29)-ene-27,28-dioic acid dimethyl ester (**7**), isolated from root of *Plumbago zeylanica* collected from India, also exhibited anti-proliferative and anti-migration activity against triple-negative breast cancer cell lines at $IC_{50}$ value of 5 µg/mL [65].

Several terpenoids have been isolated from Ethiopian plants that have claims of having anticancer activity, although these plants may have been collected from other sources. For example, sonhafouonic acid (**8**) from *Zehneria scabra*, collected from Cameroon, demonstrated potent cytotoxicity against brine shrimp assay [66], while Lin et al. [67] showed the antiproliferative activity of euphol (**9**), isolated from *Euphorbia tirucalli* from Taiwan against human gastric cancer cells. Euphol selectively promotes apoptosis by mitochondrial-dependent caspase-3 activation and growth arrest through induction of p27kip1 and inhibition of cyclin B1 in human gastric CS12 cancer cells. It also showed a selective and strong cytotoxicity against other groups of human cancer cell lines such as glioblastoma (the most frequent and aggressive type of brain tumor) [67,68]. The molecular mechanism of action of another anticancer triterpenoid, maslinic acid (**10**), isolated from the leaves of *Olea europaea* has been studied, which induced apoptosis in HT29 human colon cancer cells by directly inhibiting the expression of Bcl-2, increasing that of Bax, releasing cytochrome-C from the mitochondria and activating caspase-9 and then caspase-3 [69]. Similarly, the leaf extract of *Ricinus communis* collected from Malta was also reported for its cytotoxicity against several human tumor cells and induction of apoptosis against human breast tumors, SK-MEL-28. The monoterpenoids 1,8-cineole, camphor and α-pinene, and the sesquiterpenoid β-caryophyllene, isolated from *R. communis*, also showed cytotoxicity against similar cell lines in a dose-dependent manner [70].

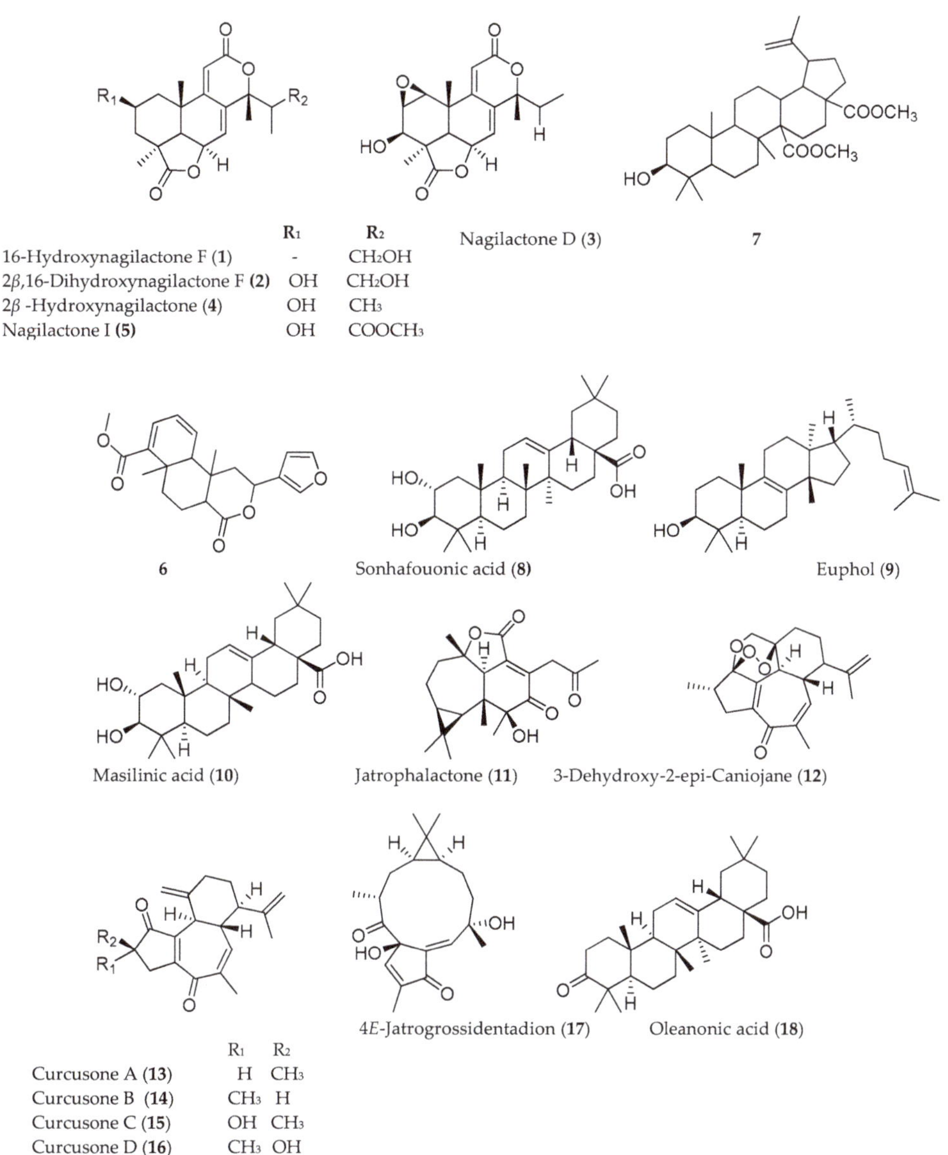


**Figure 2.** *Cont.*

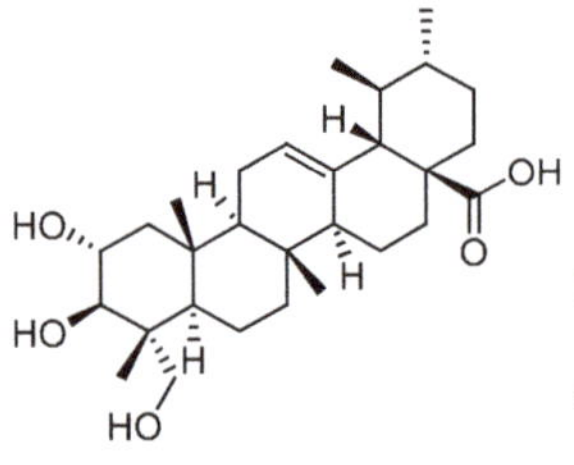

Asiatic acid (**19**)

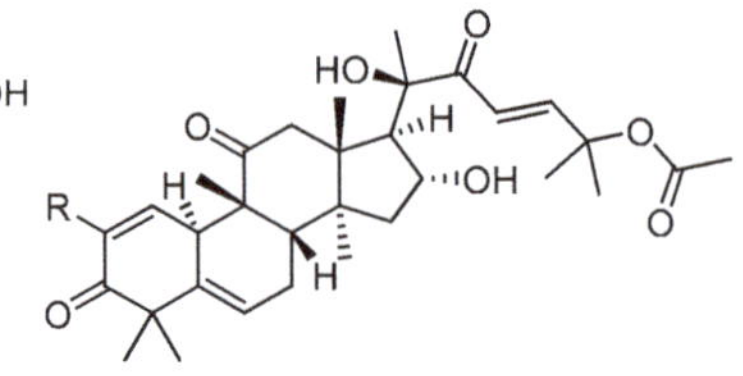

| | R |
|---|---|
| Cucurbitacin E (**20**) | OH |
| Cucurbitacin E glucoside (**23**) | Glucose |

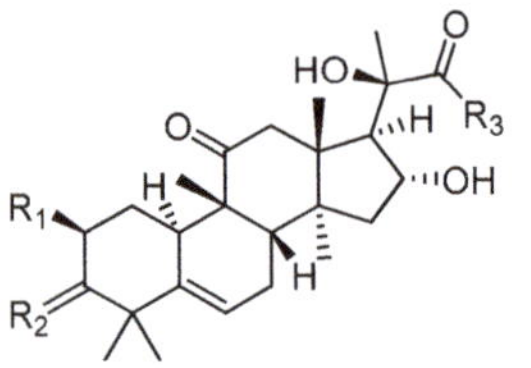

| | $R_1$ | $R_2$ | $R_3$ |
|---|---|---|---|
| Cucurbitacin B (**21**) | OH | O | $CHCH(CH_3)_2OCOCH_3$ |
| Cucurbitacin D (**22**) | OH | O | $CH_3$ |

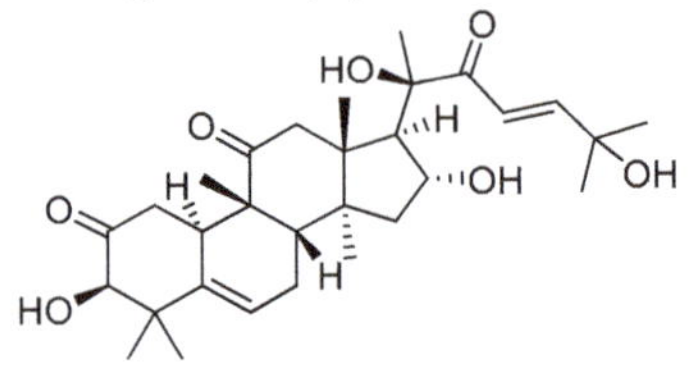

Iso Cucurbitacin D (**24**)

**Figure 2.** Structures of anticancer terpenoids reported from plants available in Ethiopia.

*Jatropha curcas* is a medicinal plant traditionally used to treat a variety of ailments in different parts of the world including Ethiopia [71]. Investigation of *J. curcas*, collected from China, resulted in the isolation of twelve phorbol esters (diterpenoids) including jatrophalactone (**11**), curcusecon A–J, 4-epi-curcusecon E, curcusone E, 3-dehydroxy-2-epi-caniojane (**12**), curcusone A (**13**), curcusone B (**14**), curcusone C (**15**), curcusone D (**16**), jatrogrosidone, 2-epi-jatrogrossidone, and 4*E*-jatrogrossidentadion (**17**) [72]. Most of these compounds showed potent cytotoxicity with $IC_{50}$ values ranging from 0.084 to 20.6 μM against HL-60, SMMC-7721, A-549, MCF-7, SW480, and HEPG2 cell lines [72,73].

The pentacyclic triterpenoid oleanonic acid (**18**), isolated from *Ekebergia capensis* [74], exhibited potent cytotoxic activity against human epithelial type 2 (HEp2) and murine mammary carcinoma (4T1) cell with $IC_{50}$ values of 1.4 and 13.3 μM, respectively. Another pentacyclic triterpenoid, asiatic acid (**19**), isolated from *Centella asiatica*, also showed 80% growth inhibition of human colorectal (SW480), human stomach (SNU668), and murine colorectal adenocarcinoma (CT26) cell lines with $IC_{50}$ values of 20 μg/mL [75]. The fresh fruit of *Cucumis prophetarum* from Saudi Arabia yielded a series of cucurbitacin and analogs (cucurbitacin E (**20**), cucurbitacin B (**21**), cucurbitacin D (**22**), cucurbitacin F 25-*O*-acetate, cucurbitacin E glucoside (**23**), dihydrocucurbitacin D, hexanor-cucurbitacin D, and isocucurbitacin D (**24**)), of which compounds **20–24** showed cytotoxic activity against MCF-7, MDA MB 231, A2780, A2780 CP, HepG2, and HCT-116 with $IC_{50}$ values ranging from 1 to 27.3 μM [76].

**Table 2.** Terpenoids isolated from medicinal plants that are traditionally used to treat cancer in Ethiopia.

| Plant Family | Class of Compounds | Cell Lines | $IC_{50}$ | Pharmacology | Isolated Active Compounds | Reference |
|---|---|---|---|---|---|---|
| *Ferula communis* L. (Apiaceae) | Daucane Sesquiterpene | Jurkat T-cells | | Ionotropism | Ferutinin | [34] |
| *Vernonia amygdalina* Delile (Asteraceae) | Sesquiterpene lactones | KB | - | - | Vernodalin and Vernomygdin | [77] |
| *Vernonia hymenolepis* A. Rich. (Asteraceae) | Sesquiterpene Dilactone | | - | - | Vernolepin | |
| *Zehneria scabra* (L.F. Sond) (Cucurbitaceae) | Triterpenoid | Brine shrimp | 10 μg/mL | | Sonhafouonicacid (**8**) | [66] |
| *Croton macrostachyus* Hochst. ex Delile* (Euphorbiaceae) | Diterpenoid | HCT116 | 50 μg/mL | Caspase mediated apoptosis | methyl 2-(furan-3-yl)-6$\alpha$,10$\beta$-dimethy-14-oxo-2,4,4$\alpha$,5,6,6$\alpha$,10$\alpha$,10$\beta$–octahydro-1H-benzo[f]isochromene-7-carboxylate) | [64] |
| *Euphorbia tirucalli* L. (Euphorbiaceae) | Triterpenoid | CS12 | 12.8 μg/mL | Apoptosis | Euphol (**9**) | [67] |
| | | AGS | 14.7 μg/mL | | | |
| | | MKN45 | 14.4 μg/mL | | | |
| *Ricinus communis* L. (Euphorbiaceae) | Monoterpenoid | SK-MEL-28 | 21.67 ± 4.74 μg/mL | Appoptosis | 1,8-Cineole, camphor, $\alpha$-pinene, β-Caryophyllene | [70] |
| | | K-562 | 24.49 ± 1.61 μg/mL | | | |
| | | COLO 679 | 20.14 ± 2.99 μg/mL | | | |
| | | OAW42 | 13.52 ± 0.20 μg/mL | | | |
| | | HT-29 | 19.86 ± 5.94 μg/mL | | | |
| | | MCF-7 | 37.87 ± 3.36 μg/mL | | | |
| | | PBMC | 13.55 ± 0.85 μg/mL | | | |
| *Jatropha curcas* L. (Euphorbiaceae) | Diterpenoid | HL-60 | 8.5 μM | | Jatrophalactone (**11**) | [72] |
| | | SMMC-7721 | 20.6 μM | | | |
| | | A-549 | 19.7 μM | | | |
| | | MCF-7 | 20.1 μM | | | |
| | | SW480 | 19.2 μM | | | |
| | | HL-60 | >40 μM | | Curcusecon A-J, 4-epi-curcusecon E, Curcusone E | |
| | | SMMC-7721 | >40 μM | | | |
| | | A-549 | >40 μM | | | |
| | | MCF-7 | >40 μM | | | |
| | | SW480 | >40 μM | | | |
| | | HL-60 | 2.86 μM | | 3-Dehydroxy-2-epi-Caniojane (**12**) | |
| | | SMMC-7721 | 3.94 μM | | | |
| | | A-549 | 3.49 μM | | | |
| | | MCF-7 | 11.69 μM | | | |
| | | SW480 | 14.05 μM | | | |

**Table 2.** *Cont.*

| Plant Family | Class of Compounds | Cell Lines | $IC_{50}$ | Pharmacology | Isolated Active Compounds | Reference |
|---|---|---|---|---|---|---|
| *Jatropha curcas* L. (Euphorbiaceae) | Diterpenoid | HL-60 | 1.63 μM | | Curcusone A (**13**) | [72] |
| | | SMMC-7721 | 3.10 μM | | | |
| | | A-549 | 3.35 μM | | | |
| | | MCF-7 | 2.47 μM | | | |
| | | SW480 | 2.10 μM | | | |
| | | HL-60 | 2.64 μM | | Curcusone B (**14**) | |
| | | SMMC-7721 | 3.30 μM | | | |
| | | A-549 | 3.88 μM | | | |
| | | MCF-7 | 3.14 μM | | | |
| | | SW480 | 2.91 μM | | | |
| | | HL-60 | 1.36 μM | | Curcusone C (**15**) | |
| | | SMMC-7721 | 2.17 μM | | | |
| | | A-549 | 3.88 μM | | | |
| | | MCF-7 | 1.61 μM | | | |
| | | SW480 | 1.99 μM | | | |
| | | HL-60 | 2.81 μM | | Curcusone D (**16**) | |
| | | SMMC-7721 | 3.58 μM | | | |
| | | A-549 | 4.70 μM | | | |
| | | MCF-7 | 2.77 μM | | | |
| | | SW480 | 2.83 μM | | | |
| | | HL-60 | 22.80 μM | | Jatrogrosidone | |
| | | SMMC-7721 | 19.49 μM | | | |
| | | A-549 | 34.93 μM | | | |
| | | MCF-7 | 21.83 μM | | | |
| | | SW480 | 20.06 μM | | | |
| | | HL-60 | 23.30 μM | | 2-epi-Jatrogrossidone | |
| | | SMMC-7721 | 18.36 μM | | | |
| | | A-549 | 36.53 μM | | | |
| | | MCF-7 | 22.72 μM | | | |
| | | SW480 | 21.08 μM | | | |

**Table 2.** *Cont.*

| Plant Family | Class of Compounds | Cell Lines | $IC_{50}$ | Pharmacology | Isolated Active Compounds | Reference |
|---|---|---|---|---|---|---|
| *Jatropha curcas* L. (Euphorbiaceae) | Diterpenoid | HEPG2 | 0.084 µM | | Curcusone C (**15**) | [73] |
| | | | 0.153 µM | | Curcusone D (**16**) | |
| | | | 0.183 µM | | 4*E*-Jatrogrossidentadion (**17**) | |
| *Premna schimperi* Engl.* (Verbenaceae) | Clerodane diterpene | L929 | 11 ± 2.3 µg/mL | - | (5*R*,8*R*,9*S*, I OR)-12-Oxo-ent-3,13(16)-clerodjen-15-oic acid | [57] |
| | | RAW264.7 | 10 ± 2.3 µg/mL | | | |
| | | SK.N.SH | 1.5 ± 0.3 µg/mL | | | |
| *Ekebergia capensis* Sparrm. (Meliaceae) | Triterpenoids | HEp2 | 1.4 µM | - | Oleanonic acid (**18**) | [74] |
| | | 4T1 | 13.3 µM | | | |
| *Olea europaea* subsp. Cuspidata (Wall. ex. G. Don) Cif. (Oleaceae) | Triterpenoids | HT-29 | 28.8 ± 0.9 µg/mL | Apoptosis | Maslinic acid (**10**) | [69] |
| *Podocarpus falcatus** (Podocarpaceae) | Terpenoids-Nagilactones (diterpenoids) | HT-29 | 0.6 ± 0.4 µM | | 16-Hydroxynagilactone F (**1**) | [63] |
| | | | 1.1 ± 0.5 µM | | 2*β*,16-Dihydroxynagilactone F (**2**) | |
| | | | 0.3 ± 0.1 µM | | 2*β*-Hydroxynagilactone F | |
| | | | >10 µM | | 7*β*-Hydroxymacrophyllic acid | |
| | | | >10 µM | | Macrophyllic acid | |
| | | | 0.9 ± 0.3 µM | | Nagilactone D (**3**) | |
| | | | 5.1 ± 0.8 µM | | 15-Hydroxynagilactone (**4**) | |
| | | | 0.5 ± 0.1 µM | | Nagilactone I (**5**) | |
| | | | >10 µM | | Inumakiol D | |
| | | | >10 µM | | Ponasterone A | |
| *Cucumis prophetarum* (Cucurbitaceae) | Triterpenoids | MCF-7 | 7.2 µM | | Cucurbitacin E (**20**) | [76] |
| | | MDA MB 231 | 2.1 µM | | | |
| | | A2780 | 5.4 µM | | | |
| | | A2780 CP | 15.9 µM | | | |
| | | HepG2 | 3.4 µM | | | |
| | | HCT-116 | 3.4 µM | | | |

**Table 2.** *Cont.*

| Plant Family | Class of Compounds | Cell Lines | $IC_{50}$ | Pharmacology | Isolated Active Compounds | Reference |
|---|---|---|---|---|---|---|
| *Cucumis prophetarum* (Cucurbitaceae) | Triterpenoids | MCF-7 | 16.0 μM | | Cucurbitacin B (**21**) | [76] |
| | | MDA MB 231 | 0.96 μM | | | |
| | | A2780 | 7.6 μM | | | |
| | | A2780 CP | 14.2 μM | | | |
| | | HepG2 | 1.7 μM | | | |
| | | HCT-116 | 1.7 μM | | | |
| | | MCF-7 | 47.9 μM | | Hexanor-Cucurbitacin D | |
| | | MDA MB 231 | 12.0 μM | | | |
| | | A2780 | >100 μM | | | |
| | | A2780 CP | >100 μM | | | |
| | | HepG2 | 37.8 μM | | | |
| | | HCT-116 | 30.7 μM | | | |
| | | MCF-7 | 26.7 μM | | Cucurbitacin D (**22**) | |
| | | MDA MB 231 | 4.0 μM | | | |
| | | A2780 | 21.6 μM | | | |
| | | A2780 CP | 6.9 μM | | | |
| | | HepG2 | 5.0 μM | | | |
| | | HCT-116 | 7.6 μM | | | |
| | | MCF-7 | 18.4 μM | | Cucurbitacin F 25-O-acetate | |
| | | MDA MB 231 | 3.4 μM | | | |
| | | A2780 | 15.8 μM | | | |
| | | A2780 CP | 15.2 μM | | | |
| | | HepG2 | 10.2 μM | | | |
| | | HCT-116 | 11.2 μM | | | |
| | | MDA MB 231 | >100 μM | | Dihydrocucurbitacin D | |
| | | | 27.3 μM | | Cucurbitacin E glucoside (**23**) | |
| | | | 1 μM | | Isocucurbitacin D (**24**) | |

**Table 2.** *Cont.*

| Plant Family | Class of Compounds | Cell Lines | $IC_{50}$ | Pharmacology | Isolated Active Compounds | Reference |
|---|---|---|---|---|---|---|
| *Centella asiatica* | Triterpenoids | SW480 | 20 µg/mL (80% growth inhibition) | Growth inhibition and apoptosis | Asiatic Acid (**19**) | [75] |
| | | SNU668 | | | | |
| | | CT26 | | | | |
| *Plumbago zeylanica* | Triterpenoids | MDA-MB-231 | 5 µg/mL | Inhibits proliferation and migration | 3$\beta$-Hydroxylup-20(29)-ene-27,28-dioic acid (**7**) | [65] |

Cell lines: HCT116 = Human colorectal carcinoma, CS12 = Human gastric carcinoma, AGS = Human gastric carcinoma, MKN-45 = Human gastric adenocarcinoma, SK-MEL-28 = Human melanoma, K562 = Human myelogenous leukemia, COLO 679 = Human melanoma, OAW42 = Human ovarian carcinoma, HT-29 = Human colorectal adenocarcinoma, MCF-7 = Human breast adenocarcinoma, PBMC = Peripheral blood mononuclear, HL-60 = Human promyelocytic leukemia, SMMC-7721 = Human hepatocarcinoma, A-549 = Human lung adenocarcinoma, SW480 = Human colorectal, HepG2 = Liver hepatocarcinoma, L929 = Murine fibroblast, RAW264.7 = murine macrophage, SK.N.SH = Human neuroblastoma, HEp-2 = Human epithelial type 2, 4T1 = Murine mammary carcinoma, HT-29 = Human colorectal adenocarcinoma, Caco-2 = Human colon carcinoma, MDA MB 231 = Triple-negative breast cancer, A2780 = Human ovarian carcinoma, A2780 CP = cisplatin-resistant ovarian carcinoma, HCT116 = Human colorectal carcinoma. $IC_{50}$ = Concentration that inhibited cell proliferation by 50%. * Plant material collected from Ethiopi.

### 3.2. Phenolic Compounds

Phenolic compounds are biosynthesized by plants through shikimate, phenylpropanoid, and flavonoid pathways, and have an aromatic ring bearing one or more hydroxyl groups. These compounds have been reported for their antioxidant, antiproliferative, and cytotoxic properties [78]. Many phenolic compounds have been identified elsewhere from the same medicinal plants that are traditionally used to manage cancer in Ethiopia. For instance, (−)-epigallocathechin (**25**) isolated from *Maytenus senegalensis* has showed potent cytotoxic activity against mouse lymphoma cell line (L5178Y) [79]. Likewise, a series phenanthrenes (5-(1-methoxyethyl)-1-methyl-phenanthren-2,7-diol (**26**); effususol A; effusol; dehydroeffusol; dehydroeffusal; 2,7-dihydroxy-1,8-dimethyl-5-vinyl-9,10-dihydrophenanthrene and juncusol; dehydrojuncusol and 1-methylpyrene-2,7-diol) from *Juncus effuses* inhibited the proliferation of five human cancer cell lines (Table 3). Among these, 5-(1-methoxyethyl)-1-methyl-phenanthren-2,7-diol (**26**) (Figure 3) was tested against MCF-7 cancer cell line and showed better cytotoxic activity [80] than all isolated compounds from *J. effuses*. Another group of phenanthrenoids (effususol A, **27**) has also demonstrated potent cytotoxicity against HT-22 cell by inducing caspase-3-mediated apoptosis [81]. Plumbagin (**28**), a naphthoquinone isolated from *Plumbago zeylanica* also induced apoptosis in human non-small cell lung ($IC_{50}$ = 6.1–10.3 μM) [82] and human pancreatic ($IC_{50}$ = 2.1 μM) [83] cancer cell lines. On the other hand, knipholone (**29**) isolated from *Kniphofia foliosa* Hochst collected from Ethiopia, induced necrotic death in mouse melanoma (B16), mouse macrophage tumor (RAW 264.7), human acute monocytic (THP-1), and promonocytic leukaemic (U937) cell lines with $IC_{50}$ values that range from 0.5 ± 0.05 to 3.3 ± 0.39 μM [15].

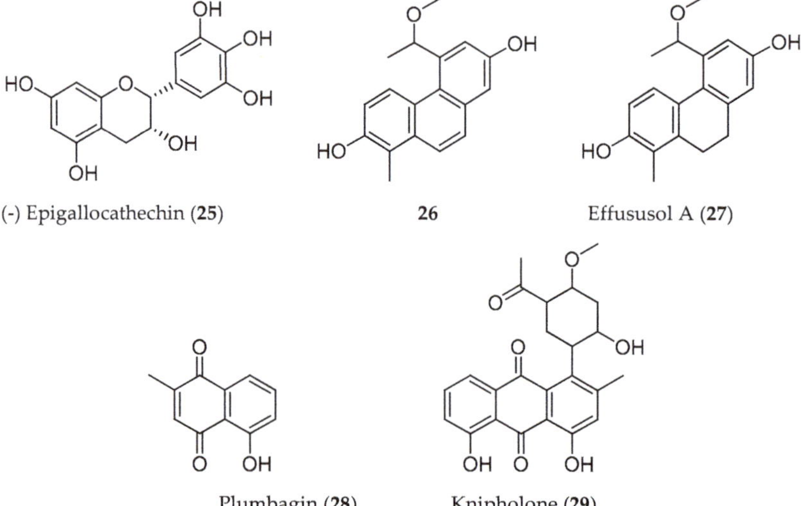


**Figure 3.** Structures of anticancer phenolic compounds reported from plants available in Ethiopia.

**Table 3.** Phenolic compounds isolated from medicinal plants that are traditionally used to treat cancer in Ethiopia.

| Plant | Class of Compounds | Cell Lines | $IC_{50}$ | Pharmacology | Isolated Active Compounds | Reference |
|---|---|---|---|---|---|---|
| *Maytenus senegalensis* (Celastraceae) | Phenolic | L5178Y | 10 µg/mL (100% inhibition) | - | (−) Epigallocathechin **(25)** | [79] |
| *Juncus effusus* L. (Juncaceae) | Phenanthrenes | MCF-7 | 10.87 ± 0.82 µM | - | 5-(1-Methoxyethyl)-1-methyl- phenanthren-2,7-diol **(26)** | [80] |
| | | | 26.68 ± 2.95 µM | | Effususol A **(27)** | |
| | | HepG-2 | 23.90 ± 3.32 µM | | Effusol | |
| | | SHSY-5Y | 22.83 ± 0.98 µM | | Dehydroeffusol | |
| | | HepG-2 | 23.13 ± 1.79 µM | | | |
| | | SMMC-7721 | 25.35 ± 2.08 µM | | Dehydroeffusal | |
| | | HepG-2 | 12.43 ± 0.41 µM | | | |
| | | Hela | 13.07 ± 2.56 µM | | | |
| | | HepG-2 | 26.04 ± 4.49 µM | | 5-Hydroxymethyl-1- methylphenanthrene-2,7-diol | |
| | | Hela | 16.35 ± 6.04 µM | | | |
| | | | 29.63 ± 0.67 µM | | 2,7-Dihydroxy-1,8-dimethyl-5-vinyl-9,10-dihydrophenanthrene and juncusol | |
| | | HepG-2 | 16.45 ± 1.12 µM | | Dehydrojuncusol | |
| | | Hela | 15.17 ± 2.47 µM | | 1-Methylpyrene-2,7-diol | |
| | | MCF-7 | 27.10 ± 1.17 µM | | | |
| | 9,10-Dihydrophenanthrene | HT22 | 100 µM | Caspase-3-mediated cytotoxicity | Effususol A **(27)** | [81] |
| *Plumbago zeylanica* | Naphthoquinones | A549 | 10.3 µM | Apoptosis | Plumbagin **(28)** | [82] |
| | | H292 | 7.3 µM | | | |
| | | H460 | 6.1 µM | | | |
| | | Panc-1 | 2.1 µM | | | [83] |
| *Kniphofia foliosa Hochst** | Phenylanthraquinones | B16 | 3.3 ± 0.39 µM | Necrotic cell death | Knipholone **(29)** | [15] |
| | | RAW 264.7 | 1.6 ± 0.25 µM | | | |
| | | U937 | 0.5 ± 0.05 µM | | | |
| | | THP-1 | 0.9 ± 0.09 µM | | | |

Cell lines: SMMC-7721 = Human hepatocarcinoma, L5178Y = Mouse lymphoma, SHSY-5Y = human neuroblastoma, MCF-7 = Human breast adenocarcinoma, SMMC-7721 = Human hepatocarcinoma, HepG2 = Liver hepatocarcinoma, Hela = Human cervical cancer, HT22 = mouse hippocampal neuronal, B16 = mouse melanoma, RAW 264.7 = mouse macrophage tumor, THP-1 = human acute monocytic leukaemic, U937 = promonocytic leukaemic;, $IC_{50}$ = Concentration that inhibited cell proliferation by 50%. * Plant material collected from Ethiopia.

### 3.3. Alkaloids

Vinblastine (**30**) and vincristine (**31**) (Figure 4) are one of the most effective bis-indole vinca alkaloids as anticancer drugs, isolated from the leaves of *Catharanthus roseus* [84]. This is one of the most precious anticancer plants indigenous to Madagascar. Previously, approximately 30 bis-indole alkaloids and over 60 monomeric indole alkaloids have been isolated from the aerial parts and roots of *C. roseus* [85,86]. Wang et al. [87] isolated three new cytotoxic dimeric indole alkaloids (**32–34**) along with other five known compounds from the whole plant of *C. roseus* collected from China (Table 4). Among the isolated compounds, leurosine (**36**) showed the most potent cytotoxic activity with $IC_{50}$ value of 0.73 ± 0.06 μM. Furthermore, the isolated three new compounds (**32–34**) also showed potent cytotoxicity against triple-negative breast cancer (MDA-MB-231) cell line with $IC_{50}$ values ranging from 0.97 ± 0.07 μM to 7.93 ± 0.42 μM. Another alkaloid, cathachunine (**40**), also showed a promising cytotoxic activity against HL-60 by inducing an intrinsic apoptotic pathway [88]. On the other hand, the monoterpenoid indole alkaloids vindoline and catharanthine, isolated from Malaysian *V. roseus*, showed weak cytotoxic activity against HCT 116 [89]. Furthermore, colchicine (**41**), isolated from the seeds of *Gloriosa superba*, demonstrated moderate activity against six human cancer cell lines (A549, MCF-7, MDA-MB231, PANC-1, HCT116, and SiHa) [90].

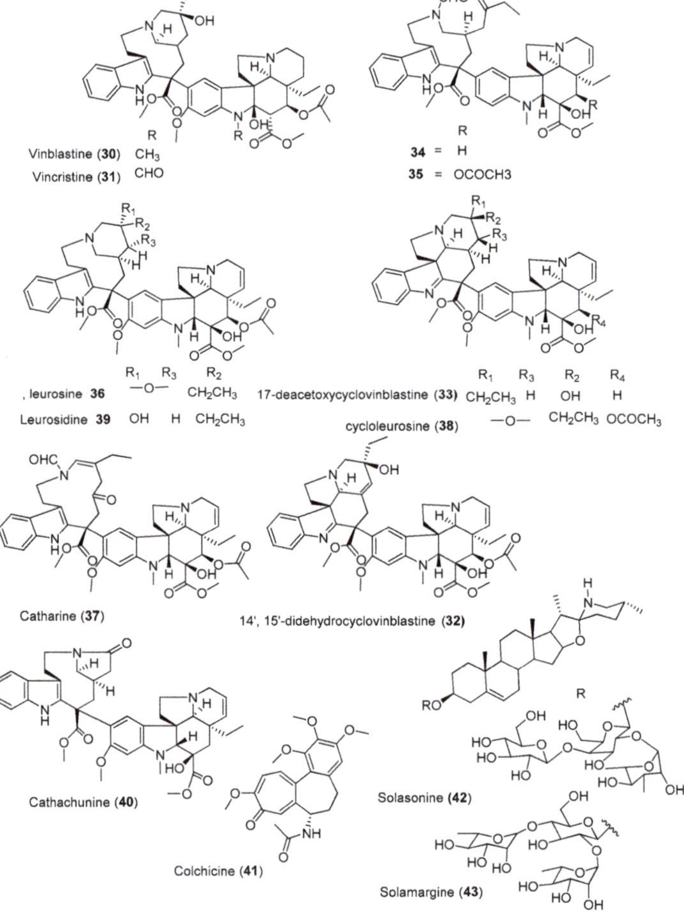


**Figure 4.** Structures of anticancer alkaloids reported from plants present in Ethiopia.

**Table 4.** Alkaloids isolated from medicinal plants that are traditionally used to treat cancer in Ethiopia.

| Plants | Class of Compounds | Cell Lines | $IC_{50}$ Values | Pharmacology | Isolated Active Compounds | Reference |
|---|---|---|---|---|---|---|
| *Catharanthus roseus* (L.) G.Don (Apocynaceae) | Bisindole alkaloid | SH-SY5Y | 0.1 μM | Mitotic arest and apoptosis | Vincristine (**31**) | [91] |
| | | MDA-MB-231 | 0.67 ± 0.03 nM | - | Vinblastine (**30**) | [87] |
| | Indole alkaloids | | 0.97 ± 0.07 μM | | 14′,15′-Didehydrocyclovinblastine (**32**) | |
| | | | 7.93 ± 0.42 μM | | 17-Deacetoxycyclovinblastine (**33**) | |
| | | | 3.55 ± 0.19 μM | | 17–Deacetoxyvinamidine (**34**) | |
| | | | 10.67 ± 0.63 μM | | Vinamidine (**35**) | |
| | | | 0.73 ± 0.06 μM | | Leurosine (**36**) | |
| | | | 8.59 ± 0.51 μM | | Catharine (**37**) | |
| | | | 1.11 ± 0.07 μM | | Cycloleurosine (**38**) | |
| | | | 4.26 ± 0.23 μM | | Leurosidine (**39**) | |
| | | HCT 116 | >200 μg/mL | | Vindoline | [89] |
| | | | 60 μg/mL | | Catharanthine | |
| | Bisindole alkaloid | HL-60 | 9.1 ± 0.7 μM | Induction of apoptosis via an intrinsic pathway | Cathachunine (**40**) | [88] |
| *Gloriosa superba* (Colchicaceae) | Alkaloid | A-549 and MDA-MB-231 | 60 nM | G2/M phase arrest | Colchicine (**41**) | [90] |
| *Solanum nigrum* (Solanaceae) | Steroidal glycoalkaloids | MGC-803 | 5.2 μg/mL | Apoptosis | Solasonine (**42**) | [92] |
| | | | 26.5 μg/mL | | $\beta$1-Solasonine | |
| | | | 8.77 μg/mL | | Solamargine (**43**) | |
| | | | 20.1 μg/mL | | Solanigroside P | |

Cell lines: MDA MB 231 = Triple-negative breast cancer, SW480 = Human colorectal, HCT116 = Human colorectal carcinoma, HL60 = Human promyelocytic leukemia, MCF-7 = Human breast adenocarcinoma, SMMC-7721 = Human hepatocarcinoma, A-549 = Human lung adenocarcinoma, MGC-803 = Human gastric cancer. $IC_{50}$ = Concentration that inhibited cell proliferation by 50%.

### *3.4. Steroids and Lignans*

Steroids and lignans, in addition to other phytochemicals, are common secondary metabolites reported from Ethiopian plants. Evidence and epidemiological studies suggest that phytosterols and lignans are protective against a wide range of diseases and possess anticancer activity [93]. Withanolides are cytotoxic steroidal lactones, reported from various plants of the family Solanaceae [94], of which withaferine-A (**44**) and 5$\beta$,6$\beta$,14$\alpha$,15$\alpha$-diepoxy-4$\beta$,27-dihydroxy-1-oxowitha-2,24-dienolide (**45**) (Figure 5), isolated from *Withania somnifera*, demonstrated anticancer activity against human lung cancer cell line (NCI-H460) with $IC_{50}$ values of 0.45 ± 0.00 and 8.3 ± 0.21 µg/mL, respectively [94]. Several buffadinolides, cardiac glycosides with steroidal nucleus, including berscillogenin, 3-epiberscillogenin, and bersenogenin [95]; hellebrigenin 3-acetate (**48**); and hellebrigenin 3,5-diacetate (**49**) [96] isolated from *Bersama abyssinica* collected from Ethiopia, demonstrated cytotoxic activities. $\beta$-Sitosterol-3-*O*-glucoside, a phytosterol from *Prunus Africana*, exhibited poor anticancer activity against three cell lines (Table 5).

Lignans and isoflavonoids are the major classes of phytoestrogens [97] which showed potential anticancer activity against various cells. Three lignans, namely, (−)-carinol (**50**), (−)-carissanol (**51**), and (−)-nortrachelogenin, isolated from *Carissa spinarum*,were found to be cytotoxic against A549, MCF-7, and WI-38 cell lines. Among these, (−)-carinol (i.e., a compound with butanediol structure) showed more potent cytotoxic activity against these three cell lines with $IC_{50}$ value of 1 µg/mL, as compared to (−)-carissanol and (−)-nortrachelogenin [98]. Secoisolariciresinol (**52**) and matairesino (**53**), two lignans isolated from *Linum usitatissimum*, exhibited cytotoxicity against MCF-7 cells with $IC_{50}$ values of 10 and 1 µM, respectively [99].

**Table 5.** Steroidal and Lignan compounds isolated from medicinal plants that are traditionally used to treat cancer in Ethiopia.

| Plant | Class of Compounds | Cell Lines | $IC_{50}$ | Isolated Active Compounds | Reference |
|---|---|---|---|---|---|
| *Prunus africana* (Hook.f.) Kalkman (Rosaceae) | Steroids | HEK293 | 937 μg/mL | *β*-Sitosterol-3-*O*-glucoside | [93] |
| | | HepG2 | 251 μg/mL | | |
| | | Caco-2 | 54 μg/mL | | |
| *Withania somnifera* (Solanaceae) | Steroidal lactone | NCI-H460 | 0.45 ± 0.00 μg/mL | Withaferin A (**44**) | [94] |
| | | | 8.3 ± 0.12 μg/mL | 5*β*,6*β*,14*α*,15*α*-Diepoxy-4*β*,27-dihydroxy-1-oxowitha-2,24-dienolide (**45**) | |
| | | | 95.6 ± 2.60 μg/mL | 27-Acetoxy-4*β*,6*α*-dihydroxy-5*β*-chloro-1-oxowitha-2,24-dienolide (**46**) | |
| | Withasteroid | MCF-7 and WRL-68 | 1.0 μg/mL | 5,6-De-epoxy-5-en-7-one-17-hydroxy withaferin A (**47**) | [100] |
| | | Caco-2 | 3.4 μg/mL | | |
| | | PC-3 | 7.4 μg/mL | | |
| *Bersama abyssinica* Fresen.* (Melianthaceae) | Steroids (bufadienolide) | KB | 0.028 μg/mL ($ED_{50}$) | Berscillogenin | [95] |
| | | | 0.62 μg/mL ($ED_{50}$) | 3-Epiberscillogenin | |
| | | | 0.0046 μg/mL ($ED_{50}$) | Bersenogenin | |
| | | | $10^{-7}$ μg/mL ($ED_{50}$) | Hellebrigenin 3-acetate (**48**) | [96] |
| | | | $10^{-3}$ μg/mL ($ED_{50}$) | Hellebrigenin 3,5-diacetate (**49**) | |
| *Carissa spinarum* (Apocynaceae) | Lignans | A549 | <1 μg/mL | (−)-Carinol (**50**) | [98] |
| | | MCF-7 | | | |
| | | WI-38 | | | |
| | | A549 | 11.0 μg/mL | (−)-Carissanol (**51**) | |
| | | MCF-7 | 17.4 μg/mL | | |
| | | WI-38 | 6.2 μg/mL | | |
| | | A549 | 29.0 μg/mL | (−)-Nortrachelogenin | |
| | | MCF-7 | 88.3 μg/mL | | |
| | | WI-38 | >100 μg/mL | | |
| *Linum usitatissimum* (Linaceae) | Lignans | MCF-7 | $1 \times 10^{-5}$ mol/L | Secoisolariciresinol (**52**) | [99] |
| | | | $1 \times 10^{-6}$ M | Matairesinol (**53**) | |

Cell lines: HEK293 = Human embryonic kidney, HepG2 = Liver hepatocarcinoma, Caco-2 = Human colon carcinoma, NCI-H460 = Human large-cell lung carcinoma, MCF-7 = Human breast adenocarcinoma, WRL-68 = human hepatic, PC-3 = Human prostate cancer, KB = Human mouth epidermal carcinoma, MGC-803 = Human gastric cancer, A-549 = Human lung adenocarcinoma, WI-38 = Normal human embryonic, $IC_{50}$ = Concentration that inhibited cell proliferation by 50%. $ED_{50}$ = Effective dose for 50% of the population * Plant material collected from Ethiopia.

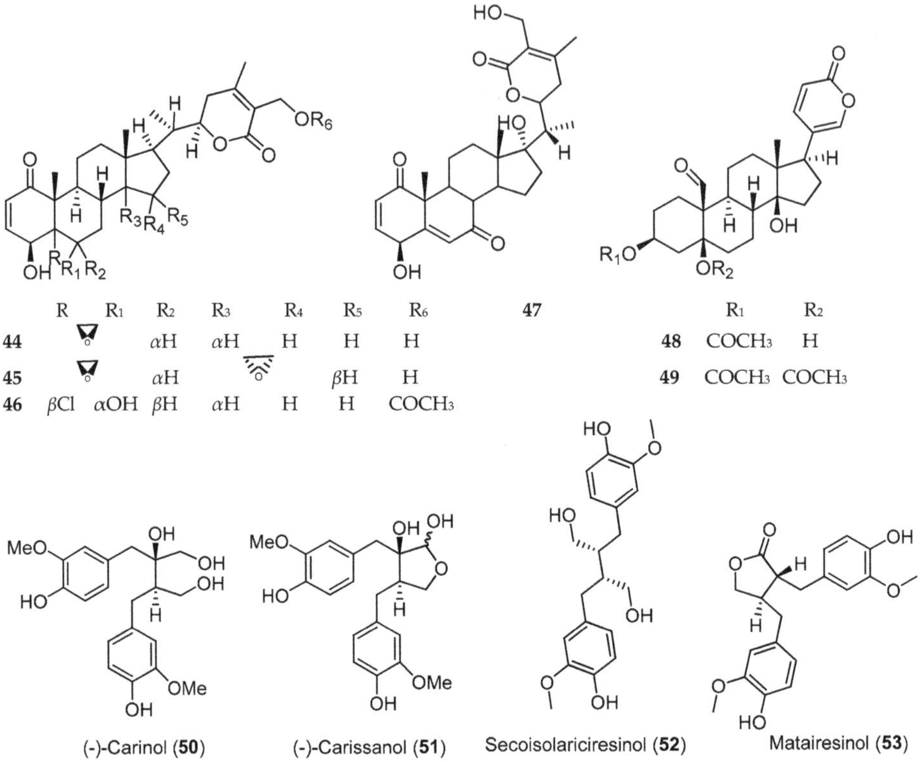

| | R | $R_1$ | $R_2$ | $R_3$ | $R_4$ | $R_5$ | $R_6$ |
|---|---|---|---|---|---|---|---|
| **44** | | | αH | αH | H | H | H |
| **45** | | | αH | | | βH | H |
| **46** | βCl | αOH | βH | αH | H | H | $COCH_3$ |

**47**

| | $R_1$ | $R_2$ |
|---|---|---|
| **48** | $COCH_3$ | H |
| **49** | $COCH_3$ | $COCH_3$ |

(-)-Carinol (**50**) (-)-Carissanol (**51**) Secoisolariciresinol (**52**) Matairesinol (**53**)

**Figure 5.** Structures of anticancer steroids and lignans reported from plants available in Ethiopia.

## 4. Preclinical, In Vivo, and Clinical Studies on Ethiopian Anticancer Plants

Preclinical studies generate data on the efficacy, safety, and pharmacokinetic properties of lead compounds, which will later be used to select better molecules for clinical trials. Assessment of the findings of preclinical in vivo animal studies supports the traditional use of plants to manage cancer in Ethiopia (Table 6). Despite the preclinical efficacy data, there are no clinically significant anticancer agents isolated from traditionally used Ethiopian plants. Moreover, there are also no clinical trials conducted on anticancer plants that are collected from Ethiopia. Among reviewed phytochemicals only ursolic acid, secoisolariciresinol (**52**), and colchicines (**41**), isolated from plants collected elsewhere, were considered further for clinical trial.

**Table 6.** Animal efficacy studies, clinical trials, and/or clinically approved agents among Ethiopian anticancer plants/compounds.

| Plants | Crude Extract | Isolated Compounds | In Vivo Studies | Clinical Trials (Status) | Clinically Approved for |
|---|---|---|---|---|---|
| *Bersama abyssinica* | | Hellebrigenin 3-acetate (**48**) | Significantly inhibits Walker intramuscular carcinosarcoma 256 in rats [96] | - | - |
| *Catharanthus roseus* | Ethanolic extract | | Significantly increased the life span and decreased the tumor volume in Ehrlich ascites carcinoma-bearing mice [101] | - | - |
| | | Vincristine (**31**) | - | - | Childhood leukaemia, Hodgkin's disease and acute panmyelosis [102] |
| | | Vinblastine (**30**) | - | - | Lymphosarcoma, choriocarcinoma, neuroblastoma and lymphocytic leukemia [103] |
| *Euphorbia tirucalli* | Hydroalcoholic extract | | Significantly enhanced survival and reduced tumor growth in Ehrlich ascites tumor-bearing mice [104] | - | - |
| | Latex | | Significantly reduced tumor growth and cachexia in Walker 256 tumor-bearing rats [105] | - | - |
| *Gloriosa superba* | Ethanolic crude extract | | Significantly reduced tumor growth in combination with gemcitabine in a murine model of pancreatic adenocarcinoma [106] | - | - |
| | | Colchicine (**41**) | - | Phase II for castrate resistant prostate cancer (Withdrawn due to funding) [107] | - |
| *Jatropha curcas* | Methanolic fractions | | Showed significant anti-metastatic and antiprolifertaive activity in C57BL/6 mice [108] | - | - |
| *Linum usitatissimum* | | Secoisolariciresinol (**52**) | - | Phase II (Completed) [109] | - |

**Table 6.** *Cont.*

| Plants | Crude Extract | Isolated Compounds | In Vivo Studies | Clinical Trials (Status) | Clinically Approved for |
|---|---|---|---|---|---|
| *Prunus Africana* | Ethanol extract | | Showed significant reduction in prostate cancer incidence in mice [110] | - | |
| | | Ursolic Acid | - | Early Phase I [111] | - |
| *Plumbago zeylanica* L. | | Plumbagin | Significantly inhibits squamous cell carcinomas in FVB/N mice [112] | | |
| *Ricinus communis* | Fruit extract | | Significantly reduced tumor volume in 4T1 syngeneic mouse model [113] | - | - |
| *Solanum nigrum* | Crude polysaccharides | | Significant growth inhibition in cervical cancer tumor-bearing mice [114] | - | - |
| | Aqueous extract | | Significantly inhibits early hepatocarcinogenesis [115] | - | - |
| *Vernonia amygdalina* | Aqueous crude extract | | Increase efficacies and optimizes treatment outcomes when given with paclitaxel in athymic mice [116] | – | - |
| *Vernonia hymenolepis* | | Vernolepin | Significantly inhibited intramuscular carcinosarcoma in walker tumor bearing rats [117] | - | - |
| *Withania somnifera* | Aqueous extract | | Decreased tumor volume in orthotopic glioma allograft rat model [118] | - | - |
| | Ethanolic extract | | Significantly improve colon cancer treatment in mice [119] | - | - |
| | | Withaferin A | Significantly inhibited HepG2-xenografts and diethylnitrosamine (DEN)-induced-hepatocellular carcinoma (HCC) in C57BL/6 mice [120] | - | - |

## 5. Conclusions

Despite the traditional use of various Ethiopian plants for the treatment of cancer by herbal medicine practitioners for many decades, only a few active anticancer crude extracts, herbal preparations, and pure compounds were tested and so far no clinical trial was conducted on them. In this review, an attempt has been made to document antiproliferative, antitumor, and cytotoxic natural products small molecules isolated from medicinal plants that are traditionally used to treat cancer in Ethiopia. However, among the reported active compounds, only few have been isolated from plants that are originated and collected from Ethiopian geographic location, despite their wider presence and traditional claim at home. The majority of compounds reported in this review are isolated from plants (corresponding to Ethiopian species) that were collected from different regions of the world. However, the comprehensive list of active compounds ($IC_{50}$ and $ED_{50}$ values) provided in this review will help to identify the most potent source(s) of these compounds, as bioactive marker(s), of local flora. Based on the higher frequency of citation *Croton macrostachyus*, *Jatropha curcas*, *Plumbago zeylanica*, and *Vernonia hymenolepsis* are potential candidates for follow-up bioassay guided investigations. Furthermore, plants with reported antiproliferative compounds such as *Podocarpus falcatus*, *Linum usitatissimum*, and *Zehneria scabra* should also be examined for additional cytotoxic compounds and evaluated against a battery of cancer cell lines.

Generally, the ecological variation has a huge impact on the biosynthesis, yield of active constituent and biological potency of secondary metabolites produced by plants of similar species from different geographical regions. Thus, Ethiopian anticancer plants might have novel active constituents to fight cancer, based on traditional medical use, than those collected from other regions due to their unique geographical location and inherent climatic condition of the diverse landscape. Unfortunately, these valuable plant resources are disappearing rapidly due to climate change, rapid urbanization, agricultural land expansion, and artificial deforestation; therefore, Ethiopian flora is facing a great challenge, and thus it is high time to examine the anticancer plants systematically with the aim to carry out chemical and biological invesigations, as well as clinical trials on promising anticancer plant extracts based on ethnopharmacological knowledge.

**Author Contributions:** I.M., K.A., S.T. and E.L.; developed the concept, analyzed the data, and wrote the manuscript. Y.A., M.K. and E.T.; performed the literature searches, and contributed to draft the manuscript. All authors have read and agreed to the published version of this manuscript.

**Funding:** This research received no external funding.

**Acknowledgments:** M.K. and I.M. would like to thanks NCNPR, University of Mississippi, for technical support in preparing the manuscript.

**Conflicts of Interest:** The authors declare no conflict of interest.

## References

1. Bray, F.; Ferlay, J.; Soerjomataram, I.; Siegel, R.L.; Torre, L.A.; Jemal, A. Global cancer statistics 2018: GLOBOCAN estimates of incidence and mortality worldwide for 36 cancers in 185 countries. *CA Cancer J. Clin.* **2018**, *68*, 394–424. [CrossRef] [PubMed]
2. Cragg, G.M.; Pezzuto, J.M. Natural products as a vital source for the discovery of cancer chemotherapeutic and chemopreventive agents. *Med. Princ. Pract.* **2016**, *25*, 41–59. [CrossRef] [PubMed]
3. Rishton, G.M. Natural products as a robust source of new drugs and drug leads: Past successes and present day issues. *Am. J. Cardiol.* **2008**, *101*, S43–S49. [CrossRef] [PubMed]
4. Newman, D.J.; Cragg, G.M. Natural products as sources of new drugs from 1981 to 2014. *J. Nat. Prod.* **2016**, *79*, 629–661. [CrossRef] [PubMed]
5. Kelbessa, E.; Demissew, S. Diversity of vascular plant taxa of the flora of Ethiopia and Eritrea. *Ethiop. J. Biol. Sci.* **2014**, *13*, 37–45.
6. Tuasha, N.; Petros, B.; Asfaw, Z. Medicinal plants used by traditional healers to treat malignancies and other human ailments in Dalle District, Sidama Zone, Ethiopia. *J. Ethnobiol. Ethnomed.* **2018**, *14*, 1–21. [CrossRef]

7. Bussa, N.F.; Belayneh, A. Traditional medicinal plants used to treat cancer, tumors and inflammatory ailments in Harari Region, Eastern Ethiopia. *S. Afr. J. Bot.* **2019**, *122*, 360–368. [CrossRef]
8. Atlabachew, M.; Chandravanshi, B.S.; Redi, M. Selected secondary metabolites and antioxidant activity of khat (*Catha edulis* Forsk) chewing leaves extract. *Int. J. Food Prop.* **2014**, *17*, 45–64. [CrossRef]
9. Worku, N.; Mossie, A.; Stich, A.; Daugschies, A.; Trettner, S.; Hemdan, N.Y.; Birkenmeier, G. Evaluation of the in vitro efficacy of *Artemisia annua*, *Rumex abyssinicus*, and *Catha edulis* Forsk extracts in cancer and *Trypanosoma brucei* cells. *ISRN Biochem.* **2013**, *2013*, 1–10. [CrossRef]
10. Tauchen, J.; Doskocil, I.; Caffi, C.; Lulekal, E.; Marsik, P.; Havlik, J.; Van Damme, P.; Kokoska, L. In vitro antioxidant and anti-proliferative activity of Ethiopian medicinal plant extracts. *Ind. Crops Prod.* **2015**, *74*, 671–679. [CrossRef]
11. Yeshak, M.Y.; Burman, R.; Asres, K.; Göransson, U. Cyclotides from an extreme habitat: Characterization of cyclic peptides from *Viola abyssinica* of the Ethiopian highlands. *J. Nat. Prod.* **2011**, *74*, 727–731. [CrossRef] [PubMed]
12. Nibret, E.; Youns, M.; Krauth-Siegel, R.L.; Wink, M. Biological activities of xanthatin from *Xanthium strumarium* leaves. *Phytother. Res.* **2011**, *25*, 1883–1890. [CrossRef] [PubMed]
13. Gebrelibanos, M.; Asres, K.; Veeresham, C. In Vitro radical scavenging activity of the leaf and bark extracts of *Senna singueana* (Del). Lock. *Ethiop. Pharm. J.* **2007**, *25*, 77–84. [CrossRef]
14. Mengesha, A.E.; Youan, B.-B.C. Anticancer activity and nutritional value of extracts of the seed of Glinus lotoides. *J. Nutr. Sci. Vitaminol.* **2010**, *56*, 311–318. [CrossRef] [PubMed]
15. Habtemariam, S. Knipholone anthrone from *Kniphofia foliosa* induces a rapid onset of necrotic cell death in cancer cells. *Fitoterapia* **2010**, *81*, 1013–1019. [CrossRef]
16. Tuasha, N.; Hailemeskel, E.; Erkö, B.; Petros, B. Comorbidity of intestinal helminthiases among malaria outpatients of Wondo Genet health centers, southern Ethiopia: Implications for integrated control. *BMC Infect. Dis.* **2019**, *19*, 659. [CrossRef]
17. Esubalew, S.T.; Belete, A.; Lulekal, E.; Gabriel, T.; Engidawork, E.; Asres, K. Review of ethnobotanical and ethnopharmacological evidences of some ethiopian medicinal plants traditionally used for the treatment of cancer. *Ethiop. J. Health Dev.* **2017**, *31*, 161–187.
18. Friis, I.; Demissew, S.; Van Breugel, P. Atlas of the potential vegetation of Ethiopia. In *Det Kongelige Danske Videnskabernes Selskab*; Nabu Press: Charleston, SC, USA, 2010.
19. Makonnen, E.; Hagos, E. Antispasmodic effect of *Bersama abyssinica* aqueous extract on guinea-pig ileum. *Phytother. Res.* **1993**, *7*, 211–212. [CrossRef]
20. Birhanu, Z. Traditional use of medicinal plants by the ethnic groups of Gondar Zuria District, North-Western Ethiopia. *J. Nat. Rem.* **2013**, *13*, 46–53.
21. Tesfaye, S.; Belete, A.; Engidawork, E.; Gedif, T.; Asres, K. Ethnobotanical study of medicinal plants used by traditional healers to treat cancer-like symptoms in eleven districts, Ethiopia. *Evid-Based Compl. Alt.* **2020**, *2020*, 1–23. [CrossRef]
22. Agize, M.; Demissew, S.; Asfaw, Z. Ethnobotany of medicinal plants in Loma and Gena bosa districts (woredas) of dawro zone, southern Ethiopia. *Topcls. J. Herb. Med.* **2013**, *2*, 194–212.
23. Giday, M.; Asfaw, Z.; Woldu, Z. Ethnomedicinal study of plants used by Sheko ethnic group of Ethiopia. *J. Ethnopharmacol.* **2010**, *132*, 75–85. [CrossRef] [PubMed]
24. Kidane, B.; van Andel, T.; van der Maesen, L.J.G.; Asfaw, Z. Use and management of traditional medicinal plants by Maale and Ari ethnic communities in southern Ethiopia. *J. Ethnobiol. Ethnomed.* **2014**, *10*, 1–15. [CrossRef] [PubMed]
25. Giday, M.; Teklehaymanot, T.; Animut, A.; Mekonnen, Y. Medicinal plants of the Shinasha, Agew-awi and Amhara peoples in northwest Ethiopia. *J. Ethnopharmacol.* **2007**, *110*, 516–525. [CrossRef]
26. Abera, B. Medicinal plants used in traditional medicine by Oromo people, Ghimbi District, Southwest Ethiopia. *J. Ethnobiol. Ethnomed.* **2014**, *10*, 1–15. [CrossRef]
27. Abebe, D.; Debella, A.; Urga, K. Medicinal plants and other useful plants of Ethiopia. *Ethiop. Health Nutr. Res. Inst.* **2003**, *156*, 194–212.
28. Regassa, R. Assessment of indigenous knowledge of medicinal plant practice and mode of service delivery in Hawassa city, southern Ethiopia. *J. Med. Plants Res.* **2013**, *7*, 517–535.

29. Kewessa, G.; Abebe, T.; Demessie, A. Indigenous knowledge on the use and management of medicinal trees and shrubs in Dale District, Sidama Zone, Southern Ethiopia. *Ethnobot. Res. Appl.* **2015**, *14*, 171–182. [CrossRef]
30. Abebe, W. A survey of prescriptions used in traditional medicine in Gondar region, northwestern Ethiopia: General pharmaceutical practice. *J. Ethnopharmacol.* **1986**, *18*, 147–165. [CrossRef]
31. Teklehaymanot, T.; Giday, M. Ethnobotanical study of medicinal plants used by people in Zegie Peninsula, Northwestern Ethiopia. *J. Ethnobiol. Ethnomed.* **2007**, *3*, 1–11. [CrossRef]
32. Birhanu, Z.; Endale, A.; Shewamene, Z. An ethnomedicinal investigation of plants used by traditional healers of Gondar town, North-Western Ethiopia. *J. Med. Plants Stud.* **2015**, *3*, 36–43.
33. Chekole, G.; Asfaw, Z.; Kelbessa, E. Ethnobotanical study of medicinal plants in the environs of Tara-gedam and Amba remnant forests of Libo Kemkem District, northwest Ethiopia. *J. Ethnobiol. Ethnomed.* **2015**, *11*, 1–38. [CrossRef] [PubMed]
34. Teklehaymanot, T. Ethnobotanical study of knowledge and medicinal plants use by the people in Dek Island in Ethiopia. *J. Ethnopharmacol.* **2009**, *124*, 69–78. [CrossRef] [PubMed]
35. Yineger, H.; Yewhalaw, D. Traditional medicinal plant knowledge and use by local healers in Sekoru District, Jimma Zone, Southwestern Ethiopia. *J. Ethnobiol. Ethnomed.* **2007**, *3*, 1–7. [CrossRef] [PubMed]
36. Kebede, A.; Ayalew, S.; Mesfin, A.; Mulualem, G. Ethnobotanical investigation of traditional medicinal plants commercialized in the markets of Dire Dawa city, eastern Ethiopia. *J. Med. Plants Stud.* **2016**, *4*, 170–178.
37. Ayele, T.T. A Review on traditionally used medicinal plants/herbs for cancer therapy in Ethiopia: Current status, challenge and future perspectives. *Org. Chem. Curr. Res.* **2018**, *7*. [CrossRef]
38. Tolossa, K.; Debela, E.; Athanasiadou, S.; Tolera, A.; Ganga, G.; Houdijk, J.G. Ethno-medicinal study of plants used for treatment of human and livestock ailments by traditional healers in South Omo, Southern Ethiopia. *J. Ethnobiol. Ethnomed.* **2013**, *9*, 1–15. [CrossRef]
39. Bekele, G.; Reddy, P.R. Ethnobotanical study of medicinal plants used to treat human ailments by Guji Oromo tribes in Abaya District, Borana, Oromia, Ethiopia. *Univers. J. Plant Sci.* **2015**, *3*, 1–8. [CrossRef]
40. Abera, B. Medicinal plants used in traditional medicine in Jimma Zone, Southwest Ethiopia. *Ethiop. J. Health Sci.* **2003**, *13*, 85–94.
41. Wabe, N.; Mohammed, M.A.; Raju, N.J. An ethnobotanical survey of medicinal plants in the Southeast Ethiopia used in traditional medicine. *Spatula DD* **2011**, *1*, 153–158. [CrossRef]
42. Zenebe, G.; Zerihun, M.; Solomon, Z. An ethnobotanical study of medicinal plants in Asgede Tsimbila district, Northwestern Tigray, northern Ethiopia. *Ethnobot. Res. Appl.* **2012**, *10*, 305–320. [CrossRef]
43. Teklay, A.; Abera, B.; Giday, M. An ethnobotanical study of medicinal plants used in Kilte Awulaelo District, Tigray Region of Ethiopia. *J. Ethnobiol. Ethnomed.* **2013**, *9*, 1–23. [CrossRef] [PubMed]
44. Yadav, R.H. Medicinal plants in folk medicine system of Ethiopia. *J. Poisonous Med. Plants Res.* **2013**, *1*, 7–11.
45. Yohannis, S.W.; Asfaw, Z.; Kelbessa, E. Ethnobotanical study of medicinal plants used by local people in menz gera midir district, north shewa zone, amhara regional state, Ethiopia. *J. Med. Plants Res.* **2018**, *12*, 296–314. [CrossRef]
46. Birhanu, A.; Ayalew, S. Indigenous knowledge on medicinal plants used in and around Robe Town, Bale Zone, Oromia Region, Southeast Ethiopia. *J. Med. Plants Res.* **2018**, *12*, 194–202. [CrossRef]
47. Gebeyehu, G.; Asfaw, Z.; Enyew, A. An ethnobotanical study of traditional use of medicinal plants and their conservation status in Mecha Wereda, west Gojjam zone of Amhara region, Ethiopia. *Int. J. Pharm. Health Care Res.* **2014**, *2*, 137–154.
48. Yineger, H.; Yewhalaw, D.; Teketay, D. Ethnomedicinal plant knowledge and practice of the Oromo ethnic group in southwestern Ethiopia. *J. Ethnobiol. Ethnomed.* **2008**, *4*, 1–11. [CrossRef]
49. Demissew, S. A description of some essential oil bearing plants in Ethiopia and their indigenous uses. *J. Essent. Oil Res.* **1993**, *5*, 465–479. [CrossRef]
50. Agisho, H.; Osie, M.; Lambore, T. Traditional medicinal plants utilization, management and threats in Hadiya zone, Ethiopia. *J. Med. Plant* **2014**, *2*, 94–108.
51. Giday, M.; Asfaw, Z.; Woldu, Z. Medicinal plants of the Meinit ethnic group of Ethiopia: An ethnobotanical study. *J. Ethnopharmacol.* **2009**, *124*, 513–521. [CrossRef]
52. Birhanu, T.; Abera, D.; Ejeta, E.; Nekemte, E. Ethnobotanical study of medicinal plants in selected Horro Gudurru Woredas, western Ethiopia. *J. Biol. Agric. Healthc.* **2015**, *5*, 83–93.

53. Gidey, M.; Beyene, T.; Signorini, M.A.; Bruschi, P.; Yirga, G. Traditional medicinal plants used by Kunama ethnic group in Northern Ethiopia. *J. Med. Plants Res.* **2015**, *9*, 494–509.
54. Giday, M.; Asfaw, Z.; Elmqvist, T.; Woldu, Z. An ethnobotanical study of medicinal plants used by the Zay people in Ethiopia. *J. Med. Plants Res.* **2015**, *9*, 494–509. [CrossRef]
55. Ragunathan, M.; Abay, S.M. Ethnomedicinal survey of folk drugs used in Bahirdar Zuria district, northwestern Ethiopia. *Indian J. Tradit. Knowl.* **2009**, *8*, 281–284.
56. Mekuanent, T.; Zebene, A.; Solomon, Z. Ethnobotanical study of medicinal plants in Chilga district, northwestern Ethiopia. *J. Nat. Remedies* **2015**, *15*, 88–112. [CrossRef]
57. Habtemariam, S. Cytotoxicity of diterpenes from *Premna schimperi* and *Premna oligotricha*. *Planta Med.* **1995**, *61*, 368–369. [CrossRef]
58. Connolly, J.D.; Hill, R.A. Triterpenoids. *Nat. Prod. Rep.* **2010**, *27*, 79–132. [CrossRef]
59. Hordyjewska, A.; Ostapiuk, A.; Horecka, A. Betulin and betulinic acid in cancer research. *J. Pre Clin. Clin. Res.* **2018**, *12*, 72–75. [CrossRef]
60. Chudzik, M.; Korzonek-Szlacheta, I.; Król, W. Triterpenes as potentially cytotoxic compounds. *Molecules* **2015**, *20*, 1610–1625. [CrossRef]
61. Phillips, D.R.; Rasbery, J.M.; Bartel, B.; Matsuda, S.P. Biosynthetic diversity in plant triterpene cyclization. *Curr. Opin. Plant Biol.* **2006**, *9*, 305–314. [CrossRef]
62. Petronelli, A.; Pannitteri, G.; Testa, U. Triterpenoids as new promising anticancer drugs. *Anticancer Drugs* **2009**, *20*, 880–892. [CrossRef] [PubMed]
63. Addo, E.M.; Chai, H.-B.; Hymete, A.; Yeshak, M.Y.; Slebodnick, C.; Kingston, D.G.; Rakotondraibe, L.H. Antiproliferative constituents of the roots of Ethiopian *Podocarpus falcatus* and structure revision of 2α-hydroxynagilactone F and nagilactone I. *J. Nat. Prod.* **2015**, *78*, 827–835. [CrossRef] [PubMed]
64. Yong, Y.; Tesso, H.; Terfa, A.; Dekebo, A.; Dinku, W.; Lee, Y.H.; Shin, S.Y.; Lim, Y. Biological evaluation of the diterpenes from *Croton macrostachyus*. *Appl. Biol. Chem.* **2017**, *60*, 615–621. [CrossRef]
65. Sathya, S.; Sudhagar, S.; Vidhya Priya, M.; Bharathi Raja, R.; Muthusamy, V.S.; Niranjali Devaraj, S.; Lakshmi, B.S. 3β-Hydroxylup-20(29)-ene-27,28-dioic acid dimethyl ester, a novel natural product from *Plumbago zeylanica* inhibits the proliferation and migration of MDA-MB-231 cells. *Chem. Biol. Interact.* **2010**, *188*, 412–420. [CrossRef] [PubMed]
66. Kongue, M.D.; Talontsi, F.M.; Lamshöft, M.; Kenla, T.J.; Dittrich, B.; Kapche, G.D.; Spiteller, M. Sonhafouonic acid, a new cytotoxic and antifungal hopene-triterpenoid from *Zehneria scabra* camerunensis. *Fitoterapia* **2013**, *85*, 176–180. [CrossRef] [PubMed]
67. Lin, M.-W.; Lin, A.-S.; Wu, D.-C.; Wang, S.S.; Chang, F.-R.; Wu, Y.-C.; Huang, Y.-B. Euphol from *Euphorbia tirucalli* selectively inhibits human gastric cancer cell growth through the induction of ERK1/2-mediated apoptosis. *Food Chem. Toxicol.* **2012**, *50*, 4333–4339. [CrossRef]
68. Silva, V.A.O.; Rosa, M.N.; Tansini, A.; Oliveira, R.J.; Martinho, O.; Lima, J.P.; Pianowski, L.F.; Reis, R.M. In vitro screening of cytotoxic activity of euphol from *Euphorbia tirucalli* on a large panel of human cancer-derived cell lines. *Exp. Ther. Med.* **2018**, *16*, 557–566. [CrossRef]
69. Reyes-Zurita, F.J.; Rufino-Palomares, E.E.; Lupiáñez, J.A.; Cascante, M. Maslinic acid, a natural triterpene from *Olea europaea* L., induces apoptosis in HT29 human colon-cancer cells via the mitochondrial apoptotic pathway. *Cancer Lett.* **2009**, *273*, 44–54. [CrossRef]
70. Darmanin, S.; Wismayer, P.S.; Camilleri Podesta, M.T.; Micallef, M.J.; Buhagiar, J.A. An extract from *Ricinus communis* L. leaves possesses cytotoxic properties and induces apoptosis in SK-MEL-28 human melanoma cells. *Nat. Prod. Res.* **2009**, *23*, 561–571. [CrossRef]
71. Burkill, H.M. *The Useful Plants of West Tropical Africa. Families E–I.*; Royal Botanic Gardens: Richmond, UK, 1994; Volume 2.
72. Liu, J.-Q.; Yang, Y.-F.; Li, X.-Y.; Liu, E.-Q.; Li, Z.-R.; Zhou, L.; Li, Y.; Qiu, M.-H. Cytotoxicity of naturally occurring rhamnofolane diterpenes from *Jatropha curcas*. *Phytochemistry* **2013**, *96*, 265–272. [CrossRef]
73. Zhang, X.-Q.; Li, F.; Zhao, Z.-G.; Liu, X.-L.; Tang, Y.-X.; Wang, M.-K. Diterpenoids from the root bark of *Jatropha curcas* and their cytotoxic activities. *Phytochem. Lett.* **2012**, *5*, 721–724. [CrossRef]
74. Irungu, B.N.; Orwa, J.A.; Gruhonjic, A.; Fitzpatrick, P.A.; Landberg, G.; Kimani, F.; Midiwo, J.; Erdélyi, M.; Yenesew, A. Constituents of the roots and leaves of *Ekebergia capensis* and their potential antiplasmodial and cytotoxic activities. *Molecules* **2014**, *19*, 14235–14246. [CrossRef]

75. Tang, X.-L.; Yang, X.-Y.; Jung, H.-J.; Kim, S.-Y.; Jung, S.-Y.; Choi, D.-Y.; Park, W.-C.; Park, H. Asiatic acid induces colon cancer cell growth inhibition and apoptosis through mitochondrial death cascade. *Biol. Pharm. Bull.* **2009**, *32*, 1399–1405. [CrossRef] [PubMed]
76. Alsayari, A.; Kopel, L.; Ahmed, M.S.; Soliman, H.S.; Annadurai, S.; Halaweish, F.T. Isolation of anticancer constituents from *Cucumis prophetarum* var. prophetarum through bioassay-guided fractionation. *BMC Complement. Altern. Med.* **2018**, *18*, 1–12. [CrossRef] [PubMed]
77. Kupchan, S.M.; Hemingway, R.J.; Karim, A.; Werner, D. Tumor inhibitors. XLVII. Vernodalin and vernomygdin, two new cytotoxic sesquiterpene lactones from *Vernonia amygdalina* Del. *J. Org. Chem.* **1969**, *34*, 3908–3911. [CrossRef] [PubMed]
78. Selassie, C.D.; Kapur, S.; Verma, R.P.; Rosario, M. Cellular apoptosis and cytotoxicity of phenolic compounds: A quantitative structure- activity relationship study. *J. Med. Chem.* **2005**, *48*, 7234–7242. [CrossRef]
79. Okoye, F.B.C.; Debbab, A.; Wray, V.; Esimone, C.O.; Osadebe, P.O.; Proksch, P. A phenyldilactone, bisnorsesquiterpene, and cytotoxic phenolics from *Maytenus senegalensis* leaves. *Tetrahedron Lett.* **2014**, *55*, 3756–3760. [CrossRef]
80. Ma, W.; Zhang, Y.; Ding, Y.-Y.; Liu, F.; Li, N. Cytotoxic and anti-inflammatory activities of phenanthrenes from the medullae of *Juncus effusus* L. *Arch. Pharm. Res.* **2016**, *39*, 154–160. [CrossRef]
81. Ishiuchi, K.; Kosuge, Y.; Hamagami, H.; Ozaki, M.; Ishige, K.; Ito, Y.; Kitanaka, S. Chemical constituents isolated from *Juncus effusus* induce cytotoxicity in HT22 cells. *J. Nat. Med.* **2015**, *69*, 421–426. [CrossRef]
82. Xu, T.-P.; Shen, H.; Liu, L.-X.; Shu, Y.-Q. Plumbagin from *Plumbago Zeylanica* L. Induces Apoptosis in Human Non-small Cell Lung Cancer Cell Lines through NF-κB Inactivation. *Asian Pac. J. Cancer Prev.* **2013**, *14*, 2325–2331. [CrossRef]
83. Chen, C.-A.; Chang, H.-H.; Kao, C.-Y.; Tsai, T.-H.; Chen, Y.-J. Plumbagin, isolated from *Plumbago zeylanica*, induces cell death through apoptosis in human pancreatic cancer cells. *Pancreatology* **2009**, *9*, 797–809. [CrossRef] [PubMed]
84. Ruszkowska, J.; Chrobak, R.; Wróbel, J.T.; Czarnocki, Z. Novel bisindole derivatives of Catharanthus alkaloids with potential cytotoxic properties. In *Developments in Tryptophan and Serotonin Metabolism*; Springer: Berlin/Heidelberg, Germany, 2003; pp. 643–646.
85. Jossang, A.; Fodor, P.; Bodo, B. A new structural class of bisindole alkaloids from the seeds of *Catharanthus roseus*: Vingramine and methylvingramine. *J. Org. Chem.* **1998**, *63*, 7162–7167. [CrossRef] [PubMed]
86. Zhang, W.-K.; Xu, J.-K.; Tian, H.-Y.; Wang, L.; Zhang, X.-Q.; Xiao, X.-Z.; Li, P.; Ye, W.-C. Two new vinblastine-type N-oxide alkaloids from *Catharanthus roseus*. *Nat. Prod. Res.* **2013**, *27*, 1911–1916. [CrossRef]
87. Wang, C.-H.; Wang, G.-C.; Wang, Y.; Zhang, X.-Q.; Huang, X.-J.; Zhang, D.-M.; Chen, M.-F.; Ye, W.-C. Cytotoxic dimeric indole alkaloids from *Catharanthus roseus*. *Fitoterapia* **2012**, *83*, 765–769. [CrossRef]
88. Wang, X.-D.; Li, C.-Y.; Jiang, M.-M.; Li, D.; Wen, P.; Song, X.; Chen, J.-D.; Guo, L.-X.; Hu, X.-P.; Li, G.-Q. Induction of apoptosis in human leukemia cells through an intrinsic pathway by cathachunine, a unique alkaloid isolated from *Catharanthus roseus*. *Phytomedicine* **2016**, *23*, 641–653. [CrossRef] [PubMed]
89. Siddiqui, M.J.; Ismail, Z.; Aisha, A.F.A.; Majid, A.M. Cytotoxic activity of *Catharanthus roseus* (Apocynaceae) crude extracts and pure compounds against human colorectal carcinoma cell line. *Int. J. Pharmacol.* **2010**, *6*, 43–47. [CrossRef]
90. Balkrishna, A.; Das, S.K.; Pokhrel, S.; Joshi, A.; Verma, S.; Sharma, V.K.; Sharma, V.; Sharma, N.; Joshi, C.S. Colchicine: Isolation, LC–MS QTof screening, and anticancer activity study of *Gloriosa superba* seeds. *Molecules* **2019**, *24*, 2772. [CrossRef]
91. Tu, Y.; Cheng, S.; Zhang, S.; Sun, H.; Xu, Z. Vincristine induces cell cycle arrest and apoptosis in SH-SY5Y human neuroblastoma cells. *Int. J. Mol. Med.* **2013**, *31*, 113–119. [CrossRef]
92. Ding, X.; Zhu, F.; Yang, Y.; Li, M. Purification, antitumor activity in vitro of steroidal glycoalkaloids from black nightshade (*Solanum nigrum* L.). *Food Chem.* **2013**, *141*, 1181–1186. [CrossRef]
93. Maiyoa, F.; Moodley, R.; Singh, M. Phytochemistry, cytotoxicity and apoptosis studies of β-sitosterol-3-oglucoside and β-amyrin from *Prunus africana*. *Afr. J. Tradit. Complement. Altern. Med.* **2016**, *13*, 105–112. [CrossRef]
94. Choudhary, M.I.; Hussain, S.; Yousuf, S.; Dar, A. Chlorinated and diepoxy withanolides from *Withania somnifera* and their cytotoxic effects against human lung cancer cell line. *Phytochemistry* **2010**, *71*, 2205–2209. [CrossRef] [PubMed]

95. Kupchan, S.M.; Moniot, J.L.; Sigel, C.W.; Hemingway, R.J. Tumor inhibitors. LXV. Bersenogenin, berscillogenin, and 3-epiberscillogenin, three new cytotoxic bufadienolides from *Bersama abyssinica*. *J. Org. Chem.* **1971**, *36*, 2611–2616. [CrossRef] [PubMed]
96. Kupchan, S.M.; Hemingway, R.J.; Hemingway, J.C. The isolation and characterization of hellebrigenin 3-acetate and hellebrigenin 3, 5-diacetate, bufadienolide tumor inhibitors from *Bersama abyssinica*. *Tetrahedron Lett.* **1968**, *9*, 149–152. [CrossRef]
97. Abarzua, S.; Szewczyk, M.; Gailus, S.; Richter, D.-U.; Ruth, W.; Briese, V.; Piechulla, B. Effects of phytoestrogen extracts from *Linum usitatissimum* on the Jeg3 human trophoblast tumour cell line. *Anticancer Res.* **2007**, *27*, 2053–2058. [PubMed]
98. Wangteeraprasert, R.; Lipipun, V.; Gunaratnam, M.; Neidle, S.; Gibbons, S.; Likhitwitayawuid, K. Bioactive compounds from *Carissa spinarum*. *Phytother. Res.* **2012**, *26*, 1496–1499. [CrossRef]
99. Abarzua, S.; Serikawa, T.; Szewczyk, M.; Richter, D.-U.; Piechulla, B.; Briese, V. Antiproliferative activity of lignans against the breast carcinoma cell lines MCF 7 and BT 20. *Arch. Gynecol. Obstet.* **2012**, *285*, 1145–1151. [CrossRef]
100. Ali, A.; Joshi, P.; Misra, L.; Sangwan, N.; Darokar, M. 5,6-De-epoxy-5-en-7-one-17-hydroxy withaferin A, a new cytotoxic steroid from *Withania somnifera* L. Dunal leaves. *Nat. Prod. Res.* **2014**, *28*, 392–398. [CrossRef]
101. Aruna, S.R. In-vitro and in-vivo antitumor activity of Catharanthus roseus. *Int. Res. J. Pharm. Appl. Sci.* **2014**, *4*, 1–4.
102. Moudi, M.; Go, R.; Yien, C.Y.S.; Nazre, M. Vinca alkaloids. *Int. J. Prev. Med.* **2013**, *4*, 1231–1235.
103. Rai, V.; Tandon, P.K.; Khatoon, S. Effect of chromium on antioxidant potential of *Catharanthus roseus* varieties and production of their anticancer alkaloids: Vincristine and vinblastine. *BioMed Res. Int.* **2014**, 1–10. [CrossRef]
104. Valadares, M.C.; Carrucha, S.G.; Accorsi, W.; Queiroz, M.L. *Euphorbia tirucalli* L. modulates myelopoiesis and enhances the resistance of tumour-bearing mice. *Int. Immunopharmacol.* **2006**, *6*, 294–299. [CrossRef] [PubMed]
105. Martins, C.G.; Appel, M.H.; Coutinho, D.S.; Soares, I.P.; Fischer, S.; de Oliveira, B.C.; Fachi, M.M.; Pontarolo, R.; Bonatto, S.J.; Iagher, F. Consumption of latex from *Euphorbia tirucalli* L. promotes a reduction of tumor growth and cachexia, and immunomodulation in Walker 256 tumor-bearing rats. *J. Ethnopharmacol.* **2020**, *255*, 1–9. [CrossRef] [PubMed]
106. Capistrano, I.R.; Vangestel, C.; Vanpachtenbeke, H.; Fransen, E.; Staelens, S.; Apers, S.; Pieters, L. Coadministration of a *Gloriosa superba* extract improves the in vivo antitumoural activity of gemcitabine in a murine pancreatic tumour model. *Phytomedicine* **2016**, *23*, 1434–1440. [CrossRef] [PubMed]
107. Drabick, J. *Phase II Trial of Oral Colchicine in Men With Castrate-Resistant Prostate Cancer Who Have Failed Taxotere-Based Chemotherapy;* Clinicaltrials Gov. National Library of Medicine: Bethesda, MD, USA, 2013.
108. Balaji, R.; Rekha, N.; Deecaraman, M.; Manikandan, L. Antimetastatic and antiproliferative activity of methanolic fraction of *Jatropha curcas* against B16F10 melanoma induced lung metastasis in C57BL/6 mice. *Afr. J. Pharm. Pharmacol.* **2009**, *3*, 547–555.
109. Fabian, C.J.; Khan, S.A.; Garber, J.E.; Dooley, W.C.; Yee, L.D.; Klemp, J.R.; Nydegger, J.L.; Powers, K.R.; Kreutzjans, A.L.; Zalles, C.M. Randomized phase IIB trial of the lignan secoisolariciresinol diglucoside in pre-menopausal women at increased risk for development of breast cancer. *Cancer Prev. Res.* **2020**, 1–32. [CrossRef]
110. Shenouda, N.S.; Sakla, M.S.; Newton, L.G.; Besch-Williford, C.; Greenberg, N.M.; MacDonald, R.S.; Lubahn, D.B. Phytosterol *Pygeum africanum* regulates prostate cancer in vitro and in vivo. *Endocrine* **2007**, *31*, 72–81. [CrossRef]
111. The University of Texas Health Science Center at San Antonio. *Phase I Clinical Trial Testing the Synergism of Phytonutrients, Curcumin and Ursolic Acid, to Target Molecular Pathways in the Prostate;* Clinicaltrials Gov, National Library of Medicine: Bethesda, MD, USA, 2020.
112. Sand, J.M.; Hafeez, B.B.; Jamal, M.S.; Witkowsky, O.; Siebers, E.M.; Fischer, J.; Verma, A.K. Plumbagin (5-hydroxy-2-methyl-1, 4-naphthoquinone), isolated from *Plumbago zeylanica*, inhibits ultraviolet radiation-induced development of squamous cell carcinomas. *Carcinogenesis* **2012**, *33*, 184–190. [CrossRef]
113. Majumder, M.; Debnath, S.; Gajbhiye, R.L.; Saikia, R.; Gogoi, B.; Samanta, S.K.; Das, D.K.; Biswas, K.; Jaisankar, P.; Mukhopadhyay, R. *Ricinus communis* L. fruit extract inhibits migration/invasion, induces apoptosis in breast cancer cells and arrests tumor progression in vivo. *Sci. Rep.* **2019**, *9*, 1–14. [CrossRef]

114. Li, J.; Li, Q.; Feng, T.; Zhang, T.; Li, K.; Zhao, R.; Han, Z.; Gao, D. Antitumor activity of crude polysaccharides isolated from *Solanum nigrum* Linne on U14 cervical carcinoma bearing mice. *Phytother. Res.* **2007**, *21*, 832–840. [CrossRef]
115. Hsu, J.-D.; Kao, S.-H.; Tu, C.-C.; Li, Y.-J.; Wang, C.-J. *Solanum nigrum* L. Extract inhibits 2-Acetylaminofluorene-induced hepatocarcinogenesis through overexpression of Glutathione S-Transferase and antioxidant enzymes. *J. Agric. Food Chem.* **2009**, *57*, 8628–8634. [CrossRef]
116. Howard, C.B.; Johnson, W.K.; Pervin, S.; Izevbigie, E.B. Recent perspectives on the anticancer properties of aqueous extracts of Nigerian *Vernonia amygdalina*. *Botanics* **2015**, *5*, 65–76. [CrossRef] [PubMed]
117. Kupchan, S.M.; Hemingway, R.J.; Werner, D.; Karim, A. Tumor inhibitors. XLVI. Vernolepin, a novel sesquiterpene dilactone tumor inhibitor from *Vernonia hymenolepis*. *J. Org. Chem.* **1969**, *34*, 3903–3908. [CrossRef] [PubMed]
118. Kataria, H.; Kumar, S.; Chaudhary, H.; Kaur, G. *Withania somnifera* Suppresses Tumor Growth of Intracranial Allograft of Glioma Cells. *Mol. Neurobiol.* **2016**, *53*, 4143–4158. [CrossRef] [PubMed]
119. Muralikrishnan, G.; Dinda, A.K.; Shakeel, F. Immunomodulatory effects of *Withania Somnifera* on Azoxymethane induced experimental colon cancer in mice. *Immunol. Investig.* **2010**, *39*, 688–698. [CrossRef]
120. Kuppusamy, P.; Nagalingam, A.; Muniraj, N.; Saxena, N.K.; Sharma, D. Concomitant activation of ETS-like transcription factor-1 and Death Receptor-5 via extracellular signal-regulated kinase in withaferin A-mediated inhibition of hepatocarcinogenesis in mice. *Sci. Rep.* **2017**, *7*, 1–13. [CrossRef]

*Review*

# Using Micropropagation to Develop Medicinal Plants into Crops

**Rita M. Moraes [1,2,*], Antonio Luiz Cerdeira [3] and Miriam V. Lourenço [1]**

[1] Santa Martha Agro Ltd.a, Rodovia Prefeito Antonio Duarte Nogueira, Km 317, Contorno Sul, Ribeirão Preto, SP 14.032-800, Brazil; mvlouren@gmail.com
[2] Fundação Fernando E. Lee, Av. Atlântica 900, Balneário, Guarujá, SP 114420-070, Brazil
[3] Embrapa Meio Ambiente, Rodovia SP-340, Km 127,5, Tanquinho Velho, Jaguariúna, SP 13918-110, Brazil; antonio.cerdeira@embrapa.br
* Correspondence: ritammoraes@santamarthaagro.com.br

**Abstract:** Medicinal plants are still the major source of therapies for several illnesses and only part of the herbal products originates from cultivated biomass. Wild harvests represent the major supply for therapies, and such practices threaten species diversity as well as the quality and safety of the final products. This work intends to show the relevance of developing medicinal plants into crops and the use of micropropagation as technique to mass produce high-demand biomass, thus solving the supply issues of therapeutic natural substances. Herein, the review includes examples of in vitro procedures and their role in the crop development of pharmaceuticals, phytomedicinals, and functional foods. Additionally, it describes the production of high-yielding genotypes, uniform clones from highly heterozygous plants, and the identification of elite phenotypes using bioassays as a selection tool. Finally, we explore the significance of micropropagation techniques for the following: a) pharmaceutical crops for production of small therapeutic molecules (STM), b) phytomedicinal crops for production of standardized therapeutic natural products, and c) the micropropagation of plants for the production of large therapeutic molecules (LTM) including fructooligosaccharides classified as prebiotic and functional food crops.

**Keywords:** medicinal plants; in vitro propagation; medicinal crops; phytomedicines

**Citation:** Moraes, R.M.; Cerdeira, A.L.; Lourenço, M.V. Using Micropropagation to Develop Medicinal Plants into Crops. *Molecules* **2021**, *26*, 1752. https://doi.org/10.3390/molecules26061752

Academic Editors: Muhammad Ilias and Dhammika Nanayakkara

Received: 24 February 2021
Accepted: 19 March 2021
Published: 21 March 2021

## 1. Introduction

For over a century, plant tissue culture technology has been an important tool in crop improvement and development: producing disease-free plant material [1], obtaining transgenic plants [2,3], breaking dormancy, and micropropagating elite plants with highly desirable chemotype [4], thereby leading to more uniform plant production [5,6]. This is the technology for conserving in vitro germplasm of elite [7,8], rare, and endangered plant species [9–11], implementing breeding programs for innumerous crops as well as encapsulated seeds [12], and studying plant biosynthesis through cell and root cultures [12,13], the interaction between endophytes and the hostplant [14,15].

High-demand plants face great challenges: Depletion of species diversity due to overharvesting and environmental pollution affecting natural populations are strong factors that support the argument for cultivating rare and elite high-yielding medicinal plants. In addition, the cultivation of medicinal plants is the most effective way of addressing the gap between supply and demand. Breeding studies are necessary both to develop pharmaceutical plants as crops and to scale up their production [16]. Still, few success stories about breeding medicinal plants such as *Artemisia annua* L. exist. Because micropropagation is the tool of producing clones—especially with high-yielding chemotypes—for industrial purposes, it solves this target-breeding problem. Moreover, as the *Echinacea* study [17] showed, micropropagation's demonstrated ability for mass selection suggests that together with bioassays it could form part of an overall strategy to screen elite phenotype lines.

Micropropagation is an in vitro technology of rapidly multiplying elite plants using modern plant tissue culture methods. It is well-known for its applications in the agro, horticultural and forestry industries, this review focuses on a less-commonly known area which is on medicinal plants and the need to develop them as medicinal crops. Li et al. [18] defined pharmaceutical crops in three distinct categories: 1) crops for the production of small therapeutic molecules (STMs), 2) standardized therapeutic extracts (STEs), and 3) large therapeutic molecules (LTMs). In addition, this review also examines micropropagation of functional food plants to ensure their development as crop.

## 2. Pharmaceutical Crops for Production of Small Therapeutic Molecules (STM)

Drug discovery programs and the formation of knowledge of different pharmacological classes of pharmaceuticals owe much to traditional medicine in countries such as China and India [19,20]. Some natural compounds are extracted or used as templates for synthesis or as a precursor for the semi-synthesis (e.g., paclitaxel, artemisinin, podophyllotoxin, cannabinoids, galantamine, vinca alkaloids, atropine, ephedrine, digoxin, morphine, quinine, reserpine, tubocurarine etc.). Many of these compounds provide therapeutic relief for several major illnesses including cancer, Alzheimer, malaria, high blood pressure, fever, and anxiety. As researchers confirm the medicinal utility of these natural resources, they suffer depletions with the increased demand.

According to McChesney et al. [20], pharmaceutical natural substances require considerations beyond supply and demand: the establishment of successful production systems must be sustainable, environmentally safe, reliable, and affordable. Thus, the development of medicinal crops is a key factor to obtaining a commercially viable source of medicinal biomass for the pharmaceutical industry. In fact, non-stable supply sources could lead to bottlenecks that limit potentially beneficial products. For example, researchers pointed to insufficiency in the biomass supply of anti-cancer pharmaceutical ingredients such as podophyllotoxin and paclitaxel, as the major limiting factor at phase III clinical trials, which led to overharvesting of the natural resources of *Podophyllum emodi* Wall ex Royle in India [21] and *Taxus baccata* L. in Europe [22].

Given the shortage of biomass supply limiting clinical phase III trials of paclitaxel and podophyllotoxin, several laboratories engaged in different research approaches that included bioprospecting studies searching for alternate sources with high yields of the active compounds [23–25]. Clippings of cultivated *Taxus* sp. became the reliable source for production of paclitaxel [18], and Sisti et al. [26] reported methods of semi-synthesis using abundant intermediates for production of paclitaxel. Majada et al. [27] reported a procedure to obtain high-yielding *T. baccata* plantlets by screening micropropagated juvenile seedlings that accumulate 10-deacetyl baccatin III. The selected genotypes of *T. baccata* grow faster and contain high taxene content.

For its part, podophyllotoxin is the starting compound for semisynthesis of etoposide and teniposide, two potent DNA topoisomerase cancer drugs utilized in the treatment of small lung and testicular cancers, lymphomas/leukemias and the water-soluble etoposide phosphate, also known as etopophos (Figure 1). To assure podophyllotoxin supply, a buffer extraction procedure using leaf biomass of mayapple plants provides a sustainable alternative source [28]. Later, we published a survey and a database of high-yielding podophyllotoxin colonies [29,30] and an in vitro propagation protocol of Podophyllum peltatum L. to rapidly produce podophyllotoxin-rich plantlets [5].

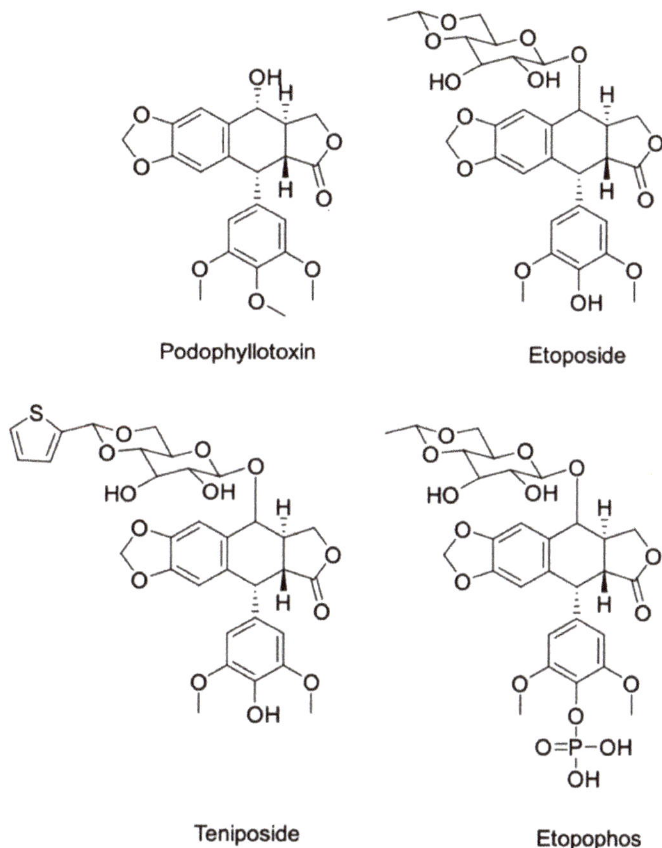


**Figure 1.** Structures of (−)-podohyllotoxin present in *Podophyllum* sp and its commercial chemotherapeutic derivatives.

*Artemisia annua* L. is the source of artemisinin, an endoperoxide sesquiterpene lactone that is very difficult to synthesize, precursor of a common anti-malarial drug (Artemether). Artemisinin production comes from cultivated plants selected for their high artemisinin content [18]. Selection of genotypes with high artemisinin concentration in wild populations resulted in lines containing up to 1.4 percent artemisinin (on dry leaves basis). The leading commercial source, 'Artemis,' exhibited extensive variation of metabolic and agronomic traits; artemisinin content on a µg/mg dry basis for individual plants ranged 22-fold, plant fresh weight varied 28-fold, and leaf area ranged 9-fold [31].

While Ferreira and Janick [32] found that the in vitro production of artemisinin will never be commercially feasible, Wetzetein et al. [33] suggested that cultivation of micropropagated high-yielding artemisinin plants with levels above 2% and improved agronomic traits (high leaf area and shoot biomass production) may reach productivity of 70 kg/ha artemisinin using a crop density of 1 plant $m^{-2}$. We include in Table 1 examples of pharmaceutical plant species classified as small therapeutic molecules STM's (18) and their micropropagation protocols to produce elite clones for higher yields. Taxol® is another success story. According to McChesney et al. [20], the path from the discovery to a pharmaceutical drug requires a viable production system (cultivation, harvest, extraction, purification and isolation) where every step of a natural product must be systematically evaluated. Micropropagation of the superior source (chemotype or variety of the species) may help to produce biomass with a high and consistent concentration of the natural product or a precursor of the natural product that can be converted economically by semi-synthesis to the final bulk active product.

**Table 1.** Commercial sources of pharmaceuticals often used in therapies of several illnesses that are micropropagated [17].

| Plant Species | Natural Substance | Therapy | Micropropagation Protocol |
|---|---|---|---|
| *Artemisia annua* L. | Artemisinin | Antimalarial | Etienne et al. [34] |
| *Catharanthus roseus* (L.) G. Don | Vincristine, Vinblastine | Anticancer | Kumar et al. [35] |
| *Campotheca acuminata* Decne | Camptothecin | Anticancer, antiviral | Nacheva et al. [36] |
| *Leucojum aestivum* L. | Galantamine | Anti-alzheimer | Zagorska et al. [37] |
| *Narcissus* sp. L. | Galantamine | Anti-alzheimer | Khonakdari et al. [38] |
| *Hyoscyamus niger* L. | Scopolamine | Parasympatholytic | Uranbey et al. [39] |
| *Pilocarpus* sp. Vahl | Pilocarpine | Anti-glaucoma | Saba et al. [40] |
| *Podophyllum emodi* Wall ex Royle | Podophyllotoxin | Anticancer, antiviral | Chakraborty et al. [41] |
| Podophyllum peltatum L. | Podophyllotoxin | Anticancer, antiviral | Moraes-Cerdeira et al. [5] |
| *Rauwolfia serpentina* (L.) Benth ex Kurz | Reserpine | Hypotensive, sedative | Bhatt et al. [42] |
| *Taxus* sp L. | Paclitaxel | Anticancer | Abbasin et al. [43] |

## 3. Phytomedicine Crops for the Production of Standardized Therapeutic Natural Products

Herein, we describe the category of phytomedicinal crops similar to what Li et al. [18] reports regarding pharmaceutical crops for production of standardized therapeutic extracts (STEs). Additionally, we relate examples wherein micropropagation proves useful as a method for ensuring the stability of biomass supply of phytomedicines by allowing breeders to select phytomedicinal crops with an eye towards maintaining genetic consistency and the sustainability of wild plant population.

Also known as botanical drugs, herbal remedies, and herbal medicines, phytomedicines are classified in the United States as dietary supplements according to the specific claim as described in the Dietary Supplement Health and Education Act (DSHEA) of 1994 [18]. In Europe, the phytomedicines are standardized and certified medicinal products and in Asia they have a status of traditional medicine. Phytomedicinal crops relate to the cultivation of medicinal species by which a mixture of multiple active compounds commercialized as standardized products. Usually, phytomedicines are evaluated for quality as the means to ensure safety, as complex mixtures of secondary compounds, to maintain consistency is fundamental to their efficacy. Thus, authenticity and uniformity and well-defined cultivation practices and postharvest processes are essential to certify safety and efficacy. Govidaraghavan and Sucher [44] reinforce that herbal productions must follow good agricultural and collection practice (GACP), good plant authentication and identification practice (GPAIP), good manufacturing practice (GMP) before and during the manufacturing process, guided by analytical tools, and micropropagation is an important tool in ensuring uniformity and consistency in open pollinated crops.

As of today, the majority of phytomedicines are still harvested from the wild, which causes habitat destruction, genetic diversity loss, as well as ingredient mislabeling, variability and contamination. In Brazil, products are sourced from the wild, as well as from cultivation in agroforest or in small gardens. They are chosen without proper guide from health-care professionals because medical schools do not include in their curriculum the disciplines of phytomedicines or phytotherapy. In 2016 the Brazilian Health Regulatory Agency, ANVISA, officially recognized twenty-eight medicinal plant species as herbal drugs and published their monographs [45] in the first edition (Memento). The monographs are a complete therapeutic guide of phytomedicinals reviewed and accepted by ANVISA as therapies used in SUS, the public health system of Brazil. The majority of the phytomedicinals included in this first edition, was introduced to Brazil by immigrants and later became part of traditional use especially by the rural communities.

The increased consumption of phytomedicine offers an opportunity to develop medicinal plant production systems as crop. Conventional plant breeding may improve agronomic traits in association with molecular markers aiding crop development. The greatest obstacles for such a program remain predicting which extracts remain active, specifically resembling all the medicinal properties described in the ones harvested in the wild [46]. In this context, micropropagation may produce clones that could be screened using bioassays to assure bioactivity. Moraes et al. [17] used tissue culture techniques to produce *Echinacea* sp. clones and later screened those using human monocytes assays to identify high and low activity. The immune response between the two selected clones after field cultivation due to bacterial endophytes was the same [47]. The selection procedure using in vitro propagation techniques, genetic markers, and bioassay work are approaches for selection of elite germplasm [17].

Micropropagation allows one to mass generate plants with genetically identical chemotype for cultivation purposes. Reinhard [48] suggested that different chemotypes in Cat's Claw (*Uncaria tometosa* (Willd. ex Schults) DC) might have different healing properties: tetracyclic oxindole alkaloid acting on the central nervous system, and the pentacyclic oxindole alkaloid affecting the immune system. The immunological effect of both alkaloid mixtures is antagonistic and therefore may be unsuitable for therapy. For Reinhard [48], the production of safe and efficacious Cat's Claw phytomedicinal requires chemical identification prior to harvesting and perhaps even before the cultivation.

Micropropagation also allows one to select plants based on the chemical profile in order to standardize a particular chemotype. Morais et al. [49] reported that the chemical composition of *Lippia sidoides* Cham. (syn. *Lippia origanoides)* varied according to cultivation sites. Thymol is the major component of essential oil extracted from crops grown in northeast Brazil [50–52], whereas carvacrol is the major component present in *L. sidoides* harvested from Lavras, Minas Gerais [53] and 1.8-cineole, isoborneol, and bornyl acetate in São Gonçalo do Abaeté, Minas Gerais, Brazil. Standardized essential oil of *L. sidoides* is recommended for topical applications on skin, mucous membranes, mouth, throat and vaginal washings as antiseptic [45]. According to Santos et al. [53], genotypes regulate chemical polymorphism thymol and carvacrol. Phenotypical variation is likely to influence biological properties and the type of industrial application. Planting thymol or carvacrol clones ensured a high-quality biomass for safe and efficacious products [54].

Finally, micropropagation proves useful to reduce consumption pressure on potentially threatened wild populations [55]. For example, bark extraction of barbatimão to produce phytomedicine has depleted genetic diversity of *Stryphnodendron polyphythum* Mart. natural resources. The bark of this Brazilian tree is widely utilized as a wound-healing phytomedicine with anti-inflammatory, antioxidant and antimicrobial activities. Souza-Moreira et al. [55] showed that proanthocyanidins present in the bark are responsible for its healing properties. França et al. [10] published an efficient micropropagation protocol to produce barbatimão plantlets, while Correa et al. [56] defined the conditions for in vitro germplasm conservation to reduce pressure on its threatened status. Table 2 includes in vitro propagation protocols to produce healthy plantlets for cultivation purposes, thus aiding the development of phytomedicinal crops.

As the above paragraphs state, micropropagation can provide an effective technique to those seeking to mold a supply chain of a product, in order to ensure the genetic homogeneity of plant clones, chemical profile, and finally sustainability of those plants harvested in the wild.

**Table 2.** Micropropagation protocols of medicinal plants considered phytomedicine by the Brazilian Regulatory Agency (ANVISA).

| Plant Species (Common Name) | Herbal Constituents | Therapy | Micropropation Protocol |
|---|---|---|---|
| *Actaea racemosa* L. (Black cohosh) | Triterpenes | Hot flashes menopause | Lata et al. [57] |
| *Aesculus hippocastanum* L. (Horse chestnut) | Coumarins (Aesculetin), Triterpenoid Saponin Glycoside | Varicose vein syndrome | Sediva et al. [58] |
| *Allium sativum* L. (Garlic) | Thiosulfinates (Allicin), Terpenes | Bronchitis, asthma, arteriosclerosis | Ayabe and Sumi [59] |
| *Aloe vera* (L.) Burm. f. (Aloe) | Polysaccharides | Laxative, healing burns and wounds | Roy and Sarka [60] |
| *Calendula officinalis* L. (Calendula) | Flavonoids, Terpenes, | Anti-inflammatory, healing wounds | Çöçü et al. [61] |
| *Cynara scolymus* L. (Artichoke) | Flavonoids, Caffeoylquinic Acids | Hepatic-biliary, dysfunction and digestive complaints | El Boullani et al. [62] |
| *Echinacea purpurea* (L.) Moench (Echinacea) | Alkamides, Cichoric Acid, Polysaccharides | Cold treatment | Jones et al. [63] |
| *Ginkgo biloba* L. (Ginkgo) | Flavonoids, Terpene lactones | Circulatory disorders | Camper et al. [64] |
| *Harpagophytum procumbens* DC. ex Meisn. (Devil's claw) | Iridoid glycosides | Anti-inflammatory | Kaliamoorthy et al. [65] |
| *Hypericum perforatum* L. (St. John's wort) | Naphthodianthrones (Hypericin, pseudohypericin) | Antidepressant | Gadzovska et al. [66] |
| *Lippia sidoides* Cham. (Pepper rosmarin) | Essential Oils | Anti-inflammatory, antifungal, antiseptic | Costa et al. [54] |
| *Matricaria chamomilla* L. (Camomile) | Flavonoids, Essential Oils | antispasmodic, anti-inflammatory | Taniguchi & Tanakano [67] |
| *Maytenus ilicifolia* Mart. (Espinheira santa) | Flavonoids, Triterpenes | Gastric disordes | Pereira et al. [68] |
| *Passiflora incarnata* L. (Passion flower) | Flavonoids, Coumarin, Umbelliferone, Indol Alkaloids | Anxiolytic | Ozarowski &Thiem [69] |
| *Paullinia cupana* Kunth (Guaraná) | Caffeine | CNS stimmulant, antioxidant | Barbosa & Mendes [70] |
| *Peumus boldus* Molina (Boldo) | Essential oils, Aporphine Alkaloid, Flavonoids | Hepatic, diuretic, laxative | Rios et al. [71] |
| *Piper methysticum* G. Forst (Kava-kava) | Kavalactones | CNS activity, antidepression, anxiolytic | Zhang et al. [72] |
| *Psidium guajava* L. (Guava) | Tannins, Flavonoids, Triterpenes | Noninfectious diarrhea | Rawls et al. [73] |
| *Stryphnodendron adstringens* (Mart.) Coville (Barbatimão) | Tannins | Wound healing | França et al. [10] |
| *Uncaria tomentosa* (Willd. ex Schults) DC. (Cat's claw) | Flavonoids, Alkaloids, Saponins, Triterpenes | Anti-inflammatory | Pereira et al. [74] |
| *Valeriana officinailis* L. (Valeriana) | Terpenes, Valepotriates, Lignans | Anxiolytic, insomnia, sedative | Abdi et al. [75] |
| *Zingiber officinale* Roscoe (Ginger) | Essential oils, Shogaol, Zingerone, Gingerol | Anti-inflammatory, anti-emetic and chemo-protective | Abbas et al. [76] |

## 4. Micropropagation of Plants for Production of Large Therapeutic Molecules (LTM) Including Fructooligosaccharides Classified as Prebiotic

Li et al. [18] has called on LTMs crop plants to be cultivated for production of large molecules such as proteins and polysaccharides and engineered crops (GM) with the ultimate goal of producing drugs or vaccines at low cost. The LTM's crops are sources of proteolytic enzymes such as papain isolated from *Carica papaya* L., bromelain from fruits and stems of pineapple, and the bioactive momordica anti-HIV from *Momordica charantia* L. We included in the LTM's species those that supply prebiotic dietary fibers that are carbon sources for fermentation pathways in the colon to support digestive health. In this section, we focus on these prebiotic fibers to highlight how micropropagation may be used to create a stable supply of crops that produce LTMs.

Fructooligosaccharides/inulin also known as FOS are universally agreed-upon prebiotics [77], and species that are rich sources of dietary fibers have tremendous effect on gut microbiome. Humans cannot digest FOS. Instead, the gut microbiome ferments these non-digested carbohydrates and produces short chain fatty acids with health benefits such as reducing the risk of cancer and increasing the absorption of both calcium and magnesium. Research on the gut microbiome has increased exponentially, revealing that the intestines greatly affect human health, especially in relation to the immune system and behavior [78,79].

FOS are present in fruits, bulbs, rhizomes, and roots of banana, onion, garlic, and species belonging to the Agavaceae and Asteraceae, which are the richest sources of FOS including chicory (*Chicorium intybus* L.), globe artichoke (*Cynara cardunculus* var. *scolymus* L. Fiori, Jerusalem artichoke (*Helianthus tuberosus* L), elecampane *(Inula helenium* L.), bear's foot (*Smallanthus uvedalia* (L.) Mack. ex Mack and yacon (*Smallanthus sonchifolius* (Poepp.) H. Rob.). According to Roberfroid [78] chicory roots provide the commercial source of FOS for industrial applications, also known as inulin, which are extracted and then processed into short-chain fructans, such as the oligofructose with 2–10 degree of polymerization by partial enzymatic hydrolysis. López and Urías-Silvas [79] reviewed the use of Agave/FOS as prebiotics called agavins whose molecular structure is composed of a complex mixture of fructans. The agavins stimulated the growth of *Bifidobacterium breve* and *Lactobacillus casei* more efficiently than most commercial inulin [80]. Melilli et al. [81] evaluated (*Cyanara cardunculus* var. *scolymus* L.) germplasm for inulin with a high degree of polymerization in the Mediterranean environment, to reduce breeding time and offer growers uniform, healthy globe artichoke plants. Ozsan and Onus [82] compared in vitro micropropagation response of open-pollinated cultivars with F1 hybrids in maturity and height. They concluded that open-pollinated cultivars are cheaper than F1 and could be used for in vitro mass propagation.

The food industry considers FOS/inulin a natural ingredient that improves sensory characteristics such as taste and texture, the stability of foams, emulsions and mouthfeel in a large range of food applications like dairy products and baked goods reducing sugar and fat content while improving health [83]. Padalino et al. [84] added inulin with different degree polymerization with whole meal flour to improve quality of functional wheat spaghetti as example of processed food. The Global Market Insights reported that inulin's (FOS) market size in 2015 was 250 kilo tons and it is expected gains of 8.5% for 2023, likely to be worth more than US$ 2.5 billion [85]. The consumers of FOS are Europe, China, Japan and North America, with Japan being the world's largest market. The COVID-19 pandemic reinforced the major role of microbiota on the immune response and well-being. We expect that more consumers will pay more attention to prebiotics that modulate the gut microbiome.

Recent studies on the traditional food yacon (*Smallanthus sonchifolius* (Poeppig & Endlicher) H. Robinson), an Andean species, demonstrated that its roots are also a rich source of FOS with a smaller degree of polymerization than chicory. It has great potential as a prebiotic and sugar substitute due to its sweet taste that is related to degree of polymerization [86,87]. The role of yacon as FOS supplementation favors a healthy microbiota while reducing pathogenic population in the gut. Furthermore, short chain fatty acids produced by the beneficial bacteria improve glucose homeostasis and lipid metabolism. Clinical studies confirm that consumption of yacon as flour or syrup prevented and treated chronic diseases [88,89]. The beneficial compounds present in storage roots of yacon classify the spices as functional food (Figure 2).

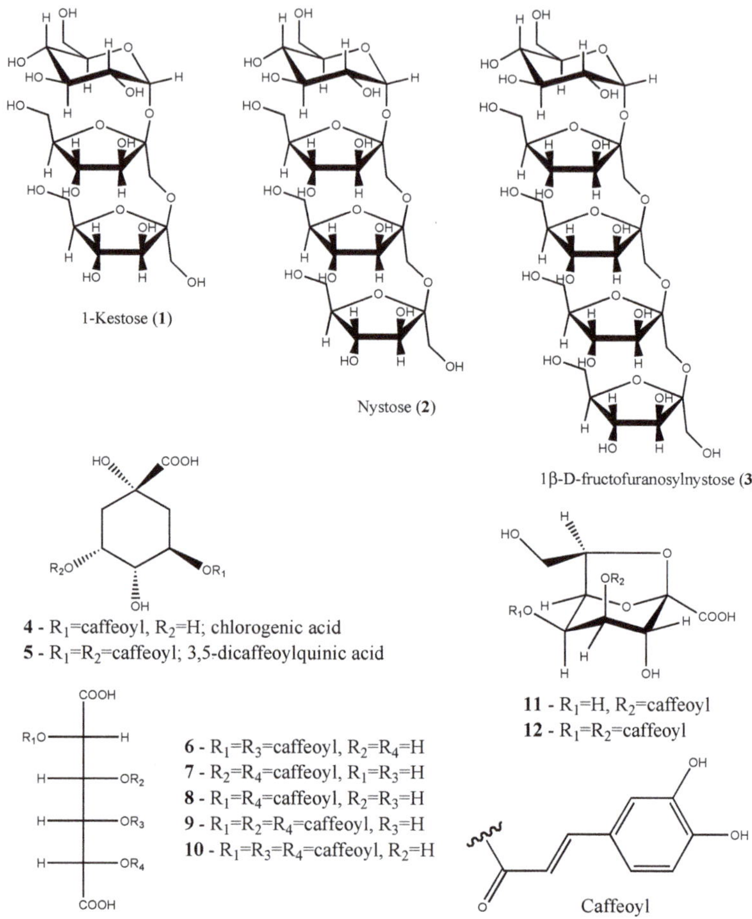


**Figure 2.** Yacon (*Smallanthus sonchifolius*) functional constituents.

Brazil is one of the largest agricultural producers in the world, but does not produce inulin/FOS from either source (chicory or artichoke) for applications in the food industry. However, in the vicinity of Sao Paulo City, yacon is produced for fresh consumption for its health benefits. Thus, to initiate any production system for supplying FOS as an ingredient with applications in the food and pharmaceutical industries, rhizospheres [90], along with storage roots may be better utilized to extract FOS. Micropropagation of yacon can still be done using axillary buds as explants of healthy plantlets for cultivation. Table 3 shows the published in vitro protocols of FOS producing plants for development of the business models of fructans as prebiotic.

Given the predicted increase in FOS/inulin consumption, supply of these LTM's crop plants will be necessary in a way such as the one suggested by McChesney et al. [20] a sustainable system to meet the demand. Yacon micropropagation is an example to stabilize the supply of crop plants as source of LTMs, thus ensuring that stability of production.

**Table 3.** Micropropagation protocols of FOS producing species.

| Plant Species | Common Name | Culture Purposes | Microprogation Protocol |
|---|---|---|---|
| *Agave* sp L. | Agave, maguey | Production of high yielding plants | Robert et al. [91] |
| *Chicorium intybus* L.<br>*Cuanara cardunculus var. scolymus* L. | Chicory<br>Globe artichoke | Germplasm conservation, Improve root quality for medicinal value<br>Propagation of open-pollinated cultivars | Previati et al. [92]<br>Dolinski and Olek [93]<br>Ozsan and Onus [82] |
| *Helianthus tuberosus* L. | Jerusalem artichoke | Large scale production of health plantlets | Abdalla [94] |
| *Smallanthus sonchifolius* (Poeppig & Endlicher) H. Robinson | Yacon | Production of healthy plantlets | Viehmannova et al. [95] |

## 5. Functional Food Crops

Metabolic syndrome is a global economic and social burden, understanding the origins, relevant factors contributing to high rates of obesity and its physiological impacts may reveal potential therapeutic targets.

Maintaining a healthy gut microbiome is one of the therapeutic goals that improve human health [83]. Dietary fibers promote wellbeing, and thus they are classified as functional food. Wildman [96] refers as functional food, the food, either natural or formulated, which will enhance physiological performance or prevent or treat disease and disorders.

Royston and Tolesfbol [97] refer to term epigenetic diet class of bioactive dietary compounds such as resveratrol in grapes, genistein in soybean, apigenin in celery, allicin in garlic, phenolic compounds in berries and omega 3 in Portulaca oleracea L. also known as purslanen [98] and other consumed foods, which have been shown to defend against the development of many different types of tumors. Compounds that act as epigenetic modulators prevent initiation and the progression of oncogenesis [97]. Micropropagation is an important tool for the propagation of selected lines in various breeding programs, as well as the recovery of pathogen-free material, or even for slow growth storage and the cryopreservation of valuable germplasm of fruit and vegetable crops.

## 6. Conclusions

Humans have long used plants to address various problems, the solutions to which often brought unintended consequences, such as overharvesting and environmental degradation. These negative consequences teach us the solution to our problems ought to be sustainable. Through a literature review, this paper argues that micropropagation can be a part of a strategy to reinforce the supply and quality of crops used for medicinal purposes: (1) small therapeutic molecules, (2) standard therapeutic extracts, (3) large therapeutic molecules, and (4) functional foods.

**Author Contributions:** Conceptualization, R.M.M. writing original draft R.M.M., A.L.C., and M.V.L. All authors have read and agreed to the published version of the manuscript.

**Funding:** This research received no external funding.

**Acknowledgments:** The authors thank Marcos Cerdeira for helping with editing.

**Conflicts of Interest:** The authors declare no conflict of interest.

## References


1. Hussain, A.; Ahmed, I.; Nazir, H.; Ullah, I. Plant Tissue Culture: Current Status and Opportunities. In *Recent Advances in Plant in vitro Culture*; Leva, A., Rinaldi, L.M.R.L., Eds.; InTech: London, UK, 2012; pp. 1–28. [CrossRef]
2. Chung, S.M.; Vaidya, M.; Tzfira, T. Agrobacterium is not alone: Gene transfer to plants by viruses and other bacteria. *Trends Plant Sci.* **2006**, *11*, 1–4. [CrossRef] [PubMed]
3. Sedaghati, B.; Haddad, R.; Bandehpour, M. Efficient plant regeneration and Agrobacterium-mediated transformation via somatic embryogenesis in purslane (*Portulaca oleracea* L.): An important medicinal plant. *Plant Cell Tissue Organ Cult.* **2019**, *136*, 231–245. [CrossRef]
4. Mohan, N.; Nikdad, S.; Singh, G. Studies on seed germination and embryo culture of *Jatropha curcas* L. under in vitro conditions. *Biotech. Bioinf. Bioenergy* **2011**, *1*, 187–194.
5. Moraes-Cerdeira, R.M.; Burandt, C.L.; Bastos, J.K.; Nanayakkara, N.P.D.; McChesney, J.D. In vitro propagation of *Podophyllum peltatum*. *Planta Med.* **1998**, *64*, 42–45. [CrossRef]
6. Mestinsek-Mubi, S.; Svetik, S.; Flajsman, M.; Murovec, J. In vitro tissue culture and genetic analysis of two high-CBD medical cannabis (*Cannabis sativa* L.) breeding lines. *Genetika* **2020**, *52*, 925–941. [CrossRef]
7. Lata, H.; Moraes, R.M.; Bertoni, B.; Pereira, A.M.S. In vitro germplasm conservation of *Podophyllum peltatum* L. under slow growth conditions. *In vitro Cell. Dev. Biol. Plant* **2010**, *46*, 22–27. [CrossRef]
8. Lata, H.; Chandra, S.; Techen, N.; Khan, I.A.; ElSohly, M.A. In vitro mass propagation of *Cannabis sativa* L.: A protocol refinement using novel aromatic cytokinin meta-topolin and the assessment of eco-physiological, biochemical and genetic fidelity of micropropagated plants. *J. Appl. Res. Med. Aromat. Plants* **2016**, *3*, 18–26. [CrossRef]
9. Thakur, J.; Dwivedi, M.D.; Sourabh, P.; Uniyal, P.L.; Pandey, A.K. Genetic Homogeneity Revealed Using SCoT, ISSR and RAPD Markers in Micropropagated *Pittosporum eriocarpum* Royle- An Endemic and Endangered Medicinal Plant. *PLoS ONE* **2016**, *11*. [CrossRef] [PubMed]
10. França, S.; Duarte, I.; Moraes, R.; Pereira, A.M.S. Micropropagation of *Stryphnodendron polyphythum* (Barbatimão). *Plant Cell Tissue Organ Cult.* **1995**, *42*, 291–293. [CrossRef]
11. Patel, A.K.; Lodha, D.; Shekhawat, N.S. An improved micropropagation protocol for the ex situ conservation of *Mitragyna parvifolia* (Roxb.) Korth. (Rubiaceae): An endangered tree of pharmaceutical importance. *In vitro Cell. Dev. Biol. Plant* **2020**, *56*, 817–826. [CrossRef] [PubMed]
12. Rout, G.R.; Samantaray, S.; Das, P. In vitro manipulation and propagation of medicinal plants. *Biotechnol. Adv.* **2000**, *18*, 91–120. [CrossRef]
13. Halder, M.; Roychowdhury, D.; Jha, S. A Critical Review on Biotechnological Interventions for Production and Yield Enhancement of Secondary Metabolites in Hairy Root Cultures. In *Hairy Roots*; Srivastava, V., Mehrotra, S., Mishra, S., Eds.; Springer: Singapore, 2018; pp. 21–44.
14. Lata, H.; Li, X.C.; Silva, B.; Moraes, R.M.; Halda-Alija, L. Identification of IAA-producing endophytic bacteria from micropropagated Echinacea plants using 16S rRNA sequencing. *Plant Cell Tissue Organ Cult.* **2006**, *85*, 353–359. [CrossRef]
15. Rosa, L.; Tabanca, N.; Techen, N.; Wedge, D.; Pan, Z.; Bernier, U.; Becnel, J.; Agramonte, N.; Walker, L.; Moraes, R. Diversity and Biological Activities of Endophytic Fungi Associated with Micropropagated Medicinal Plant Echinacea purpurea (L.) Moench. *Am. J. Plant Sci.* **2012**, *3*, 1105–1114. [CrossRef]
16. Wang, W.; Xu, J.; Fang, H.; Li, Z.; Li, M. Advances and challenges in medicinal plant breeding. *Plant Sci.* **2020**, *298*, 110573. [CrossRef] [PubMed]
17. Moraes, R.; Lata, H.; Sumyanto, J.; Pereira, A.; Bertoni, B.; Joshi, V.; Pugh, N.; Khan, I.; Pasco, D. Characterization and pharmacological properties of in vitro propagated clones of *Echinacea tennesseensis* (Beadle) Small. *Plant Cell Tissue Organ Cult.* **2011**, *106*, 309–315. [CrossRef]
18. Li, S.; Yuan, W.; Yang, P.; Antoun, M.; Balick, M.; Cragg, G. Pharmaceutical crops: An overview. *Pharm. Crop* **2010**, *1*, 1–17. [CrossRef]
19. Cragg, G.; Newman, D. Plants as a source of anti-cancer and anti-HIV agents. *Ann. Appl. Biol.* **2003**, *143*, 127–133. [CrossRef]
20. McChesney, J.; Venkataraman, S.; Henri, J. Plant natural products: Back to the future or into extinction? *Phytochemistry* **2007**, *68*, 2015–2022. [CrossRef] [PubMed]
21. Foster, S. Medicinal plant conservation and genetic resources: Examples from the temperate Northern hemisphere. *Acta Hortic.* **1993**, *330*, 67–74. [CrossRef]
22. Cragg, G.; Schepartz, S.; Suffness, M.; Grever, M. The Taxol Supply Crisis. New NCI Policies for Handling the Large-Scale Production of Novel Natural Product Anticancer and Anti-HIV Agents. *J. Nat. Prod.* **1993**, *56*, 1657–1668. [CrossRef]
23. Bastos, J.K.; Burandt, C.L.; Nanayakkara, N.P.D.; Bryant, L.; McChesney, J.D. Quantitation of Aryltetralin Lignans in Plant Parts and among Different Populations of *Podophyllum peltatum* by Reversed-Phase High-Performance Liquid Chromatography. *J. Nat. Prod.* **1996**, *59*, 406–408. [CrossRef]
24. Bedir, E.; Khan, I.; Moraes, M.R. Bioprospecting for podophyllotoxin. In *Trends in New Crops and New uses*; Janik, J., Whipkey, A., Eds.; ASHS Press: Alexandria, VA, USA, 2001; pp. 545–549.
25. Elsohly, H.; Croom, E.; Kopycki, W.; Joshi, A.; Elsohly, M.; McChesney, J. Concentrations of taxol and related taxanes in the needles of different *Taxus cultivars*. *Phytochem. Anal.* **1995**, *6*, 149–156. [CrossRef]

26. Sisti, N.J.; Brinkman, H.R.; McChesney, J.D.; Chander, M.D.; Liang, X.; Zygmunt, J. Methods and useful intermediates for paclitaxel synthesis from C-7, C-10 di-cbz 10-deacetylbaccatin III. Patent US 6448417B1, 10 September 2002.
27. Majada, J.; Sierra, M.; Sánchez-Tamés, R. One step more towards taxane production through enhanced Taxus propagation. *Plant Cell Rep.* **2000**, *19*, 825–830. [CrossRef] [PubMed]
28. Canel, C.; Dayan, F.; Ganzera, M.; Khan, I.; Rimando, A.; Burandt, C.; Moraes, R. High Yield of Podophyllotoxin from Leaves of *Podophyllum peltatum* by In situ Conversion of Podophyllotoxin 4-O-β-D-Glucopyranoside. *Planta Med.* **2001**, *67*, 97–99. [CrossRef] [PubMed]
29. Moraes, R.; Burandt, C.; Ganzera, M.; Li, X.; Khan, I.; Canel, C. The American mayapple revisited—*Podophyllum peltatum*—Still a potential cash crop? *Econ. Bot.* **2000**, *54*, 471–476. [CrossRef]
30. Moraes, R.; Momm, H.; Silva, B.; Maddox, V.; Easson, G.; Lata, H.; Ferreira, D. Geographic Information System Method for Assessing Chemo-Diversity in Medicinal Plants. *Planta Med.* **2006**, *71*, 1157–1164. [CrossRef]
31. Graham, I.; Besser, K.; Blumer, S.; Branigan, C.; Czechowski, T.; Elias, L.; Guterman, I.; Harvey, D.; Isaac, P.; Khan, A.; et al. The Genetic Map of *Artemisia annua* L. Identifies Loci Affecting Yield of the Antimalarial Drug Artemisinin. *Science* **2010**, *327*, 328–331. [CrossRef]
32. Ferreira, J.F.S.; Janick, J. Roots as an enhancing factor for the production of artemisinin in shoot cultures of *Artemisia annua*. *Plant Cell Tissue Organ Cult.* **1996**, *44*, 211–217. [CrossRef]
33. Wetzstein, H.; Porter, J.; Janick, J.; Ferreira, J.F.S.; Mutui, T.M. Selection and Clonal Propagation of High Artemisinin Genotypes of *Artemisia annua*. *Front Plant Sci.* **2018**, *9*, 358. [CrossRef] [PubMed]
34. Etienne, H.; Berthoulym, M. Temporary immersion systems in plant micropropagation. *Plant Cell Tissue Organ Cult.* **2002**, *69*, 215–231. [CrossRef]
35. Kumar, A.; Prakash, K.; Sinha, R.; Kumar, N. In Vitro Plant Propagation of *Catharanthus roseus* and Assessment of Genetic Fidelity of Micropropagated Plants by RAPD Marker Assay. *Appl. Biochem. Biotechnol.* **2013**, *169*, 804–900. [CrossRef] [PubMed]
36. Nacheva, L.; Dimitrova, N.; Ivanova, V.; Cao, F.; Zhu, Z. Micropropagation of Camptotheca Acuminata Decne (Nyssaceae)—Endangered Ornamental and Medicinal Tree. *Acta Univ. Agric. Silvic.* **2020**, *68*, 679–686.
37. Zagorska, N.; Stanilova, M.; Ilcheva, V.; Gadeva, P. Micropropagation of *Leucojum aestivum* L. (Summer Snowflake). In *High-Tech and Micropropagation VI*; Bajaj, Y.P.S., Ed.; Springer: Heidelberg, Germany, 1997; Volume 40, pp. 178–192. [CrossRef]
38. Khonakdari, M.; Rezadoost, H.; Heydari, R.; Mirjalili, M. Effect of photoperiod and plant growth regulators on in vitro mass bulblet proliferation of *Narcissus tazzeta* L. (*Amaryllidaceae*), a potential source of galantamine. *Plant Cell Tissue Organ Cult.* **2020**, *142*, 187–199. [CrossRef]
39. Uranbey, S. Thidiazuron induced adventitious shoot regeneration in *Hyoscyamus niger*. *Biol. Plantarum.* **2005**, *49*, 427–430. [CrossRef]
40. Sabá, R.; Lameira, O.; Luz, J.; Gomes, A.; Innecco, R. Micropropagation of the jaborandi. *Hortic. Bras.* **2002**, *20*, 106–109.
41. Chakraborty, A.; Bhattacharya, D.; Ghanta, S.; Chattopadhyay, S. An efficient protocol for in vitro regeneration of *Podophyllum hexandrum*, a critically endangered medicinal plant. *Indian J. Biotech.* **2010**, *9*, 217–220.
42. Bhatt, R.; Arif, M.; Gaur, A.K.; Rao, P.B. Rauwolfia serpentina: Protocol optimization for in vitro propagation. *Afr. J. Biotechnol.* **2009**, *7*, 4265–4268.
43. Abbasin, Z.; Zamani, S.; Movahedi, S.; Khaksar, G.; Tabatabaei, B.E.S. In vitro micropropagation of yew (*Taxus baccata*) and Production of Plantlets. *Biotechnology* **2010**, *9*, 48–54. [CrossRef]
44. Govidaraghavan, S.; Sucher, N.J. Quality Assessment of medicinal herbs and their extracts: Criteria and prerequisites for consistent safety and efficacy of herbal medicine. *Epilepsy Behav.* **2015**, *52*, 363–371. [CrossRef] [PubMed]
45. ANVISA. *Memento Fitoterápico da Farmacopéia Brasileira*; Agência Nacional de Vigilância Sanitária: Brasília, Brazil, 2016.
46. Canter, P.; Thomas, H.; Ernst, E. Bringing medicinal plants into cultivation: Opportunities and challenges for biotechnology. *Trends Biotechnol.* **2005**, *23*, 180–185. [CrossRef] [PubMed]
47. Tamta, H.; Pugh, N.D.; Balachandran, P.; Moraes, R.; Sumiyanto, J.; Pasco, D. Variability of In Vitro Macrophage Activation by Commercially Diverse Bulk Echinacea Plant Material is Due Predominantly to Bacterial Lipoproteins and Lipopolysaccharides. *J. Agr. Food Chem.* **12009**, *56*, 10552–10556. [CrossRef] [PubMed]
48. Reinhard, K.H. Uncaria tomentosa (Willd) D.D. Cat's Claw, Una de Gato or Saventraro. *J. Altern. Complement. Med.* **1999**, *5*, 143–151. [CrossRef]
49. Morais, S.R.; Oliveira, T.L.S.; Oliveira, L.P.; Tresvenzol, L.M.F.; da Conceicao, E.C.; Rezende, M.H.; Fiuza, T.d.S.; Costa, E.A.; Ferri, P.H.; de Paula, J.R. Essential Oil Composition, Antimicrobial and Pharmacological Activities of Lippia sidoides Cham. (Verbenaceae) From São Gonçalo do Abaeté, Minas Gerais. *Brazil. Pharmacogn. Mag.* **2016**, *12*, 262–270.
50. Cavalcanti, S.; Niculau, E.S.; Blank, A.; Camara, C.; Araújo, I.N.; Alves, P. Composition and acaricidal activity of Lippia sidoides essential oil against two-spotted spider mite (*Tetranychus urticae* Koch). *Bioresour. Technol.* **2009**, *101*, 829–832. [CrossRef]
51. Marco, C.; Teixeira, E.; Simplicio, A.A.F.; Oliveira, C.; Costa, J.; Feitosa, J. Chemical composition and allelopathyc activity of essential oil of *Lippia sidoides* Cham. *Chil. J. Agr. Res.* **2012**, *72*, 157–160. [CrossRef]
52. Lima, R.; Cardoso, M.; Moraes, J.C.; Carvalho, S.; Rodrigues, V.; Guimarães, L. Chemical composition and fumigant effect of essential oil of Lippia sidoides Cham. and monoterpenes against Tenebrio molitor (L.) (Coleoptera: Tenebrionidae). *Cienc. Agrotec.* **2011**, *35*, 664–671. [CrossRef]

53. Santos, C.P.; Oliveira, T.C.; Pinto, J.A.O.; Fontes, S.S.; Cruz, E.M.O.; Arrigoni-Blank, M.F.; Andrade, T.M.A.; Matos, I.L.; Alves, P.B.; Innecco, R.; et al. Chemical diversity and influence of plant age on the essential oil from *Lippia sidoides* Cham. *Germplasm. Ind. Crop. Prod.* **2015**, *76*, 416–421. [CrossRef]
54. Costa, A.; Arrigoni-Blank, M.; Blank, A.; Mendonça, A.; Amancio, V.; Lédo, A. In vitro establishment of *Lippia sidoides* Cham. *Hortic. Bras.* **2007**, *25*, 68–72. [CrossRef]
55. Souza-Moreira, T.; Queiroz-Fernandes, G.; Pietro, R. Stryphnodendron Species Known as "Barbatimão": A Comprehensive Report. *Molecules* **2018**, *23*, 910. [CrossRef]
56. Corrêa, V.; Cerdeira, A.; Fachin, A.; Bertoni, B.; Pereira, P.; França, S.; Momm, H.; Moraes, R.; Pereira, A. Geographical variation and quality assessment of *Stryphnodendron adstringens* (Mart.) Coville within Brazil. *Genet. Resour. Crop. Evol.* **2012**, *59*, 1349–1356. [CrossRef]
57. Lata, H.; Bedir, E.; Hosick, A.; Ganzera, M.; Khan, I.; Moraes, R. In vitro Plant Regeneration from Leaf-Derived Callus of *Cimicifuga racemosa*. *Planta Med.* **2002**, *68*, 912–915. [CrossRef]
58. Sedivá, J.; Vlašínová, H.; Mertelík, J. Shoot regeneration from various explants of horse chestnut (*Aesculus hippocastanum* L.). *Sci. Hortic.* **2013**, *161*, 223–227. [CrossRef]
59. Ayabe, M.; Sumi, S. Establishment of a novel tissue culture method, stem-disc culture, and its practical application to micropropagation of garlic (*Allium sativum* L.). *Plant Cell Rep.* **1998**, *17*, 773–779. [CrossRef] [PubMed]
60. Roy, S.; Sarkar, A. In vitro regeneration and micropropagation of *Aloe vera* L. *Sci. Hortic.* **1991**, *47*, 107–113. [CrossRef]
61. Çöçü, S.; Uranbey, S.; İpek, A.; Khawar, K.M.; Sarihan, E.O.; Kaya, M.D.; Parmaksız, I.; Özcan, S. Adventitious Shoot Regeneration and Micropropagation in *Calendula officinalis* L. *Biol. Plant.* **2004**, *48*, 449–451. [CrossRef]
62. El Boullani, R.; Elmoslih, A.; Elfinti, A.; Abdelhamid, E.M.; Serghini, M. Improved in Vitro Micropropagation of Artichoke (*Cynara cardunculus* var. *scolymus* L.). *Eur. J. Sci. Res.* **2012**, *80*, 430–436.
63. Jones, A.; Yi, Z.; Murch, S.; Saxena, P. Thidiazuron-induced regeneration of *Echinacea purpurea* L.: Micropropagation in solid and liquid culture systems. *Plant Cell Rep.* **2007**, *26*, 13–19. [CrossRef]
64. Camper, N.; Coker, P.; Wedge, D.; Keese, R. In vitro culture of Ginkgo. *In Vitro Cell. Dev. Biol. Plant* **1997**, *33*, 125–127. [CrossRef]
65. Kaliamoorthy, S.; Naidoa, G.; Achar, P. Micropropagation of *Harpagophytum procumbens*. *Biol. Plantarum* **2008**, *52*, 191–194. [CrossRef]
66. Gadzovska Simic, S.; Maury, S.; Saida, O.; Righezza, M.; Kascakova, S.; Refregiers, M.; Spasenoski, M.; Joseph, C.; Hagège, D. Identification and quantification of hypericin and pseudohypericin in different *Hypericum perforatum* L. in vitro cultures. *Plant Physiol. Biochem.* **2005**, 591–601. [CrossRef]
67. Taniguchi, H.; Takano, H.T. Micropropagation of *Matricaria chamomilla* (*Camomile*). In *High-Tech and Micropropagation VI*; Biotechnology in Agriculture and Forestry; Bajaj, Y.P.S., Ed.; Springer: Berlin/Heidelberg, Germany, 2012; pp. 332–342.
68. Pereira, A.; Moro, J.; Cerdeira, R.; França, S. Effect of phytoregulators and physiological characteristics of the explants on micropropagation of *Maytenus ilicifolia*. *Plant Cell Tissue Organ Cult.* **1995**, *42*, 295–297. [CrossRef]
69. Ożarowski, M.; Thiem, B. Progress in micropropagation of *Passiflora* spp. to produce medicinal plants: A mini-review. *Rev. Bras. Farmacogn.* **2013**, *23*, 937–947. [CrossRef]
70. Barbosa, C.B.; Mendes, L.A.C. Estabelecimento de protocolo para o cultivo in vitro do guarana (*Paullinia cupana* (Mart.) Ducke). In *Anais da 1 Jornada de Iniciação Científica da Embrapa Amazonia*; Embrapa: Brasilia, Brazil, 2004.
71. Rios, D.; Sandoval, D.; Gomes, C. In vitro culture of Peumus boldus Molina via direct organogenesis. *J. Med. Chem.* **2010**, *21*, 70–72.
72. Zhang, Z.; Zhao, L.; Chen, X.; Zheng, X. Sucessful micropropagation protocol of *Piper methysticum*. *Biol. Plantarum.* **2008**, *52*, 110–112. [CrossRef]
73. Rawls, B.; Harris-Shultz, K.; Dhekney, S.; Forrester, I.; Sitther, V. Clonal Fidelity of Micropropagated *Psidium guajava* L. Plants Using Microsatellite Markers. *Am. J. Plant Sci.* **2015**, *6*, 2385–2392. [CrossRef]
74. Pereira, R.; Valente, L.; Pinto, J.E.; Bertolucci, S.; Bezerra, G.; Alves, F.; Santos, P.; Benevides, P.; Siani, A.; Rosario, S.; et al. In Vitro Cultivated *Uncaria tomentosa* and *Uncaria guianensis* with Determination of the Pentacyclic Oxindole Alkaloid Contents and Profiles. *J. Braz. Chem. Soc.* **2008**, *19*, 1193–1200. [CrossRef]
75. Abdi, G.; Salehi, H.; Khosh-Khui, M. Nano silver: A novel nanomaterial for removal of bacterial contaminants in valerian (*Valeriana officinalis* L.) tissue culture. *Acta Physiol. Plant.* **2008**, *30*, 709–714. [CrossRef]
76. Abbas, M.; Taha, H.; Aly, U.; El-Shabrawi, H.; Gaber, E.-S. In vitro propagation of ginger (*Zingiber officinale* Rosco). *J. Gene. Eng. Biotechnol.* **2011**, *9*, 165–172. [CrossRef]
77. Carlson, J.; Erickson, J.; Lloyd, B.; Slavin, J. Health Effects and Sources of Prebiotic Dietary Fiber. *Curr. Dev. Nutr.* **2018**, 1–8. [CrossRef]
78. Roberfroid, M. Prebiotics and probiotics: Are they functional foods? *Am. J. Clin. Nutr.* **2000**, *71*, 1682–1688. [CrossRef]
79. López, M.G.; Urias-Silvas, J. Agave fructans as prebiotics. In *Recent Advances in Fructooligosaccharides Research*; Norio, S., Noureddine, B., Shuichi, O., Eds.; Research Signpost: Trivandrum, Kerala, India 2007; pp. 297–310.
80. Al-Sheraji, S.H.; Ismail, A.; Manap, M.Y.; Mustafa, S.; Yusof, R.M.; Hassan, F.A. Prebiotics as functional foods. *J. Funct. Food* **2013**, 1542–1553. [CrossRef]
81. Melilli, M.G.; Branca, F.; Sillitti, C.; Scandurra, S.; Calderaro, P.; Di Stefano, V. Germplasm evaluation to obtain inulin with high degree of polymerization in Mediterranean environment. *Nat. Prod. Res.* **2020**, *34*, 187–191. [CrossRef] [PubMed]

82. Ozsan, T.; Onus, A.N. Comparative study on in vitro micropropagation response of seven globe artichoke *Cynara cardunculus* var. *scolymus* (L.) Fiori cultivars: Open-pollinated cultivars vs F-1 hybrids. *Not. Bot. Horti. Agrobot. Cluj Napoca* **2020**, *48*, 1210–1220. [CrossRef]
83. Caetano, B.; Sivieri, K.; Antunes de Moura, N. Yacon (Smallanthus sonchifolius) as a Food Supplement: Health-Promoting Benefits of Fructooligosaccharides. *Nutrients* **2016**, *8*, 463. [CrossRef] [PubMed]
84. Padalino, L.; Costa, C.; Conte, A.; Melilli, M.G.; Sillitti, C.; Bognanni, R.; Raccuia, S.A.; Del Nobile, M.A. The quality of functional whole-meal durum wheat spaghetti as affected by inulin polymerization degree. *Carbohydr. Polym.* **2017**, *173*, 84–90. [CrossRef]
85. Global Market Insights. Available online: https://www.gminsights.com/industry-analysis/inulin-market#:~{}:text=Global%20inulin%20market%20size%20was,USD%202.5%20billion%20by%202023 (accessed on 3 August 2020).
86. Genta, S.; Habib, N.; Pons, J.; Manrique, I.; Grau, A.; Sanchez, S. Yacon syrup: Beneficial effects on obesity and insulin resistance in humans. *Clin. Nutr. ESPEN* **2009**, *28*, 182–187. [CrossRef] [PubMed]
87. Geyer, M.; Manrique, I.; Degen, L.; Beglinger, C. Effect of Yacon (*Smallanthus sonchifolius*) on Colonic Transit Time in Healthy Volunteers. *Digestion* **2008**, *78*, 30–33. [CrossRef] [PubMed]
88. Adriano, L.; Dionísio, A.; Abreu, F.; Carioca, A.A.F.; Zocolo, G.; Wurlitzer, N.; Pinto, C.; Oliveira, A.; Sampaio, H. Yacon syrup reduces postprandial glycemic response to breakfast: A randomized, crossover, double-blind clinical trial. *Int. Food. Res. J.* **2019**, *126*, 108682. [CrossRef] [PubMed]
89. Machado, A.; Silva, N.; Chaves, J.; Alfenas, R. Consumption of yacon flour improves body composition and intestinal function in overweight adults: A randomized, double-blind, placebo-controlled clinical trial. *Clin. Nutr. ESPEN* **2018**, *29*, 22–29. [CrossRef] [PubMed]
90. Sumiyanto, J.; Dayan, F.; Cerdeira, A.; Wang, Y.-H.; Khan, I.; Moraes, R. Oligofructans content and yield of yacon (*Smallanthus sonchifolius*) cultivated in Mississippi. *Sci. Hortic.* **2012**, *148*, 83–88. [CrossRef]
91. Robert, M.L.; Herrera-Herrera, J.L.; Castilho, C.; Ojeda, G.; Herrera-Amillo, M.A. An Efficient method for the micropropagation of Agave Species. *Methods Mol. Biol.* **2006**, *318*, 165–178. [PubMed]
92. Prevati, A.; Benelli, C.; Re, F.D.; Carlo, A.; Vettori, C.; Lambardi, M. In vitro propagation and conservation of red chicory germplasm. In Proceedings of the The Role of Biotechnology, Turim, Italy, 5–7 March 2005; pp. 207–208.
93. Dolinski, R.; Olek, A. Micropropagation of wild chicory (*Cichorium intybus* L. var. silvestre bisch.) from leaf explants. *Acta Sci. Pol. Hortorum Cultus* **2013**, *12*, 33–44.
94. Abdalla, N.A. Micropropagation of Jerusalem Artichoke (*Helianthus tuberosus* L.) Plant. Ph.D. Thesis, Department of Horticulture Faculty of Agriculture, Ain Shams University, Cairo, Egypt, 2020.
95. Viehmannova, I.; Bortlova, Z.; Vitamvas, J.; Cepkova, P.H.; Eliasova, K.; Svobodova, E.; Travnickova, M. Assessment of somaclonal variation in somatic embryo-derived plants of yacon [*Smallanthus sonchifolius* (Poepp. And Endl.) H. Robinson] using inter simple sequence repeat analysis and flow cytometry. *Electron. J. Biotechnol.* **2014**, 102–106. [CrossRef]
96. Wildman, R.E.C. Nutraceuticals: A Brief review of historical and teleological aspects. In *Hand Book of Nutraceuticals and Functional Foods*; Wildman, R.E.C., Ed.; CRC Press: London, UK, 2001.
97. Royston, K.I.; Tollefsbol, T.O. The epigenetic impact of cruciferous vegetables on cancer prevention. *Curr. Pharmacol. Rep.* **2015**, *1*, 46–51. [CrossRef]
98. Melilli, M.G.; Pagliaro, A.; Scandurra, S.; Gentile, C.; Di Stefano, V. Omega-3 rich foods: Durum wheat spaghetti fortified with *Portulaca oleracea*. *Food Biosci.* **2020**, 37. [CrossRef]

Review

# Approaches toward the Separation, Modification, Identification and Scale up Purification of Tetracyclic Diterpene Glycosides from *Stevia rebaudiana* (Bertoni) Bertoni

**Wilmer H. Perera** [1,*] **and James D. McChesney** [2,*]

1 CAMAG Scientific, Inc., 515 Cornelius Harnett Drive, Wilmington, NC 28401, USA
2 Ironstone Separations, Inc., Etta, MS 38627, USA
* Correspondence: wilmer.perera@gmail.com (W.H.P.); jdmcchesney@yahoo.com (J.D.M.); Tel.: +1-800-334-3909 or +1-303-808-4104 (W.H.P.); Fax: +1-910-343-1834 (W.H.P.)

**Abstract:** *Stevia rebaudiana* (Bertoni) Bertoni is a plant species native to Brazil and Paraguay well-known by the sweet taste of their leaves. Since the recognition of rebaudioside A and other steviol glycosides as generally recognized as safe by the United States Food and Drug Administration in 2008 and grant of marketing approval by the European Union in 2011, the species has been widely cultivated and studied in several countries. Several efforts have been dedicated to the isolation and structure elucidation of minor components searching for novel non-caloric sugar substitutes with improved organoleptic properties. The present review provides an overview of the main chemical approaches found in the literature for identification and structural differentiation of diterpene glycosides from *Stevia rebaudiana*: High-performance Thin-Layer Chromatography, High-Performance Liquid Chromatography, Electrospray Ionization Mass Spectrometry and Nuclear Magnetic Resonance Spectroscopy. Modification of diterpene glycosides by chemical and enzymatic reactions together with some strategies to scale up of the purification process saving costs are also discussed. A list of natural diterpene glycosides, some examples of chemically modified and of enzymatically modified diterpene glycosides reported from 1931 to February 2021 were compiled using the scientific databases Google Scholar, ScienceDirect and PubMed.

**Keywords:** *Stevia rebaudiana*; HPTLC; HPLC methods; mass spectrometry dissociation pattern; NMR; endocyclic and exocyclic diterpene glycosides; scale up

**Citation:** Perera, W.H.; McChesney, J.D. Approaches toward the Separation, Modification, Identification and Scale up Purification of Tetracyclic Diterpene Glycosides from *Stevia rebaudiana* (Bertoni) Bertoni. *Molecules* **2021**, *26*, 1915. https://doi.org/10.3390/molecules26071915

Academic Editor: Muhammad Ilias

Received: 3 March 2021
Accepted: 24 March 2021
Published: 29 March 2021



## 1. Introduction

*Stevia rebaudiana*, a native plant species to Brazil and Paraguay, was discovered by Mosè Giacomo Bertoni in 1899, an Italian-speaking Swiss naturalist and described as *Eupatorium rebaudianum* Bertoni. In 1905 the plant was grouped in a different genus and named as *Stevia rebaudiana* (Bertoni) Bertoni. *S. rebaudiana* has been widely used to sweeten teas and potions for centuries by Guarani indigenous due to the high content in diterpene glycosides (DGs) in their leaves [1]. Essential oils have been reported from *S. rebaudiana*, being spathulenol (13.4–40.9%), caryophyllene oxide (1.3–18.7%), β-caryophyllene (2.1–16.0%) and β-pinene (5.5–21.5%) the major volatile compounds [2]. Non-glycosidic labdane-type diterpenes have also been isolated e.g., sterebins I -N [3] and several hydroxycinnamate and flavonoid derivatives have been identified [4,5]. There is no doubt, that the DGs is the most significant group of secondary metabolites from *S. rebaudiana* due to their applications in the Sweeteners Industry. Stevioside was the first steviol glycoside isolated from this Asteraceae in 1931, reported by M. Bridel and R. Lavielle, two French Chemist [6]. Then, rebaudiosides A, B and steviolbioside were discovered by a Japanese group 45 years later, followed by dulcosides A, B and rebaudiosides C (same chemical structure as dulcoside B), D and E in 1977 [7–10]. In 2002 rebaudioside F and its saponification products were reported [11]. In 2008, the United States Food and Drug Administration granted generally recognized as safe (GRAS) regulatory acceptance of rebaudioside A while Food Standards Australia New Zeeland accepted the use of steviol glycosides [12,13]. Other steviol glycosides were recognized as safe in USA in 2010, and the European Union approved DGs from *S. rebaudiana* for marketing in November 2011 [14,15]. A great scientific contribution was made by Ohta et al. [16], where 21 steviol glycosides were isolated and characterized. Since that time, multiple groups have been working intensively on searching for new steviol glycosides. Rebaudioside A and stevioside, the major steviol glycosides (accounting for more than 40% of the DGs present in leaf extract) have shown more potent natural sweetness than sucrose which make them good candidates as non-caloric sugar-substitutes [1]. Hence, *S. rebaudiana* is being widely commercialized in mainland China, Japan, Korea, India, and elsewhere (Midmore and Rank, 2012) and several rebaudioside A-based products as well as table-top sweeteners are commercially available in the US consumer market.

It was recently discovered that rebaudioside A and stevioside, potentiate the activity of TRPM5 (a $Ca^{2+}$-activated cation channel expressed in type II taste receptor cells and pancreatic β-cells) enhancing glucose-induced insulin secretion in a TRPM5-dependent manner. TRPM5 has been suggested as a potential target to prevent and treat type 2 diabetes [17]. It has also been described that steviol glycosides stimulates insulin secretion in a dose- and glucose-dependent manner from isolated mouse islets of Langerhans. Steviol glucuronide has been identified in human urine, thus steviol is the main active metabolite after oral intake of steviol glycosides [18,19]. The antihyperglycemic and blood pressure-reducing effects of stevioside in the diabetic Goto-Kakizaki rat and the mechanism of the hypoglycemic effect of stevioside has been described [20,21].

Despite the potential positive human health benefits of rebaudioside A and the relatively easy access to this steviol glycoside, its lingering aftertaste makes it less desirable as a sweetener. Therefore, several efforts have been dedicated to preparing commercial *Stevia* extracts rich in DGs to find next generation non-caloric DGs with improved organoleptic properties and thus, attract consumers to a better health lifestyle.

Important advances have been made over the last several years in the methods of analysis and purification of these complex mixtures of glycosides. All these developments have allowed the discovery of several new natural and slightly modified DGs with remarkable potential as sugar substitutes e.g., several rebaudioside A isomers, a stevioside like-compound with a disaccharide linked at position C-12 of the aglycone [22], rebaudioside R (unique example with a non-glucose monosaccharide linked directly at position C-13 (xylose) [23], rebaudioside U and similar compounds, the only steviol glycosides with an arabinopyranosyl moiety [24,25], rebaudioside W (unique hexa-glycoside from *S. rebaudiana* with three different sugar monosaccharides attached (glucose, xylose and rhamnose) [26]. Additionally, rebaudioside IX series are the largest natural DGs composed of nine glucose units linked to steviol [27–29], only one natural example of rebaudioside M isomer has been reported [30] and a few new compounds with an *ent*-atisene core have also been reported [31,32]. Nowadays, many natural, chemically, and enzymatically modified diterpene glycosides with diverse sugar interlinkages and cores have been discovered, hence, a wide spectrum of structures are now available to evaluate for their sweetener potential and to understand better the relationship between structure-sweetness/lingering aftertaste. Rebaudioside M, has been pointed out as the next generation non-caloric sweetener but in our non-professional tasters experience there are other DGs with interesting organoleptic properties that should be explored e.g., rebaudiosides U and Y.

Herein, we summarize the most noteworthy chemical approaches for the separation, identification and structural elucidation of diterpene glycosides from *Stevia rebaudiana*: High-performance Thin-Layer Chromatography, High-Performance Liquid Chromatography, High-Resolution Electrospray Ionization Mass Spectrometry and Nuclear Magnetic Resonance Spectroscopy. The modification of diterpene glycosides by chemical and enzymatic reactions together with some strategies to achieve the scale up of the purification process economically sustainable are also discussed. An updated listing of the natural and chemically modified diterpene glycosides has been included.

## 2. Results

### 2.1. Chemical and Enzymatic Reactions of Diterpene Glycosides

#### 2.1.1. Alkaline Hydrolysis

*Stevia* DGs are thermostable natural compounds in a neutral solution, although in an alkaline medium with some increase in temperature can be hydrolyzed and yield the free acid compounds at C-19 [16,26].

DGs with one sugar and few examples with two sugars (less hindered as rebaudiosides I and U) attached at position C-19 cleave using mild alkaline conditions. However, DGs with a hindered disaccharide or longer oligomers at C-19 only cleave using stronger alkaline conditions. This difference in saponification conditions for DGs together with the preparation and characterization of some saponified DGs has been useful for the development of a simple and rapid reversed-phase C-18 high-performance liquid chromatography method (HPLC) to identify known and detect novel C-13 oligomer arrangements, see Section 4.1 [26]. In Table 1 are shown all the DGs from *S. rebaudiana* that have been modified by using alkaline conditions.

#### 2.1.2. Acid Hydrolysis

The DGs under acidic conditions may undergo double bond isomerization, sugar cleavage from the glycosides, and Wagner-Meerwein rearrangement of the diterpene aglycone portion depending on temperature. The acid hydrolysis of DGs has been studied since several soft drinks and juices prepared with DGs from *S. rebaudiana* as sweeteners can produce undesirable products under inappropriate temperature storage that may affect the taste of these acidic beverages. Even under different mild acid conditions a double bond isomerization process occurs yielding some by-products that have been isolated and characterized [33–39]. However, proper high-resolution analytical methods need to be developed to differentiate natural DGs from their endocyclic isomer compounds since they elute very closely in HPLC [40]. Additionally, a cleavage of all the sugars in DGs occurs under strong acid conditions producing a mixture of at least three main aglycones from this reaction: isosteviol (major), the endo steviol isomer and steviol (as very minor component). Hence, acid hydrolysis is not a good method to produce steviol in quantities. Recently, these three structurally close aglycones have been purified in their intact form by a two-step gradient high-performance silica gel chromatography [38]. In Table 1 are shown all the DGs formed under basic or acidic conditions while all the aglycones produced from steviol glycosides under acidic conditions and others reported from *S. rebaudiana* are compiled in Figure 1.

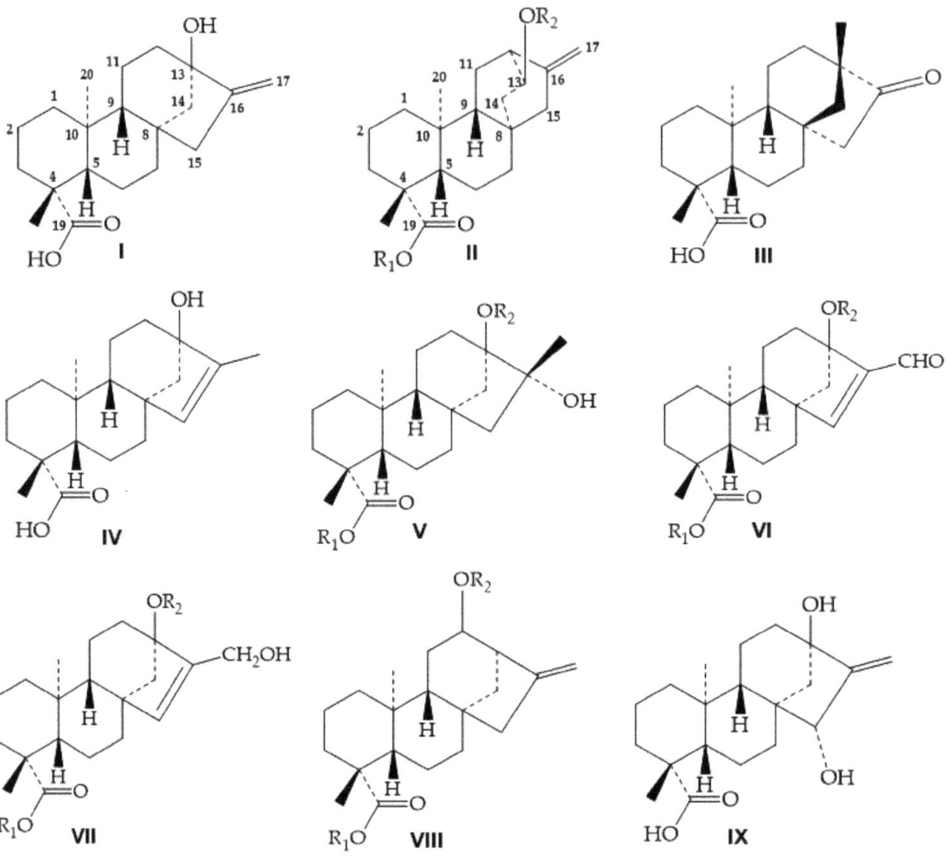

**Figure 1.** Aglycone cores reported from *S. rebaudiana* as natural forms, degradation products or forming part of a glycoside compound ($R_1$ and $R_2$). steviol (**I**); *ent*-atisene (**II**); isosteviol (**III**); endo steviol (**IV**); $CH_3$ and OH at C-16 (**V**); CHO at C-16 (**VI**); $CH_2OH$ at C-16 (**VII**); C-12 linkage (**VIII**); 15-α-hydroxy-rebaudioside M (**IX**).

**Table 1.** Compounds reported as degradation products from natural diterpene glycosides by alkaline or acid conditions. **a** 13-[(2-*O*-*β*-D-xylopyranosyl-*β*-D-glucopyranosyl-)oxy]*ent*-kaur-16-en-19-oic acid *β*-D-glucopyranosyl ester; **b** 13-[(2-*O*-6-deoxy-*β*-D-glucopyranosyl-3-*O*-*β*-D-glucopyranosyl-*β*-D-glucopyranosyl)oxy]*ent*-kaur-16-en-19-oic acid *β*-D-glucopyranosyl ester.

| **DGs from Alkaline Hydrolysis** | | | | | | | |
|---|---|---|---|---|---|---|---|
| **Common Name** | **Oligosaccharide Moieties** | | **AS** | **Chemical Formula** | **Accurate MW** | **Starting Material** | **Ref** |
| | **C-13** | **C-19** | | | | | |
| - | Xylβ(1-2)Glcβ1- | - | I | $C_{31}H_{48}O_{12}$ | 612.3146 | a | [26] |
| Dulcoside $A_1$ | Rhaα(1-2)Glcβ1- | - | I | $C_{32}H_{50}O_{12}$ | 626.3303 | Dulcoside A | [26] |
| Rebaudioside $G_1$ | Glcβ(1-3)Glcβ1- | - | I | $C_{32}H_{50}O_{13}$ | 642.3252 | Rebaudioside G | [26] |
| Rebaudioside $F_1$ | Xylβ(1-2)[Glcβ(1-3)]Glcβ1- | - | I | $C_{37}H_{58}O_{17}$ | 774.3675 | Rebaudioside F | [11,26] |
| Rebaudioside $R_1$ | Glc(1-2)[Glcβ(1-3)]Xylβ1- | - | I | $C_{37}H_{58}O_{17}$ | 774.3675 | Rebaudioside R | [26] |
| Rebaudioside $Z_1$ | Glcβ(1-6)[Glcβ(1-2)]Glcβ1- | - | I | $C_{38}H_{60}O_{18}$ | 804.3781 | Rebaudioside Z | [31] |
| - | Glcβ(1-2)[Glcβ(1-3)]Glcβ1- | | II | $C_{38}H_{60}O_{18}$ | 804.3781 | - | [31] |
| - | 6-deoxyGlcβ(1-2)[Glcβ(1-3)]Glcβ1- | - | I | $C_{38}H_{60}O_{17}$ | 788.3831 | b | [26] |
| Rebaudioside $H_1$ | Glcβ(1-6)Glcβ(1-3)[Glcβ(1-3)]Glcβ1- | - | I | $C_{44}H_{70}O_{23}$ | 966.4309 | Rebaudioside H | [26] |

**Table 1.** *Cont.*

| Common Name | Oligosaccharide Moieties | | AS | Chemical Formula | Accurate MW | Starting Material | Ref |
|---|---|---|---|---|---|---|---|
| | C-13 | C-19 | | | | | |
| **DGs from Alkaline Hydrolysis** | | | | | | | |
| Rebaudioside $L_1$ | Glcβ(1-3)Rhaα(1-2)[Glcβ(1-3)]Glcβ$_1$- | - | I | $C_{44}H_{70}O_{23}$ | 966.4309 | Rebaudioside L | [26] |
| **DGs from acid hydrolysis** | | | | | | | |
| Isosteviol | - | - | III | $C_{20}H_{30}O_3$ | 318.2195 | Rebaudioside A/Stevioside | [34,38] |
| *Endo*-steviol | - | - | IV | $C_{20}H_{30}O_3$ | 318.2195 | Rebaudioside A/Stevioside | [34,38] |
| *Endo*-steviolmonoside | Glcβ$_1$- | - | IV | $C_{26}H_{40}O_8$ | 480.2724 | Rebaudioside A/Rubusoside | [38,39] |
| *Endo*-rebaudioside $G_1$ | Glcβ(1-3)Glcβ$_1$- | - | IV | $C_{32}H_{50}O_{13}$ | 642.3252 | Rebaudioside A | [38] |
| *Endo*-steviolbioside | Glcβ(1-2)Glcβ$_1$- | - | IV | $C_{32}H_{50}O_{13}$ | 642.3252 | Rebaudioside A | [38] |
| *Endo*-rubusoside | Glcβ$_1$- | Glcβ$_1$- | IV | $C_{32}H_{50}O_{13}$ | 642.3252 | Rebaudioside A/Rubusoside | [38,39] |
| *Iso-stevioside/Endo-*stevioside | Glcβ(1-2)Glcβ$_1$- | Glcβ$_1$- | IV | $C_{38}H_{60}O_{18}$ | 804.3781 | Rebaudioside A | [38,41] |
| *Iso-rebaudioside B/Endo*-rebaudioside B | Glcβ(1-2)[Glcβ(1-3)]-Glcβ$_1$- | - | IV | $C_{38}H_{60}O_{18}$ | 804.3781 | Rebaudioside A | [38,42] |
| - | - | Glcβ(1-2)[Glcβ(1-3)]Glcβ$_1$- | III | $C_{38}H_{60}O_{18}$ | 804.3781 | Rebaudioside M | [35] |
| *Iso-rebaudioside AEndo*-rebaudioside A | Glcβ(1-2)[Glcβ(1-3)]Glcβ$_1$- | Glcβ$_1$- | IV | $C_{44}H_{70}O_{23}$ | 966.4309 | Rebaudioside A | [38] |
| - | Glcβ(1-2)[Glcβ(1-3)]Glcβ$_1$- | Glcβ(1-2)[Glcβ(1-3)]Glcβ$_1$- | IV | $C_{56}H_{90}O_{33}$ | 1290.5366 | Rebaudioside M | [35] |
| - | Glcβ(1-2)[Glcβ(1-3)]Glcβ$_1$- | Glcβ(1-2)[Glcβ(1-3)]Glcβ$_1$- | V | $C_{56}H_{92}O_{34}$ | 1308.5472 | Rebaudioside M | [35] |

### 2.1.3. Hydrogenation of the Exocyclic Double Bond

Catalytic hydrogenation of the exocyclic olefinic bond has also been studied in detail for some DGs and the sensory evaluation of their derivatives has also been reported. In this reaction, both isomers, α-$CH_3$ and β-$CH_3$ at position C-16 of the aglycone have been obtained and characterized Figure 2.

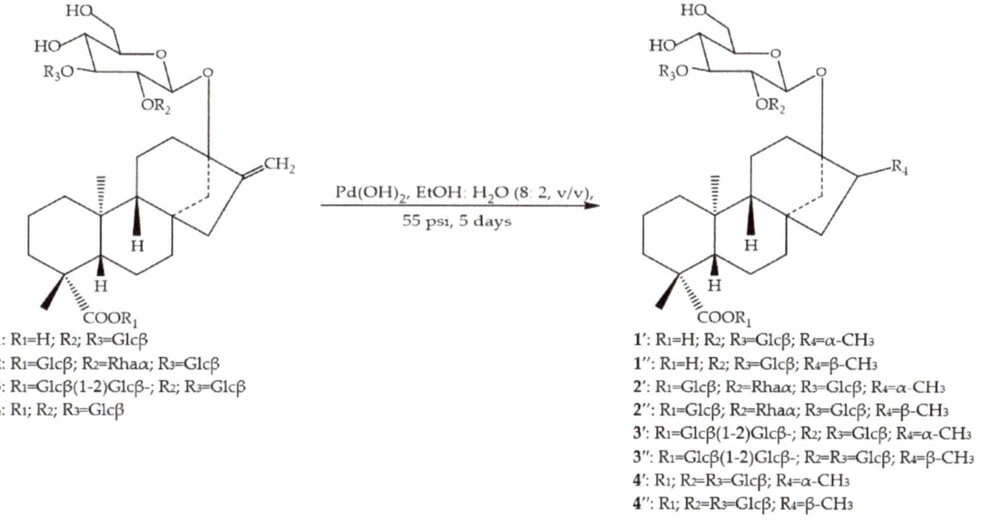


**Figure 2.** Catalytic hydrogenation of the exocyclic double bond in some steviol glycosides.

Some diterpene glycoside compounds have been prepared from rebaudiosides A, B, C and D using different catalysts and reaction conditions [43,44]. The organoleptic properties of the mixture of the reduced α-$CH_3$ and β-$CH_3$ diterpene glycoside isomers have been assayed. The reduction of the exocyclic olefinic bond at C-16 decreased the sweet taste of the natural DGs from 25 to 75%, hence the reductive modification of this site of the aglycone steviol is not a good target for improving the organoleptic properties of natural DGs.

#### 2.1.4. Enzymatic Reactions

Steviol glycosides and gibberellins share the first steps of the biosynthetic pathway where the pathway splits after the formation of *ent*-kaurenoic acid to yield gibberellins and steviol. Steviol glycosides are formed after glycosylation by means of the Uridine 5′-diphospho-glycosylltransferase enzymes producing glucosylation, xylosylation, rhamnosylation, etc. in different positions contributing to the formation of wide spectrum in monosaccharide arrangements [45,46].

The preparation of known and new DGs by using glycosyltransferase enzymes and/or genetically modified microorganisms (e.g., *Saccharomyces cerevisiae* strains) have been used as an approach for producing DGs with improved taste. Glycosyltransferase enzymes transfer glycosyl moieties from an activated nucleoside monosaccharide into mono or oligosaccharide acceptor containing molecules. Some glycosyltransferase enzymes retain or invert the stereochemistry of the substrates. Some scientific articles have been recently published and various patents have been granted.

Some of the examples found in the literature retaining the stereochemistry of the substrate are the bioconversion of rebaudioside A into rebaudioside I and M2 (rebaudioside M isomer with Glcβ(1-6)[Glcβ(1-2)]Glc$\beta_1$- at C-19) and the bioconversion of rebaudioside D3 into (Rebaudioside D isomer with Glcβ(1-6)[Glcβ(1-2)]-Glc$\beta_1$- at C-13) from rebaudioside E employing UDP-glycosyltransferases using uridine 5-diphosphoglucose (UDP-glucose) as a donor of the sugar moiety.

In the same way, some new minor DGs with inversion of the stereochemistry of some anomeric protons were isolated from a cyclodextrin glycosyltransferase glucosylated *Stevia* extract containing a high percentage of DGs. Interestingly, α-glucosyl linkages were mainly detected at position 4 of some sugar monosaccharide of moieties attached at position C-13 and/or C-19 of the steviol aglycone: 13-[(2-*O*-β-D-glucopyranosyl-3-*O*-(4-*O*-α-D-glucopyranosyl)-β-D-glucopyranosyl-β-D-glucopyranosyl)oxy] ent-kaur-16-en-19-oic acid-[(4-*O*-α-D-glucopyranosyl-β-D-glucopyranosyl) ester]; 13-[(2-*O*-β-D-glucopyranosyl-β-D-glucopyranosyl)oxy] ent-kaur-16-en-19-oic acid-[(4-*O*-(4-*O*-(4-*O*-α-D-glucopyranosyl)-α-D-glucopyranosyl)-α-D-glucopyranosyl)-β-D-glucopyranosyl ester]; 13-[(2-*O*-β-D-glucopyranosyl-3-*O*-(4-*O*-(4-*O*-(4-*O*-α-D-glucopyranosyl)-α-D-glucopyranosyl)-α-D-glucopyranosyl)-β-D-glucopyranosyl-β-D-glucopyranosyl)oxy] ent-kaur16-en-19-oic acid β-D-glucopyranosyl ester and 13-[(2-*O*-β-D-glucopyranosyl-3-*O*-(4-*O*- (4-*O*-(4-*O*-α-D-glucopyranosyl)-α-D-glucopyranosyl)-α-D-glucopyranosyl)-β-D-glucopyranosyl-β-D-glucopyranosyl)oxy] ent-kaur-16-en-19-oic acid-[(4-*O*-α-D-glucopyranosyl-β-D-glucopyranosyl) ester]. The different enzymes, cyclodextrin glycosyl transferase, α and β-Glucosidases, α and β-Galactosidase and β-fructosidase transglycosylation systems together with β-Glycosyltransferase glycosylation systems using UDP-sugars used in the biotransformation of steviol glycosides have been summarized [47].

Additionally, enzymatic hydrolysis seems to be the most efficient method to produce steviol in quantities. Several enzymes have been used for this purpose e.g., juices of the snail *Helix pomatia*, pectinase or hesperidinase affords the aglycone, steviol [6,48,49].

## 3. Methods of Separation for Diterpene Glycosides

### *3.1. Analytical Methods*

#### 3.1.1. High-Performance Thin-Layer Chromatography

High-Performance Thin-Layer Chromatography (HPTLC) is a standardized methodology that has been widely used in the Botanical Industry [50] but has also found multiple applications in different fields [51–53]. Several HPTLC methods have been developed to identify or quantify selected steviol glycosides (stevioside, steviolbioside, dulcoside A and rebaudiosides A–D) in food and Stevia formulations, and also to detect degradation products like steviol and isosteviol [54]. The main stationary phases used have been silica gel 60, silica gel 60 diol, LiChrospher silica gel 60 F254 S and Proteochrom although the best separation has been achieved on HPTLC silica gel 60 plates. Multiple mobile phases have been assessed to optimize the separation of selected steviol glycosides but ethyl acetate, methanol and formic acid (93:40:1 *v*/*v*/*v*) and ethyl acetate, methanol and acetic acid (3:1:1, *v*/*v*/*v*) seem to be the mixtures where the best separation has been achieved so far. However, with these mobile phases the migration of very polar steviol glycosides like rebaudioside M, N and O is not guaranteed and the separation of the endocyclic steviol isomer, a degradation product formed under acidic conditions from steviol glycosides, needs to be further studied. Steviol glycosides have been visualized on the silica plates by derivatization mainly with a solution of 2-naphthol or primuline for aglycones. Hence, densitometric analysis using absorption and fluorescence modes have been used [54,55] for quantitation. Steviol glycosides can be identified by HPTLC-ESI-MS using a TLC-MS interface and specific structural information can be obtained if collision energy is carefully studied as discussed in 4.3. In Figure 3 the HPTLC separation of selected steviol glycosides is presented.

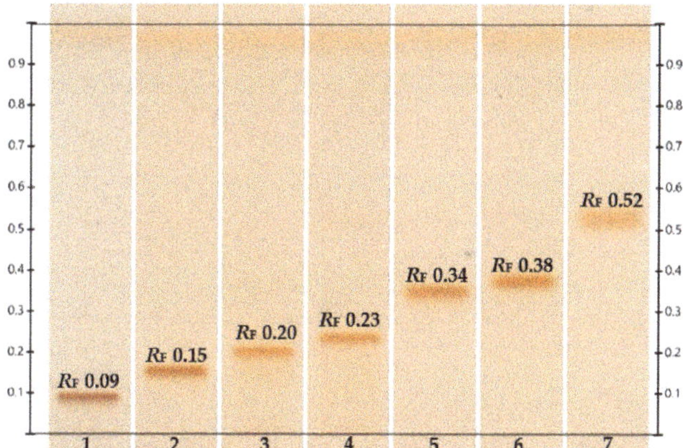


**Figure 3.** Silica gel HPTLC plate for rebaudiosides D, A, C, stevioside, rebaudioside B, dulcoside A and steviolbioside (tracks 1 to 7). Developed with ethyl acetate, methanol and formic acid (93:40:1 *v*/*v*/*v*) with no saturation but activated with $MgCl_2$ at a relative humidity of 33%. The detection was performed after derivatization with a solution of 2-naphthol under white light.

### 3.1.2. High-Performance Liquid Chromatography

High performance liquid chromatography (HPLC) is the usual methodology for the analyses of DGs with UV, PDA and MS as the most common detectors [56,57]. Several methods have been developed to separate DGs from complex *S. rebaudiana* extract mixtures. Since the most common aglycone is steviol, novel potential sugar-substitutes could be found based on the number of monosaccharides, type of sugar units and their arrangements. These slight structural differences often cause co-elution of DGs in a single HPLC method.

An overview of the advantages and disadvantages of some HPLC methods using different stationary phases (HILIC, $NH_2$, RP-C18, Sepaxdiol, Synergi, silica gel) is given herein, although detailed information can be found [40,58]. Several purified DGs containing different numbers of sugars and linkage arrangements to steviol were analyzed for this purpose. Interestingly, in all the methods reported, acetonitrile: water acidified elution gradients were described for the analyses of DGs [40,58]. The use of a common binary mobile phase is a practical methodology for the analyses of DGs with simple changes in the column adsorbent chemistry (Figure 4). There is no change in selectivity between the silica gel and the amino method. In silica gel, stevioside and rebaudioside C coelute while in an amino column, a good separation is achieved, but rubusoside and steviolbioside are better resolved in silica than in amino column. When a Synergi column is utilized, there is a big room for the separation of very polar steviol glycosides (more polar than rebaudioside O) and also between rebaudiosides A and N. There is no change in selectivity if HILIC column is compared with Sepax-diol one, although sharper peaks are gotten with a Sepax-diol column. The appropriate combination of HPLC methods which best resolved components in the range of retention times of interest give a good understanding of the DGs in the extract and chromatography fractions and provide guidance for further preparative methods to be used.

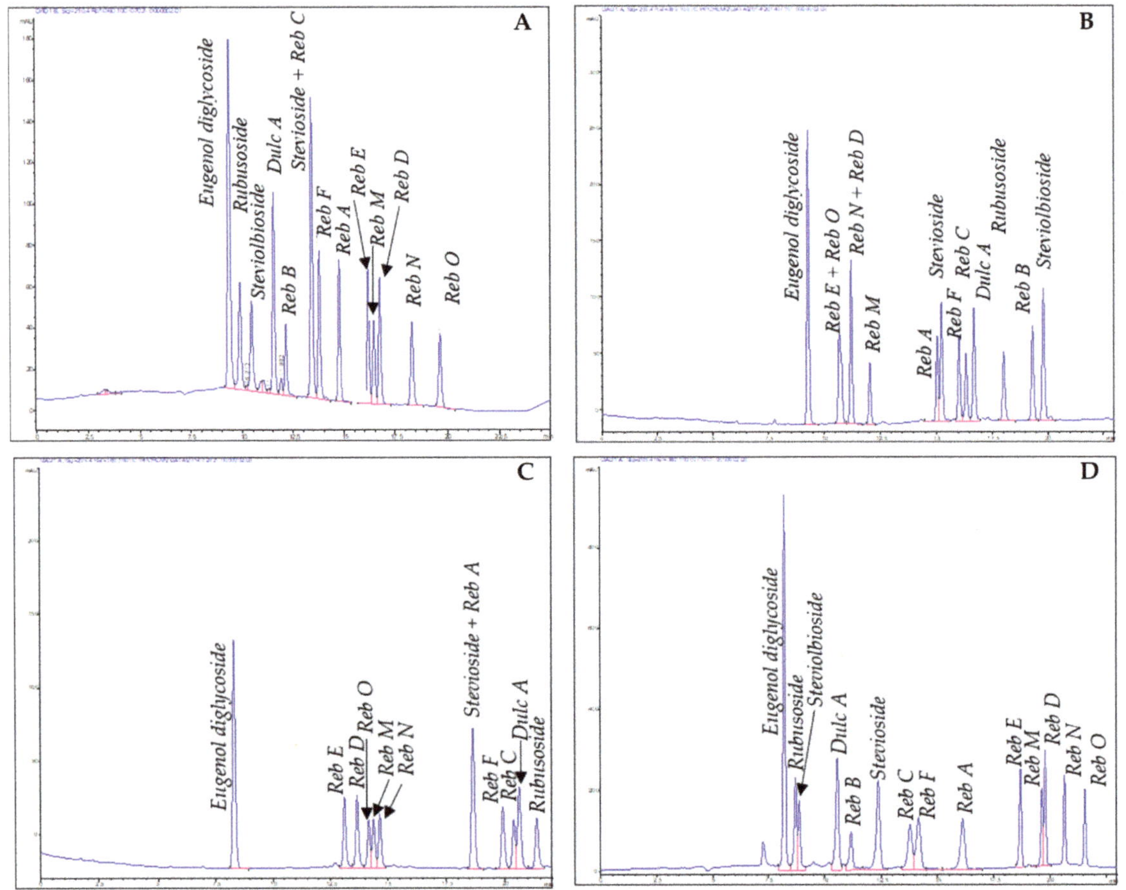


**Figure 4.** *Cont.*

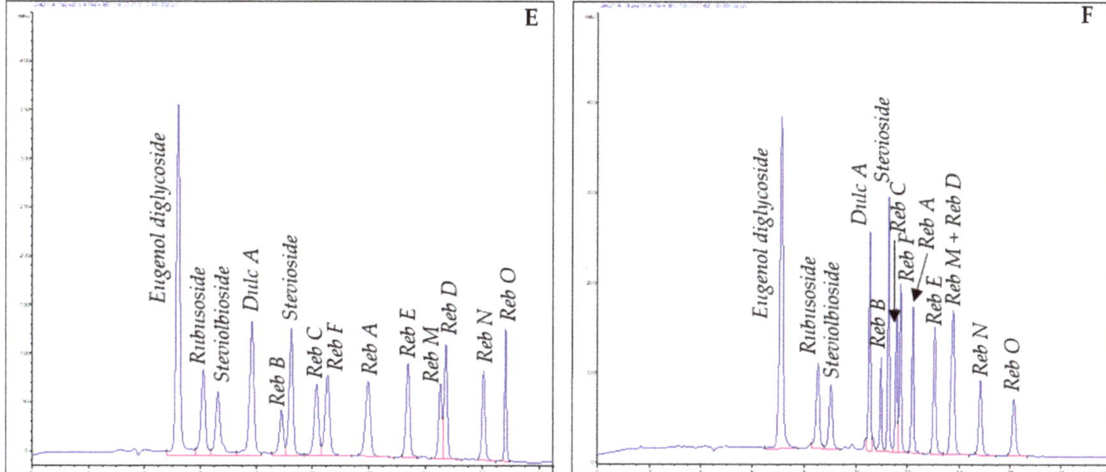


**Figure 4.** Selected HPLC methods used for the analysis of steviol glycosides: silica method (**A**), RP-C18 (**B**), Synergi (**C**), amino (**D**), HILIC (**E**) and Sepax-diol (**F**).

*3.2. Preparative Methods*

Several approaches have been described in the literature to isolate and purify known and novel DGs. However, most of the preparative methods described allowed the purification of some few milligrams of DGs. Since novel DGs with refined organoleptic properties need to be found to enhance *Stevia* DGs consumption as a non-caloric sweetener and thus improve human health, the purification process of DGs needs to be scaled-up to access for assay the very minor glycosides as potential sugar-substitutes still undiscovered. Herein we describe some strategies using our own technology to get known compounds in gram to kilogram quantities. Using this technology [59], several novel DGs were purified in hundreds of milligrams and gram quantities [40]. This technology also provides an economic and convenient approach to the isolation and purification of larger quantities of potentially interesting natural products for confirmation of their bioactivities.

Two main approaches have been used for scaling-up the purification process of DGs: utilization of reverse phase (C-18) and silica gel stationary phases. Several sized columns were utilized for the isolation and purification of DGs: (1 × 25 cm, 5 µm), preparative (2.1 × 40 cm, 10 µm) and large-scale high-performance chromatography (7.5 × 50 cm, 10 µm) [59]. Reverse phase chromatography is perceived to have economic advantages over normal phase chromatography due to the usual practice of replacement of the normal phase adsorbent after one or at most a few uses whereas the reverse phase adsorbent can be used for hundreds of separations. The most used mobile phase for purifying DGs in reverse phase chromatography is acidified mixtures of acetonitrile: water in different ratios depending on the analytes. The solvent mixtures after processing the fractions are recovered (c.a 70% in acetonitrile as an aqueous azeotrope) and re-used in the next chromatography after adjusting the ratios based on the density of the mixtures. Thus, we can save solvents and make the purification process of these compounds at large scale more economic.

On the other hand, silica gel stationary phase for chromatography is perceived to not be reusable or of a very limited useful life. Consequently, less expensive poor-quality normal phase adsorbents are typically employed in preparative column packings. The poorer quality normal phase adsorbent is usually of irregular shaped particles and possesses a wide particle size distribution which together provides overall poor chromatographic performance for the packed bed. High quality normal phase adsorbents are available with spherical particles and narrow particle size distributions. As quality normal phase adsorbent costs about $5000 or more per kilogram and as the adsorbent is perceived not to be reusable, development of silica gel preparative chromatographic processes has largely been avoided or not considered. In our own experience a high-performance preparative silica gel column (10 μm, spherical silica gel) was packed in 2009 and hundreds of separations have been performed until now without re-packing. Appropriate column regeneration and re-equilibration to the next mobile phase composition increases the lifetime of the stationary phase and restores the performance qualities of the column [60].

With the technologies outlined above, silica gel chromatographic processes can be developed providing cost savings to users through better performance, higher capacity, easier product recovery, less costly solvent recovery, and less costly solvent disposal. Compound recovery from the organic solvent of the mobile phase used in silica gel chromatography is easier and less expensive than from the water containing mobile phase used in reverse phase chromatography. At production scale, the energy required to recycle the normal phase organic solvents is significantly less than that of reversed phase aqueous solvents. The waste disposal costs are reduced for the normal phase organic solvents because of their usually higher BTU content.

One of the most common mobile phases for separating non-glycosylated steviol and related aglycones using this technology are mixtures of *n*-heptane: wet acidified EtOAc and/or *n*-heptane: *wa*EtOAc: MeOH [40]. However, the detection of these compounds is generally set at 205 or 210 nm, below the UV absorbance cut off wavelength of ethyl acetate. An alternative mobile phase to solve this problem is the use of methyl tert-butyl ether (MTBE). This mobile phase was useful for the separation of isosteviol from a mixture of steviol and other isomers [40].

Obviously, normal phase separation of the glycosides requires a more polar mobile phase. Mixtures of EtOAc: MeOH: $H_2O$ with 0.1% AcOH or MTBE: MeOH: $H_2O$ with 0.1% AcOH in different ratios have been widely used in our experience with good separation of several DGs. Both organic mobile phases are easily recovered, and the solvent composition ratios could be adjusted by scouting by TLC for subsequent chromatography.

All methodologies have been useful to purify several steviol glycosides most of them in gram quantities and could be applied for any natural product research. In Table 2 are shown all the tetracyclic diterpene reported as natural compounds from *S. rebaudiana*.

## 4. Structure Elucidation for Diterpene Glycosides

The diversity of DGs has been considerably enhanced with the discovery of several new compounds in the last decade. Herein, we summarize all the DGs isolated from *S. rebaudiana* and reported as natural compounds from 1931 to 2021 (Table 2). The main structural differences have been found in the number of sugar units linked to the aglycone at positions C-13 and C-19 and the different arrangements described, thus some rebaudioside families have been described. Nevertheless, some other DGs with an aglycone core different from steviol have also been reported, and chemical structures are compiled in Figure 1.

The isolation and structure elucidation of minor DGs in quantities not only requires appropriate instrumentation and several chromatographic steps but also high-quality spectroscopic and spectrometric equipment. NMR analysis is the most extensively used technique to elucidate unambiguously natural compounds. However, as an aid for structure elucidation of new steviol glycosides some alternative methods have been reported in the literature and described herein.

**Table 2.** Diterpene glycosides purified from *Stevia rebaudiana*.

| Common Name | Oligosaccharide Moieties | | AS | Chemical Formula | MW | Ref |
|---|---|---|---|---|---|---|
| | C-13 | C-19 | | | | |
| Steviol | H | H | **I** | $C_{20}H_{30}O_3$ | 318.2195 | [61] |
| Steviol monoacetate | H | $COCH_3$ | **I** | $C_{22}H_{32}O_4$ | 360.2192 | [62] |
| Steviolmonoside | Glcβ$_1$- | H | **I** | $C_{26}H_{40}O_8$ | 480.2724 | [16] |
| Steviol-19-O-β-D glucoside | H | Glcβ$_1$- | **I** | $C_{26}H_{40}O_8$ | 480.2724 | [63] |
| Rubusoside | Glcβ$_1$- | Glcβ$_1$- | **I** | $C_{32}H_{50}O_{13}$ | 642.3252 | [16] |
| **Oxidized core family** | | | | | | |
| Isosteviol-19-*O*-β-glucoside | - | Glcβ$_1$- | **III** | $C_{26}H_{40}O_8$ | 480.2724 | [42] |
| - | Glcβ(1-2)Glcβ$_1$- | Glcβ$_1$- | **VI** | $C_{38}H_{58}O_{19}$ | 818.3573 | [37] |
| - | Glcβ(1-2)Glcβ$_1$- | Glcβ$_1$- | **VII** | $C_{38}H_{60}O_{19}$ | 820.3728 | [37] |
| - | Glcβ(1-2)[Glcβ(1-3)]Glcβ$_1$- | Glcβ$_1$- | **VII** | $C_{44}H_{70}O_{24}$ | 982.4258 | [64] |
| - | Glcβ(1-2)[Glcβ(1-3)]Glcβ$_1$- | Glcβ$_1$- | **IV** | $C_{44}H_{72}O_{24}$ | 984.4415 | [41] |
| **Rebaudioside A family** | | | | | | |
| Steviolbioside | Glcβ(1-2)Glcβ$_1$- | H | **I** | $C_{32}H_{50}O_{13}$ | 642.3252 | [7] |
| - | Glcβ$_1$- | Glcβ(1-2)Glcβ$_1$- | **I** | $C_{38}H_{60}O_{18}$ | 804.3781 | [22] |
| Rebaudioside KA | Glcβ(1-2)Glcβ$_1$- | Glcβ$_1$- | **VIII** | $C_{38}H_{60}O_{18}$ | 804.3781 | [22] |
| Stevioside | Glcβ(1-2)Glcβ$_1$- | Glcβ$_1$- | **I** | $C_{38}H_{60}O_{18}$ | 804.3781 | [6] |
| Rebaudioside B | Glcβ(1-2)[Glcβ(1-3)]Glcβ$_1$- | H | **I** | $C_{38}H_{60}O_{18}$ | 804.3781 | [7] |
| Rebaudioside E | Glcβ(1-2)Glcβ$_1$- | Glcβ(1-2)Glcβ$_1$- | **I** | $C_{44}H_{70}O_{23}$ | 966.4309 | [9] |
| Rebaudioside A | Glcβ(1-2)[Glcβ(1-3)]Glcβ$_1$- | Glcβ$_1$- | **I** | $C_{44}H_{70}O_{23}$ | 966.4309 | [7] |
| - | Glcβ(1-6)Glcβ(1-2)Glcβ$_1$- | Glcβ$_1$- | **I** | $C_{44}H_{70}O_{23}$ | 966.4309 | [65] |
| Rebaudioside Y | Glcβ(1-2)Glcβ$_1$- | Glcβ(1-6)Glcβ$_1$- | **I** | $C_{44}H_{70}O_{23}$ | 966.4309 | [58] |
| Rebaudioside Z | Glcβ(1-6)[Glcβ(1-2)]Glcβ$_1$- | Glcβ$_1$- | **I** | $C_{44}H_{70}O_{23}$ | 966.4309 | [31] |
| Rebaudioside D | Glcβ(1-2)[Glcβ(1-3)]Glcβ$_1$- | Glcβ(1-2)Glcβ$_1$- | **I** | $C_{50}H_{80}O_{28}$ | 1128.4838 | [9] |
| Rebaudioside I | Glcβ(1-2)[Glcβ(1-3)]Glcβ$_1$- | Glcβ(1-3)Glcβ$_1$- | **I** | $C_{50}H_{80}O_{28}$ | 1128.4838 | [16] |
| Rebaudioside M | Glcβ(1-2)[Glcβ(1-3)]Glcβ$_1$- | Glcβ(1-2)[Glcβ(1-3)]Glcβ$_1$- | **I** | $C_{56}H_{90}O_{33}$ | 1128.4838 | [16] |
| Rebaudioside L | Glcβ(1-6)Glcβ(1-2)[Glcβ(1-3)]Glcβ$_1$- | Glcβ$_1$- | **I** | $C_{50}H_{80}O_{28}$ | 1128.4838 | [16] |
| - | Glcβ(1-6)[Glcβ(1-3)]Glcβ$_1$- | Glcβ(1-2)Glcβ$_1$- | **I** | $C_{50}H_{80}O_{28}$ | 1128.4838 | [41] |
| - | Glcβ(1-2)Glcβ$_1$- | Glcβ(1-6)[Glcβ(1-2)Glcβ$_1$- | **I** | $C_{50}H_{80}O_{28}$ | 1128.4838 | [66] |
| 15α-hydroxy-Reb M | Glcβ(1-2)[Glcβ(1-3)]Glcβ$_1$- | Glcβ(1-2)[Glcβ(1-3)]Glcβ$_1$- | **IX** | $C_{56}H_{90}O_{34}$ | 1306.5313 | [30] |

Table 2. *Cont.*

| Common Name | Oligosaccharide Moieties | | AS | Chemical Formula | MW | Ref |
|---|---|---|---|---|---|---|
| | C-13 | C-19 | | | | |
| **Rebaudioside VIII and IX families** | | | | | | |
| Rebaudioside VIII | Glcβ(1-2)[Glcβ(1-3)]-Glcα(1-6)-Glcβ(1-2)[Glcβ(1-3)]Glcβ$_1$- | Glcβ(1-2)Glcβ$_1$- | I | $C_{68}H_{110}O_{43}$ | 1614.6420 | [29] |
| Rebaudioside IX | Glcβ(1-2)[Glcβ(1-3)]-Glcα(1-6)-Glcβ(1-2)[Glcβ(1-3)]Glcβ$_1$- | Glcβ(1-2)[Glcβ(1-3)]Glcβ$_1$- | I | $C_{74}H_{119}O_{48}$ | 1776.6949 | [27] |
| Rebaudioside IXa | Glcβ(1-2)[Glcβ(1-3)]Glcβ$_1$- | Glcβ(1-2)[Glcβ(1-3)]Glcβ(1-2)[Glcβ(1-3)]-Glcα(1-6)-Glcβ$_1$- | I | $C_{74}H_{119}O_{48}$ | 1776.6949 | [28] |
| Rebaudioside IXb | Glcβ(1-2)[Glcβ(1-3)]Glcβ$_1$- | Glcβ(1-2)[Glcβ(1-3)]-Glcα(1-6)-Glcβ(1-2)[Glcβ(1-3)]Glcβ$_1$- | I | $C_{74}H_{119}O_{48}$ | 1776.6949 | [28] |
| Rebaudioside IXc | Glcβ(1-2)[Glcβ(1-3)]Glcβ$_1$- | Glcβ(1-2)[Glcβ(1-2)[Glcβ(1-3)]-Glcα(1-3)-Glcβ(1-3)]Glcβ$_1$- | I | $C_{74}H_{119}O_{48}$ | 1776.6949 | [28] |
| Rebaudioside IXd | Glcβ(1-2)[Glcβ(1-3)]Glcβ(1-2)[Glcβ(1-3)]-Glcα(1-6)-Glcβ$_1$- | Glcβ(1-2)[Glcβ(1-3)]Glcβ$_1$- | I | $C_{74}H_{119}O_{48}$ | 1776.6949 | [29] |
| **Rebaudioside C family** | | | | | | |
| Dulcoside A | Rhaα(1-2)Glcβ$_1$- | Glcβ$_1$- | I | $C_{38}H_{60}O_{17}$ | 788.3831 | [8] |
| Dulcoside B | Rhaα(1-2)[Glcβ(1-3)]Glcβ$_1$- | H | I | $C_{38}H_{60}O_{17}$ | 788.3831 | [16] |
| Rebaudioside C/Dulcoside B | Rhaα(1-2)[Glcβ(1-3)]Glcβ$_1$- | Glcβ$_1$- | I | $C_{44}H_{70}O_{22}$ | 950.4356 | [8,10] |
| Rebaudioside S | Glcα(1-2)Glcβ$_1$- | Rhaα(1-2)Glcβ$_1$- | I | $C_{44}H_{70}O_{22}$ | 950.4356 | [23] |
| - | Glcβ(1-2)Glcβ$_1$- | Rhaα(1-2)Glcβ$_1$- | I | $C_{44}H_{70}O_{22}$ | 950.4358 | [25] |
| Rebaudioside H | Glcβ(1-3)Rhaα(1-2)[Glcβ(1-3)]Glcβ$_1$- | Glcβ$_1$- | I | $C_{50}H_{80}O_{27}$ | 1112.4888 | |
| Rebaudioside K | Rhaα(1-2)[Glcβ(1-3)]Glcβ$_1$- | Glcβ(1-2)Glcβ$_1$- | I | $C_{50}H_{80}O_{27}$ | 1112.4888 | [16] |
| Rebaudioside J | Glcβ(1-2)[Glcβ(1-3)]Glcβ$_1$- | Rhaα(1-2)Glcβ$_1$- | I | $C_{50}H_{80}O_{27}$ | 1112.4888 | [16] |
| **Rebaudioside C family** | | | | | | |
| - | Rhaα(1-2)[Glcβ(1-3)]Glcβ$_1$- | Glcβ(1-6)Glcβ$_1$- | I | $C_{50}H_{80}O_{27}$ | 1112.4887 | [25] |
| Rebaudioside N | Glcβ(1-2)[Glcβ(1-3)]Glcβ$_1$- | Rhaα(1-2)[Glcβ(1-3)]Glcβ$_1$- | I | $C_{56}H_{90}O_{32}$ | 1274.5417 | [16] |
| Rebaudioside O | Glcβ(1-2)[Glcβ(1-3)]Glcβ$_1$- | Glcβ(1-3)Rhaα(1-2)[Glcβ(1-3)]Glcβ$_1$- | I | $C_{62}H_{100}O_{37}$ | 1436.5945 | [16] |
| **Glcβ(1-3)Glcβ$_1$- family** | | | | | | |
| Rebaudioside G | Glcβ(1-3)Glcβ$_1$- | Glcβ$_1$- | I | $C_{38}H_{60}O_{18}$ | 788.3831 | [16] |

**Table 2.** *Cont.*

| Common Name | Oligosaccharide Moieties | | AS | Chemical Formula | MW | Ref |
|---|---|---|---|---|---|---|
| | C-13 | C-19 | | | | |
| | | **Rebaudioside F family** | | | | |
| - | Xylβ(1-2)Glcβ1- | Glcβ1- | I | $C_{37}H_{58}O_{17}$ | 774.3675 | [67] |
| Rebaudioside F | Xylβ(1-2)[Glcβ(1-3)]Glcβ1- | Glcβ1- | I | $C_{43}H_{68}O_{22}$ | 936.4203 | [11] |
| - | Glcβ(1-2)[Xylβ(1-3)]Glcβ1- | Glcβ1- | I | $C_{43}H_{68}O_{22}$ | 936.4203 | [67] |
| - | Glcβ(1-2)Glcβ1- | Xylβ(1-6)Glcβ1- | I | $C_{43}H_{68}O_{22}$ | 936.4203 | [37] |
| Rebaudioside R | Glcβ(1-2)[Glcβ(1-3)]Xylβ1- | Glcβ1- | I | $C_{43}H_{68}O_{22}$ | 936.4203 | [23] |
| Rebaudioside V | Xylβ(1-2)[Glcβ(1-3)]Glcβ1- | Glcβ(1-2)Glcβ1- | I | $C_{49}H_{78}O_{27}$ | 1098.4730 | [26] |
| - | Glcβ(1-2)[Glcβ(1-3)]Glcβ1- | Xylβ(1-2)Glcβ1- | I | $C_{49}H_{78}O_{27}$ | 1098.4730 | [25] |
| - | Glcβ(1-2)Glcβ1- | Xylβ(1-2)[Glcβ(1-4)]Glcβ1- | I | $C_{49}H_{78}O_{27}$ | 1098.4730 | [25] |
| Rebaudioside T | Glcβ(1-2)[Glcβ(1-3)]Glcβ1- | Xylβ(1-2)[Glcβ(1-3)]Glcβ1- | I | $C_{55}H_{88}O_{32}$ | 1260.5260 | [24,68] |
| - | Glcβ(1-2)[Xylβ (1-3)]Glcβ1- | Glcβ(1-2)[Glcβ(1-3)]Glcβ1- | I | $C_{55}H_{88}O_{32}$ | 1260.5260 | [25] |
| | | **Fruβ family** | | | | |
| - | Glcβ(1-2)[Fruβ(1-3)]Glcβ1- | Glcβ1- | I | $C_{44}H_{70}O_{23}$ | 966.4309 | [65] |
| | | **Glcα family** | | | | |
| - | Glcβ(1-2)Glcβ1- | Glcα(1-2)[Glcα(1-4)]-Glcβ1- | I | $C_{50}H_{80}O_{28}$ | 1128.4838 | [69] |
| - | Glcα(1-3)Glcβ(1-2)[Glcβ(1-3)]Glcβ1- | Glcβ1- | I | $C_{50}H_{80}O_{28}$ | 1128.4838 | [70] |
| - | Glcα(1-4)Glcβ(1-3)[Glcβ(1-2)]Glcβ1- | Glcβ1- | I | $C_{50}H_{80}O_{28}$ | 1128.4838 | [70] |
| | | 6-deoxyGlcβ family | | | | |
| - | 6-deoxyGlcβ(1-2)Glcβ1- | Glcβ1- | I | $C_{38}H_{60}O_{17}$ | 788.3831 | [71] |
| - | 6-deoxyGlcβ(1-2)[Glcβ(1-3)]Glcβ1- | Glcβ1- | I | $C_{44}H_{70}O_{22}$ | 950.4356 | [71] |
| - | Glcβ(1-2)[Glcβ(1-3)]Glcβ1- | 6-deoxyGlcβ1- | I | $C_{44}H_{70}O_{22}$ | 950.4356 | [72] |
| | | **Rebaudioside W family (three different sugar moieties)** | | | | |
| Rebaudioside W | Xylβ(1-2)[Glcβ(1-3)]-Glcβ1- | Rhaα(1-2)[Glcβ(1-3)]Glcβ1- | I | $C_{55}H_{88}O_{31}$ | 1244.5311 | [26] |
| | | **Rebaudioside U Araα family** | | | | |
| Rebaudioside U/6 | Glcβ(1-2)[Glcβ(1-3)]Glcβ1- | Araα(1-6)Glcβ1- | I | $C_{49}H_{78}O_{27}$ | 1098.4730 | [24,25] |
| - | Glcβ(1-2)[Glcβ(1-3)]Glcβ1- | Araα(1-6)[Glcβ(1-2)]Glcβ1- | I | $C_{55}H_{88}O_{32}$ | 1260.5258 | [25] |

**Table 2.** *Cont.*

| Common Name | Oligosaccharide Moieties | | AS | Chemical Formula | MW | Ref |
|---|---|---|---|---|---|---|
| | C-13 | C-19 | | | | |
| | | *Ent*-atisene-core family | | | | |
| - | Glcβ(1-2)[Glcβ(1-3)]Glc$β_1$- | Glc$β_1$- | II | $C_{44}H_{70}O_{23}$ | 966.4309 | [31] |
| Stevatisene J | Glcβ(1-2)[Glcβ(1-3)]Glc$β_1$- | Rhaα(1-2)Glc$β_1$- | II | $C_{50}H_{80}O_{27}$ | 1112.4888 | [32] |
| Stevatisene K | Rhaα(1-2)[Glcβ(1-3)]Glc$β_1$- | Glcβ(1-2)Glc$β_1$- | II | $C_{50}H_{80}O_{27}$ | 1112.4888 | [32] |
| Stevatisene T | Glcβ(1-2)[Glcβ(1-3)]Glc$β_1$- | Xylβ(1-2)[Glcβ(1-3)]Glc$β_1$- | II | $C_{55}H_{88}O_{32}$ | 1260.5260 | [32] |
| Stevatisene N | Glcβ(1-2)[Glcβ(1-3)]Glc$β_1$- | Rhaα(1-2)[Glcβ(1-3)]Glc$β_1$- | II | $C_{56}H_{90}O_{32}$ | 1274.5417 | [32] |
| Stevatisene O | Glcβ(1-2)[Glcβ(1-3)]Glc$β_1$- | Glcβ(1-3)Rhaα(1-2)[Glcβ(1-3)]Glc$β_1$- | II | $C_{62}H_{100}O_{37}$ | 1436.5945 | [32] |

*4.1. Reversed-Phase High-Performance Liquid Chromatography*

High-performance liquid chromatograph is a very common instrument in most of the laboratories today. Thus, a method based on using reversed-phase C18 column retention times of several previously characterized natural DGs and others with no sugar at position C-19 was described as an aid for structure elucidation of new and known DGs [26]. Specifically, the method is helpful to identify known and detect novel aglycone-C13 oligosaccharide moieties and give some hints about C-19 linkages. Elution order of several DGs and their aglycone-C13 oligosaccharide substituted with different sugar arrangements are summarized in Table 3.

**Table 3.** Diterpene glycosides organized by increasing retention times in an RP-C18 HPLC method.

| DG | $R_T$ (min) | $\Delta_{RT}$ (min) | Oligosaccharide Positions | |
|---|---|---|---|---|
| | | | C-13 | C-19 |
| Reb O | 2.92 | 0.16 | Glcβ(1-2)[Glcβ(1-3)]Glc$\beta_1$- | Glcβ(1-3) Rhaα(1-2)[Glcβ(1-3)]-Glc$\beta_1$- |
| Reb N | 3.08 | - | Glcβ(1-2)[Glcβ(1-3)]Glc$\beta_1$- | Rhaα(1-2) [Glcβ(1-3)]-Glc$\beta_1$- |
| Reb E | 3.08 | 0.18 | Glc$\beta_1$(1-2)Glc$\beta_1$- | Glc$\beta_1$(1-2)Glc$\beta_1$ - |
| Reb D | 3.26 | 0.23 | Glcβ(1-2)[Glcβ(1-3)]Glc$\beta_1$- | Glc$\beta_1$(1-2)Glc$\beta_1$ - |
| Reb J | 3.49 | 0.07 | Glcβ(1-2)[Glcβ(1-3)]Glc$\beta_1$- | Rhaα (1-2)Glc$\beta_1$- |
| Reb W | 3.56 | 0.20 | Xylβ(1-2)[Glcβ(1-3)]Glc$\beta_1$- | Rhaα(1-2) [Glcβ(1-3)]-Glc$\beta_1$- |
| Reb M | 3.76 | 0.04 | Glcβ(1-2)[Glcβ(1-3)]Glc$\beta_1$- | Glcβ(1-2)[Glcβ(1-3)]-Glc$\beta_1$- |
| Reb Y | 3.80 | 0.09 | Glcβ(1-2)Glc$\beta_1$- | Glc$\beta_1$(1-6)Glc$\beta_1$- |
| Reb V | 3.89 | 0.05 | Xylβ(1-2)[Glcβ(1-3)]Glc$\beta_1$- | Glc$\beta_1$(1-2)Glc$\beta_1$ - |
| Reb Z | 3.94 | 0.24 | Glcβ(1-2)[Glcβ(1-6)]-Glcβ1- | Glc$\beta_1$- |
| Reb U | 4.18 | 0.17 | Glcβ(1-2)[Glcβ(1-3)]Glc$\beta_1$- | Araα(1-6)-Glc$\beta_1$- |
| a | 4.35 | 0.50 | Glcβ(1-2)[Glcβ(1-3)]Glc$\beta_1$- | Glc$\beta_1$- |
| Reb T | 4.85 | 0.61 | Glcβ(1-2)[Glcβ(1-3)]Glc$\beta_1$- | Xylβ(1-2)[Glcβ(1-3)]-Glc$\beta_1$- |
| Reb H | 5.46 | 0.18 | Glcβ(1-3)Rhaα(1-2)[Glcβ(1-3)]Glc$\beta_1$- | Glc$\beta_1$- |
| Reb L | 5.64 | 0.71 | Glcβ(1-6)Glcβ(1-3)[Glcβ(1-3)]Glc$\beta_1$- | Glc$\beta_1$- |
| Reb I | 6.35 | 0.76 | Glcβ(1-2)[Glcβ(1-3)]Glc$\beta_1$- | Glc$\beta_1$(1-3)Glc$\beta_1$- |
| Reb A | 7.11 | 0.42 | Glcβ(1-2)[Glcβ(1-3)]Glc$\beta_1$- | Glc$\beta_1$- |
| Stev | 7.53 | 0.38 | Glcβ(1-2)Glc$\beta_1$- | Glc$\beta_1$- |
| Iso-Reb A | 7.91 | 0.65 | Glcβ(1-2)[Glcβ(1-3)]Glc$\beta_1$- | Glc$\beta_1$- |
| Iso-Stev | 8.56 | 0.52 | Glcβ(1-2)Glc$\beta_1$- | Glc$\beta_1$- |
| Reb F | 9.08 | 0.24 | Xylβ(1-2)[Glcβ(1-3)]Glc$\beta_1$- | Glc$\beta_1$- |
| Reb $Z_1$ | 9.32 | 0.47 | Glcβ(1-2)[Glcβ(1-6)]Glcβ1- | H- |
| Reb C | 9.79 | 0.13 | Rhaα(1-2)[Glcβ(1-3)]Glc$\beta_1$- | Glc$\beta_1$- |
| Reb R | 9.92 | 0.34 | Glc(1-2)[Glcβ(1-3)]Xylβ1- | Glc$\beta_1$- |
| b | 10.26 | 0.16 | Xylβ(1-2)Glc$\beta_1$- | Glc$\beta_1$- |
| Dulc A | 10.42 | 0.08 | Rhaα(1-2)Glc$\beta_1$- | Glc$\beta_1$- |
| c | 10.50 | 0.69 | Glcβ(1-2)[Glcβ(1-3)]Glc$\beta_1$- | - |
| Reb G | 11.19 | 0.26 | Glcβ(1-3)Glc$\beta_1$- | Glc$\beta_1$- |
| d | 11.45 | 0.56 | 6-deoxyGlcβ(1-2)[Glcβ(1-3)]Glc$\beta_1$- | Glc$\beta_1$- |
| Reb $L_1$ | 12.01 | 0.28 | Glcβ(1-3) Rhaα(1-2)[Glcβ(1-3)]Glc$\beta_1$- | H- |
| Reb $H_1$ | 12.29 | 0.23 | Glcβ(1-6) Glcβ(1-3)[Glcβ(1-3)]Glc$\beta_1$- | H- |
| Rub | 12.52 | 1.80 | Glc$\beta_1$- | Glc$\beta_1$- |
| Reb B | 14.32 | 0.12 | Glcβ(1-2)[Glcβ(1-3)]Glc$\beta_1$- | H- |
| Iso-Reb B | 14.44 | 0.44 | Glcβ(1-2)[Glcβ(1-3)]Glc$\beta_1$- | H- |
| Stev-bio | 14.88 | 0.29 | Glcβ(1-2)Glc$\beta_1$- | H- |
| Iso-Stevbio | 15.17 | 0.15 | Glcβ(1-2)Glc$\beta_1$- | H- |
| Dulc B | 15.31 | 0.52 | Rhaα(1-2)[Glcβ(1-3)]Glc$\beta_1$- | H- |
| Reb $F_1$ | 15.83 | 0.61 | Xylβ(1-2)[Glcβ(1-3)]Glc$\beta_1$- | H- |
| Dulc $A_1$ | 16.45 | 0.12 | Rhaα(1-2)Glc$\beta_1$- | H- |
| Reb $R_1$ | 16.57 | 0.70 | Glc(1-2)[Glcβ(1-3)]Xylβ1- | H- |
| e | 17.26 | 0.02 | Xylβ(1-2)Glc$\beta_1$- | H- |
| Reb $G_1$ | 17.28 | 0.1 | Glcβ(1-3)Glc$\beta_1$- | H- |
| f | 17.36 | 2.19 | 6-deoxyGlcβ(1-2)[Glcβ(1-3)]Glc$\beta_1$- | H- |
| Stev-mono | 19.55 | | Glc$\beta_1$- | H- |

Steviolmonoside (Stev-mono); Iso-steviolbioside (Iso-Stevbio); Rubusoside (Rub); Steviolbioside (Stev-bio), Stevioside (Stev); Iso-Steviolbioside (Iso-Stevbio); Iso-Stevioside (Iso-Stev); Rebaudioside $G_1$ (Reb $G_1$); Rebaudioside G (Reb G); Dulcoside $A_1$ (Dulc $A_1$); Dulcoside A (Dulc A); Rebaudioside B (Reb B); Rebaudioside A (Reb A); Iso-Rebaudioside $A_1$ (Iso-Reb $A_1$); Iso-Rebaudioside A (Iso-Reb A); Dulcoside B (Dulc B); Rebaudioside C (Reb C); Rebaudioside $F_1$ (Reb $F_1$); Rebaudioside F (Reb F); Rebaudioside $L_1$ (Reb $L_1$); Rebaudioside L (Reb L); Rebaudioside H (Reb H); Rebaudioside $H_1$ (Reb $H_1$); **a:** 13-[(2-*O*-β-D-glucopyranosyl-3-*O*-β-D-glucopyranosyl-β-D-glucopyranosyl) oxy]*ent*-hydroxyatis-16-en-19-oic acid -β-D-glucopyranosy ester; **b:** 13-[(2-*O*-*β*-D-xylopyranosyl-*β*-D-glucopyranosyl-)oxy]*ent*-kaur-16-en-19-oic acid *β*-D-glucopyranosyl ester; **c:** 13-[(2-*O*-β-D-glucopyranosyl-3-*O*-β-D-glucopyranosyl-β-D-glucopyranosyl) oxy]*ent*-hydroxyatis-16-en-19-oic acid; **d:** 13-[(2-*O*-6-deoxy-*β*-D-glucopyranosyl-3-*O*-*β*-D-glucopyranosyl-*β*-D-glucopyranosyl)oxy]*ent*-kaur-16-en-19-oic acid *β*-D-glucopyranosyl ester; **e:** 13-[(2-*O*-β-D-xylopyranosyl-β-D-glucopyranosyl-)oxy]kaur-16-en-19-oic acid; **f:** 13-[(2-*O*-6-deoxy-β-D-glucopyranosyl-3-*O*-β-D-glucopyranosyl-β-D-glucopyranosyl)oxy]kaur-16-en-19-oic acid.

The number of sugar components may be inferred by retention time ranges: *DGs* with retention times < 7 min (oligosaccharide with five to seven units), although rebaudioside E or similar isomers, with two monosaccharide units at C-13 and C-19, four glucose units with a 1-6 linkage in C-13 or four glucose units attached to a an *ent*-atisene core show retention times under 7 min, *DGs* with retention times between 7 to 11.5 min (with a di or trisaccharide at C-13 and a monosaccharide at C-19, although three glucose units at C-13 with a 1-6 linkage or linked to an *ent*-atisene core also elute in this range), *DGs* with retention times between 11.6- 12.5 min (a tetrasaccharide oligomer is attached at C-13 and no sugar at C-19) while rubusoside, the only steviol glycosides reported with one sugar unit at both C-13 and C-19, elutes at 12.5 min and *DGs* with retention times greater than 12.5 *min* (a mono, di or trisaccharide at C-13 and a free carboxylic acid at C-19). However, to confirm the identity of highly substituted DGs, a normal phase chromatographic method should be run.

The method is based on comparing retention times of a pure natural DG and its saponification product with those reported in Table 3. If the retention time of a natural DG and its saponification product match with one of those reported in Table 3, most probably the DG has previously been identified. During the analyses of novel DGs few possibilities may be found: (1) DGs with a novel C-13 and known C-19 arrangement; (2) DGs with known C-13 and new C-19 arrangement; (3) DGs with unknown C-13 and C-19 arrangements or (4) DGs with known C-13 and C-19 moieties but with a unique combination that makes the DGs become an undescribed structure e.g., rebaudioside V [26].

For identifying the sugar component numbers and arrangement at the C-13 moiety, the C-19 portion is cleaved under alkaline hydrolysis and clean up following acidification. HPLC retention time of the aglycone-C13 oligosaccharide moiety is obtained using specific concentration and chromatographic conditions [26]. Comparing retention times with those reported in Table 3 will provide rapid information about the C-13 oligosaccharide arrangement or its novelty.

The strength of the alkaline hydrolysis is also important to get hints about the number of saccharide units and infer their positions of the C-19 oligomer. Mild alkaline hydrolysis over 30 min easily cleaves the monosaccharide linked at position C-19 (rebaudioside A, stevioside, rebaudioside F, etc.). Ready cleavage also occurs when less hindered oligomers are attached at position C-19 e.g., rebaudioside I (Glcβ(1-3)Glcβ-) and rebaudioside U (Araα(1-6)Glc-). However, the same easy cleavage with a disaccharide hexose/pentose (1-4) hexose is expected. We also believe that an easy cleavage could happen in a trisaccharide or a tetrasaccharide with position 2 of the sugar directly linked at C-19 unsubstituted [26]. DGs with hindered disaccharide Glcβ(1-2)Glcβ- at position C-19 (Rebaudiosides D, E, J) and also trisaccharides and tetrasaccharides (with position 2 of the sugar directly bonded to C-19 substituent) as rebaudiosides M and O, respectively, only cleave completely using strong alkaline conditions (reflux, 1 h).

### 4.2. High-Resolution Electrospray Ionization Tandem Mass Spectrometry

As glycosidic compounds, DGs give greater ionization efficiencies in negative ion mode. The dissociation pattern of DGs has been carefully described by ranging collision energies (CE) using high-resolution electrospray ionization tandem mass spectrometry (HRESIMS) [73,74]. The understanding of the HRESIMS dissociation pattern of DGs is of great importance since this allows the rapid detection and identification of known and novel DGs. Obviously, chemical structures of new DGs should be confirmed by NMR experiments.

*Stevia* DGs are a good model to study the sequence of cleavage, by tandem mass spectrometry, of each sugar monomer from the parent ions due to the variety of types of sugar monomers and their arrangement in the oligosaccharide.

#### 4.2.1. Cleavage of the C-19 Moiety

It is well-known that monosaccharides and oligomers attached at position C-19 (ester portion) is the first moiety cleaved from DGs parent ions. Only glycosides with glucose directly attached at C-19 have been found so far. Ranging of collision energies (CE) allows one to infer important structural information about the C-19 moiety: number of sugar units at C-19 and position of sugar linkages. DGs containing a monosaccharide at C-19 position cleave easily at CE of 10 eV supported with the presence of two major ions in the MS spectrum, the deprotonated molecule $[M-H]^-$ and the major product ion [M-H-162]- Da. All the DGs containing a monosaccharide at position C-19 analyzed by HRESIMS showed the same trend e.g., rebaudiosides A, G, H, L and stevioside among others. In the same way, DGs containing less hindered disaccharides at position C-19 also cleave at low collision energy. Rebaudiosides I (Glc(1-3)Glc-) and U (Ara(1-6)Glc-) are a couple of DGs with a less hindered disaccharide attached at C-19. Both compounds show cleavage of the C-19 moiety at 10 eV but produce lower product ion intensities: rebaudioside I-steviol-C13$^-$, 14.4% intensity and rebaudioside U-steviol-C13$^-$, 54% intensity. Product ion intensities of these less hindered C-19 disaccharides should be studied carefully since most probably the position of the second monosaccharide could be also inferred.

However, DGs with two sugars at both positions, C-13 and C-19, show different product ion intensities of the steviol with the C-13 moiety $[steviol-C13]^-$ at 10 eV. Rebaudiosides E and S, the only examples at hand, showed both the deprotonated molecule ion $[M-H]^-$ and the product ion $[steviol-C13]^-$. Nevertheless, significantly lower product ion intensities were observed: rebaudioside E, $[M-H-324]^-$ Da with $m/z = 641$ Da and intensity of 5.9% due to the loss of two glucoses and rebaudioside S, $[M-H-308]^-$ Da with $m/z = 641$ Da and intensity of 7.5% with loss of one glucose and one rhamnose.

On the other hand, DGs containing hindered disaccharides (Rebaudioside D, E, K, etc.) and longer oligomer chains (Rebaudioside s M, N, O, W, etc.) at position C-19 with a C-13 moiety of more than two sugar units do not cleave at low CE. Due to this highly hindered ester, 40 eV needed to be applied to cleave the C-19 portion and produce detectable [steviol-C13 glycoside]$^-$ product ions. We also expect a relatively easy cleavage at low CE of the C-19 moiety in DGs with a tri and tetrasaccharide oligomer at C-19 with the C-2 of the glucose directly linked to C-19 unsubstituted.

#### 4.2.2. Cleavage of the C-13 Moiety

In contrast to the C-19 moiety, a broad cleavage of the C-13 moieties in DGs occurs at 70 eV. In branched trisaccharides at C-13, sugars at C-3 cleave first, followed by sugars at C-2 and finally sugars directly linked at C-13. The dissociation pattern of three tetraglycosides with a glucose attached at C-19 and deoxyGlc$\beta$(1-2)[Glc$\beta$(1-3)]Glc$\beta_1$- (13-[(2-*O*-6-deoxy-$\beta$-D-glucopyranosyl-3-*O*-$\beta$-D-glucopyranosyl-$\beta$-D-glucopyranosyl) oxy]*ent*-kaur-16-en-19-oic acid $\beta$-D-glucopyranosyl ester), Xyl$\beta$(1-2)[Glc$\beta$(1-3)]Glc$\beta_1$- (rebaudioside F) and Glc$\beta$(1-

2)[Glc$\beta$(1-3)]Xyl$\beta_1$- (rebaudioside R) are three DG examples that allow the understanding of the sequential loss of each sugar unit from C-13 (Figure 5).

Rebaudiosides H, L, VIII, IX and IXd are the only reported examples of natural DGs with more than three monosaccharide units at C-13. Rebaudiosides L, VIII and series IX contain only glucoses which makes it difficult to understand the sequential cleavage of C-13 moieties. However, it was found that rebaudioside H [Glc(1-3)Rha$\alpha$(1-2)[Glc$\beta$(1-3)]Glc$\beta_1$-] showed a sequential loss of four sugar units at CE (70 eV) following the C-19 cleavage as follows: first two glucoses, $m/z$ = 787.3729 Da and 625.3197 Da, followed by a rhamnose $m/z$ = 479.2637 Da and a further glucose loss $m/z$ = 317.2106 Da. Probably, the glucose attached at C-3 of the rhamnose is the first cleaved followed by glucose at C-3 of the glucose directly attached at C-13. DGs have been grouped into 11 groups by molecular weights or structure similarities for an ease of understanding of the different dissociation pattern of several DGs and isomers. MS data of several natural and prepared DGs have been recently reported [74].

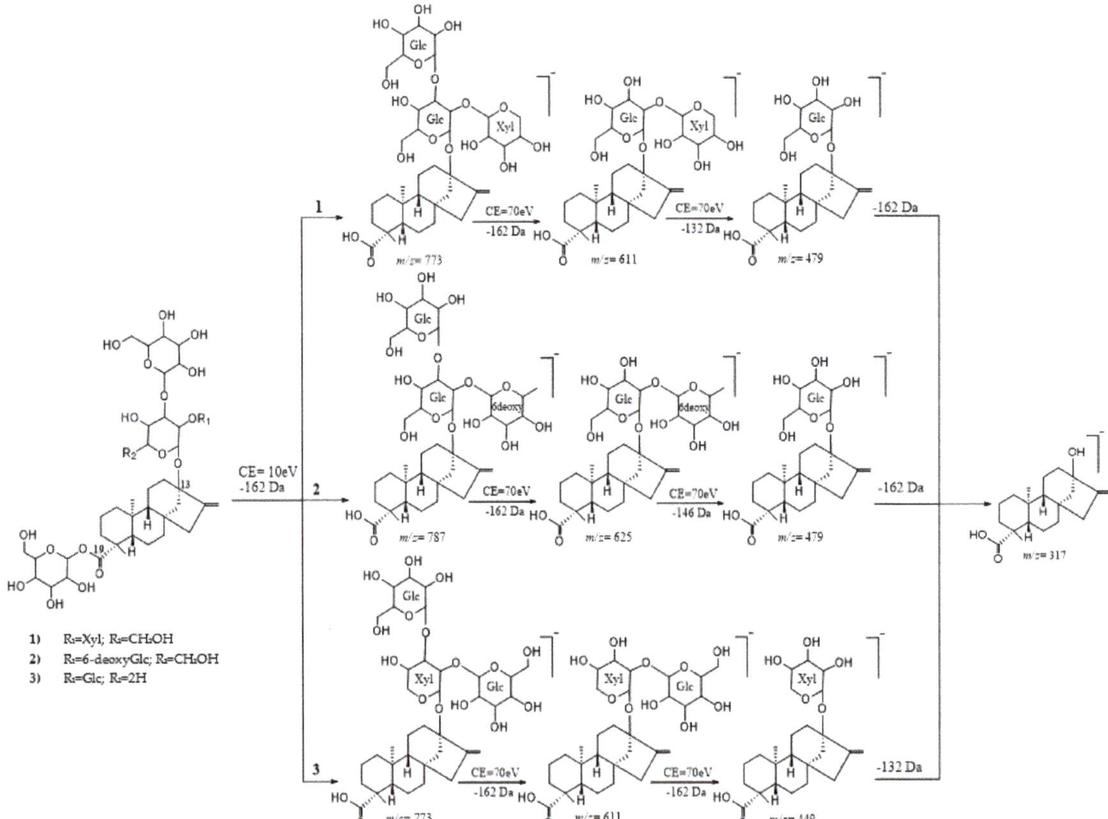


**Figure 5.** Fragmentation pattern of steviol glycosides.

### 4.3. Determination of the Absolute Configuration of the Monosaccharides.

A simple and rapid reversed-phase C18 high-performance and ultra-high-performance liquid chromatography methods have been developed to determine the absolute configuration of several monosaccharides linked to different kind aglycones [75,76]. This method is not only appropriate to DGs from *S. rebaudiana*, but it can also be applied for several classes of glycosides: flavonoid glycosides, alkaloid glycosides and triterpene glycosides [77,78] among others. The method is based on cleavage of all the monosaccharides linked to the aglycone under acid hydrolysis. Further liquid-liquid partitions with adequate organic solvent are needed to separate the monosaccharides from the aglycone. It is well-known in steviol glycosides that steviol is not the main aglycone recovered but rather a mixture of isosteviol, endo-steviol and steviol. Basically, two step reactions are needed to yield monosaccharide derivatives traceable by UV with enough difference in retention times between D and L monosaccharide enantiomers. The first step reaction is between the monosaccharides and L-cysteine methyl ester to yield the thiazolidine derivatives followed with further addition of phenylisothiocyanate to yield the thiocarbamoyl thiazolidine monosaccharide derivatives. Both step reactions are performed at 60–70 °C in pyridine. The identity and absolute configuration of the monosaccharide could be identified by comparison of the retention times of the derivatives with appropriate standards [75,76]. Thiohydantoin is a by-product yielded in the reaction which has earlier retention times in RP-C18 HPLC method than the thiocarbamoyl-thiazolidine sugar derivatives [75].

## 5. NMR Experiments

Several approaches have been discussed in previous sections, all of them provide important information for the structure elucidation of the DGs but still those approaches do not fully elucidate the structures of several DGs. The oligosaccharides at position C-19 and isomeric aglycones are not easily identified using previous methods. There is no doubt that NMR is the most powerful technique for the structure elucidation of any compound if an appropriate amount and purity are in hand. Steviol is the main aglycone in DGs from *S. rebaudiana*, although other natural and degradation cores have also been reported (Figure 1). As far as we know, only steviol, isosteviol, endo steviol and the aglycone with the hydrated double bond at 16,17 (Figure 1V) have been isolated as the aglycone forms. The $^{13}$C-NMR spectra for the aglycones of isosteviol, steviol and endocyclic steviol isomer are shown in Figure 6. The $^{1}$H and $^{13}$C NMR chemical shifts of the main aglycones found in *S. rebaudiana* or in DGS from leaves of this plant are listed (Table 4).

Herein, we discuss the key NMR chemical shifts to differentiate the aglycone cores from *S. rebaudiana*. Steviol has present an exocyclic double bond at position C-16 and the glycosylation sites at position C-13 and C-19 (Figure 1I). The double bond is characterized by a quaternary (158.3 ppm) and a methylene olefinic carbon resonance (103.5 ppm, $^{13}$C resonances); and proton resonances (5.48 and 5.04 ppm) while endo steviol, has an endocyclic double bond at position C-15 (135.0, $^{13}$C) and 5.24 ppm $^{1}$H) and C-16 (145.9 ppm $^{13}$C) with an additional methyl group at C-17 (12.7 $^{13}$C and 1.59 ppm $^{1}$H) (Figure 1IV). Isosteviol is easily differentiated from the other aglycones due to the presence of a ketone group (221.3 ppm $^{13}$C) at C-16 and an additional methyl group (29.9 $^{13}$C and 1.40 ppm $^{1}$H) (Figure 1III). Compounds 7 with a molecular weight of (332 Da) (Figure 1VI) and 8 with 334 Da (Figure 1VII) possess an endocyclic double bond like endo steviol. The slight structural differences are found at position C-17 where a remarkable difference in chemical shifts is observed, compound 7 has an -CHO (191.4 $^{13}$C and 9.61 ppm $^{1}$H) and 8 a -$CH_2OH$ group (59.2 $^{13}$C and 4.11; 4.29 $^{1}$H). Different from the other cores, compound 4 with a molecular weight of 336 Da (Figure 1V), does not present any double bond, the main difference is found in the groups linked at position C-16, -$CH_3$ (22.2 $^{13}$C; 1.32 ppm $^{1}$H) and -OH (77.1 ppm $^{1}$H).

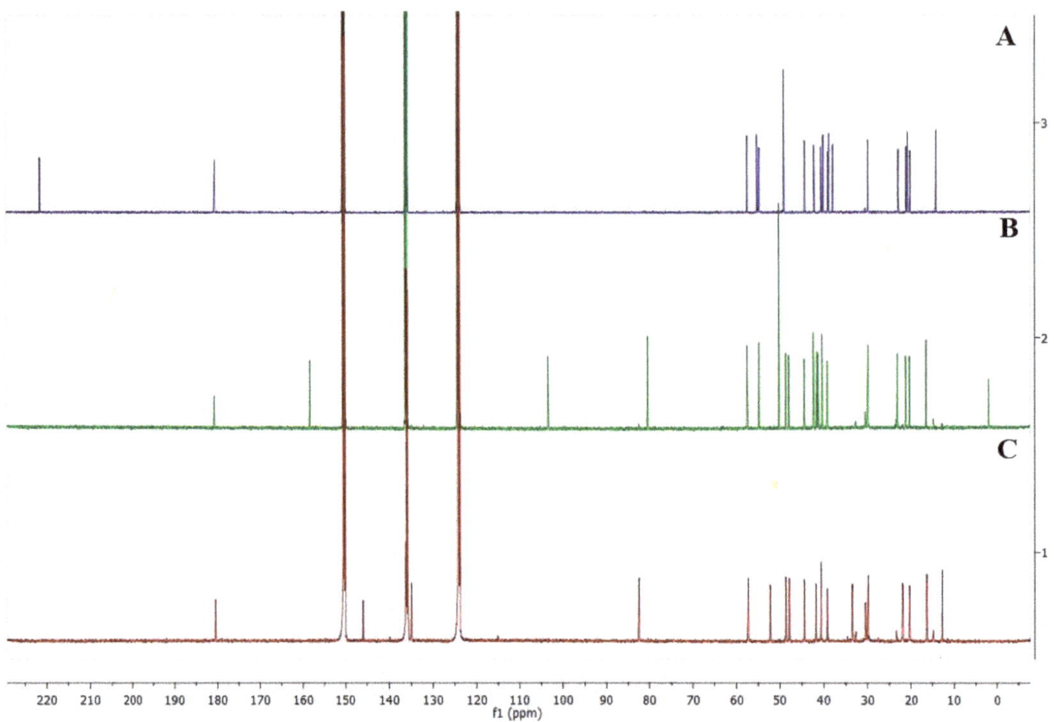


**Figure 6.** Comparison of the $^{13}$C-NMR of isosteviol (**A**), steviol (**B**) and endocyclic steviol isomer (**C**).

**Table 4.** $^{1}$H and $^{13}$C chemical shifts of the compounds **1–9**.

| Position | **1** [(a)] | | **2** [(b)] | | **3** [(a)] | | **4** [(a)] | | **5** [(a)] | |
|---|---|---|---|---|---|---|---|---|---|---|
| | $\delta_H$ | $\delta_C$ | $\delta_H$ | $\delta_C$ | $\delta_H$ | $\delta_C$ | $\delta_C$ | | $\delta_H$ | $\delta_C$ |
| 1 | 0.87; 1.87 | 41.5 | 0.79 | 40.2 | 0.92; 1.92 | 41.8 | 0.78; 1.75 | 40.3 | 0.88; 1.67 | 40.6 |
| 2 | 2.28; 1.52 | 20.3 | 2.13; 1.38 | 19.7 | 2.32; 1.56 | 20.4 | 1.34; 2.23 | 19.3 | 2.23; 1.46 | 20.1 |
| 3 | 2.05; 1.87 | 41.3 | 2.36 | 39.1 | 2.50; 1.11 | 39.2 | 1.00; 2.32 | 38.4 | 1.61; 1.33 | 37.9 |
| 4 | - | 44.4 | - | 44.6 | - | 44.4 | - | 43.8 | - | 44.3 |
| 5 | 1.11 | 57.5 | 1.03 | 57.9 | 1.14 | 57.3 | 1.04 | 57.1 | 1.14 | 57.4 |
| 6 | 2.07; 2.21 | 23.2 | 2.41; 1.84 | 21.1 | 2.23; 2.07 | 22.1 | 2.11; 2.43 | 23.1 | 2.04; 2.04 | 22.9 |
| 7 | 1.55; 1.45 | 42.4 | 2.78 | 39.6 | 1.90; 1.79 | 33.4 | 1.37; 1.88 | 42.8 | 1.66; 1.66 | 42.1 |
| 8 | - | 42.3 | - | 34.9 | - | 48.7 | - | *NA* | - | 49.1 |
| 9 | 1.02 | 54.8 | 1.01 | 51.7 | 0.97 | 47.9 | 0.84 | 54.8 | 1.14 | 55.2 |
| 10 | - | 40.3 | - | 44.6 | - | 40.6 | - | *NA* | - | 38.8 |
| 11 | 1.79; 1.55 | 21.3 | 1.59 | 27.3 | 1.72; 1.72 | 21.9 | 1.52; 1.71 | 19.8 | 1.60; 1.20 | 21.1 |
| 12 | 2.48; 1.09 | 39.2 | 4.11 | 78.4 | 1.75; 1.60 | 40.6 | 1.85; 2.67 | 31.6 | 2.45; 1.09 | 39.0 |
| 13 | - | 80.3 | 2.94 | 42.0 | - | 82.4 | - | 87.6 | - | 40.1 |
| 14 | 2.35; 1.58 | 48.7 | 1.42; 1.05 | 39.9 | 2.47; 1.77 | 52.3 | 2.44; 2.58 | 40.3 | 1.49; 1.38 | 54.7 |
| 15 | 2.25; 2.22 | 48.0 | 2.23; 1.91 | 49.0 | 5.24 | 135.0 | 1.41; 1.83 | 54.3 | 2.71; 2.66 | 49.1 |
| 16 | - | 158.3 | - | 148.4 | - | 145.9 | - | 77.1 | - | 221.3 |
| 17 | 5.48; 5.04 | 103.5 | 5.24; 4.90 | 109.5 | 1.89 | 12.7 | 1.32 | 22.2 | 1.04 | 20.7 |
| 18 | 1.28 | 29.9 | 1.26 | 29.1 | 1.38 | 29.8 | 1.28 | 27.7 | 1.40 | 29.9 |
| 19 | - | 180.7 | - | 177.4 | - | 180.6 | - | 176.9 | - | 180.7 |
| 20 | 1.20 | 16.4 | 1.04 | 13.3 | 1.24 | 16.3 | 1.31 | 16.0 | 0.98 | 14.1 |

**Table 4.** *Cont.*

| Position | 6 [(b)] | | 7 [(b)] | | 8 [(b)] | | 9 [(b)] | |
|---|---|---|---|---|---|---|---|---|
| | $\delta_H$ | $\delta_H$ | $\delta_H$ | $\delta_C$ | $\delta_H$ | $\delta_C$ | $\delta_C$ | $\delta_H$ |
| 1 | 0.92; 1.59 | 39.5 | 0.89; 1.88 | 41.7 | 0.86; 1.86 | 41.8 | 0.86; 1.89 | 40.9 |
| 2 | 1.92; 1.40 | 18.4 | 1.42; 1.95 | 19.7 | 1.41; 1.96 | 19.8 | 1.45; 2.15 | 20.1 |
| 3 | 2.20; 1.09 | 37.7 | 1.07; 2.17 | 38.6 | 1.06; 2.14 | 38.8 | 1.09; 2.06 | 39.0 |
| 4 | 1.12 | 43.7 | - | 44.3 | - | 44.7 | - | 45.1 |
| 5 | 1.78; 2.05 | 57.1 | 1.18 | 57.6 | 1.12 | 58.0 | 1.07 | 58.2 |
| 6 | 1.57; 1.16 | 19.9 | 1.88; 2.05 | 21.1 | 1.83; 2.00 | 21.5 | 1.72; 2.08 | 23.6 |
| 7 | 1.09 | 38.8 | 1.66; 1.75 | 39.0 | 1.55; 1.65 | 40.2 | 1.33; 1.68 | 36.7 |
| 8 | 1.63; 1.45 | 33.9 | - | 50.6 | - | 49.5 | - | 46.4 |
| 9 | 2.51 | 51.0 | 1.00 | 46.1 | 0.90 | 47.7 | 0.94 | 54.3 |
| 10 | 4.04 | 38.1 | - | 40.5 | - | 40.1 | - | 40.9 |
| 11 | 1.15; 2.51 | 26.3 | 1.77; 1.82 | 20.6 | 1.63; 1.72 | 21.8 | 1.40; 1.76 | 20.0 |
| 12 | 1.89; 2.11 | 41.0 | 1.74; 1.87 | 31.4 | 1.69; 1.81 | 30.4 | 1.45; 2.16 | 38.7 |
| 13 | 4.70; 4.84 | 77.8 | - | 88.5 | - | 90.2 | - | 88.7 |
| 14 | 1.24 | 37.7 | 1.95; 2.32 | 47.7 | 1.73; 2.32 | 49.6 | 1.66; 2.16 | 40.8 |
| 15 | 0.89 | 47.4 | 6.57 | 158.0 | 5.36 | 136.6 | 3.70 | 80.8 |
| 16 | 0.92; 1.59 | 146.7 | - | 147.7 | - | 146.9 | - | 157.2 |
| 17 | 1.92; 1.40 | 108.0 | 9.61 | 191.4 | 4.11; 4.29 | 59.2 | 5.31; 5.54 | 110.2 |
| 18 | 2.20; 1.09 | 27.5 | - | 178.0 | - | 178.4 | 1.25 | 28.5 |
| 19 | 1.12 | 176.8 | 1.22 | 28.5 | 1.21 | 28.6 | - | 178.8 |
| 20 | 1.78; 2.05 | 11.9 | 1.02 | 16.0 | 1.00 | 15.8 | 0.97 | 17.3 |

[(a)] NMR spectra were recorded in Pyr-$d_5$. [(b)] MeOH-$d_4$, **1:** steviol; **2:** C-12 aglycone; **3:** endo steviol isomer; **4:** aglycone with a $CH_3$ and OH groups linked at C-16; **5:** isosteviol; **6:** *ent*-atisene core from 13-[(2-O-β-D-glucopyranosyl-3-O-β-D-glucopyranosyl-β-d-glucopyranosyl) oxy] ent-hydroxyatis-16-en-19-oic acid -β-D-glucopyranosy ester; **7:** aglycone with a CHO group linked at C-16; **8:** aglycone with a $CH_2OH$ group linked at C-16; **9:** aglycone from 15α-hydroxy-rebaudioside M. Chemical structures of the aglycone cores are presented in Figure 1: steviol (I); *ent*-atisene (II); isosteviol (III); endo steviol (IV); $CH_3$ and OH at C-16 (V); CHO group at C-16 (VI); $CH_2OH$ at C-16 (VII); C-12 linkage (VIII); 15-α-hydroxy-rebaudioside M (IX).

Compounds 2 (Figure 1VII) and 6 (Figure 1II) share very similar chemical shifts, although different structures have been assigned. Compound 6 was unambiguously elucidated using appropriate 1D and 2D NMR experiments and aglycone was defined as ***ent***-13(***S***)-hydroxyatisenoic acid core. If compared with steviol, the main differences were found in C-13 (77.8 ppm $^{13}$C), C-14 (37.7 ppm $^{13}$C), C-16 (146.7 ppm $^{13}$C) and C-17 (108.0 ppm $^{13}$C). Structure of compound 2 should be reanalyzed. Compound 9 shows similar chemical shifts that steviol aglycone with a main difference in the chemical shifts of H-15 and C-15 with 3.70 ppm and 80.8, respectively. The NMR chemical shifts of selected DGs with similar structures are compiled from Tables 5–8. In Table 5 are shown the NMR data for DGs with three glucose units while in Table 6, DGs with four glucose units.

**Table 5.** $^1$H and $^{13}$C chemical shifts of the compounds **10–13**.

| Moiety | Position | 10 [(a)] | | 11 [(b)] | | 12 [(b)] | | 13 [(a)] | |
|---|---|---|---|---|---|---|---|---|---|
| | | $\delta_H$ | $\delta_C$ | $\delta_H$ | $\delta_C$ | $\delta_H$ | $\delta_C$ | $\delta_H$ | $\delta_C$ |
| Aglycone | 1 | 1.71; 0.73 | 40.9 | 0.76; 1.75 | 40.8 | 0.79 | 40.2 | *NA* | 40.7 |
| | 2 | 2.21; 1.42 | 19.5 | 1.70; 2.17 | 20.2 | 1.38; 2.13 | 19.7 | *NA* | 19.2 |
| | 3 | 2.33; 1.02 | 38.5 | 1.82; 2.14 | 38.9 | 2.36 | 39.1 | *NA* | 38.1 |
| | 4 | – | 44.1 | – | 44.5 | – | 44.6 | *NA* | 43.9 |
| | 5 | 1.05; 2.45 | 57.5 | 0.99 | 57.6 | 1.03 | 57.9 | *NA* | 57.3 |
| | 6 | 2.45; 1.90 | 22.3 | 1.91; 2.20 | 22.2 | 1.84; 2.41 | 21.1 | *NA* | 22.0 |
| | 7 | 1.26 | 41.8 | 1.31; 1.51 | 41.9 | 2.78 | 39.6 | *NA* | 41.5 |
| | 8 | – | 42.7 | – | 42.2 | – | 34.9 | *NA* | 42.5 |
| | 9 | 0.86 | 54.0 | 0.93 | 54.2 | 1.01 | 51.7 | *NA* | 53.8 |
| | 10 | – | 39.9 | – | 39.8 | – | 44.6 | *NA* | 39.7 |

**Table 5.** *Cont.*

| Moiety | Position | 10 [(a)] | | 11 [(b)] | | 12 [(b)] | | 13 [(a)] | |
|---|---|---|---|---|---|---|---|---|---|
| | | $\delta_H$ | $\delta_C$ | $\delta_H$ | $\delta_C$ | $\delta_H$ | $\delta_C$ | $\delta_H$ | $\delta_C$ |
| | 11 | 1.60 | 20.7 | 1.48 | 20.7 | 1.59 | 27.3 | NA | 20.6 |
| | 12 | 2.22; 1.92 | 36.8 | 2.75; 1.10 | 38.0 | 4.11 | 78.4 | NA | 36.6 |
| | 13 | – | 86.2 | – | 87.2 | 2.94 | 42.0 | NA | 85.9 |
| | 14 | 2.70; 1.77 | 44.6 | 1.94; 2.45 | 44.8 | 1.42; 1.05 | 39.9 | NA | 44.3 |
| | 15 | 2.09; 2.02 | 47.7 | 2.10 | 48.6 | 1.91; 2.23 | 49.0 | NA | 47.5 |
| | 16 | – | 154.5 | – | 153.8 | – | 148.4 | NA | 154.3 |
| | 17 | 5.68; 5.04 | 104.7 | 5.10; 5.64 | 105.5 | 4.90; 5.40 | 109.5 | NA | 104.5 |
| | 18 | 1.22 | 28.4 | 1.42 | 29.4 | 1.26 | 29.1 | NA | 28.2 |
| | 19 | – | 177.2 | – | 176.1 | – | 177.4 | NA | 177.0 |
| | 20 | 1.27 | 15.7 | 0.99 | 16.5 | 1.04 | 13.3 | NA | 15.4 |
| Glcβ-C19 | 1′ | 6.08 | 95.9 | 6.23 | 93.8 | 6.18 | 96.4 | 6.16 | 95.6 |
| | 2′ | 4.15 | 74.0 | NA | NA | NA | NA | NA | NA |
| | 3′ | 4.17 | 79.0 | NA | NA | NA | NA | NA | NA |
| | 4′ | 4.27 | 71.1 | NA | NA | NA | NA | NA | NA |
| | 5′ | 3.93 | 79.3 | NA | NA | NA | NA | NA | NA |
| | 6′ | 4.39; 4.30 | 62.2 | NA | NA | NA | NA | NA | NA |
| Glcβ(1-X) | 1″ | – | – | 5.10 | 105.8 | – | – | – | – |
| | 2″ | – | – | NA | NA | – | – | – | – |
| | 3″ | – | – | NA | NA | – | – | – | – |
| | 4″ | – | – | NA | NA | – | – | – | – |
| | 5″ | – | – | NA | NA | – | – | – | – |
| | 6″ | – | – | NA | NA | – | – | – | – |
| Glcβ-C13 | 1‴ | 5.12 | 98.0 | 5.12 | 99.7 | 5.01 | 103.0 | 5.08 | 97.6 |
| | 2‴ | 4.14 | 84.6 | – | – | NA | NA | NA | NA |
| | 3‴ | 4.25 | 78.24 | – | – | NA | NA | NA | NA |
| | 4‴ | 4.00 | 72.2 | – | – | NA | NA | NA | NA |
| | 5‴ | 3.88 | 77.9 | – | – | NA | NA | NA | NA |
| | 6‴ | 4.55; 4.19 | 62.9 | – | – | NA | NA | NA | NA |
| Glcβ(1-Y) | 1⁗ | 5.27 | 106.8 | – | – | 5.24 | 106.4 | 95.6 | 104.7 |
| | 2⁗ | 4.18 | 77.0 | – | – | NA | NA | NA | NA |
| | 3⁗ | 4.23 | 78.16 | – | – | NA | NA | NA | NA |
| | 4⁗ | 4.39 | 71.6 | – | – | NA | NA | NA | NA |
| | 5⁗ | 3.94 | 78.6 | – | – | NA | NA | NA | NA |
| | 6⁗ | 4.46; 4.41 | 62.7 | – | – | NA | NA | NA | NA |

[(a)] NMR spectra recorded in Pyr-$d_5$, [(b)] MeOH-$d_4$. **10:** stevioside (Y = 2) [79]; **11**: rebaudioside KA (X = 2) [22]; **12:** 12-α-[(2-O-β-D-glucopyranosyl-β-D-glucopyranosyl)oxy]ent-kaur-16-en-19-oic acid β-D-glucopyranosyl ester (Y = 2) [22]; **13:** rebaudioside G (Y = 3) [16] NA: not assigned.

In Table 7 are presented DGs with four sugar units, one xylose and three glucoses with different arrangements whereas in Table 8, a few examples of DGs with four sugar units, one rhamnose or 6-deoxyglucose together with three glucose with different arrangements.

**Table 6.** $^{1}H$ and $^{13}C$ chemical shifts of the compounds **14–19**.

| Moiety | Position | **14** [(a)] | | **15** [(a)] | | **16** [(a)] | | **17** [(b)] | | **18** [(b)] | | **19** [(b)] | |
|---|---|---|---|---|---|---|---|---|---|---|---|---|---|
| | | $\delta_H$ | $\delta_C$ | $\delta_H$ | $\delta_C$ | $\delta_H$ | $\delta_C$ | $\delta_H$ | $\delta_C$ | $\delta_H$ | $\delta_C$ | $\delta_H$ | $\delta_C$ |
| Aglycone | 1 | 2.34; 0.98 | 39.1 | 0.78; 1.77 | 41.2 | 0.73; 1.73 | 41.2 | 0.92; 1.59 | 39.5 | 0.87; 1.87 | 41.4 | 0.86; 1.87 | 41.5 |
| | 2 | 2.18; 1.14 | 20.1 | 2.22; 1.45 | 19.9 | 2.20; 1.43 | 19.9 | 1.92; 1.40 | 18.4 | 1.41; 1.92 | 19.8 | 1.41; 1.94 | 19.7 |
| | 3 | 2.21; 1.95 | 37.4 | 1.03; 2.35 | 37.3 | 2.28; 1.79 | 36.8 | 2.20; 1.09 | 37.7 | 1.06; 2.14 | 38.8 | 1.05; 2.17 | 38.8 |
| | 4 | – | 43.4 | – | 44.5 | – | 44.5 | – | 43.7 | – | 43.8 | – | 44.6 |
| | 5 | 1.01 | 57.9 | 1.05 | 57.8 | 1.03 | 57.8 | 1.12 | 57.1 | 1.12 | 58.1 | 1.12 | 58.2 |
| | 6 | 2.47; 1.90 | 22.9 | 2.46; 1.92 | 22.6 | 1.91; 2.50 | 22.7 | 1.78; 2.05 | 19.9 | 1.84; 1.97 | 22.6 | 1.84; 2.04 | 22.8 |
| | 7 | 1.25; 1.25 | 42.4 | 1.30 | 42.2 | 1.28; 1.28 | 42.2 | 1.57; 1.16 | 38.8 | 1.41; 1.54 | 42.2 | 1.42; 1.55 | 42.4 |
| | 8 | – | 44.7 | – | 43.1 | – | 43.3 | – | 33.9 | – | 43.0 | – | 43.4 |
| | 9 | 0.86 | 54.6 | 0.88 | 54.5 | 0.87 | 54.3 | 1.09 | 51.0 | 0.97 | 54.7 | 0.98 | 55.0 |
| | 10 | – | 40.5 | – | 40.3 | – | 40.3 | – | 38.1 | – | 39.4 | – | 40.5 |
| | 11 | 1.60; 1.60 | 21.4 | 1.68 | 21.1 | 1.68; 1.68 | 21.1 | 1.63; 1.45 | 26.3 | 1.64; 1.80 | 21.0 | 1.65; 1.81 | 21.1 |
| | 12 | 1.70; 0.73 | 41.4 | 2.25; 2.00 | 37.3 | 1.02; 2.35 | 38.9 | 2.51 | 41.0 | 1.52; 1.97 | 37.9 | 1.55; 1.98 | 37.8 |
| | 13 | – | 86.8 | – | 86.9 | – | 86.4 | – | 77.8 | – | 87.8 | – | 87.7 |
| | 14 | 2.72; 1.78 | 45.1 | 2.66; 1.81 | 45.0 | 1.97; 2.74 | 45.1 | 4.04 | 37.7 | 1.50; 2.27 | 45.0 | 1.51; 2.26 | 45.0 |
| | 15 | 2.04; 2.04 | 48.3 | 2.05 | 48.2 | 2.06; 2.06 | 48.0 | 1.15; 2.51 | 47.4 | 2.04; 2.12 | 48.2 | 2.04; 2.14 | 48.4 |
| | 16 | – | 155.2 | – | 154.7 | – | 155.0 | – | 146.7 | – | 154.5 | – | 154.4 |
| | 17 | 5.70; 5.07 | 105.4 | 5.01; 5.64 | 105.1 | 5.10; 5.73 | 105.4 | 1.89; 2.11 | 108.0 | 4.85; 5.18 | 105.3 | 4.84; 5.17 | 105.2 |
| | 18 | 1.25 | 28.9 | 1.25 | 28.8 | 1.23 | 28.8 | 4.70; 4.84 | 27.5 | 1.21 | 28.6 | 1.23 | 28.5 |
| | 19 | – | 178.0 | – | 177.5 | – | 177.7 | 1.24 | 176.8 | – | 178.4 | – | 178.4 |
| | 20 | 1.28 | 16.3 | 1.32 | 16.0 | 1.31 | 16.0 | 0.89 | 11.9 | 0.97 | 16.0 | 0.98 | 16.0 |
| Glcβ-C19 | 1′ | 6.01 | 96.5 | 6.12 | 96.3 | 6.12 | 96.3 | 5.43 | 94.2 | 5.37 | 95.4 | 5.34 | 95.5 |
| | 2′ | 4.09 | 74.4 | 4.13 | 79.8 | 4.17 | 74.4 | 3.39 | 72.6 | 3.36 | 73.8 | 3.35 | 74.0 |
| | 3′ | 4.13 | 79.5 | 4.18 | 78.7 | 3.98 | 79.5 | 3.43 | 77.3 | 3.46 | 78.5 | 3.44 | 78.4 |
| | 4′ | 4.26 | 72.0 | 4.28 | 71.5 | 4.33 | 71.4 | 3.39 | 69.7 | 3.36 | 70.9 | 3.45 | 70.9 |
| | 5′ | 4.02 | 78.8 | 3.97 | 79.2 | 4.22 | 79.9 | 3.39 | 77.3 | 3.36 | 78.2 | 3.45 | 77.5 |
| | 6′ | 4.68; 4.31 | 70.0 | 4.40 | 63.4 | 4.43; 4.57 | 62.6 | 3.85; 3.71 | 61.0 | 3.68; 3.82 | 62.2 | 3.58; 3.61 | 62.1 |
| Glcβ -C13 | 1″ | 5.03 | 105.9 | 5.08 | 98.8 | 5.19 | 98.3 | 4.56 | 100.5 | 4.59 | 97.2 | 4.61 | 97.4 |
| | 2″ | 4.00 | 75.9 | 4.38 | 81.3 | 4.23 | 84.4 | 3.65 | 79.0 | 3.58 | 81.5 | 3.45 | 82.4 |
| | 3″ | 3.88 | 79.1 | 4.15 | 88.6 | 4.27 | 78.6 | 3.71 | 86.0 | 3.55 | 77.9 | 3.56 | 78.1 |
| | 4″ | 4.22 | 77.8 | 3.87 | 71.2 | 4.45 | 71.8 | 3.38 | 68.8 | 3.30 | 71.6 | 3.25 | 72.0 |
| | 5″ | 3.88 | 79.2 | 3.79 | 77.9 | 4.08 | 77.8 | 3.34 | 76.1 | 3.22 | 77.5 | 3.26 | 77.7 |
| | 6″ | 4.57; 4.38 | 63.5 | 4.50; 4.40 | 63.2 | 4.51; 4.78 | 70.3 | 3.91; 3.68 | 61.4 | 3.65; 3.83 | 62.2 | 3.63; 3.85 | 63.0 |

**Table 6.** *Cont.*

| Moiety | Position | 14 [(a)] | | 15 [(a)] | | 16 [(a)] | | 17 [(b)] | | 18 [(b)] | | 19 [(b)] | |
|---|---|---|---|---|---|---|---|---|---|---|---|---|---|
| | | $\delta_H$ | $\delta_C$ | $\delta_H$ | $\delta_C$ | $\delta_H$ | $\delta_C$ | $\delta_H$ | $\delta_C$ | $\delta_H$ | $\delta_C$ | $\delta_H$ | $\delta_C$ |
| Glcβ(1-2) | 1‴ | 5.15 | 98.7 | 5.58 | 105.3 | 5.30 | 106.8 | 4.81 | 102.2 | 4.61 | 104.3 | 4.57 | 105.1 |
| | 2‴ | 4.18 | 85.4 | 4.20 | 76.8 | 4.11 | 75.7 | 3.19 | 74.6 | 3.28 | 75.6 | 3.27 | 76.2 |
| | 3‴ | 3.92 | 78.7 | 4.31 | 74.4 | 4.23 | 78.8 | 3.35 | 76.5 | 3.33 | 77.8 | 3.36 | 78.0 |
| | 4‴ | 4.29 | 78.9 | 4.28 | 72.0 | 4.28 | 72.1 | 3.22 | 70.6 | 3.33 | 71.5 | 3.30 | 71.6 |
| | 5‴ | 4.03 | 72.8 | 3.97 | 79.0 | 3.91 | 78.3 | 3.33 | 76.6 | 3.44 | 77.5 | 3.25 | 78.4 |
| | 6‴ | 4.57; 4.22 | 63.2 | 4.40 | 62.5 | 4.22; 4.57 | 63.3 | 3.87; 3.66 | 61.9 | 3.78; 4.10 | 69.3 | 3.63; 3.91 | 63.0 |
| Sugar*(1-X) | 1⁗ | 5.30 | 107.6 | 5.33 | 105.3 | 5.14 | 105.9 | 4.65 | 103.1 | 4.54 | 104.0 | 3.61; 3.67 | 63.8 |
| | 2⁗ | 4.22 | 78.8 | 4.08 | 76.8 | 4.06 | 77.2 | 3.28 | 73.9 | 3.19 | 74.8 | – | 105.1 |
| | 3⁗ | 4.27 | 71.5 | 4.20 | 74.4 | 4.27 | 78.9 | 3.39 | 76.8 | 3.46 | 77.9 | 4.09 | 78.8 |
| | 4⁗ | 4.45 | 72.1 | 4.23 | 72.4 | 4.12 | 72.6 | 3.31 | 70.1 | 3.29 | 71.5 | 4.00 | 76.3 |
| | 5⁗ | 3.97 | 79.4 | 4.05 | 78.8 | 4.03 | 78.5 | 3.37 | 76.8 | 3.30 | 77.8 | 3.72 | 83.6 |
| | 6⁗ | 4.49; 4.38 | 63.2 | 4.57; 4.32 | 62.8 | 4.22; 4.57 | 63.3 | 3.90; 3.66 | 61.2 | 3.61; 3.86 | 62.6 | 3.72; 3.98 | 61.6 |

[(a)] NMR spectra recorded in Pyr-$d_5$, [(b)] MeOH-$d_4$. **14**: rebaudioside Y (X = 6) [58]; **15**: rebaudioside A (*Glc, X = 3) [79] **16**: rebaudioside Z (*Glc, X = 6) [31]; **17**: 13-[(2-O-β-D-glucopyranosyl-3-O-β-D-glucopyranosyl-β-D-glucopyranosyl) oxy]ent-hydroxyatis-16-en-19-oic acid -β-D-glucopyranosy ester (*Glc, X = 3) [31]; **18**: 13-[(2-*O*-(6-*O*-β-D-glucopyranosyl)-3-O-β-D-glucopyranosyl-β-D-glucopyranosyl)oxy] kaur-16-en-18-oic acid β-D-glucopyranosyl ester (*Glc, X = 6 as follows Glcβ(1-6)Glcβ(1-2)Glc$\beta_1$- [65]; **19**: 13-[(2-*O*-β-D-glucopyranosyl-3-*O*-β-D-fructofuranosyl-β-D-glucopyranosyl)oxy] kaur-16-en-18-oic acid β-D-glucopyranosyl ester (*Fru, X = 3) [65]. NA: not assigned.

**Table 7.** $^1$H and $^{13}$C chemical shifts of the compounds **20–23**.

| Moiety | Position | **20** $^{(a)}$ | | **21** $^{(a)}$ | | **22** $^{(b)}$ | | **23** $^{(b)}$ | |
|---|---|---|---|---|---|---|---|---|---|
| | | $\delta_H$ | $\delta_C$ | $\delta_H$ | $\delta_C$ | $\delta_H$ | $\delta_C$ | $\delta_H$ | $\delta_C$ |
| Aglycone | 1 | 0.77; 1.78 | 40.8 | 0.75; 1.77 | 40.1 | 0.85; 1.87 | 41.6 | 0.85; 1.88 | 41.5 |
| | 2 | 1.41; 2.20 | 19.4 | 1.42; 2.22 | 19.9 | 1.44; 1.93 | 19.8 | 1.41; 1.94 | 19.8 |
| | 3 | 1.00; 2.32 | 38.4 | 1.89; 2.38 | 38.9 | 1.06; 2.17 | 38.7 | 1.05; 2.15 | 38.7 |
| | 4 | – | 44.0 | – | 44.4 | – | 44.3 | – | 44.8 |
| | 5 | 1.02 | 57.3 | 1.05 | 57.8 | 1.12 | 58.4 | 1.12 | 58.2 |
| | 6 | 1.89; 2.44 | 22.1 | 1.98; 2.40 | 22.6 | 1.83; 2.08 | 22.8 | 1.83; 2.03 | 22.8 |
| | 7 | 1.28 | 41.7 | 1.36; 1.43 | 42.2 | 1.43; 1.55 | 42.5 | 1.44; 1.55 | 42.4 |
| | 8 | – | 42.5 | – | 44.7 | – | 43.2 | – | 43.3 |
| | 9 | 0.89 | 54.0 | 0.91 | 54.7 | 0.98 | 54.7 | 0.97 | 55.0 |
| | 10 | – | 39.8 | – | 40.2 | – | 38.8 | – | 40.7 |
| | 11 | 1.66 | 20.6 | 1.68 | 20.9 | 1.66; 1.82 | 21.0 | 1.63; 1.80 | 21.0 |
| | 12 | 1.85; 2.28 | 37.0 | 2.19; 1.89 | 38.4 | 1.46; 2.01 | 37.6 | 1.54; 1.99 | 37.9 |
| | 13 | – | 86.4 | – | 87.2 | – | 87.1 | – | 88.0 |
| | 14 | 1.79; 2.65 | 44.3 | 1.74; 2.55 | 44.7 | 1.54; 2.27 | 45.0 | 1.51; 2.26 | 45.0 |
| | 15 | 2.03; 2.14 | 47.7 | 2.06; 2.14 | 48.5 | 2.04 | 48.3 | 2.04; 2.14 | 48.3 |
| | 16 | – | 154.2 | – | 154.7 | – | 153.6 | – | 153.7 |
| | 17 | 4.99; 5.63 | 104.6 | 5.20; 5.60 | 105.4 | 4.82; 5.11 | 104.7 | 4.84; 5.18 | 105.2 |
| | 18 | 1.20 | 28.3 | 1.27 | 29.0 | 1.21 | 28.6 | 1.20 | 28.6 |
| | 19 | – | 177.0 | – | 177.3 | – | 178.2 | – | 178.5 |
| | 20 | 1.31 | 15.5 | 1.25 | 16.1 | 1.00 | 16.0 | 0.97 | 16.0 |
| Glcβ-C19 | 1′ | 6.10 | 95.8 | 6.16 | 96.3 | 5.34 | 95.7 | 5.34 | 95.4 |
| | 2′ | 4.19 | 73.9 | *NA* | *NA* | 3.32 | 73.8 | 3.35 | 73.7 |
| | 3′ | 4.10 | 79.2 | *NA* | *NA* | 3.44 | 78.4 | 3.42 | 78.2 |
| | 4′ | 4.24 | 71.0 | *NA* | *NA* | 3.34 | 70.9 | 3.39 | 70.9 |
| | 5′ | 3.93 | 78.5 | *NA* | *NA* | 3.36 | 78.2 | 3.52 | 77.6 |
| | 6′ | 4.39; 4.30 | 62.1 | *NA* | *NA* | 3.60; 3.82 | 62.5 | 3.76; 4.06 | 68.8 |
| Xyl(1-6) | 1″ | – | – | – | – | – | – | 4.30 | 104.6 |
| | 2″ | – | – | – | – | – | – | 3.58 | 72.2 |
| | 3″ | – | – | – | – | – | – | 3.53 | 74.0 |
| | 4″ | – | – | – | – | – | – | 3.79 | 69.1 |
| | 5″ | – | – | – | – | – | – | 3.50; 3.84 | 66.3 |
| | 6″ | – | – | – | – | – | – | – | – |
| Sugar*-C13 | 1‴ | 5.02 | 97.9 | 4.96 | 98.7 | 4.59 | 97.8 | 4.60 | 97.4 |
| | 2‴ | 4.22 | 80.7 | *NA* | *NA* | 3.59 | 81.5 | 3.45 | 82.4 |
| | 3‴ | 4.06 | 88.3 | *NA* | *NA* | 3.68 | 87.3 | 3.54 | 77.9 |
| | 4‴ | 3.83 | 70.6 | *NA* | *NA* | 3.34 | 70.9 | 3.25 | 71.6 |
| | 5‴ | 3.72 | 77.3 | *NA* | *NA* | 3.30 | 77.2 | 3.25 | 77.9 |
| | 6‴ | 4.42; 4.02 | 62.6 | – | – | 3.60; 3.82 | 62.3 | 3.62; 3.83 | 62.6 |
| Sugar**(1-2) | 1⁗ | 5.42 | 105.4 | 5.54 | 105.0 | 4.63 | 104.0 | 4.58 | 105.1 |
| | 2⁗ | 4.10 | 75.9 | *NA* | *NA* | 3.25 | 73.4 | 3.27 | 77.8 |
| | 3⁗ | 4.12 | 78.6 | *NA* | *NA* | 3.42 | 78.6 | 3.36 | 77.9 |
| | 4⁗ | 4.23 | 71.2 | *NA* | *NA* | 3.32 | 71.1 | 3.29 | 71.4 |
| | 5⁗ | 4.35; 3.65 | 67.5 | *NA* | *NA* | 3.36 | 78.4 | 3.24 | 77.7 |
| | 6⁗ | – | – | *NA* | *NA* | 3.62; 3.80 | 62.7 | 3.62; 3.83 | 62.6 |
| Sugar***(1-3) | 1′′′′′ | 5.26 | 104.8 | 5.33 | 105.1 | 4.61 | 105.3 | – | – |
| | 2′′′′′ | 4.01 | 75.2 | *NA* | *NA* | 3.54 | 73.0 | – | – |
| | 3′′′′′ | 4.17 | 79.0 | *NA* | *NA* | 3.50 | 74.3 | – | – |
| | 4′′′′′ | 4.13 | 71.5 | *NA* | *NA* | 3.46 | 69.9 | – | – |
| | 5′′′′′ | 4.02 | 78.6 | *NA* | *NA* | 3.62; 3.86 | 67.6 | – | – |
| | 6′′′′′ | 4.51; 4.26 | 62.3 | *NA* | *NA* | – | – | – | – |

$^{(a)}$ NMR spectra recorded in Pyr-$d_5$, $^{(b)}$ MeOH-$d_4$. **20**: rebaudioside F (*Glc, **Xyl and ***Glc) [11]; **21**: rebaudioside R (*Xyl, **Glc and ***Glc) [23]; **22**: 13-[(2-*O*-β-D-glucopyranosyl-3-*O*-β-D-xylopyranosyl-β-D-glucopyranosyl)oxy] *ent*-kaur-16-en-19-oic acid β-D-glucopyranosyl ester (*Glc, **Glc and *** Xyl) [68]; **23**: 13-[(2-O-β-D-glucopyranosyl-β-D-glucopyranosyl) oxy]-kaur-16-en-18-oic acid-(6-O-β-D-xylopyranosyl-β-D-glucopyranosyl) ester (*Glc and **Glc) [37].

**Table 8.** $^1$H and $^{13}$C chemical shifts of the compounds **24–27**.

| Moiety | Position | **24** $^{(b)}$ | | **25** $^{(b)}$ | | **26** $^{(b)}$ | | **27** $^{(a)}$ | |
|---|---|---|---|---|---|---|---|---|---|
| | | $\delta_H$ | $\delta_C$ | $\delta_H$ | $\delta_C$ | $\delta_H$ | $\delta_C$ | $\delta_H$ | $\delta_C$ |
| Aglycone | 1 | 0.76; 1.66 | 41.1 | 0.83; 1.85 | 40.9 | 0.85; 1.86 | 41.6 | *NA* | 40.8 |
| | 2 | 1.70; 2.11 | 20.3 | 1.40; 1.92 | 19.2 | 1.40; 1.94 | 19.8 | *NA* | 19.4 |
| | 3 | 1.91; 2.15 | 38.1 | 1.55; 1.96 | 37.0 | 1.05; 2.14 | 38.9 | *NA* | 38.5 |
| | 4 | – | 44.8 | – | 43.5 | – | 44.8 | *NA* | 43.9 |
| | 5 | 1.00 | 58.8 | 1.07 | 57.5 | 1.12 | 58.3 | *NA* | 57.5 |
| | 6 | 1.86; 2.11 | 22.5 | 1.86; 1.90 | 21.8 | 1.84; 2.02 | 22.7 | *NA* | 22.0 |
| | 7 | 1.32; 1.62 | 42.1 | 1.43; 1.54 | 41.7 | 1.41; 1.54 | 42.5 | *NA* | 41.8 |
| | 8 | – | 43.1 | – | 54.0 | - | 43.2 | *NA* | 42.2 |
| | 9 | 0.91 | 54.4 | 0.98 | 54.1 | 0.97 | 55.0 | *NA* | 54.1 |
| | 10 | – | 40.2 | – | 39.0 | – | *NA* | *NA* | 39.9 |
| | 11 | 1.70 | 21.1 | 1.63; 1.80 | 19.8 | 1.63; 1.77 | 21.0 | *NA* | 20.6 |
| | 12 | 2.68; 1.15 | 38.1 | 1.07; 2.27 | 37.5 | 1.51; 1.97 | 38.1 | *NA* | 38.5 |
| | 13 | – | 86.6 | – | 87.6 | - | 88.7 | *NA* | 87.3 |
| | 14 | 1.70; 2.50 | 45.1 | 1.50; 2.24 | 44.6 | 1.52; 2.24 | 44.9 | *NA* | 43.2 |
| | 15 | 2.10 | 48.3 | 2.04; 2.13 | 47.7 | 2.03; 2.14 | 48.5 | *NA* | 48.7 |
| | 16 | – | 155.1 | – | 152.5 | – | 153.2 | *NA* | 152.9 |
| | 17 | 5.10; 5.72 | 105.6 | 4.84; 2.13 | 104.7 | 4.83; 5.18 | 105.1 | *NA* | 106.1 |
| | 18 | 1.50 | 29.6 | 1.25 | 28.4 | 1.21 | 28.6 | *NA* | 28.2 |
| | 19 | – | 176.0 | – | 177.3 | – | 178.4 | *NA* | 177.1 |
| | 20 | 1.14 | 17.3 | 0.93 | 16.1 | 0.99 | 16.1 | *NA* | 15.2 |
| Glcβ-C19 | 1′ | 6.24 | 94.1 | 5.60 | 93.5 | 5.38 | 95.5 | 5.98 | 96.0 |
| | 2′ | *NA* | *NA* | 3.60 | 77.7 | 3.34 | 73.6 | *NA* | *NA* |
| | 3′ | *NA* | *NA* | 3.57 | 78.2 | 3.44 | 78.4 | *NA* | *NA* |
| | 4′ | *NA* | *NA* | 3.40 | 70.1 | 3.34 | 70.9 | *NA* | *NA* |
| | 5′ | *NA* | *NA* | 3.39 | 77.4 | 3.36 | 78.2 | *NA* | *NA* |
| | 6′ | *NA* | *NA* | 3.70; 3.79 | 62.3 | 3.60; 3.82 | 62.3 | *NA* | *NA* |
| Rha(1-2) | 1″ | 6.40 | 102.0 | 5.30 | 100.9 | – | – | – | – |
| | 2″ | *NA* | *NA* | 3.90 | 71.0 | – | – | – | – |
| | 3″ | *NA* | *NA* | 3.61 | 71.1 | – | – | – | – |
| | 4″ | *NA* | *NA* | 3.37 | 72.7 | – | – | – | – |
| | 5″ | *NA* | *NA* | 3.75 | 69.3 | – | – | – | – |
| | 6″ | *NA* | *NA* | 1.24 | 17.1 | – | – | – | – |
| Glcβ-C13 | 1‴ | 5.10 | 98.5 | 4.61 | 96.5 | 4.56 | 97.7 | 5.04 | 97.7 |
| | 2‴ | *NA* | *NA* | 3.47 | 81.2 | 3.54 | 80.9 | *NA* | *NA* |
| | 3‴ | *NA* | *NA* | 3.57 | 79.2 | 3.67 | 87.5 | *NA* | *NA* |
| | 4‴ | *NA* | *NA* | 3.31 | 70.7 | 3.34 | 70.9 | *NA* | *NA* |
| | 5‴ | *NA* | *NA* | 3.25 | 77.2 | 3.30 | 77.2 | *NA* | *NA* |
| | 6‴ | *NA* | *NA* | 3.63; 3.83 | 62.0 | 3.60; 3.82 | 62.3 | *NA* | *NA* |
| Sugar*(1-2) | 1⁗ | 5.22 | 107.0 | 4.60 | 104.4 | 4.71 | 104.0 | 6.34 | 104.2 |
| | 2⁗ | *NA* | *NA* | 3.24 | 75.0 | 3.22 | 75.8 | *NA* | *NA* |
| | 3⁗ | *NA* | *NA* | 3.35 | 77.7 | 3.30 | 74.0 | *NA* | *NA* |
| | 4⁗ | *NA* | *NA* | 3.21 | 70.7 | 3.02 | 76.9 | *NA* | *NA* |
| | 5⁗ | *NA* | *NA* | 3.21 | 76.5 | 3.26 | 81.2 | *NA* | *NA* |
| | 6⁗ | *NA* | *NA* | 3.64; 3.84 | 61.8 | 1.25 | 18.0 | *NA* | 18.8 |
| Glcβ(1-3) | 1″‴′ | – | – | – | – | 4.63 | 104.2 | 4.94 | 102.1 |
| | 2″‴′ | – | – | – | – | 3.24 | 73.4 | *NA* | *NA* |
| | 3″‴′ | – | – | – | – | 3.42 | 78.6 | *NA* | *NA* |
| | 4″‴′ | – | – | – | – | 3.32 | 71.1 | *NA* | *NA* |
| | 5″‴′ | – | – | – | – | 3.36 | 78.4 | *NA* | *NA* |
| | 6″‴′ | – | – | – | – | 3.62; 3.80 | 62.7 | *NA* | *NA* |

$^{(a)}$ NMR spectra recorded in Pyr-$d_5$, $^{(b)}$ MeOH-$d_4$. **24**: rebaudioside S [23]; **25**: no trivial or systematic name was assigned (*Glc) [25]; **26**: 13-[(2-O-6-deoxy-β-D-glucopyranosyl-3-O-β-D-glucopyranosyl-β-D-glucopyranosyl)oxy] ent-kaur-16-en19-oic acid β-D-glucopyranosyl ester (*6deoxyGlc) [71]; **27**: rebaudioside C (*Rha) [16].

## 6. Conclusions

*Stevia rebaudiana* and its steviol glycosides have been widely studied, and in the continuous search for non-caloric sugar substitutes with improved taste, multiple new tetracyclic diterpene glycosides have been isolated from leaf of *S. rebaudiana* or have been prepared by chemical or enzymatic reactions from selected steviol glycosides. Herein, an updated list of natural DGs isolated from *S. rebaudiana* has been compiled, along with some chemically modified DGs. Some approaches for the rapid detection of new DG structures in fractions rich in DGs have also been presented. Thus, the fragmentation pattern for DGs by High-Resolution Electrospray Ionization Mass Spectrometry has been presented. The modification of DGs by a simple saponification reaction with further analysis of the products by Reverse-Phase High-Performance Liquid Chromatography for the detection of new moieties at position C-13 is another approach summarized in this review. Several HPLC methods with diverse stationary phases, but with a unique mobile phase used for the analysis of fractions rich of specific DGs were discussed. However, a major drawback in the search of novel structures in natural products, is the isolation of quantities that are insufficient to allow bioassays or in the case of DGs, conducting tasting assays to better understand the relationship structure-sweetness/bitterness of the DGs, and in turn followed by toxicological evaluation. Some strategies to save cost in the scale-up of the purification process were also shared. With the described strategies, several very minor DGs were purified in quantities of hundreds of milligrams to multiple grams. Currently, several DGs with different aglycone cores, numbers and types of sugar units, and arrangements have been well documented so far. Hence, the number of DG structures available is vast, although the reported tasting results are infrequent or at least not visible in the scientific literature. Further studies need to be pointed out for the tasting evaluation of the DGs already discovered. With the availability of this information, adequate strategies could be followed to overexpress specific groups of DGs in *S. rebaudiana* and/or to use appropriate blends of DGs with improved taste.

**Author Contributions:** Both W.H.P. and J.D.M. contributed to writing the manuscript. All authors have read and agreed to the published version of the manuscript.

**Funding:** This research received no external funding.

**Institutional Review Board Statement:** Not applicable.

**Informed Consent Statement:** Not applicable.

**Conflicts of Interest:** The authors declare no conflict of interest.

**Sample Availability:** Samples of the compounds described in this review are available from the authors.

## References

1. Kinghorn, A.D. *Stevia: The Genus Stevia; Medicinal and Aromatic Plants—Industrial Profiles*; Kinghorn, A.D., Ed.; Taylor & Francis: London, UK, 2002; Volume 19, pp. 1–17.
2. Cioni, P.L.; Morelli, I.; Andolfi, L.; Macchia, M.; Ceccarini, L. Qualitative and quantitative analysis of essential oils of five lines stevia rebaudiana bert. genotypes cultivated in pisa (italy). *J. Essent. Oil. Res.* **2006**, *18*, 76–79. [CrossRef]
3. McGarvey, B.D.; Attygalle, A.B.; Starratt, A.N.; Xiang, B.; Schroeder, F.C.; Brandle, J.E.; Meinwald, J. New Non-Glycosidic Diterpenes from the Leaves of Stevia r Ebaudiana. *J. Nat. Prod.* **2003**, *66*, 1395–1398. [CrossRef]
4. Rajbhandari, A.; Roberts, M.F. The Flavonoids of Stevia Rebaudiana. *J. Nat. Prod.* **1983**, *46*, 194–195. [CrossRef]
5. Karaköse, H.; Jaiswal, R.; Kuhnert, N. Characterization and quantification of hydroxycinnamate derivatives in stevia rebaudiana leaves by LC-MSn. *J. Agric. Food Chem.* **2011**, *59*, 18. [CrossRef] [PubMed]
6. Bridel, M.; Lavielle, R. Sur Le Principe Sucre Des Feuilles de Kaa-He-e (Stevia Rebaundiana B). *Acad. Sci. Paris Comptes Rendus* **1931**, *192*, 1123–1125.
7. Kohda, H.; Kasai, R.; Yamasaki, K.; Murakami, K.; Tanaka, O. New sweet diterpene glucosides from stevia rebaudiana. *Phytochemistry* **1976**, *15*, 981–983. [CrossRef]
8. Kobayashi, M.; Horikawa, S.; Degrandi, I.H.; Ueno, J.; Mitsuhashi, H. Dulcosides A and B, New diterpene glycosides from stevia rebaudiana. *Phytochemistry* **1977**, *16*, 1405–1408. [CrossRef]

9. Sakamoto, I.; Yamasaki, K.; Tanaka, O. Application of 13C NMR spectroscopy to chemistry of plant glycosides: Rebaudiosides-d and-e, new sweet diterpene-glucosides of stevia rebaudiana bertoni. *Chem. Pharm. Bull.* **1977**, *25*, 3437–3439. [CrossRef]
10. Sakamoto, I.; Yamasaki, K.; Tanaka, O. Application of 13C NMR spectroscopy to chemistry of natural glycosides: Rebaudioside-c, a new sweet diterpene glycoside of stevia rebaudiana. *Chem. Pharm. Bull.* **1977**, *25*, 844–846. [CrossRef]
11. Starratt, A.N.; Kirby, C.W.; Pocs, R.; Brandle, J.E. Rebaudioside F, A diterpene glycoside from stevia rebaudiana. *Phytochemistry* **2002**, *59*, 367–370. [CrossRef]
12. Andress, S. *Agency Response Letter GRAS Notice No. GRN 000252, CFSAN/Office of Food Additive Safety*; Whole Earth Sweeten Company LLC: Chicago, IL, USA, 2008.
13. Application A540—Steviol Glycosides as Intense Sweeteners. Available online: https://www.foodstandards.gov.au/code/applications/pages/applicationa540stevi3096.aspx (accessed on 21 March 2021).
14. The European Commission. Commission Regulation (EU) No. 1131/2011: Amending Annex II to Regulation (EC) No. 1333/2008 of the European Parliament and of the Council with Regard to Steviol Glycosides. *Off. J. Eur. Union* **2011**, 205–295. Available online: https://eur-lex.europa.eu/LexUriServ/LexUriServ.do?uri=OJ:L:2011:295:0205:0211:EN:PDF (accessed on 21 March 2021).
15. McQuate, R.S. *Agency Response Letter GRAS Notice No. GRN000304, CFSAN/Office of Food Additive Safety*; GRAS Associates LLC: Bend, OR, USA, 2010.
16. Ohta, M.; Morita, K.; Inoue, A.; Tamai, T.; Fujita, I.; Ohta, M.; Sasa, S. Characterization of novel steviol glycosides from leaves of stevia rebaudiana morita. *J. Appl. Glycosci.* **2010**, *57*, 199–209. [CrossRef]
17. Philippaert, K.; Pironet, A.; Mesuere, M.; Sones, W.; Vermeiren, L.; Kerselaers, S.; Pinto, S.; Segal, A.; Antoine, N.; Gysemans, C.; et al. Steviol glycosides enhance pancreatic Beta-Cell function and taste sensation by potentiation of TRPM5 channel activity. *Nat. Commun.* **2017**, *8*, 14733. [CrossRef]
18. Gu, W.; Rebsdorf, A.; Anker, C.; Gregersen, S.; Hermansen, K.; Geuns, J.M.C.; Jeppesen, P.B. Steviol glucuronide, a metabolite of steviol glycosides, potently stimulates insulin secretion from isolated mouse islets: Studies in vitro. *Endocrinol. Diabetes Metab.* **2019**, *2*, e00093. [CrossRef] [PubMed]
19. Geuns, J.M.C.; Buyse, J.; Vankeirsbilck, A.; Temme, E.H.M.; Compernolle, F.; Toppet, S. Identification of Steviol Glucuronide in Human Urine. *J. Agric. Food Chem.* **2006**, *54*, 2794–2798. [CrossRef]
20. Chen, T.-H.; Chen, S.-C.; Chan, P.; Chu, Y.-L.; Yang, H.-Y.; Cheng, J.-T. Mechanism of the hypoglycemic effect of stevioside, a glycoside of stevia rebaudiana. *Planta Med.* **2005**, *71*, 108–113. [CrossRef]
21. Jeppesen, P.B.; Gregersen, S.; Rolfsen, S.E.D.; Jepsen, M.; Colombo, M.; Agger, A.; Xiao, J.; Kruhøffer, M.; Orntoft, T.; Hermansen, K. Antihyperglycemic and blood pressure-reducing effects of stevioside in the diabetic goto-kakizaki rat. *Metabolism* **2003**, *52*, 372–378. [CrossRef] [PubMed]
22. Ibrahim, M.A.; Rodenburg, D.L.; Alves, K.; Fronczek, F.R.; McChesney, J.D.; Wu, C.; Nettles, B.J.; Venkataraman, S.K.; Jaksch, F. Minor diterpene glycosides from the leaves of stevia rebaudiana. *J. Nat. Prod.* **2014**, *77*, 1231–1235. [CrossRef] [PubMed]
23. Ibrahim, M.A.; Rodenburg, D.L.; Alves, K.; Perera, W.H.; Fronczek, F.R.; Bowling, J.; McChesney, J.D. Rebaudiosides R and S, Minor diterpene glycosides from the leaves of stevia rebaudiana. *J. Nat. Prod.* **2016**, *79*, 1468–1472. [CrossRef]
24. Perera, W.H.; Ghiviriga, I.; Rodenburg, D.L.; Alves, K.; Bowling, J.J.; Avula, B.; Khan, I.A.; McChesney, J.D. Rebaudiosides T and U, Minor C-19 xylopyranosyl and arabinopyranosyl steviol glycoside derivatives from stevia rebaudiana (bertoni) bertoni. *Phytochemistry* **2017**, *135*, 106–114. [CrossRef]
25. Purkayastha, S.; Clos, J.F.; Prakash, I.; Yin, C.S.; Markosyan, A. Additional minor diterpene glycosides from stevia rebaudiana. *IOSR J. Appl. Chem.* **2019**, *12*, 48–55.
26. Perera, W.H.; Ghiviriga, I.; Rodenburg, D.L.; Carvalho, R.; Alves, K.; McChesney, J.D. Development of a high-performance liquid chromatography procedure to identify known and detect novel c-13 oligosaccharide moieties in diterpene glycosides from stevia rebaudiana (bertoni) bertoni (asteraceae): Structure elucidation of rebaudiosides V and W. *J. Sep. Sci.* **2017**, *40*, 3771–3781. [CrossRef] [PubMed]
27. Prakash, I.; Ma, G.; Bunders, C.; Charan, R.D.; Ramirez, C.; Devkota, K.P.; Snyder, T.M. A novel diterpene glycoside with nine glucose units from stevia rebaudiana bertoni. *Biomolecules* **2017**, *7*, 10. [CrossRef] [PubMed]
28. Prakash, I.; Hong, S.; Ma, G.; Bunders, C.; Devkota, K.P.; Charan, R.D.; Ramirez, C.; Snyder, T.M. Complete structure elucidation of new steviol glycosides possessing 9 glucose units isolated from stevia rebaudiana. *Nat. Prod. Commun.* **2017**, *12*. [CrossRef]
29. Ma, G.; Bechman, A.; Bunders, C.; Devkota, K.P.; Charan, R.D.; Ramirez, C.; Snyder, T.M.; Priedemann, C.; Prakash, I. New diterpene glycosides from stevia rebaudiana bertoni: Rebaudioside VIII and Rebaudioside IXd. *Nat. Prod. Commun.* **2018**, *13*. [CrossRef]
30. Prakash, I.; Ma, G.; Bunders, C.; Devkota, K.P.; Charan, R.D.; Ramirez, C.; Snyder, T.M.; Priedemann, C. A new diterpene glycoside: 15α-hydroxy-rebaudioside m isolated from stevia rebaudiana–Indra prakash. *Nat. Prod. Commun.* **2015**, *10*, 1177. [CrossRef]
31. Perera, W.H.; Ghiviriga, I.; Rodenburg, D.L.; Alves, K.; Wiggers, F.T.; Hufford, C.D.; Fronczek, F.R.; Ibrahim, M.A.; Muhammad, I.; Avula, B.; et al. Tetra-Glucopyranosyl diterpene Ent-Kaur-16-En-19-Oic acid and Ent-13(S)-Hydroxyatisenoic acid derivatives from a commercial extract of stevia rebaudiana (bertoni) bertoni. *Molecules* **2018**, *23*, 3328. [CrossRef]
32. Devkota, K.P.; Charan, R.D.; Priedemann, C.; Donovan, R.; Snyder, T.M.; Ramirez, C.; Harrigan, G.; Ma, G.; Prakash, I. Five new ent-atisene glycosides from stevia rebaudiana. *Nat. Prod. Commun.* **2019**, *14*. [CrossRef]
33. Lee, T. Steviol Glycoside Isomers 2009. U.S. Patent No. 7964232 B2, 19 March 2009.

34. Avent, A.G.; Hanson, J.; Oliveira, B.H. Hydrolysis of the diterpenoid glycoside, stevioside. *Phytochemistry* **1990**, *29*, 2712–2715. [CrossRef]
35. Prakash, I.; Chaturvedula, V.S.P.; Markosyan, A. Structural characterization of the degradation products of a minor natural sweet diterpene glycoside rebaudioside m under acidic conditions. *Int. J. Mol. Sci.* **2014**, *15*, 1014–1025. [CrossRef] [PubMed]
36. Prakash, I.; Clos, J.F.; Chaturvedula, V.S.P. Stability of rebaudioside a under acidic conditions and its degradation products. *Food Res. Int.* **2012**, *48*, 65–75. [CrossRef]
37. Chaturvedula, V.S.P.; Clos, J.F.; Rhea, J.; Milanowski, D.; Mocek, U.; DuBois, G.E.; Prakash, I. Minor diterpenoid glycosides from the leaves of stevia rebaudiana. *Phytochem. Lett.* **2011**, *4*, 209–212. [CrossRef]
38. Perera, W.H.; Docampo, M.L.; Wiggers, F.T.; Hufford, C.D.; Fronczek, F.R.; Avula, B.; Khan, I.A.; McChesney, J.D. Endocyclic double bond isomers and by-products from rebaudioside a and stevioside formed under acid conditions. *Phytochem. Lett.* **2018**, *25*, 163–170. [CrossRef]
39. Prakash, I.; Bunders, C.; Devkota, K.P.; Charan, R.D.; Hartz, R.M.; Sears, T.L.; Snyder, T.M.; Ramirez, C. Degradation products of rubusoside under acidic conditions. *Nat. Prod. Commun.* **2015**, *10*. [CrossRef]
40. Rodenburg, D.L.; Alves, K.; Perera, W.H.; Ramsaroop, T.; Carvalho, R.; McChesney, J.D. Development of hplc analytical techniques for diterpene glycosides from stevia rebaudiana (bertoni) bertoni: Strategies to scale-up. *J. Braz. Chem. Soc.* **2016**, *27*, 1406–1412.
41. Chaturvedula, V.S.P.; Prakash, I. Utilization of RP-HPLC fingerprinting analysis for the identification of diterpene glycosides from stevia rebaudiana. *Int. J. Res. Phytochem. Pharmacol.* **2011**, *1*, 88–92.
42. Prakash Chaturvedula, V.S.; Upreti, M.; Prakash, I. Diterpene glycosides from stevia rebaudiana. *Molecules* **2011**, *16*, 3552–3562. [CrossRef] [PubMed]
43. Chaturvedula, V.S.P.; Prakash, I. Hydrogenation of the exocyclic olefinic bond at C-16/C-17 position of Ent-Kaurane diterpene glycosides of stevia rebaudiana using various catalysts. *Int. J. Mol. Sci.* **2013**, *14*, 15669–15680. [CrossRef]
44. Prakash, I.; Campbell, M.; Chaturvedula, V.S.P. Catalytic hydrogenation of the sweet principles of stevia rebaudiana, rebaudioside b, rebaudioside c, and rebaudioside d and sensory evaluation of their reduced derivatives. *Int. J. Mol. Sci.* **2012**, *13*, 15126–15136. [CrossRef]
45. Brandle, J.E.; Telmer, P.G. Steviol glycoside biosynthesis. *Phytochemistry* **2007**, *68*, 1855–1863. [CrossRef]
46. Ceunen, S.; Geuns, J.M. Steviol glycosides: Chemical Diversity, metabolism, and function. *J. Nat. Prod.* **2013**, *76*, 1201–1228. [CrossRef]
47. Gerwig, G.J.; Te Poele, E.M.; Dijkhuizen, L.; Kamerling, J.P. Stevia glycosides: Chemical and enzymatic modifications of their carbohydrate moieties to improve the sweet-tasting quality. *Adv. Carbohydr. Chem. Biochem.* **2016**, *73*, 1–72. [CrossRef] [PubMed]
48. Ruddat, M.; Heftmann, E.; Lang, A. biosynthesis of steviol. *Arch. Biochem. Biophys.* **1965**, *110*, 496–499. [CrossRef]
49. Mizukami, H.; Shiiba, K.; Ohashi, H. Enzymatic determination of stevioside in stevia rebaudiana. *Phytochemistry* **1982**, *21*, 1927–1930. [CrossRef]
50. Reich, E.; Schibli, A. *High.-Performance Thin-Layer Chromatography for the Analysis of Medicinal Plants*; Thieme Medical Publisher Inc.: New York, NY, USA, 2007; ISBN 978-1-58890-409-6.
51. Jarne, C.; Savirón, M.; Lapieza, M.P.; Membrado, L.; Orduna, J.; Galbán, J.; Garriga, R.; Morlock, G.E.; Cebolla, V.L. High-Performance thin-layer chromatography coupled with electrospray ionization tandem mass spectrometry for identifying neutral lipids and sphingolipids in complex samples. *J. AOAC Int.* **2018**, *101*, 1993–2000. [CrossRef]
52. Jarne, C.; Cebolla, V.L.; Membrado, L.; Galbán, J.; Savirón, M.; Orduna, J.; Garriga, R. Separation and profiling of monoglycerides in biodiesel using a hyphenated technique based on high-performance thin-layer chromatography. *Fuel* **2016**, *177*, 244–250. [CrossRef]
53. Barret, L.-A.; Polidori, A.; Bonneté, F.; Bernard-Savary, P.; Jungas, C. A new high-performance thin layer chromatography-based assay of detergents and surfactants commonly used in membrane protein studies. *J. Chromatogr. A* **2013**, *1281*, 135–141. [CrossRef] [PubMed]
54. Wald, J.P.; Morlock, G.E. Quantification of steviol glycosides in food products, stevia leaves and formulations by planar chromatography, including proof of absence for steviol and isosteviol. *J. Chromatogr. A* **2017**, *1506*, 109–119. [CrossRef]
55. Morlock, G.E.; Meyer, S.; Zimmermann, B.F.; Roussel, J.-M. High-Performance thin-layer chromatography analysis of steviol glycosides in stevia formulations and sugar-free food products, and benchmarking with (ultra) high-performance liquid chromatography. *J. Chromatogr. A* **2014**, *1350*, 102–111. [CrossRef]
56. Wölwer-Rieck, U. The leaves of stevia rebaudiana (bertoni), their constituents and the analyses thereof: A review. *J. Agric. Food Chem.* **2012**, *60*, 886–895. [CrossRef]
57. Kolb, N.; Herrera, J.L.; Ferreyra, D.J.; Uliana, R.F. Analysis of sweet diterpene glycosides from stevia rebaudiana: Improved HPLC method. *J. Agric. Food Chem.* **2001**, *49*, 4538–4541. [CrossRef] [PubMed]
58. Perera, W.H.; Ramsaroop, T.; Carvalho, R.; Rodenburg, D.L.; McChesney, J.D. A silica gel orthogonal high-performance liquid chromatography method for the analyses of steviol glycosides: Novel tetra-glucopyranosyl steviol. *Nat. Prod. Res.* **2019**, *33*, 1876–1884. [CrossRef] [PubMed]
59. McChesney, J.D.; Rodenburg, D.L. Preparative chromatography and natural products discovery. *Curr. Opin. Biotechnol.* **2014**, *25*, 111–113. [CrossRef] [PubMed]
60. McChesney, J.D.; Rodenburg, D.L. Chromatography Methods 2014. U.S Patent 8801924 B2, 12 August 2014.

61. Minne, V.J.; Compernolle, F.; Toppet, S.; Geuns, J.M. Steviol quantification at the picomole level by high-performance liquid chromatography. *J. Agric. Food Chem.* **2004**, *52*, 2445–2449. [CrossRef] [PubMed]
62. Chaturvedula, V.S.P.; Kinger, K.P.; Campbell, M.R.; Prakash, I. NMR spectral assignments of steviol and steviol monoacetate. *J. Chem. Pharm. Res.* **2012**, *4*, 2666–2670.
63. Gardana, C.; Scaglianti, M.; Simonetti, P. Evaluation of steviol and its glycosides in stevia rebaudiana leaves and commercial sweetener by ultra-high-performance liquid chromatography-mass spectrometry. *J. Chromatogr. A* **2010**, *1217*, 1463–1470. [CrossRef]
64. Chaturvedula, V.S.P.; Prakash, I. Structure elucidation of three new diterpene glycosides from stevia rebaudiana. *Int. J. Phys. Sci.* **2011**, *6*, 6698–6705.
65. Chaturvedula, V.S.P.; Rhea, J.; Milanowski, D.; Mocek, U.; Prakash, I. Two minor diterpene glycosides from the leaves of stevia rebaudiana. *Nat. Prod. Commun.* **2011**, *6*. [CrossRef]
66. Chaturvedula, V.S.P.; Meneni, S.R. A new penta β-D-Glucopyranosyl diterpene from stevia rebaudiana. *Int. J. Org. Chem.* **2017**, *7*, 91–98. [CrossRef]
67. Chaturvedula, V.S.P.; Prakash, I. Additional minor diterpene glycosides from stevia rebaudiana. *Nat. Prod. Commun.* **2011**, *6*. [CrossRef]
68. Prakash, I.; Chaturvedula, V.S.P. Additional minor diterpene glycosides from stevia rebaudiana bertoni. *Molecules* **2013**, *18*, 13510–13519. [CrossRef]
69. Chaturvedula, V.S.P. Indra prakash diterpene glycosides from stevia rebaudiana. *J. Med. Plants Res.* **2011**, *5*, 4838–4842.
70. Chaturvedula, V.S.P.; Upreti, M.; Prakash, I. Structures of the novel α-glucosyl linked diterpene glycosides from stevia rebaudiana. *Carbohydr. Res.* **2011**, *346*, 2034–2038. [CrossRef]
71. Chaturvedula, V.S.P.; Prakash, I. Structures of the novel diterpene glycosides from stevia rebaudiana. *Carbohydr. Res.* **2011**, *346*, 1057–1060. [CrossRef]
72. Chaturvedula, V.; Rhea, J.; Milanowski, D.; Mocek, U.; Prakash, I. Isolation and structure elucidation of two new minor diterpene glycosdies from stevia rebaudiana. *Org. Chem. Curr. Res.* **2011**, *1*, 2161-0401.
73. Zimmermann, B.F. Tandem mass spectrometric fragmentation patterns of known and new steviol glycosides with structure proposals—zimmermann—2011—Rapid communications in mass spectrometry. *Rapid Commun. Mass. Spectrom.* **2011**, *25*, 1575–1582. [CrossRef] [PubMed]
74. Perera, W.H.; Avula, B.; Khan, I.A.; McChesney, J.D. Assignment of sugar arrangement in branched steviol glycosides using electrospray ionization quadrupole time-of-flight tandem mass spectrometry. *Rapid Commun. Mass. Spectrom.* **2017**, *31*, 315–324. [CrossRef] [PubMed]
75. Tanaka, T.; Nakashima, T.; Ueda, T.; Tomii, K.; Kouno, I. Facile discrimination of aldose enantiomers by reversed-phase HPLC. *Chem. Pharm. Bull.* **2007**, *55*, 899–901. [CrossRef]
76. Wang, Y.-H.; Avula, B.; Fu, X.; Wang, M.; Khan, I.A. Simultaneous determination of the absolute configuration of twelve monosaccharide enantiomers from natural products in a single injection by a UPLC-UV/MS method. *Planta Med.* **2012**, *78*, 834–837. [CrossRef] [PubMed]
77. Perera, W.H.; Shivanagoudra, S.R.; Pérez, J.L.; Kim, D.M.; Sun, Y.K.; Jayaprakasha, G.S.; Patil, B. Anti-Inflammatory, antidiabetic properties and in silico modeling of cucurbitane-type triterpene glycosides from fruits of an indian cultivar of *Momordica charantia* L. *Molecules* **2021**, *26*, 1038. [CrossRef]
78. Tangpaisarnkul, N.; Tuchinda, P.; Wilairat, P.; Siripinyanond, A.; Shiowattana, J.; Nobsathian, S. Development of pure certified reference material of stevioside. *Food Chem.* **2018**, *255*, 75–80. [CrossRef] [PubMed]
79. Steinmetz, W.E.; Lin, A. NMR studies of the conformation of the natural sweetener rebaudioside A. *Carbohydr. Res.* **2009**, *344*, 2533–2538. [CrossRef] [PubMed]

Article

# Insecticidal and Biting Deterrent Activities of *Magnolia grandiflora* Essential Oils and Selected Pure Compounds against *Aedes aegypti*

**Abbas Ali [1,*], Nurhayat Tabanca [1,2], Betul Demirci [3], Vijayasankar Raman [1], Jane M. Budel [4], K. Hüsnü Can Baser [5] and Ikhlas A. Khan [1]**

1 National Center for Natural Products Research, The University of Mississippi, Oxford, MS 38677, USA; nurhayat.tabanca@usda.gov (N.T.); vraman@olemiss.edu (V.R.); ikhan@olemiss.edu (I.A.K.)

2 United States Department of Agriculture, Agricultural Research Service (USDA-ARS), Subtropical Horticulture Research Station (SHRS), Miami, FL 33158, USA

3 Department of Pharmacognosy, Faculty of Pharmacy, Anadolu University, 26470 Eskisehir, Turkey; betuldemirci@gmail.com

4 Departamento de Ciências Farmacêuticas, Universidade Estadual de Ponta Grossa (UEPG), Ponta Grossa, PR 84030-900, Brazil; janemanfron@hotmail.com

5 Department of Pharmacognosy, Faculty of Pharmacy, Near East University, 99138 Nicosia, Northern Cyprus; khcbaser@gmail.com

* Correspondence: aali@olemiss.edu or dr_aliabbas@hotmail.com

Academic Editor: Maria Carla Marcotullio

Received: 26 February 2020; Accepted: 14 March 2020; Published: 17 March 2020

**Abstract:** In our natural products screening program for mosquitoes, we tested essential oils extracted from different plant parts of *Magnolia grandiflora* L. for their insecticidal and biting deterrent activities against *Aedes aegypti*. Biting deterrence of seeds essential oil with biting deterrence index value of 0.89 was similar to *N,N*-diethyl-3-methylbenzamide (DEET). All the other oils were active above the solvent control but the activity was significantly lower than DEET. Based on GC-MS analysis, three pure compounds that were only present in the essential oil of seed were further investigated to identify the compounds responsible for biting deterrent activity. 1-Decanol with PNB value of 0.8 was similar to DEET (PNB = 0.8), whereas 1-octanol with PNB value of 0.64 showed biting deterrence lower than 1-decanol and DEET. The activity of 1-heptanol with PNB value of 0.36 was similar to the negative control. Since 1-decanol, which was 3.3% of the seed essential oil, showed biting deterrence similar to DEET as a pure compound, this compound might be responsible for the activity of this oil. In in vitro A & K bioassay, 1-decanol with MED value of 6.25 showed higher repellency than DEET (MED = 12.5). Essential oils of immature and mature fruit showed high toxicity whereas leaf, flower, and seeds essential oils gave only 20%, 0%, and 50% mortality, respectively, at the highest dose of 125 ppm. 1-Decanol with $LC_{50}$ of 4.8 ppm was the most toxic compound.

**Keywords:** Magnoliaceae; GC-FID; GC-MS; mosquito control; 1-decanol; 1-octanol; larvicidal activity; deterrent; biopesticides

---

## 1. Introduction

Insect disease vectors transmit many disease pathogens and are important in global public health. *Aedes aegypti* (L.) and *Ae. albopictus* (Skuse) are the primary and secondary vectors of Zika and dengue as well as other viruses [1]. The use of synthetic insecticides in mosquito control has proven to be one of the major approaches for the prevention and reduction of mosquito-borne disease incidence [2]. Insect repellents also play an important role in the reduction of disease incidence by preventing infected mosquitoes from biting humans [3]. Moreover; repellents have always been used against host-seeking vectors as they provide immediate; localized; personal protection. *N*,*N*-Diethyl-3-methylbenzamide (DEET) has been in use for more than 60 years and is the gold standard to which all repellents are measured in the marketplace [4]. The discovery of novel insecticides and repellents against disease vectors from non-toxic and biodegradable plant sources continues to be the focus of recent research efforts [5–8].

*Magnolia grandiflora* L. (Magnoliaceae) is a large evergreen tree native to North America [9] that has medicinal and ornamental values. Medicinal use of various parts of *M. grandiflora* is reported in American Indian medicine and also listed as a bitter tonic and antimalarial. Several biologically active compounds have been reported from *Magnolia* species [10–12]. As a part of our natural product screening program for mosquitoes, we tested essential oils from various parts of *M. grandiflora* for their larvicidal and biting deterrent activities. This paper reports insecticidal and biting deterrent activities of essential oils and select pure compounds from various parts of *M. grandiflora* against yellow fever mosquito, *Aedes aegypti*.

## 2. Results

Water-distilled essential oils of the leaves, flowers, immature fruits, mature fruits, and seeds of *M. grandiflora* were analyzed by GC-FID and GC-MS. Chemical compositions of the essential oils and total ion current (TIC) chromatogram are given in Table 1 and Figure S1 (Suplementary Material). Chemical profiles of the oils varied among essential oils. Sesquiterpene hydrocarbons (31.5%) were dominant in the seeds and immature fruit essential oils (32%) whereas the leaf and flower oils were rich in monoterpene hydrocarbons (30.9% and 43.8%, respectively). Oxygenated monoterpenes (36.9%) were the major components of the mature fruit oil followed by monoterpene hydrocarbons (15.2%), sesquiterpene hydrocarbons (15.5%), and oxygenated sesquiterpenes (16.2%). The α- and β-pinenes and 1,8-cineole were the major contents of leaf, flower, immature fruit, and mature fruit whereas these compounds were either very low or absent in seed essential oil (Table 1). The seed oil was differentiated from other essential oils because of the presence of fatty acid; hexadecanoic acid (2.9%) and fatty acid esters (2.2%). The saturated aliphatic esters (10.9%) and two phenolic compounds; methyl chavicol (2.6%) and eugenol (1.3%) were also only found in the seed essential oil. Major compounds, α- and β-pinenes and 1,8-cineole present in leaf, flower, immature fruit, and fruit essential oils were either in very low concentration or absent in seed essential oil (Table 1). Fatty acids and esters (5.1%) were high and aliphatic esters (10.9%) were present only in the seed essential oil. Hexadecanoic acid (2.9%), 1-decanol (3.3%), 1-octanol (6.2%), and 1-heptanol were also present only in the seed essential oil.

**Table 1.** The chemical composition of essential oils of the *Magnolia grandiflora*.

| RRI | Compound | Leaf (%) | Flower (%) | Immature Fruit (%) | Mature Fruit (%) | Seed (%) | IM |
|---|---|---|---|---|---|---|---|
| 1032 | α-Pinene | 6.3 | 8.0 | 4.8 | 3.8 | 1.0 | RRI, MS |
| 1063 | Ethyl 2-methylbutyrate | - | - | - | - | 1.5 | MS |
| 1076 | Camphene | 0.1 | 0.7 | 1.1 | 1.6 | - | RRI, MS |
| 1100 | Isobutyl isobutyrate | - | - | - | - | 0.7 | MS |
| 1118 | β-Pinene | 23.0 | 32.3 | 12.7 | 6.9 | 1.2 | RRI, MS |
| 1132 | Sabinene | - | 0.3 | - | - | - | RRI, MS |
| 1151 | Propyl 2-methylbutyrate | - | - | - | - | 1.0 | MS |
| 1174 | Myrcene | - | 0.4 | - | - | 0.6 | RRI, MS |
| 1176 | α-Phellandrene | - | - | - | - | 1.1 | RRI, MS |
| 1185 | Isobutyl 2-methylbutyrate | - | - | - | - | 2.4 | MS |
| 1198 | Isobutyl 3-methylbutyrate | - | - | - | - | 1.9 | MS |
| 1203 | Limonene | 1.1 | 1.0 | 1.7 | 1.4 | 1.0 | RRI, MS |
| 1213 | 1,8-Cineole | 4.1 | 4.4 | 4.5 | 12.2 | - | RRI, MS |
| 1218 | β-Phellandrene | - | - | - | - | 7.3 | RRI, MS |
| 1241 | Butyl-2-methylbutyrate | - | - | - | - | 1.3 | MS |

**Table 1.** *Cont.*

| RRI | Compound | Leaf (%) | Flower (%) | Immature Fruit (%) | Mature Fruit (%) | Seed (%) | IM |
|---|---|---|---|---|---|---|---|
| 1246 | (*Z*)-β-Ocimene | - | 0.8 | - | - | - | MS |
| 1280 | *p*-Cymene | 0.4 | 0.3 | 0.8 | 1.3 | 5.5 | RRI, MS |
| 1290 | Terpinolene | - | - | 0.5 | 0.2 | - | RRI, MS |
| 1286 | 2-Methyl butyl 2-methylbutyrate | - | - | - | - | 1.3 | MS |
| 1299 | 2-Methylbutyl isovalerate | - | - | - | - | 0.8 | MS |
| 1429 | Perillene | - | - | - | - | 0.4 | MS |
| 1450 | *trans*-Linalool oxide (*Furanoid*) | - | 0.8 | - | - | - | MS |
| 1452 | α,*p*-Dimethylstyrene | 0.2 | - | - | 0.3 | - | MS |
| 1463 | 1-Heptanol | - | - | - | - | 0.5 | MS |
| 1493 | α-Ylangene | - | - | - | - | 0.7 | MS |
| 1497 | α-Copaene | - | - | 0.2 | 0.2 | 1.6 | RRI, MS |
| 1532 | Camphor | - | - | - | 0.3 | - | RRI, MS |
| 1553 | Linalool | 0.7 | 4.7 | 0.5 | 0.8 | - | RRI, MS |
| 1562 | 1-Octanol | - | - | - | - | 6.2 | MS |
| 1586 | Pinocarvone | 0.3 | - | 0.3 | 1.8 | - | RRI, MS |
| 1591 | Bornyl acetate | - | 0.2 | 2.8 | 4.1 | 0.4 | RRI, MS |
| 1594 | *trans*-β-Bergamotene | 0.3 | 1.2 | 0.7 | 0.4 | 1.7 | MS |
| 1600 | β-Elemene | 13.6 | 7.7 | 12.9 | 5.7 | - | MS |
| 1611 | Terpinen-4-ol | 0.8 | 0.6 | 1.1 | 0.8 | - | RRI, MS |
| 1612 | β-Caryophyllene | 3.4 | 1.1 | 7.9 | 2.9 | 8.8 | RRI, MS |
| 1648 | Myrtenal | 1.1 | 0.9 | 0.7 | 4.0 | - | MS |
| 1661 | *trans*-Pinocarvyl acetate | 3.8 | 3.3 | 1.5 | 2.3 | - | MS |
| 1669 | Sesquisabinene | - | - | - | - | 1.7 | MS |
| 1670 | *trans*-Pinocarveol | 0.9 | 0.8 | 0.5 | 2.4 | - | RRI, MS |
| 1687 | α-Humulene | 0.8 | 0.4 | 1.4 | 0.6 | 1.0 | RRI, MS |
| 1687 | Methyl chavicol | - | - | - | - | 2.6 | RRI, MS |
| 1688 | Selina-4,11-diene | 0.4 | 0.3 | 1.1 | - | - | MS |
| 1695 | (*E*)-β-Farnesene | - | - | - | - | 0.4 | MS |
| 1704 | Myrtenyl acetate | - | - | - | - | - | MS |
| 1704 | γ-Muurolene | - | - | 0.6 | 0.5 | 2.7 | MS |
| 1706 | α-Terpineol | 2.4 | 2.5 | 5.1 | 3.9 | - | RRI, MS |
| 1719 | Borneol | 0.2 | - | 0.7 | 1.2 | - | RRI, MS |
| 1725 | Verbenone | - | - | - | 0.7 | - | RRI, MS |
| 1726 | Germacrene D | - | 0.3 | - | - | - | RRI, MS |
| 1740 | α-Muurolene | - | - | - | - | 1.6 | MS |
| 1742 | Geranial | - | 0.5 | - | - | - | RRI, MS |
| 1742 | β-Selinene | 1.5 | 1.2 | 2.9 | 1.6 | 0.9 | MS |
| 1744 | α-Selinene | 1.4 | 0.9 | 2.3 | 1.5 | 0.7 | MS |
| 1766 | 1-Decanol | - | - | - | - | 3.3 | MS |
| 1773 | δ-Cadinene | - | 0.3 | 1.3 | 0.3 | 4.0 | MS |
| 1776 | γ-Cadinene | - | 0.1 | - | 0.5 | 2.0 | MS |
| 1784 | (*E*)-α-Bisabolene | 0.4 | 0.6 | 1.2 | 0.4 | 0.8 | MS |
| 1799 | Cadina-1,4-diene | - | - | - | - | 0.3 | MS |
| 1804 | Myrtenol | 1.5 | 1.4 | 0.5 | 2.2 | - | MS |
| 1808 | Nerol | - | 0.1 | - | - | - | RRI, MS |
| 1849 | Calamenene | - | - | 0.5 | 0.4 | 1.8 | MS |
| 1857 | Geraniol | - | 2.5 | - | - | - | RRI, MS |
| 1864 | *p*-Cymen-8-ol | 1.0 | 0.4 | - | 0.6 | - | RRI, MS |
| 1872 | *cis*-Myrtanol | - | tr | - | - | - | MS |
| 1879 | *trans*-Myrtanol | - | 0.2 | - | - | - | MS |

**Table 1.** *Cont.*

| RRI | Compound | Leaf (%) | Flower (%) | Immature Fruit (%) | Mature Fruit (%) | Seed (%) | IM |
|---|---|---|---|---|---|---|---|
| 1941 | α-Calacorene | - | - | 0.2 | 0.2 | 0.9 | MS |
| 1948 | *trans*-Jasmone | - | 1.0 | - | - | - | MS |
| 2008 | Caryophyllene oxide | 3.9 | 0.9 | 1.8 | 7.2 | 1.9 | RRI, MS |
| 2029 | Perilla alcohol | - | - | - | 0.3 | - | MS |
| 2050 | (*E*)-Nerolidol | 1.3 | 1.7 | 0.6 | 0.4 | 1.2 | RRI, MS |
| 2071 | Humulene epoxide-II | 0.6 | 0.2 | 0.3 | 1.2 | - | MS |
| 2080 | Junenol (=*Eudesm-4(15)-en-6-ol*) | - | 0.2 | - | - | - | MS |
| 2100 | Heneicosane | - | 0.5 | - | - | - | RRI, MS |
| 2186 | Eugenol | - | - | - | - | 1.3 | RRI, MS |
| 2187 | T-Cadinol | 0.3 | 0.5 | 0.7 | 0.2 | 0.3 | MS |
| 2209 | T-Muurolol | 0.3 | 0.8 | 0.9 | 0.7 | 0.1 | MS |
| 2226 | Methyl hexadecanoate | 0.4 | - | 0.6 | - | 0.3 | RRI, MS |
| 2219 | δ-Cadinol | - | - | 0.2 | - | - | MS |
| 2255 | α-Cadinol | 0.3 | 1.3 | 1.4 | 0.7 | 0.3 | MS |
| 2256 | Cadalene | - | - | - | - | 0.7 | MS |
| 2262 | Ethyl hexadecanoate | - | - | - | - | 0.7 | MS |
| 2269 | Guaia-6,10(14)-dien-4β-ol | 0.3 | - | 0.6 | 0.7 | - | MS |
| 2273 | Selin-11-en-4α-ol | 1.0 | 1.8 | 4.0 | 1.5 | - | MS |
| 2300 | Tricosane | - | 0.6 | - | - | - | RRI, MS |
| 2316 | Caryophylla-2(12),6(13)-dien-5β-ol (=Caryophylladienol I) | - | - | - | 1.4 | - | MS |
| 2353 | Chavicol | - | - | - | - | 0.7 | MS |
| 2369 | (2*E*,6*E*)-Farnesol | - | 2.3 | - | - | - | MS |
| 2389 | Caryophylla-2(12),6-dien-5β-ol (=Caryophyllenol I) | - | - | - | 1.8 | - | MS |
| 2456 | (Z)-9-Methyl octadecanoate (=Methyl oleate) | - | - | 1.7 | - | 0.5 | RRI, MS |
| 2509 | (Z.Z)-9,12-methyl octadecadienoate (=Methyl linoleate) | - | - | - | - | 0.7 | RRI, MS |
| 2931 | Hexadecanoic acid | - | - | 0.5 | - | 2.9 | RRI, MS |
| | Monoterpene hydrocarbons | 30.9 | 43.8 | 21.6 | 15.2 | 17.7 | |
| | Oxygenated monoterpenes | 16.8 | 23.3 | 18.2 | 36.9 | 0.4 | |
| | Sesquiterpene hydrocarbons | 21.4 | 13.5 | 32.0 | 15.5 | 31.5 | |
| | Oxygenated Sesquiterpenes | 8.4 | 10.3 | 11.7 | 16.2 | 4.6 | |
| | Fatty acids and their esters | 0.4 | - | 2.8 | - | 5.1 | |
| | Aliphatic esters | - | - | - | - | 10.9 | |
| | others | 0.2 | 2.1 | - | 0.3 | 15.0 | |
| | Total | 78.1 | 93.0 | 86.3 | 84.1 | 85.2 | |

RRI: relative retention indices calculated against n-alkanes; %: calculated from FID data; tr: trace (< 0.1 %); IM: identification method based on the relative retention indices (RRI) of authentic compounds on the HP Innowax column; MS, identified based on computer matching of the mass spectra with those of the Wiley and MassFinder libraries and comparison with literature data. % calculated from FID data.; -: not detected.

The essential oils obtained from five different plant parts of *M. grandiflora* were investigated for their biting deterrent activity against *Ae. aegypti*. All the essential oils showed biting deterrence above the negative control. Seeds essential oil produced significantly higher biting deterrence than the essential oils from the other parts (Figure 1). Seed essential oil with high minimum effective dose BDI value (0.89) showed biting deterrent activity similar to DEET whereas all the other essential oils had activity lower than DEET. 1-Decanol with PNB value of 0.8 showed biting deterrence similar to DEET (PNB = 0.8) whereas the activity of 1-octanol was above negative control but lower than DEET and 1-decanol (Figure 2).

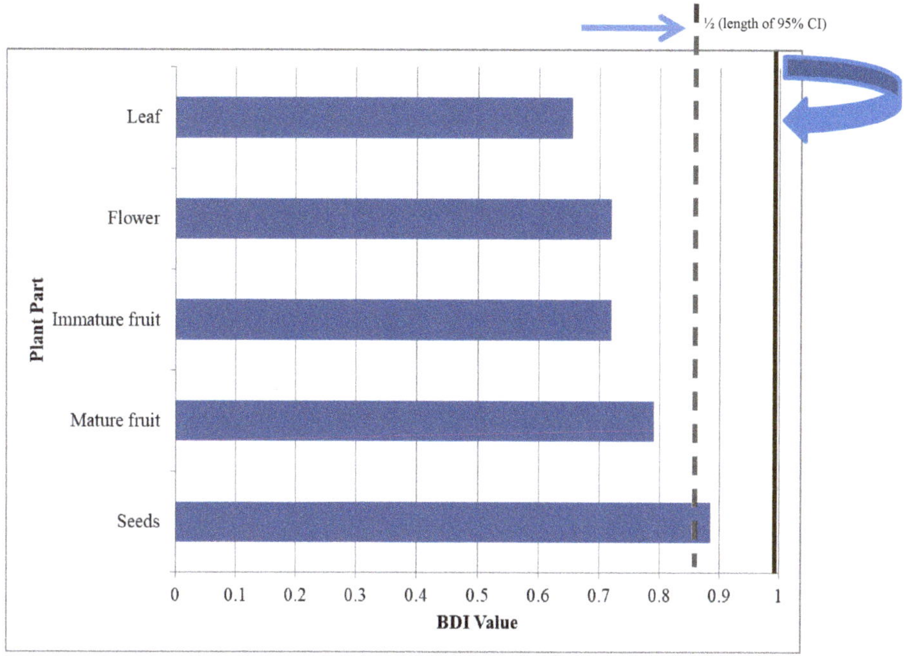


**Figure 1.** Mean BDI (± SEM) values of the essential oils from different parts of *M. grandiflora* against *Ae. aegypti*. All the essential oils were tested at 10 μg/cm$^2$ whereas DEET (*N,N*-diethyl-3-methylbenzamide) was tested at 25 nmol/cm$^2$ and ethanol a solvent control. A BDI value greater than 0 indicates biting deterrence relative to ethanol and a BDI value not significantly different from 1 shows deterrence similar to DEET.

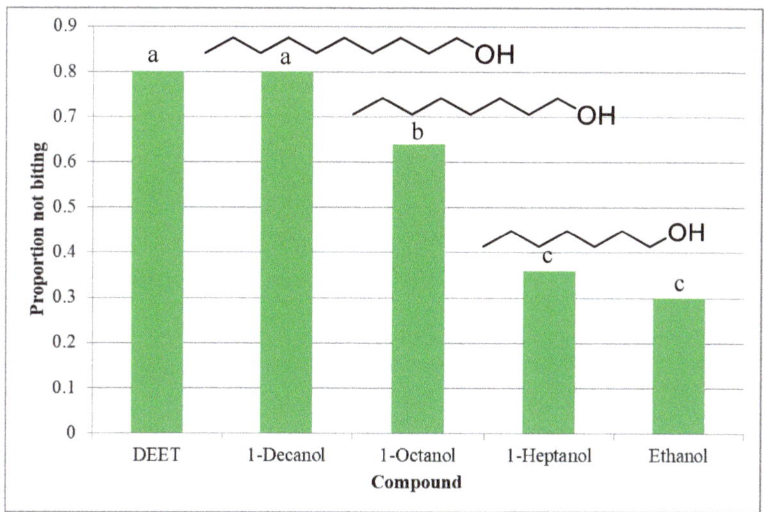


**Figure 2.** Mean proportion not biting values of the essential oils from different parts of *M. grandiflora* against *Ae. aegypti*. Essential oils were tested at 10μg/cm$^2$ while DEET at 25 nmol/cm$^2$ was tested as a positive control. Mean proportions sharing the same letter are not significantly different.

Biting deterrence of 1-heptanol was similar to the negative control. 1-Decanol with 3.3% of seed essential oil appears to be the major compound responsible for the biting deterrent activity of the seed essential oil. In in vitro A & K bioassay, MED value of 1-decanol (6.25) was lower than DEET (12.5) which indicated better repellency of 1-decanol than DEET (Figure 3).

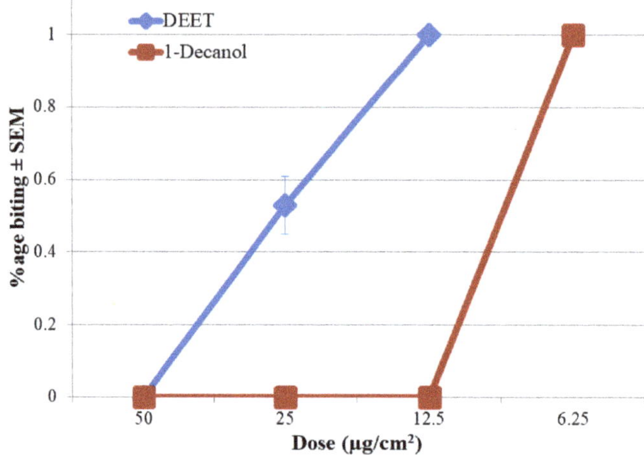


**Figure 3.** Data are %age (mean ± SEM) females biting. Minimum effective dosage (MED) values in this bioassay were ≤ 1% biting in 1 min which is two females out of 200 in this cage.

The toxicity of essential oils from *M. grandiflora* against 1-d-old larvae of *Ae. aegypti* is given in Table 2. In initial screening, essential oils of immature and mature fruits showed high toxicity whereas leaf, flower, and seeds essential oils gave only 20%, 0%, and 50% mortality, respectively, at the highest screening dose of 125 ppm. Therefore dose-response bioassays were not conducted on leaf, flower, and seeds essential oils. Immature fruit and mature fruit essential oils with $LC_{50}$ values of 49.4 and 48.9 ppm, respectively at 24-h post-treatment showed similar toxicity.

**Table 2.** Toxicity of essential oils from *M. grandiflora* and its select pure compounds against 1-day-old *Ae. aegypti* at 24-h post-treatment.

| Essential oil | $LC_{50}$ (95%CI) * | $LC_{90}$ (95%CI) | $\chi^2$ | DF |
|---|---|---|---|---|
| Immature fruit | 49.4 (39.4–64.2) | 135.9 (96.5–244.2) | 43.0 | 48 |
| Mature fruit | 48.9 (42.3–56.9) | 116.9 (94.6–158.3) | 83.3 | 48 |
| 1-Decanol | 4.8 (4.2–5.5) | 10.2 (8.5–13.2) | 83.2 | 48 |
| 1-Octanol | 34.3 (30.3–38.7) | 63.9 (54.4–80.5) | 73.7 | 48 |
| Leaf | 20% ** | | | |
| Flower | 0% | | | |
| Seed | 50% | | | |

* $LC_{50}$ and $LC_{90}$ values are in ppm and 95% C.I are confidence intervals. ** Leaf, flower and seeds essential oils gave only 20%, 0 and 50% mortality, respectively, at the highest dose of 125 ppm.

Pure compounds 1-decanol, 1-octanol, and 1-heptanol, present in seed essential oil were also screened for larvicidal activity. Both 1-decanol and 1-octanol were active in screening bioassays whereas 1-heptanol did not show any mortality at the highest dose of 125 ppm. 1-Decanol and 1-octanol were further evaluated to observe the dose response. 1-Decanol with $LC_{50}$ of 4.8 ppm was the most toxic compound followed by 1-octanol ($LC_{50}$ = 34.3 ppm) at 24-h post-treatment. 1-Decanol was very toxic ($LC_{50}$ = 4.8 ppm) as a pure compound, the seed essential oil that contained 3.3% of this compound showed 50% mortality at the highest dose of 125 ppm. 1-Decanol amounted to be 4.07 ppm as a part of seed essential oil at 125 ppm which caused mortality similar to the pure compound. 1-Decanol and 1-octanol amounted to be 4.07 and 7.75 ppm, respectively, as a part of the essential oil dose of 125 ppm. Since the toxicity of 1-octanol as a pure compound was low, the main compound responsible for the toxicity of the seed essential oil appears to be 1-decanol.

## 3. Discussion

Schuhly et al. [13] reported β-elemene as the major compound in the fruit essential oil of *M. grandiflora* which corroborates the finding of this study having β-elemene (6–14%) in all the essential oils of *M. grandiflora* except seed oil. Guerra-Boone et al. [14] reported bornyl acetate (20.9%) as the major compound in *M. grandiflora* leaf oil, however, we did not detect this compound in the leaf oil. Garg and Kumar [15] reported β-caryophyllene as the major compound (34.8%) in flower essential oil whereas only a small amount (1.1%) was detected in the present study. Farag and Al-Mahdy [16] reported variation in the contents of *M. grandiflora* flower oil volatiles obtained through headspace and water distillation techniques indicating the effects of the isolation technique on the yield of different compounds. Such differences in chemical compositions of essential oils are expected and can be attributed to many factors including geographic location, genetic factors, climate, crop stage, harvesting time, and processing method [17,18].

In our previous studies, some of the compounds that were present in these essential oils exhibited very insignificant biting deterrence. α-Phellandrene (BDI = 0.52), (+)-α-pinene (BDI = 0.47), (-)-α-pinene (BDI = 0.41), (+)-β-pinene (BDI = 0.57), (-)-β-pinene (BDI = 0.51), *p*-cymene (BDI = 0.48), *trans*-sabinene hydrate (BDI = 0.61) showed biting deterrent activity lower than DEET. β-caryophyllene and caryophyllene oxide with BDI values of 0.54 and 0.66, respectively, were also significantly lower than DEET at 25 nmol/cm$^2$ and these two sesquiterpenes also did not repel mosquitoes up to the highest dose of 1.5 mg/cm$^2$ in cloth patch assay [6,19–22]. Hexadecanoic acid that was only present in seed essential oil was reported to have biting deterrence lower (PNB = 0.72) than DEET [23]. In our previous study, we found that mid-chain length acids ($C_{10:0}$ to $C_{13:0}$) showed the highest biting deterrent activity against Ae. aegypti as compared to short-chain length acids ($C_{6:0}$ to $C_{9:0}$) [22]. The current study reveals similar pattern of medium chain length fatty alcohol ($C_{10:0}$) had higher biting deterrent activity than short-chain length alcohol ($C_{8:0}$). However, we shall work on this hypothesis and confirm the activity toward short, med, and long-chain fatty alcohols. Many methylbutyrates present in the seed essential oil were tested in our screening program and found not active as biting deterrents (Ali personal communications). Since most of the major compounds that were present only in seed essential oil did not show any significant activity, 1-decanol might be the main compound responsible for the biting deterrent activity of *M. grandiflora* seed essential oil.

The toxicity of many natural compounds present in plant essential oils against mosquitoes has been reported in the literature. α-Pinene ($LC_{50}$ = 49.5 ppm), β-pinene ($LC_{50}$ = 35.9 ppm), β-caryophyllene ($LC_{50}$ = 26.0 ppm), and caryophyllene oxide ($LC_{50}$ = 29.8 ppm) were active as larvicides against *Ae. aegypti* whereas 1,8-cineole did not show any mortality at the highest screening dose of 125 ppm [19,23]. Monoterpenes that were present in variable concentrations in these essential oils showed larvicidal activity. These higher percentages of monoterpenes (α- and β-pinenes) in combination with other compounds may be responsible for the high toxicity of immature- and mature fruit essential oils. We will further explore other compounds present in *M. grandiflora* essential oils for their potential as larvicides against mosquitoes. Ethanolic extracts of sarcotesta of the seeds of *M. dealbata* were reported to have 96.4% mortality at 0.1 mg/mL against the Mexican fruit fly (*Anastrepha ludens* Loew) whereas the extracts from the other parts were inactive [24].

## 4. Materials and Methods

### 4.1. Chemicals

Individual compounds such as 1-decanol (CAS # 112-30-1), 1-octanol (CAS# 111-87-5), and 1-heptanol (CAS # 111-70-6) were obtained from the National Center for Natural Products Research (NCNPR) Repository of The University of Mississippi, University, MS, USA Repository. These compounds were previously purchased from Sigma-Aldrich Co., St. Louis, MO, USA.

### 4.2. Plant Materials

Whole samples of leaves, flowers, immature and mature fruits, and seeds (Figure 4) were freshly collected from an identified *M. grandiflora* tree at the University of Mississippi campus in 2018. Voucher specimens of all the samples—leaves (NCNPR # 20286), stem-bark (# 20874), flowers (# 20316), immature fruit (# 20871), mature fruit with seeds removed (# 20872), and seeds (# 20873)—were deposited in the Repository of Botanicals at NCNPR, University of Mississippi.

**Figure 4.** Study material. **A**: flowers; **B**: leaves, immature fruits, mature fruits and seeds. Photos courtesy of V.R.

### 4.3. Extraction of Essential Oils

For the extraction of essential oils, leaves, flowers, immature and mature fruits, and seeds of *M. grandiflora* were separately subjected to hydrodistillation for 3 h, using a modified Clevenger-type apparatus. Seeds were crushed in a mortar and pestle before hydrodistillation (Figure 5). The resultant oils were stored in glass vials at 4 ± 0.5 °C with no light. The yields were calculated on a moisture-free basis for mature fruits, flowers, seeds, immature fruits and leaves at 0.02, 0.06, 0.1, 0.1, 1.5% respectively whereas there was no oil present in the stem-bark sample.

**Figure 5.** Study material. **A**: Seeds were separated from mature fruits; **B**: seeds were prepared and gently crushed in a mortar and pestle (**C**), and were subsequently hydrodistilled. Photos courtesy of V.R and N.T.

### 4.4. GC-MS Analysis

The GC-MS analysis was carried out with an Agilent 5975 GC-MSD system Agilent 5975 (SEM Ltd., Istanbul, Turkey). Innowax FSC column (60 m × 0.25 mm, 0.25 μm film thickness) was used with helium as the carrier gas (0.8 mL/min). GC oven temperature was kept at 60 °C for 10 min and programmed to 220 °C at a rate of 4 °C/min, and kept constant at 220 °C for 10 min and then programmed to 240 °C at a rate of 1 °C/min. The split ratio was adjusted at 40:1. The injector temperature was set at 250 °C. Mass spectra were recorded at 70 eV. The mass ranged from *m/z* 35 to 450.

### 4.5. GC Analysis

The GC analysis was carried out using an Agilent 6890N GC system Agilent 5975 (SEM Ltd., Istanbul, Turkey). The FID detector temperature was 300 °C. To obtain the same elution order with GC-MS, simultaneous auto-injection was done on a duplicate of the same column applying the same operational conditions. Relative percentage amounts of the separated compounds were calculated from FID chromatograms. The analysis results are given in Table 1.

Identification of the essential oil components was carried out by comparison of their relative retention times with those of authentic samples or by comparison of their relative retention index (RRI) to series of *n*-alkanes [25,26]. Computer matching against commercial (Wiley GC/MS Library, MassFinder Software 4.0) and in-house "Başer Library of Essential Oil Constituents" which includes over 3200 genuine compounds with MS and retention data from pure standard compounds and components of known oils as.

### 4.6. Insects

*Aedes aegypti* used in these studies were from a laboratory colony maintained at the Mosquito and Fly Research Unit, Center for Medical, Agricultural and Veterinary Entomology, USDA-ARS, Gainesville, Florida since 1952. We received the eggs and stored them in our laboratory until needed. Mosquitoes were reared to the adult stage by feeding the larvae on a larval diet of 2% slurry of 3:2 beef liver powder (now Foods, Bloomingdale, Illinois) and Brewer's yeast (Lewis Laboratories Ltd., Westport, CT, USA). The eggs were hatched and reared to the pupal stage in an environment-controlled room at a temperature of 27 °C ± 2 °C and 60 ± 10% RH in a photoperiod regimen of 12:12 (L: D) h. The adults were fed on cotton pads moistened with a 10% sucrose solution placed on the top of screens of 4-L cages.

### 4.7. Mosquito Biting Bioassay

Bioassays were conducted using a six-celled in vitro Klun and Debboun (K & D) module bioassay system developed by Klun et al. [27] for quantitative evaluation of biting deterrent properties of compounds. The K & D system consists of a six-well reservoir with each of the 4 × 3 cm wells containing 6 mL of feeding solution. We used the CPDA-1 ± ATP solution instead of human blood [22]. CPDA-1 and ATP preparations were freshly made on the day of the test and contained a green fluorescent tracer dye (fluorescent water-based tracer "Green"; www.blacklightworld.com) that allowed for the identification of mosquitoes that were fed on the solution. The squashed mosquitoes were observed under black light (FEIT, BPESL15T/BLB 13W 120VAC 60Hz 200mA, Ul#E170906) for feeding. DEET (97% purity *N,N*-diethyl-3-methylbenzamide) was used as a positive control (Sigma-Aldrich Co., St. Louis, MO, USA) and ethanol (Fisher Scientific Chemical Co. Fairlawn, NJ, USA) was used as solvent control. Stock and dilutions of all extracts and DEET were prepared in ethanol. All essential oils were evaluated at dosages of 10 μg/cm$^2$ and DEET along with the pure compounds was tested at a concentration of 25 nmol/cm$^2$.

The temperature of the solution was maintained at 37 °C by using a circulatory bath. The test compounds and controls were randomly applied to six 4 × 3 cm marked portions of nylon organdy strip, which was positioned over the six, membrane-covered wells. A six-celled K & D module containing five 6–15-day-old females per cell was positioned over the six wells, trap doors were opened and mosquitoes allowed access for 3 minutes, after which they were collected back into the module. Mosquitoes were squashed and the presence of green dye (or not) in the gut was used as an indicator of feeding. A replicate consisted of six treatments: four samples, DEET, and ethanol as solvent control. Five replicates were conducted per day using new batches of mosquitoes in each replication.

### 4.8. In vitro A & K Repellent Bioassay

Bioassays were conducted using Ali and Khan (A & K) bioassay system developed by Ali et al. [28] for quantitative evaluation of repellency against mosquitoes. Minimum effective dosage (MED) values in this bioassay were determined using a method described by Katritzky et al. [29]. Briefly, the bioassay system consists of a 30 × 30 × 30 cm collapsible aluminum cage having one penal of clear transparent acrylic sheet with 120 × 35 mm slit through which the blood box containing a removable feeding device was attached. The top of the blood box had a sliding door used to expose the females to the treatment during the bioassay. Rectangular areas of 4 × 7.5-cm were marked on the collagen sheet that matched the measurement of the rectangular liquid reservoirs. Treatments were applied in a volume of 107 µL using a micropipette. Treated collagen was secured on the feeding reservoir containing the feeding solution using a thin layer of grease (Dow Coming Corp., Midland, MI, USA). The feeding device was then pushed inside the blood box and the sliding door was opened to expose the females to the treatment. The numbers of females landing and biting were recorded visually for 1 min. To ensure proper landing and biting, we used 3-4 cages at a time and only one treatment replication of individual compounds was completed in a single cage. The data are presented as %age biting as a function of concentration. MED is ≤ 1% biting out of 200 females in the cage. A total of five replicates were conducted.

### 4.9. Larval Bioassay

Bioassays were conducted using the bioassay system described by Pridgeon et al. [30] to determine the larvicidal activity of essential oils from different parts of *Magnolia grandiflora* against *Ae. aegypti*. Eggs were hatched and larvae were held overnight in the hatching cup in a temperature-controlled room maintained at a temperature of 27 ± 2 °C and 60 ± 10% RH. Five 1-day-old larvae were transferred in each of 24-well tissue culture plates in a 40–50 µL droplet of water. Total of 50 µL of larval diet (2% slurry of 3:2 beef liver powder and brewer's yeast) and 1 mL of deionized water were added to each well by using a Finnpipette stepper (Thermo Fisher, Vantaa, Finland). All the essential oils and pure compounds were diluted in DMSO. After the treatment, the plates were swirled in clock-wise and counter-clockwise motions and front and back and side to side five times to ensure even mixing of the chemicals. Larval mortality was recorded 24-h post-treatment. Larvae that showed no movement in the well after manual disturbance were recorded as dead. A series of 4-5 dosages were used in each treatment to get a range of mortality. Treatments were replicated ten times for each extract/compound.

### 4.10. Statistical Analyses

Proportion not biting (PNB) was calculated using the procedure described by Ali et al. [22]. As the K & D module bioassay system can handle only four treatments along with negative and positive controls to make direct comparisons among more than four test compounds and to compensate for variation in overall response among replicates, biting deterrent activity was quantified as biting deterrence index (BDI) [22]. The BDI's were calculated using the following formula:

$$\left[BDI_{i,j,k}\right] = \left[\frac{PNB_{i,j,k} \;-\; PNB_{c,j,k}}{PNB_{d,j,k} \;-\; PNB_{c,j,k}}\right]$$

where $PNB_{i,j,k}$ denotes the proportion of females not biting when exposed to test compound $i$ for replication $j$ and day $k$ ($i$ = 1–4, $j$ = 1–5, $k$ = 1–2), $PNB_{c,j,k}$ denotes the proportion of females not biting the solvent control "c" for replication $j$ and day $k$ ($j$ = 1–5, $k$ = 1–2) and $PNB_{d,j,k}$ denotes the proportion of females not biting in response to DEET *"d"*(positive control) for replication *j* and day *k* (*j* = 1–5, *k* = 1–2). This formula adjusts for inter-day variation in response and incorporates information from the solvent control as well as the positive control.

A BDI value of 0 indicates an effect similar to ethanol, while any value greater than 0 indicates biting deterrent effect relative to ethanol. BDI values not significantly different from 1, are statistically similar to DEET. BDI values were analyzed using SAS Proc ANOVA [31]. To determine whether confidence intervals include the values of 0 or 1 for treatments, Scheffe's multiple comparison procedure with the option of CLM was used in SAS [31]. $LC_{50}$ values for larvicidal data were calculated by using SAS, Proc Probit [31].

## 5. Conclusions

The essential oil of *M. grandiflora* seeds exhibited biting deterrent activity activity similar to DEET. All the major compounds (concentration >1%) except 1-decanol that were present only in seed essential oil were not active biting deterrents which indicated that the major activity of this essential oil might be due to 1-decanol. 1-Decanol also showed promising larvicidal activity. This high activity of 1-decanol indicated the potential of this compound to be developed as an effective mosquito population management tool. Further studies will be continued to evaluate these natural products in different formulations in large scale laboratory bioassays and field trials.

**Supplementary Materials:** The following are available online at http://www.mdpi.com/1420-3049/25/6/1359/s1. The total ion current (TIC) chromatogram of essential oils of the leaves, flowers, immature fruits, mature fruits and seeds of *Magnolia grandiflora* are available as supporting information (Figure S1).

**Author Contributions:** Conceptualization, A.A. and I.A.K.; sample collection and identification, V.R.; methodology, A.A., N.T., B.D., J.M.B., K.H.C.B.; formal analysis, N.T. and A.A.; resources, J.M.B. and V.R.; writing—original draft preparation, A.A. and N.T.; writing—review and editing, A.A., N.T., and V.R.; funding acquisition, I.A.K. All authors have read and agreed to the published version of the manuscript.

**Funding:** This study was supported in part by USDA/ARS grant No. 58-6066-6-043.

**Acknowledgments:** This study was supported in part by USDA/ARS grant No. 58-6066-6-043. We thank Dan Kline, Mosquito and Fly Research Unit, Center for Medical, Agricultural and Veterinary Entomology, USDA-ARS, Gainesville, for supplying mosquito eggs.

**Conflicts of Interest:** The authors declare no conflict of interest. Mention of trade names or commercial products in this publication is solely for the purpose of providing specific information and does not imply recommendation or endorsement by the U.S. Department of Agriculture. USDA is an equal opportunity provider and employer.

## References

1. Ali, A.; Abbas, A.; Debboun, M. Zika virus: Epidemiology, vector and sexual transmission neurological disorders and vector management—A review. *Int. J. Curr. Res.* **2017**, *10*, 58721–58737.
2. Bhatt, S.; Weiss, D.J.; Cameron, E.; Bisanzio, D.; Mappin, B.; Dalrymple, U.; Battle, K.E.; Moyes, C.L.; Henry, A.; Eckhoff, P.A.; et al. The effect of malaria control on *Plasmodium falciparum* in Africa between 2000 and 2015. *Nature* **2015**, *526*, 207–211. [CrossRef]
3. Leal, W.S. The enigmatic reception of DEET—The gold standard of insect repellents. *Currt. Opin. Insect Sci.* **2006**, *6*, 93–98. [CrossRef]
4. Frances, S.P. Efficacy and safety of repellents containing DEET. In *Insect Repellents: Principles, Methods and Uses*; Debboun, M., Frances, S.P., Stickman, D., Eds.; CRC Press: New York, NY, USA, 2007; pp. 311–325.
5. Wedge, D.E.; Klun, J.A.; Tabanca, N.; Demirci, B.; Ozek, T.; Baser, K.H.C.; Liu, Z.; Zhang, S.; Cantrell, C.L.; Zhan, J. Bioactivity-guided fractionation and GC/MS fingerprinting of *Angelica sinensis* and *Angelica archangelica* root components for antifungal and mosquito deterrent activity. *J. Agric. Food Chem.* **2006**, *57*, 464–470. [CrossRef] [PubMed]
6. Tabanca, N.; Avonto, C.; Wang, M.; Parcher, J.F.; Ali, A.; Demirci, B.; Raman, V.; Khan, I.A. Comparative investigation of *Umbellularia californica* and *Laurus nobilis* leaf essential oils and identification of constituents active against *Aedes aegypti*. *J. Agric Food Chem.* **2013**, *61*, 12283–12291. [CrossRef] [PubMed]

7. Ali, A.; Tabanca, N.; Demirci, B.; Blythe, E.K.; Baser, K.H.C.; Khan, I.A. Chemical composition and biological activity of essential oils from four *Nepeta* species and hybrids against *Aedes aegypti* (L.) (Diptera: Culicidae). *Rec. Nat. Prod.* **2016**, *10*, 137–147.
8. Cantrell, C.L.; Maxwell, A.P.; Ali, A. Isolation and identification of mosquito (*Aedes aegypti*) biting-deterrent compounds from the native American ethnobotanical remedy plant *Hierochloe odorata* (Sweetgrass). *J. Agric. Food Chem.* **2016**, *63*, 447–456. [CrossRef]
9. Wang, Y.; Mu, R.; Wang, X.; Liu, S.; Fan, Z. Chemical composition of volatile constituents of *Magnolia grandiflora*. *Chem. Nat. Compounds* **2009**, *45*, 257–258. [CrossRef]
10. Rao, K.V.; Wu, W.N. Glycosides of *Magnolia grandiflora*. III: Structural elucidation of magnolenin C. *J. Nat. Prod.* **1978**, *41*, 56–62.
11. El-Feraly, F.S. Melampolides from *Magnolia grandiflora*. *Phytochemistry* **1984**, *23*, 2372–2374. [CrossRef]
12. Rao, K.V.; Davis, T.L. Constituents of *Magnolia grandiflora*. III. Toxic principle of the wood. *J. Nat. Prod.* **1982**, *45*, 283–287. [CrossRef] [PubMed]
13. Schuhly, W.; Ross, S.A.; Mehmedic, Z.; Fischer, N.H. Essential oils of the follicles of four North American *Magnolia* Species. *Nat. Prod. Commun.* **2008**, *3*, 1117–1119. [CrossRef]
14. Guerra-Boone, L.; Alvarez-Román, R.; Salazar-Aranda, R.; Torres-Cirio, A.; Rivas-Galindo, V.M.; Waksman de Torres, N.; González González, G.M.; Pérez-López, L.A. Chemical compositions and antimicrobial and antioxidant activities of the essential oils from *Magnolia grandiflora*, *Chrysactinia mexicana*, and *Schinus molle* found in northeast Mexico. *Nat. Prod. Commun.* **2013**, *8*, 135–138. [CrossRef]
15. Garg, S.N.; Kumar, S. Volatile constituents from the flowers of *Magnolia grandiflora* L. From Lucknow, India. *J. Essent. Oils Res.* **1999**, *11*, 633–634. [CrossRef]
16. Farag, M.A.; Al-Mahdy, D.A. Comparative study of the chemical composition and biological activities of *Magnolia grandiflora* and *Magnolia virginiana* flower essential oils. *Nat. Prod. Res.* **2013**, *27*, 1091–1097. [CrossRef] [PubMed]
17. Rao, B.R.R.; Sastry, K.P.; Saleem, S.M.; Rao, E.V.S.P.; Syamasundar, K.V.; Ramesh, S. Volatile flower oils of three genotypes of rose-scented geranium (*Pelargonium* sp.). *Flavour Fragr. J.* **2000**, *15*, 105–107.
18. Ali, A.; Tabanca, N.; Demirci, B.; Baser, K.H.C.; Ellis, J.; Gray, S.; Lackey, B.R.; Murphy, C.; Khan, I.A.; Wedge, D.E. Composition, mosquito larvicidal, biting deterrent and antifungal activity of essential oils of different plant parts of *Cupressus arizonica* var. *glabra* (Sudw.) Little ('Carolina Sapphire'). *Nat. Prod. Commun.* **2013**, *8*, 257–260. [CrossRef]
19. Ali, A.; Tabanca, N.; Kurkcuoglu, M.; Duran, A.; Blyhthe, E.K.; Khan, I.A.; Baser, K.H.C. Chemical composition, larvicidal and biting deterrent activity of essential oils of two subspecies of *Tanacetum argenteum* (Lam.) Willd. and individual constituents against *Aedes aegypti* (L.) (Diptera: Culicidae). *J. Med. Entomol.* **2014**, *51*, 824–830. [CrossRef]
20. Tabanca, N.; Gao, Z.; Demirci, B.; Techen, N.; Wedge, D.E.; Ali, A.; Sampson, B.J.; Werle, C.; Bernier, U.R.; Khan, I.A.; et al. Molecular and phytochemical investigation of *Angelica dahurica* and *A. pubescens* essential oils and their biological activity against *Aedes aegypti*, *Stephanitis pyrioides* and *Colletotrichum* Species. *J. Agric. Food Chem.* **2014**, *62*, 8848–8857. [CrossRef]
21. Ali, A.; Tabanca, N.; Ozek, G.; Ozek, T.; Aytac, Z.; Bernier, U.R.; Agramonte, N.M.; Baser, K.H.C.; Khan, I.A. Essential oils of *Echinophora lamondiana* (Apiales: Umbelliferae): A relationship between chemical profile and biting deterrence and larvicidal activity against mosquitoes (Diptera: Culicidae). *J. Med. Entomol.* **2015**, *52*, 93–100. [CrossRef]
22. Ali, A.; Cantrell, C.L.; Bernier, U.R.; Duke, S.O.; Schneider, J.C.; Khan, I. *Aedes aegypti* (Diptera: Culicidae) Biting deterrence: Structure-activity relationship of saturated and unsaturated fatty acids. *J. Med. Entomol.* **2012**, *49*, 1370–1378. [CrossRef] [PubMed]
23. Tabanca, N.; Bernier, U.R.; Ali, A.; Wang, M.; Demirci, B.; Blythe, E.K.; Khan, S.I.; Baser, K.H.C.; Khan, I.A. Bioassay-guided investigation of two *Monarda* essential oils for repellent activity against yellow fever mosquito *Aedes aegypti*. *J. Agric. Food Chem.* **2013**, *61*, 8573–8580. [CrossRef] [PubMed]
24. Flores-Estevez, N.; Vasquez-Morales, S.G.; Cano-Medina, T.; Sanchez-Velasquez, L.R.; Noa-Carrazana, J.C.; Diaz-Fleischer, F. Insecticidal activity of raw ethanolic extracts from *Magnolia dealbata* Zucc. on a tephritid pest. *J. Environ. Sci. Health B* **2013**, *48*, 582–586. [CrossRef] [PubMed]
25. McLafferty, F.W.; Stauffer, D.B. *The Wiley/NBS Registry of Mass Spectral Data*; J. Wiley and Sons: New York, NY, USA, 1989.

26. Hochmuth, D.H. *MassFinder 4.0, Hochmuth Scientific Consulting*, Hamburg, Germany, 2008.
27. Klun, J.A.; Kramer, M.; Debboun, M. A new *in vitro* bioassay system for discovery of novel human-use mosquito repellents. *J. Am. Mosq. Cont. Assoc.* **2005**, *21*, 64–70. [CrossRef]
28. Ali, A.; Cantrell, C.L.; Khan, I.A. A new in vitro bioassay system for the discovery and quantitative evaluation of mosquito repellents. *J. Med. Entomol.* **2017**, *54*, 1328–1336. [CrossRef]
29. Katritzky, A.R.; Wang, W.; Slavov, S.; Dobchev, D.A.; Hall, C.D.; Tsikolia, M.; Bernier, U.R.; Elejalde, N.M.; Clark, G.G.; Linthicum, K.J. Novel carboxamides as potential mosquito repellents. *J. Med. Entomol.* **2010**, *47*, 924–938. [CrossRef]
30. Pridgeon, J.W.; Becnel, J.J.; Clark, G.G.; Linthicum, K.J. A high-throughput screening method to identify potential pesticides for mosquito control. *J. Med. Entomol.* **2009**, *46*, 335–341. [CrossRef]
31. SAS Institute. *SAS Online Doc., version 9.2*; SAS Institute: Cary, NC, USA, 2007.


**Sample Availability:** Samples of the compounds 1-decanol, 1-octanol, and 1-heptanol. are available from A.A..

*Article*

# Characterization, Quantification and Quality Assessment of Avocado (*Persea americana* Mill.) Oils

**Mei Wang [1], Ping Yu [2,3,4], Amar G. Chittiboyina [1], Dilu Chen [5], Jianping Zhao [1], Bharathi Avula [1], Yan-Hong Wang [1] and Ikhlas A. Khan [1,6,*]**

1 National Center for Natural Products Research, School of Pharmacy, University of Mississippi, University, MS 38677, USA; meiwang@olemiss.edu (M.W.); amar@olemiss.edu (A.G.C.); jianping@olemiss.edu (J.Z.); bavula@olemiss.edu (B.A.); wangyh@olemiss.edu (Y.-H.W.)
2 State Key Laboratory of Food Science and Technology, Nanchang University, Nanchang 330031, China; cpu_yuping@126.com
3 Jiangxi Province Key Laboratory of Edible and Medicinal Resources Exploitation, Nanchang University, Nanchang 330031, China
4 School of Resource and Environmental and Chemical Engineering, Nanchang University, Nanchang 330031, China
5 School of Pharmacy, Hunan University of Chinese Medicine, Changsha 410208, China; cdl918@126.com
6 Division of Pharmacognosy, Department of BioMolecular Sciences, School of Pharmacy, University of Mississippi, University, MS 38677, USA
* Correspondence: ikhan@olemiss.edu; Tel.: +1-662-915-7821

Academic editor: Muhammad Ilias
Received: 14 February 2020; Accepted: 18 March 2020; Published: 24 March 2020

**Abstract:** Avocado oil is prized for its high nutritional value due to the substantial amounts of triglycerides (TGs) and unsaturated fatty acids (FAs) present. While avocado oil is traditionally extracted from mature fruit flesh, alternative sources such as avocado seed oil have recently increased in popularity. Unfortunately, sufficient evidence is not available to support the claimed health benefit and safe use of such oils. To address potential quality issues and identify possible adulteration, authenticated avocado oils extracted from the fruit peel, pulp and seed by supercritical fluid extraction (SFE), as well as commercial avocado pulp and seed oils sold in US market were analyzed for TGs and FAs in the present study. Characterization and quantification of TGs were conducted using UHPLC/ESI-MS. Thirteen TGs containing saturated and unsaturated fatty acids in avocado oils were unambiguously identified. Compared to traditional analytical methods, which are based only on the relative areas of chromatographic peaks neglecting the differences in the relative response of individual TG, our method improved the quantification of TGs by using the reference standards whenever possible or the reference standards with the same equivalent carbon number (ECN). To verify the precision and accuracy of the UHPLC/ESI-MS method, the hydrolysis and transesterification products of avocado oil were analyzed for fatty acid methyl esters using a GC/MS method. The concentrations of individual FA were calculated, and the results agreed with the UHPLC/ESI-MS method. Although chemical profiles of avocado oils from pulp and peel are very similar, a significant difference was observed for the seed oil. Principal component analysis (PCA) based on TG and FA compositional data allowed correct identification of individual avocado oil and detection of possible adulteration.

**Keywords:** *Persea americana* Mill.; avocado oil; triglyceride; fatty acid; UHPLC/ESI-MS; GC/MS; quality evaluation

## 1. Introduction

Avocado (*Persea americana* Mill.) is a member of the Lauraceae family. Although avocado trees are native to Central America, they are also widely distributed in tropical and subtropical countries.

Anatomically, the avocado fruit can be distinguished into three regions - the innermost seed that constitutes 20% of the fruit, the pulp covering the major portion (65%) and the outermost peel (15%) [1,2]. Popularly known as "vegetable butter" or "butter pear", the fruit contains a substantial amount of triglycerides (TGs) along with a high content of unsaturated fatty acids. It is also rich in many other bioactive phytochemicals such as carotenoids, tocopherols, phytosterols, aliphatic alcohols and hydrocarbons [3,4].

Unlike oil extracted from other fruits, the oil from avocado fruit is often extracted from the mature fruit flesh [4], and its lipid content has been reported as the highest among all known fruit and vegetable varieties [5–7]. Avocado oil has a multitude of applications such as a culinary oil and as an ingredient in healthcare products, cosmetics, pharmaceuticals and nutraceuticals. The consumption of avocado oil has become popular owing to its high nutritional value and potential benefit to human health, including the management of hypercholesterolemia [8,9], hypertension [10], diabetes and fatty liver disease [11]. The oil can also reduce cardio-metabolic risk [12] and possesses anti-cancer and antimicrobial properties [13,14]. Over the last decade, the production of avocado oil worldwide has grown steadily and currently accounts for about 4.4 million tons of fresh fruit [15,16].

TGs are the most important nutritive group of compounds in avocado oil and represent a significant amount (~90%) of the entire oil composition. Chemically, TGs are complex hydrophobic molecular species formed by the esterification of three fatty acids (FAs) with a glycerol backbone under enzymatic catalysis. The complexity of TGs is due to a large number of possible FA combinations attached to the glycerol skeleton, which can differ in the number of acyl carbon atoms (CNs), the degree of unsaturation, and the position and configuration (cis/trans) of the double bonds (DBs) in each FA. Furthermore, the TG molecule demonstrates optical activity (enantiomers) when the two primary hydroxyl groups are esterified with different FAs, and the stereo-specific distribution (regioisomers) can vary when stereo-chemical positions (sn-1, 2 or 3) on the glycerol skeleton are attached by various combinations of FAs. Several analytical techniques have been employed for the qualitative and quantitative determination of TGs in edible oils, ranging from spectroscopy methods such as infrared spectroscopy [17,18] and nuclear magnetic resonance [19] to chromatographic techniques including gas chromatography (GC), liquid chromatography (LC) [20] and supercritical fluid chromatography (SFC) coupled with mass spectrometry/tandem mass spectrometry [21]. Non-aqueous reversed-phase liquid chromatography coupled with positive-ion atmospheric pressure chemical ionization (APCI) mass spectrometry has become increasingly popular and currently is the most widely used separation technique for TGs analysis. By using this technique, the separation of TGs is governed by the equivalent carbon number (ECN) defined as ECN = CN – 2DB. Separations of TGs within the same ECN group [22,23], cis/trans isomers and isomers with different positional DB have been reported [24]. In contrast, GC is the most commonly used method for the analysis of FAs, but it requires transesterification to convert TGs to its corresponding fatty acid methyl esters (FAMEs). Although high-temperature GC for the direct determination of intact TGs has been reported [25], samples subjected to this technique must be thermally stable and resistant to isomerization.

In recent years, the popularity of avocado oil in the US market has been promoted with oils extracted from alternative sources such as avocado seed. Some manufacturers and consumers have considered avocado seed oil as a source of fatty acids, carbohydrates, dietary fiber and a broad range of phytochemicals. Unfortunately, there is no sufficient evidence to support the claimed health benefits and safe use of such oils. In addition, vegetable oils are among the top 25 ingredients that are most susceptible to adulteration and represent 24% of reported fraud cases [18]. Thus, avocado oil could be a target for fraudulent practices such as adulteration with low-cost oils. Therefore, the development of accurate and reproducible methods for TG and FA analysis in avocado oil is needed for characterization and quality control of this valuable commodity.

As part of an ongoing research program on the authentication, safety and biological evaluation of phytochemicals and dietary supplements, an in-depth chemical investigation of avocado oil was performed. The current study aimed to establish the comprehensive profile of TGs in oils extracted from avocado peel, pulp and seed. A UHPLC/ESI-MS method was developed for the identification and quantification of 13 TGs present in authenticated and commercial avocado oils. Furthermore, the hydrolysis and transesterification products of avocado oils were analyzed for FAMEs using a GC/MS method. To verify the precision and accuracy of the developed methods, the results from GC/MS and UHPLC/ESI-MS were compared. Finally, the TG and FA compositional data, along with chemometric analysis, was used for quality evaluation and identification of possible adulteration in commercial oils.

## 2. Results and Discussion

### 2.1. Supercritical Fluid Extraction of Avocado Oil

Generally, avocado oil is extracted from avocado pulp by centrifugation, cold pressing or solvent extraction [26]. These extraction methods are time-consuming and economically unfavorable. Supercritical fluid extraction (SFE), on the other hand, is an environmentally friendly and cost-effective extraction method with a multitude of applications in the food, pharmaceutical and fine chemical industries. In the present study, SFE was used for the extraction of oils from avocado peel, pulp and seed. The yields of the oils along with their physicochemical properties determined by the standard method of the American Oil Chemists' Society [27] were compared with the oils extracted with the AOAC method [28]. The results are given in Table 1. The color of the oils extracted from the SFE method was much lighter than that of the solvent extraction, indicating a smaller amount of chlorophyll being extracted. The TG and FA profiles for these two extraction methods were similar as determined by UHPLC/MS and GC/MS analysis. Therefore, the oils extracted by SFE were used for further studies.

**Table 1.** Extraction conditions, oil yields and physicochemical properties (mean ± SD, $n = 3$).

| Extraction Method | | SFE | | | Solvent | | |
|---|---|---|---|---|---|---|---|
| Extraction Conditions | Solvent | $CO_2$ | | | *n*-Hexane | | |
| | Time (min) | 40 | | | 480 | | |
| | Temperature (°C) | 50 | | | 70 | | |
| | Pressure (bar) | 250 | | | atmospheric | | |
| Sample | | Peel | Pulp | Seed *,b | Peel | Pulp | Seed *,b |
| Oil Yield *,a (%) | | 16.89 ± 0.56 | 56.10 ± 0.66 | 1.61 ± 0.31 | 17.32 ± 0.30 | 58.35 ± 0.27 | 1.66 ± 0.61 |
| Physicochemical Properties | Acid Value (g/oleic acid 100 g) | 1.60 ± 0.13 | 0.94 ± 0.03 | N/A | 0.69 ± 0.03 | 0.38 ± 0.00 | N/A |
| | Saponification Value (mg KOH/g) | 186.90 ± 0.74 | 193.71 ± 1.05 | N/A | 182.15 ± 0.11 | 186.15 ± 0.88 | N/A |
| | Iodine Value (g $I_2$/100 g) | 80.62 ± 0.38 | 90.76 ± 0.46 | N/A | 72.38 ± 0.41 | 88.40 ± 0.91 | N/A |
| | Peroxide Value | 2.36 ± 0.13 | 1.18 ± 0.02 | N/A | 3.53 ± 0.09 | 2.41 ± 0.13 | N/A |

*,a: Oil yields were calculated based on the dry weight. *,b: Not enough quantity to perform the measurements.

### 2.2. TGs Profile of Avocado Oils

#### 2.2.1. Method Development and Optimization

Avocado oils are characterized by a high content of TGs. The separation and unambiguous identification of structurally similar TGs in avocado oil pose great analytical challenges. In this study, 12 commercially available TG reference standards were purchase (Table 2) and used for method development and quantification. The method was optimized regarding the chromatographic column, eluent and gradient program. Different reversed-phase columns including several Agilent ZORBAX columns such as Eclipse Plus $C_{18}$, SB-$C_{18}$, XDB-$C_{18}$, SB-$C_8$ (with the same dimensions of 2.1 × 100 mm × 1.8 μm), and ACQUITY UPLC BEH $C_{18}$ (Waters, 2.1 × 100 mm × 1.7 μm) as a standalone column or in a combination were investigated. Finally, three ACQUITY UPLC BEH $C_{18}$ connected in series were used to provide the best separation of the targeted TGs. For the evaluation of eluents, unlike non-aqueous eluents used by the majority of TGs analyses [21,29–31], eluent consisting of acetonitrile with 0.1% water (v/v) and isopropanol with 5 mM ammonium formate was used, and the chromatographic peak shapes were greatly improved. The chromatograms showing the method optimization using a mixed solution containing LLO, LLP, OLO, PLO, PPoO, OOO, OOP and PPO standards are illustrated in Figure 1. The chromatograms of authenticated oils extracted from avocado peel, pulp and seed, as well as one of the commercial products claimed as avocado seed oil, are shown in Figure 2. It is worth noting, as the ECNs increased from 44 to 50, the retention times for the corresponding TGs also increased (Figure 2). Although similar TG profiles of avocado peel and pulp oils were observed, a significant difference was present in avocado seed oil. The profile for the commercial avocado seed oil appeared to be dramatically different with any of the other avocado oils, suggesting that this particular oil was possibly adulterated with other oils.

**Table 2.** Information of triglyceride reference standards.

| No. | Compound | Abbr. | ECN | MW | Formula | CAS # |
|---|---|---|---|---|---|---|
| 1 | 1,3-linolein-2-olein | LOL | 44 | 881.40 | $C_{57}H_{100}O_6$ | 2190-22-9 |
| 2 | 1,2-linolein-3-olein | LLO | 44 | 881.40 | $C_{57}H_{100}O_6$ | 2190-21-8 |
| 3 | 1,2- linolein-3-palmitin | LLP | 44 | 855.36 | $C_{55}H_{98}O_6$ | 2190-15-0 |
| 4 | 1,3-olein-2-linolein | OLO | 46 | 883.42 | $C_{57}H_{102}O_6$ | 2190-19-4 |
| 5 | 1-olein-2-palmitin-3-linolein | OPL | 46 | 857.38 | $C_{55}H_{100}O_6$ | 2534-97-6 |
| 6 | 1-palmitin-2-linolein-3-olein | PLO | 46 | 857.38 | $C_{55}H_{100}O_6$ | 2680-59-3 |
| 7 | 1-palmitin-2-palmitolein-3-olein | PPoO | 46 | 831.34 | $C_{53}H_{98}O_6$ | 81637-60-7 |
| 8 | 1-palmitin-2-olein-3-linolein | POL | 46 | 857.38 | $C_{55}H_{100}O_6$ | 2680-59-3 |
| 9 | 1,2-olein-3-palmitin | OOP | 48 | 859.39 | $C_{55}H_{102}O_6$ | 2190-30-9 |
| 10 | 1,2-palmitin-3-olein | PPO | 48 | 833.36 | $C_{53}H_{100}O_6$ | 1867-91-0 |
| 11 | 1,3-olein-2-palmitin | OPO | 48 | 859.39 | $C_{55}H_{102}O_6$ | 1716-07-0 |
| 12 | triolein | OOO | 48 | 885.43 | $C_{57}H_{104}O_6$ | 122-32-7 |

ECN: equivalent carbon number, calculated as ECN = CN (number of carbon atoms) – 2DB (double bonds).

### 2.2.2. Identification of TGs

Unambiguous identification of complex TGs in avocado oil is desirable for the accurate quantification of TGs. Many HPLC detection techniques including refractive index [32], UV-Vis and evaporative light-scattering (ELSD) [21], have been applied for the qualitative analysis of TGs in plant oils. Although each of these detection methods has its own advantages, it may not be possible to confidently identify TGs with complete or even partial chromatographic resolution in complex plant oils that contain numerous species with the same ECNs. In our study, both APCI and ESI with positive/negative ion modes were evaluated, and ESI(+) provided better sensitivity for all the TGs with the optimized solvent system (Figure 1D). The ESI(+) mass spectra and notation of fragment ions for representative TGs are illustrated in Figure 3. As described before, TGs in plant oils usually exist as a mixture of positional isomers differing by the acyl attachment, sn-1, sn-2 and sn-3. The $[M + H]^+$ and $[M + NH_4]^+$ ions were detected in all TGs with relatively low abundances compared to the corresponding $[M + H\text{-}RCOOH]^+$ ions. Except for these two ions, single-acid type ($R_1R_1R_1$) provided only one ion such as $[OO]^+$ in OOO. Mixed-acid type ($R_1R_2R_1$ or $R_1R_1R_2$) always provided two different ions, such as $[LL]^+$ and $[LO]^+$ for both LLO and LOL types TGs. Conversely, three different ions were observed for mix-acid type ($R_1R_2R_3$), such as $[LP]^+$, $[LO]^+$ and $[OP]^+$ for OLP and OPL. All the examples are shown in Figure 3. Although the theoretical ion abundance ratios of $[LO]^+/[LL]^+$ should be 2:1 for both LLO and LOL, different values (1.2 for LLO and 3.0 for LOL, respectively) were observed. A similar observation was achieved for OLP and OPL. The theoretical ion abundance ratios of $[LP]^+/[OP]^+/[LO]^+$ should be 1:1:1 for both OLP and OPL, whereas the measured values were 0.62:0.38:1 for OLP, and 1.8:2.1:1 for OPL. This observation indicated that the loss of FA from the equivalent sn-1 and sn-3 positions is preferred rather than the middle position sn-2. Figure 4 proposed the possible mechanism for the cleavage of fatty acids from TGs at different positions. Based on observed fragments, the formation of a 5-member ring as a result of losing FA group in the sn-2 position is less favorable than the formation of a more stable 6-member ring in position sn-1 or sn-3 [33].

The relative abundances of ions formed from the loss of FAs in different positions in the MS afford the confident identification of the acyl position on the glycerol backbone.

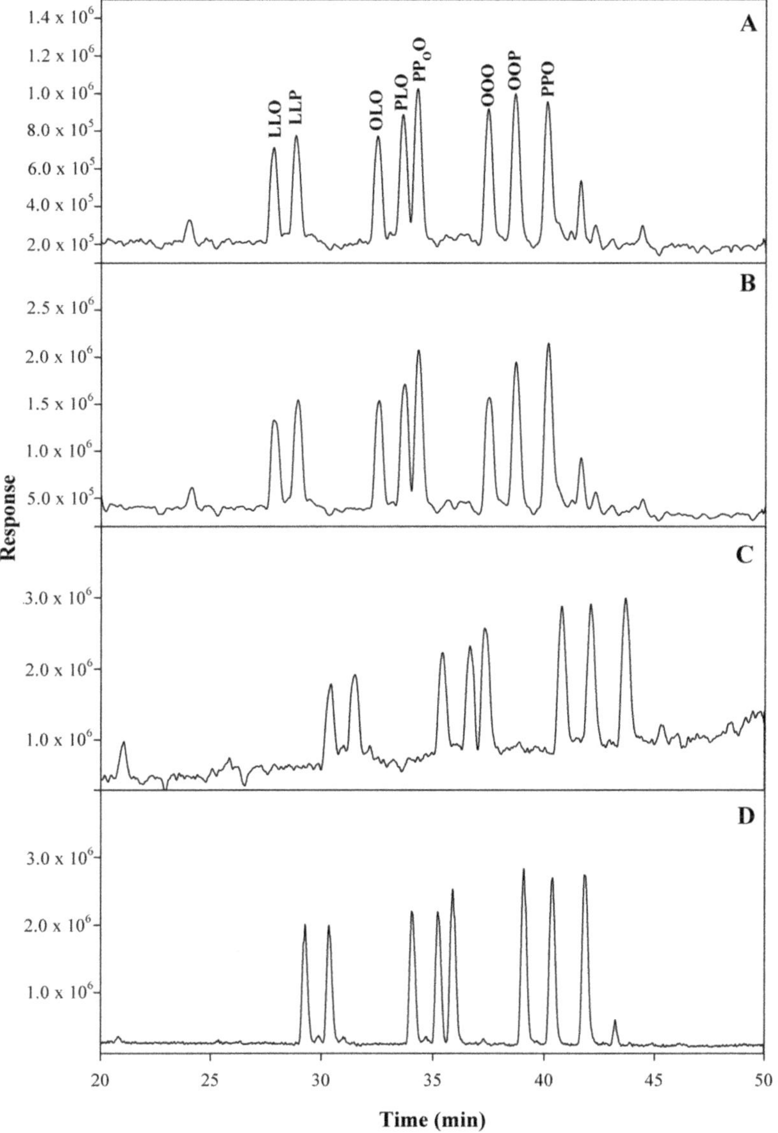


**Figure 1.** Total ion chromatograms of (**A**) APCI(+), eluent: acetonitrile/isopropanol; (**B**) ESI(+), eluent: acetonitrile/isopropanol; (**C**) ESI(+), eluent: acetonitrile with 0.05% formic acid/isopropanol with 0.05% formic acid and 5 mM ammonium formate; (**D**) ESI(+), eluent: acetonitrile with 0.1% water and 0.05% formic acid/isopropanol with 0.05% formic acid and 5 mM ammonium formate.

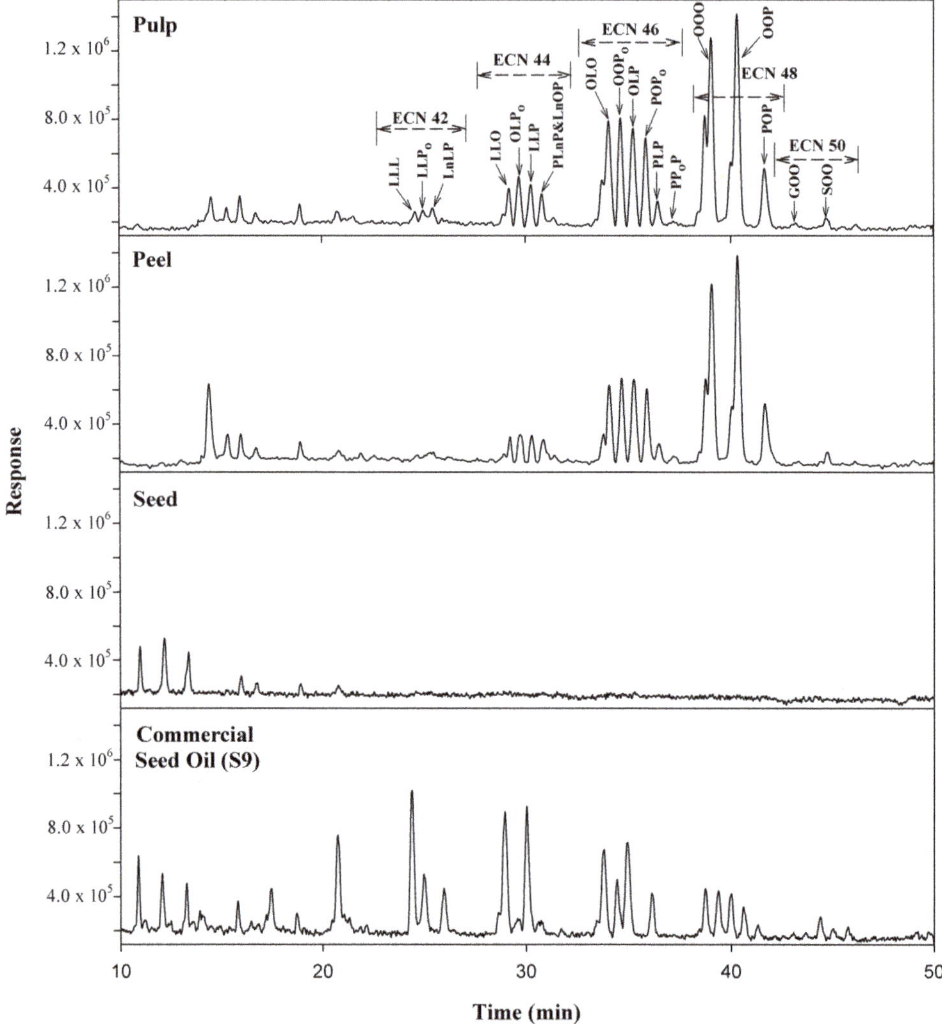


**Figure 2.** Total ion chromatograms of authenticated avocado oils extracted from pulp, peel, seed and commercial avocado seed oil. ECNs: the equivalent carbon numbers.

An alternative approach for the identification of TGs in avocado oils is to measure the ion survival yield (ISY). As demonstrated in Figure 3, various types of ions were formed by in-source collision-induced dissociation (IS-CID). The production of information-rich fragments by IS-CID can materially aid in the identification of components, elucidation of structures, and distinction between isomers and chemically similar components in complex mixtures [34]. Ion distribution in IS-CID has been commonly measured by ISY defined as $ISY = \frac{I_a}{I_a + \sum I_b}$ where $I_a$ represents the measured intensity of the monitored ion and $I_b s$ are the intensities of the additional ions formed in the source. Accurate qualitative and quantitative analysis requires that the ISYs of reference standards should be independent of the concentrations of standards, and the ISYs should be equivalent to the standards and analyzed samples. The ISYs for the representative TG standards were measured over the range of 5–400 µg/mL (Figure 5). The ISYs for the fragment ion of [LO]+ after the loss of one FA from different positions (sn-1, sn-2 or sn-3) were calculated from the representative TGs. For example, LLO and LOL both displayed a pseudo-molecular ion $[M + H]^+$ at 881 with a fragment ion $[LO]^+$ at 601. However, the ISY for $[LO]^+$ generated from LLO was 0.38 (±0.01), whereas that from LOL was 0.50 (±0.01). Similarly, ISYs for $[LO]^+$ from PLO and OPL were 0.43 (±0.01) and 0.21 (±0.01), respectively, suggesting that the measured ISYs can be used for the identification and characterization of TGs in complex samples [35].

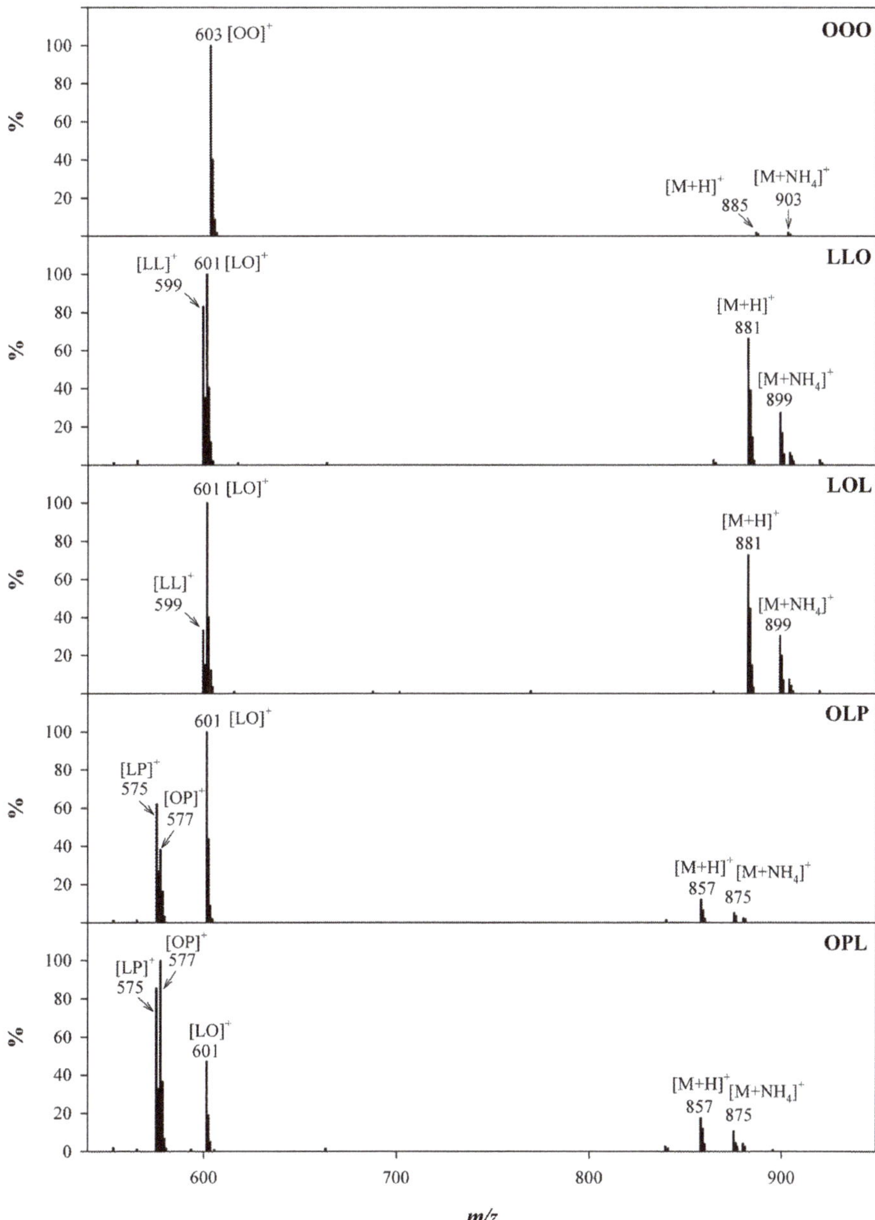


**Figure 3.** Positive ion ESI spectra of TGs containing different acyls on the glycerol backbone. OOO, LLO and LOL, and OLP and OPL were used as representative examples for single-acid type ($R_1R_1R_1$), mixed-acid type ($R_1R_2R_1$ or $R_1R_1R_2$) and mixed-acid type ($R_1R_2R_3$), respectively.

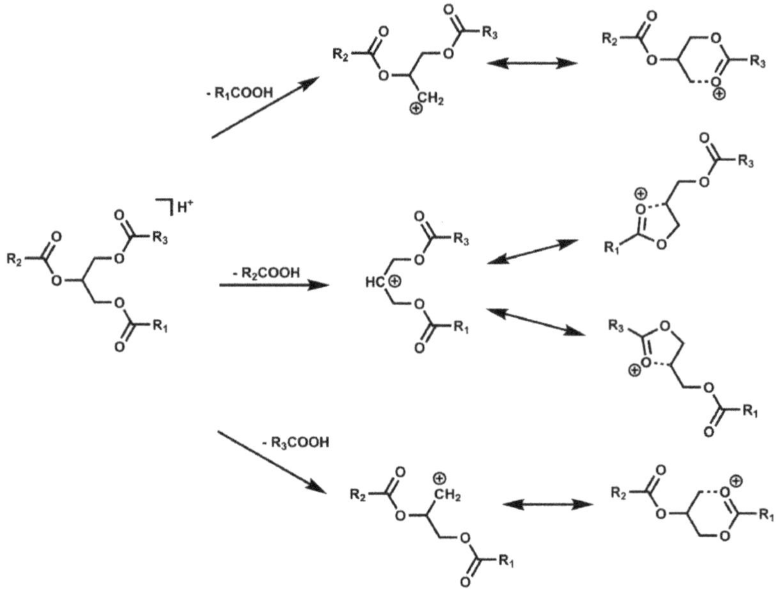

**Figure 4.** Schematic representation of fatty acids elimination from sn-1, sn-2 and sn-3 positions by ESI(+) MS.

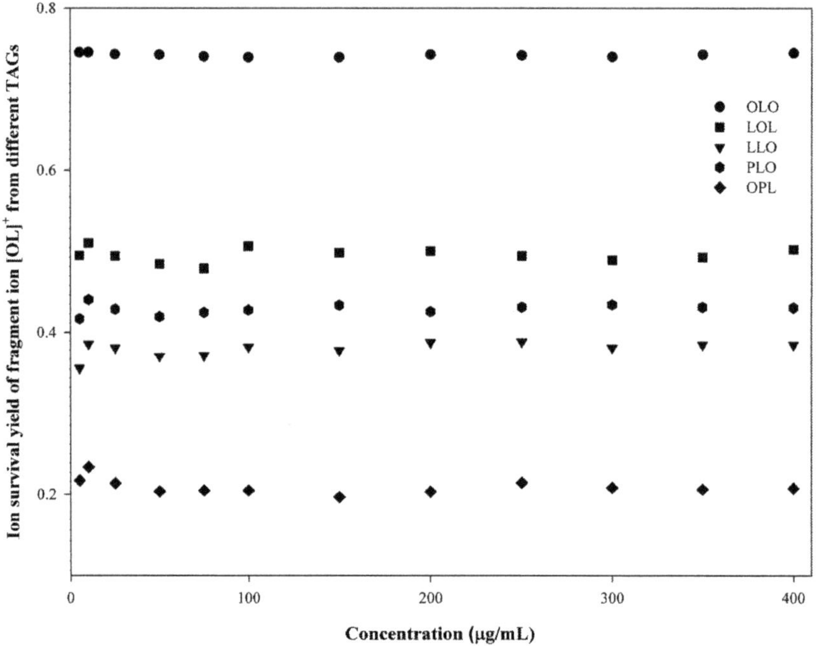


**Figure 5.** Ion survival yields (ISYs) for the fragment ion $[OL]^+$ from representative TGs over the concentration range.

### 2.2.3. Quantification of TGs

A prolonged challenge in the accurate quantification of TGs is the lack of commercially available reference standards. Natural TGs are complex, and commercial standards are available only for a limited number, mostly single-acid ($R_1R_1R_1$) type. Thus, quantification based on the calibration curve from each individual TG reference standard is practically impossible. Previously, the quantification of TGs was based on the relative peak areas neglecting the differences in the relative response for each individual TG. Later on, a more sophisticated approach using response factors (RFs) was reported by Holcapek, et al. [31], in which the calibration curves of single-acid TGs were measured, and the RFs of mixed-acid TGs expressed relative to the most common OOO were calculated. In our study, 12 commercially available TG reference standards were used for the TGs quantification. Calibration curves for the 12 standards were realized by plotting the logarithms of the sum peak areas of all ions from IS-CID for each TG versus logarithms of analyte concentrations as shown in Figure 6. The results exhibited good linearity ($R^2 > 0.99$) over the concentration range of 1–400 μg/mL, and the limits of quantification were 1 μg/mL for all the analytes. When the total content of TGs was calculated, the recovery values were all in the range of 95%–107% (RSD < 7%), and the RSD values of precision, including intra-day and inter-day, were determined to be < 7%. Interestingly, the calibration curves for TGs within the same ECN group were nearly overlapped as shown in Figure 6, suggesting that the quantification method could select one single standard from each ECN group, and simultaneously determine multiple compounds in the same ECN group when reference standards are commercially unavailable. This single standard method could significantly lower the cost and time of the experiment [36].

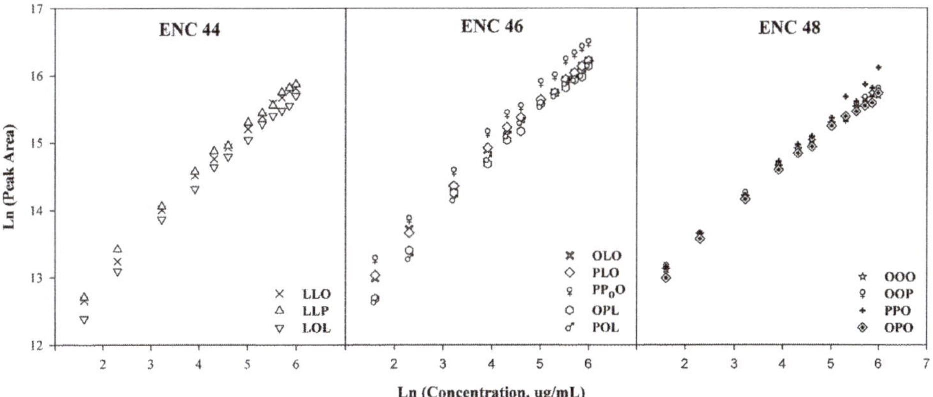


**Figure 6.** Calibration curves (grouped by ECN) for 12 commercially available reference standards.

The authenticated avocado peel, pulp and seed oils, sesame oil and soybean oil which have been reported as potential adulterants [17], along with 19 commercial avocado pulp or seed oils (Table S1) were quantified for the 13 characteristic TGs identified in avocado oils. The quantification results are given in Table 3. The quantification of all the TGs in the seed oil was below the detection limit of the current method. In addition, the yield from avocado seed oil was less than 2%, demonstrating that any commercial avocado seed oil sold in US market would be questionable due to the poor economic viability. Avocado peel and pulp oil showed very similar chemical profiles. The total compositions of the 13 TGs were 67.4% in the peel oil and 86.3% in the pulp oil, respectively. OOO and OOP are the most prominent TGs, accounting for ~25% of total TG contents in both peel and pulp oils, followed by OLO (~18%) and OOPo (~7%). On the other hand, soybean and sesame oils showed significant differences (Table 3). OLO is the most abundant compound in both sesame oil (~42%) and soybean oil

(~29%), whereas OOO and OOP are relatively low, accounting for 10% and 2.5% in sesame oil and 6% and 3% in soybean oil, respectively. Other TGs, such as LLL have been detected in both sesame and soybean oils but were not identified in avocado oil. Among the analyzed 19 (S1–S19) commercial avocado pulp and seed oils purchased in the US market, S10 and S19, claimed as avocado seed oils, demonstrated very similar chemical profiles as avocado pulp oil. S4 and S9, claimed to be avocado seed oil, along with S6, S7 and S16 (plant parts were not specified), showed significant differences to avocado pulp oil, but exhibited similar profiles to soybean oil. These samples could possibly be adulterated with soybean oil. S17 was claimed as avocado pulp oil but demonstrated a profile close to a combination of avocado and sesame oils.

**Table 3.** Concentrations (mg/g) of TGs quantified in authenticated and commercial oils.

| Sample | LLO | OLPo | LLP | PLnP | OLO | OOPo | OLP | POPo | PLP | OOO | OOP | POP | SOO | Total |
|---|---|---|---|---|---|---|---|---|---|---|---|---|---|---|
| | ENC 44 | | | | ENC 46 | | | | | ENC 48 | | | ENC 50 | |
| Authenticated Sample | | | | | | | | | | | | | | |
| Avocado Peel | 6.11 | 9.88 | 8.90 | 3.03 | 114.56 | 48.36 | 46.24 | 45.04 | 4.87 | 184.01 | 168.52 | 32.22 | 2.54 | 674.27 |
| Avocado Pulp | 12.13 | 23.00 | 17.71 | 5.36 | 163.65 | 64.59 | 57.27 | 53.46 | 7.14 | 215.97 | 213.03 | 26.88 | 2.37 | 862.55 |
| Avocado Seed | ND | ND | ND | ND | ND | ND | ND | ND | ND | ND | ND | ND | ND | |
| Sesame | 153.08 | ND | 53.07 | ND | 360.28 | ND | 96.16 | ND | 8.04 | 87.04 | 51.14 | 6.97 | 39.92 | 855.69 |
| Soybean | 105.65 | ND | 100.94 | ND | 130.00 | ND | 65.01 | ND | 18.16 | 11.16 | 14.07 | 2.10 | 4.57 | 451.67 |
| Commercial Sample | | | | | | | | | | | | | | |
| S1 | 20.08 | 10.19 | 14.80 | 3.52 | 196.21 | 38.00 | 51.55 | 36.95 | 6.57 | 286.11 | 179.71 | 27.78 | 17.39 | 888.87 |
| S2 | 27.56 | 8.12 | 18.09 | 2.88 | 172.84 | 41.11 | 44.14 | 33.82 | 3.78 | 246.88 | 159.06 | 22.50 | 23.77 | 804.52 |
| S3 | 13.27 | 8.47 | 11.56 | 4.12 | 137.46 | 38.66 | 40.19 | 39.87 | 6.87 | 265.76 | 179.46 | 29.03 | 22.59 | 797.32 |
| S4 | 24.77 | ND | ND | 0.34 | 96.29 | ND | ND | ND | ND | ND | 20.10 | 3.27 | 18.80 | 163.56 |
| S5 | 44.51 | ND | 13.41 | ND | 165.75 | 6.64 | 27.82 | ND | 1.45 | 377.75 | 97.94 | 4.35 | 51.84 | 791.44 |
| S6 | 107.34 | ND | 93.88 | 0.58 | 137.12 | ND | 61.63 | ND | 14.79 | 14.30 | 13.24 | 1.96 | 6.40 | 451.24 |
| S7 | 96.36 | ND | 55.88 | ND | 190.11 | ND | 53.50 | ND | 5.76 | 250.13 | 57.13 | 4.27 | 33.76 | 746.91 |
| S8 | 8.66 | 22.47 | 18.53 | 8.29 | 120.27 | 55.97 | 64.96 | 68.08 | 12.79 | 155.42 | 214.73 | 46.91 | 3.49 | 800.57 |
| S9 | 106.36 | ND | 104.00 | 0.68 | 134.13 | ND | 60.83 | ND | 15.26 | 15.09 | 13.43 | 2.05 | 5.30 | 457.14 |
| S10 | 43.90 | ND | 8.32 | ND | 342.05 | ND | 24.27 | ND | 0.49 | 243.62 | 33.42 | 0.69 | 13.73 | 710.50 |
| S11 | 47.24 | ND | 13.44 | ND | 166.03 | 5.03 | 29.59 | ND | 2.99 | 372.85 | 97.72 | 9.37 | 40.82 | 785.08 |
| S12 | 9.16 | 20.91 | 20.09 | 8.69 | 139.01 | 54.32 | 53.33 | 68.14 | 11.85 | 204.35 | 213.48 | 35.05 | 4.10 | 842.47 |
| S13 | 18.04 | 22.61 | 21.84 | 6.08 | 183.92 | 43.22 | 64.18 | 36.20 | 9.24 | 190.67 | 182.76 | 22.59 | 2.23 | 803.58 |
| S14 | 26.41 | 2.20 | 12.33 | 0.67 | 121.38 | 15.54 | 28.78 | 10.28 | 2.16 | 330.82 | 98.37 | 10.09 | 36.80 | 695.84 |
| S15 | 41.38 | ND | 8.15 | ND | 302.47 | ND | 21.06 | ND | 0.41 | 219.81 | 27.92 | 0.45 | 12.43 | 634.07 |
| S16 | 97.55 | ND | 89.99 | ND | 120.85 | ND | 58.65 | ND | 13.95 | 9.48 | 11.62 | 1.62 | 4.53 | 408.24 |
| S17 | 68.92 | 1.41 | 70.08 | 1.68 | 109.23 | 20.16 | 60.72 | 13.79 | 12.50 | 66.07 | 62.91 | 13.00 | 4.47 | 504.95 |
| S18 | 44.27 | ND | 12.57 | ND | 145.54 | 5.11 | 23.75 | 0.75 | 1.39 | 364.13 | 82.06 | 4.46 | 43.27 | 727.28 |
| S19 | 10.69 | ND | 10.13 | 1.41 | 72.82 | 23.30 | 40.16 | 17.31 | 16.65 | 278.46 | 129.25 | 37.57 | 26.84 | 664.59 |

ND: Not detected.

### 2.3. *FAs Profile of Avocado Oils*

Except for the most abundant compounds of TG in avocado oil, the derived FA composition can also be exploited as a peculiar fingerprint indicative of the oil's quality and authenticity [37]. FAs are a group of very complex compounds, including monounsaturated FAs, polyunsaturated FAs and saturated FAs. In this study, the selected HP-88 with (88% cyanopropy)aryl-polysiloxane stationary phase GC capillary column is a high-polarity column designed for the separation of FAMEs including those positional cis/trans isomers. All the authenticated and commercial oils were transesterified and analyzed by GC/MS. The FA profile comprised a total of seven FAs in avocado oils, viz. palmitic acid, palmitoleic acid, stearic acid, oleic acid, vaccenic acid, linoleic acid and linolenic acid. Oleic acid was the major FA (~60%), followed by palmitic acid (~15%), linoleic acid (~10%), palmitoleic acid (~7%) and vaccenic acid (~6%). This agreed with the individual FA moiety identified in the TGs by UHPLC/ESI-MS. The quantification results are summarized in Table 4. Again, the FA profiles of soybean and sesame oils showed significant differences from the avocado oils. Oleic acid was the most abundant FA (~44%) in sesame oil, followed by linoleic acid (~40%), palmitic acid (~9%) and stearic acid (~5%). In soybean oil, linoleic acid (~57%) was the major FA, followed by oleic acid (~20%), palmitoleic acid (~11%) and linolenic acid (~7%). Interestingly, avocado oils contain much less stearic acid (~0.3%) compared to soybean oil (~4%) and sesame oil (~5%). The FA compositions of the commercial samples were evaluated. Similar to the TGs analysis, S4, S6, S9, S16 and S17 contained a relatively high concentration of stearic acid (3.0%–4.8%), like sesame and soybean oils, and the FA profiles were significantly different with the authenticated avocado oil. Therefore, these five samples are likely adulterated.

**Table 4.** Concentrations (mg/g) of FAs quantified in authenticated and commercial oils.

| Sample | | Palmitic | Palmitoleic | Stearic | Oleic | Vaccenic | Linoleic | Linolenic | Total |
|---|---|---|---|---|---|---|---|---|---|
| | | * C16:0 | C16:1 | C18:0 | C18:1 n9 | C18:1 n11 | C18:2 n9,12 | C18:3 n9,12,15 | |
| Authenticated Sample | | | | | | | | | |
| Avocado | Peel | 101.76 | 37.26 | 1.79 | 416.79 | 39.82 | 55.09 | 7.24 | 659.76 |
| | Pulp | 138.42 | 67.19 | 1.36 | 538.68 | 52.66 | 89.38 | 5.61 | 893.30 |
| | Seed | 5.72 | 1.46 | ND | 11.21 | 1.84 | 14.34 | 3.11 | 37.67 |
| Sesame | | 87.76 | 2.91 | 49.91 | 428.78 | 8.91 | 384.17 | 2.38 | 964.83 |
| Soybean | | 103.50 | ND | 36.05 | 192.76 | 12.07 | 551.96 | 71.20 | 967.54 |
| Commercial Sample | | | | | | | | | |
| S1 | | 140.54 | 45.56 | 9.75 | 628.24 | 35.28 | 104.96 | 5.12 | 969.44 |
| S2 | | 135.90 | 44.55 | 17.57 | 600.69 | 32.53 | 122.65 | 4.59 | 958.48 |
| S3 | | 149.64 | 52.67 | 11.02 | 636.63 | 32.51 | 81.85 | 3.86 | 968.19 |
| S4 | | 96.24 | ND | 36.00 | 176.21 | 11.83 | 529.57 | 68.28 | 918.14 |
| S5 | | 57.99 | 2.87 | 18.56 | 780.24 | 10.99 | 124.76 | 2.36 | 997.76 |
| S6 | | 99.11 | 1.46 | 45.46 | 195.88 | 12.15 | 524.70 | 75.54 | 954.29 |
| S7 | | 70.61 | 1.70 | 19.69 | 555.30 | 7.45 | 263.92 | 6.65 | 925.31 |
| S8 | | 216.42 | 68.96 | 4.18 | 540.09 | 39.63 | 100.33 | 7.17 | 976.78 |
| S9 | | 103.67 | 1.39 | 37.24 | 189.92 | 12.12 | 539.78 | 73.01 | 957.13 |
| S10 | | 31.11 | 2.36 | 12.81 | 584.51 | 24.07 | 157.89 | 58.09 | 870.84 |
| S11 | | 69.93 | 2.80 | 16.78 | 763.75 | 11.01 | 125.40 | 2.46 | 992.14 |
| S12 | | 176.64 | 86.07 | 2.61 | 492.78 | 46.63 | 86.01 | 5.51 | 896.25 |
| S13 | | 140.03 | 60.34 | 1.39 | 472.40 | 54.00 | 126.68 | 7.34 | 862.18 |
| S14 | | 75.76 | 16.95 | 18.62 | 659.98 | 20.40 | 100.36 | 3.29 | 895.36 |
| S15 | | 30.87 | 2.45 | 13.00 | 570.62 | 23.34 | 155.03 | 58.48 | 853.80 |
| S16 | | 101.52 | ND | 38.11 | 177.06 | 11.93 | 523.32 | 77.03 | 928.98 |
| S17 | | 128.68 | 22.63 | 28.12 | 296.92 | 24.61 | 381.71 | 51.09 | 933.77 |
| S18 | | 59.78 | 3.24 | 18.18 | 758.07 | 10.49 | 123.22 | 2.86 | 975.85 |
| S19 | | 144.53 | 31.56 | 19.40 | 621.16 | 24.22 | 77.97 | 6.96 | 925.79 |

*: The formula is expressed as CN (carbon number): DB (double bond) with the position of double bond.

To verify the precision and accuracy of the developed UHPLC/ESI-MS method, the composition data of TG and FA obtained from UHPLC and GC were compared to the authenticated avocado peel and pulp oils. The concentration of individual FA was calculated for all the identified TGs. Oleic acid and vaccenic acid are structurally similar and only differed in double bond positions and these two FAs could not be separated with UHPLC/MS. Thus, they were combined when compared with the TG data. The comparison of TG and FA data is summarized in Table 5. The measurement of TGs by the UHPLC/MS method might be affected by several factors, such as: i) one FA may be distributed among many different combinations in TGs, resulting in a TG concentration below the detection limit [31]; ii) the trace amount of mono/diacylglycerols were not calculated in the TGs method; and iii) the coelution of TGs may complicate the identification of trace FAs. All these factors may result in making the total compositions of TGs slightly lower than FAs. Taking all the factors into account, the data from LC and GC methods were consistent and within an acceptable experimental error range.

### *2.4. Identification of Adulteration Using Chemometric Method*

To further identify possible adulteration, both UHPLC/MS and GC/MS data were subjected to multivariate data reduction chemometric analysis consisting of the principal component analysis (PCA). The PCA score plots are shown in Figure 7A,B for the UHPLC/MS and GC/MS data, respectively. Distinctive groups were observed in both techniques. In each plot, the avocado peel and pulp oils were grouped together, whereas the avocado seed oil was much further away from the peel and pulp oil (Figure 7A,B). None of the samples claimed as avocado seed oils (S4, S9, S10 and S19) were clustered with the seed oil. Instead, S10 and S19 were grouped with pulp and peel oils, and S4 and S9 were grouped closely with soybean oil. S6 and S16 were labeled as avocado oil, but PCA indicated soybean oil as a possible adulterant. S17 might be a mixture of avocado oil with soybean or sesame oil. The adulteration identified by PCA further confirmed the results from both UHPLC/MS and GC/MS analyses.

**Table 5.** Comparison of concentrations (weight %) of individual FA calculated from UHPLC/MS and GC/MS.

| Fatty Acid | Avocado Peel | | Avocado Pulp | |
|---|---|---|---|---|
| | LC/MS | GC/MS | LC/MS | GC/MS |
| Palmitic C16:0 | 16.31 | 15.42 | 15.35 | 15.59 |
| Palmitoleic C16:1 | 4.81 | 5.65 | 5.13 | 7.57 |
| Stearic C18:0 | 0.36 | 0.27 | 0.21 | 0.15 |
| Oleic & Vaccenic C18:1 | 68.15 | 69.21 | 66.98 | 66.21 |
| Linoleic C18:2 | 10.21 | 8.35 | 12.11 | 10.07 |
| Linolenic C18:3 | 0.16 | 1.10 | 0.22 | 0.63 |

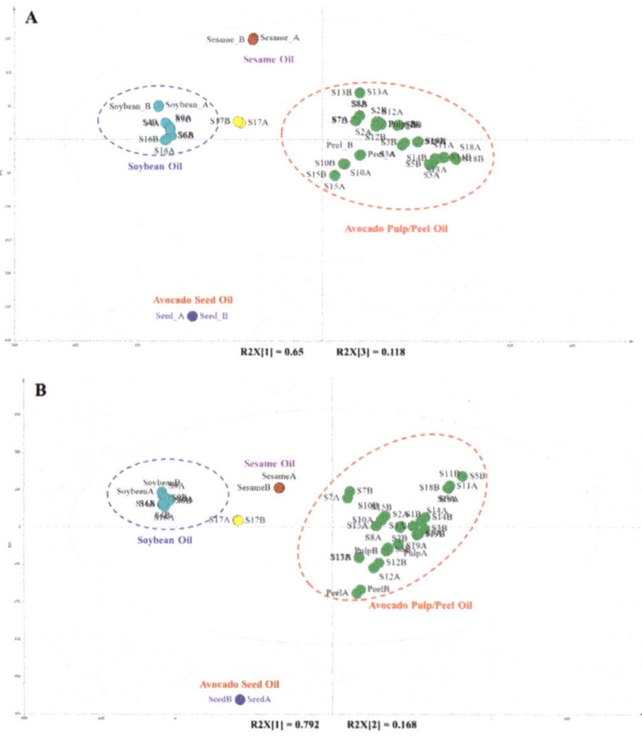


**Figure 7.** PCA score plots of authenticated and commercial avocado oils, sesame oil and soybean oil. (**A**) UHPLC/MS TG data and (**B**) GC/MS FA data. Each sample was prepared in duplicate.

## 3. Materials and Methods

### 3.1. Chemicals and Materials

*n*-Hexane (GC grade), 2-propanol (Optima LC/MS grade) and ammonium formate (HPLC grade) were purchased from Sigma-Aldrich (St. Louis, MO, USA). Formic acid obtained from Honeywell (Waltham, MA, USA) was of HPLC grade. Water was obtained from a Milli-Q system (Millipore, Burlington, MA, USA). The reference standards of soybean oil and sesame oil (Analytical grade) were also obtained from Sigma-Aldrich.

TG standards: OOO was purchased from Nu-Chek-Prep, Inc. (Elysian, MN, USA). LLO, LLP, OLO, PLO, OOP, PPO, $PPO_O$, LOL, OPL, OPO and POL (Table 2) were purchased from Larodan (Monroe, MI, USA). The purities of all the TGs standards were >99.0% as per the label and further confirmed by peak area normalization with LC/MS analysis.

FAMEs standards: the methyl esters of undecanoic acid, palmitic acid, palmitoleic acid, steric acid, oleic acid, vaccenic acid, linoleic acid and linolenic acid were obtained from Nu-Chek-Prep, Inc. The purities of all the FAMEs standards were >99.0% by the label and further confirmed by peak area normalization with GC/MS analysis. The FAMEs mixture ($C_4$–$C_{24}$) consisting of 36 compounds with the positional DB and cis/trans configuration isomers was purchased from Sigma-Aldrich and used for further compound identification.

Mature fresh fruits of avocado (*P. americana*) were purchased from different local grocery stores located in Oxford, MS, USA. All the fruits were selected manually with good morphological integrity. The authenticity of the avocado fruits was confirmed by Dr. John Sabestian, a taxonomist at the National Center of Natural Products Research (NCNPR), University of Mississippi. After washing and drying at room temperature, the peel, pulp and seed were separated and cut into small pieces. The isolated parts were freeze-dried for 24 h until constant weights were obtained. The dehydrated samples were then kept in sealed containers and stored at –20 °C to avoid any possible degradation and content loss.

Nineteen avocado oil products claimed to contain avocado pulp or seed oil were purchased from different grocery stores in the US or via various online commercial vendors (Table S1). Each of these 19 commercial samples, along with the authenticated avocado fruits, was assigned a unique identification code, and representative voucher samples were deposited in the Botanical Repository of NCNPR at the University of Mississippi.

### 3.2. Sample Preparation

Two extraction methods, viz. solvent extraction and SFE, were performed and evaluated. The solvent extraction was conducted by following the method described in AOAC 920.39 [28]. Five grams of each avocado peel, pulp or seed (dried powder) was extracted using a Soxhlet apparatus with 200 mL n-hexane at 70 °C for 4 h, and the procedure was repeated once. For the SFE, the same amount of each sample was loaded into the extraction vessel and mixed with glass beads. The extraction parameters, such as $CO_2$ flow rate, extraction time, pressure and temperature as well as co-solvent were optimized to obtain the highest oil yields possible. Finally, the $CO_2$ flow rate of 10 mL/min, 250 bar, 50 °C and 40 min were adopted. For both extraction methods, the oil yields were calculated as the percentage of oil obtained based on the weight of the sample used. The solvent extraction was used as a reference method for the comparison of oil yields obtained from SFE method.

### 3.3. Determination of TGs Using UHPLC/ESI-MS

Stock solutions of LLO, LLP, OLO, PLO, OOP, PPO, PPoO, OOO, LOL, OPL, OPO and POL at the concentration of 5 mg/mL were prepared in *n*-hexane:isopropanol (1:1, *v*/*v*). These solutions were diluted with the same solvent mixture yielding 12 calibration working solutions between 1–400 μg/mL. All the oil samples were prepared in 1 mg/mL and 500 μg/mL prior to chromatographic analysis.

The chromatographic system consisted of an Agilent 1290 Infinity series UHPLC with a diode array detector, binary pump, auto-liquid sampler and thermostated column compartment (Santa Clara, CA, USA). The UHPLC conditions were: three ACQUITY UPLC BEH $C_{18}$ columns (3.0 ×100 mm, 1.7 µm, Waters, Milford, MA, USA) were connected in series with the column temperature set to 30 °C. The eluents consisted of acetonitrile with 0.1% water (A), isopropanol with 5 mM ammonium formate (B), and both contained 0.05% formic acid. The gradient elution started at 20% B, programmed to 50% B in 45 min, and then 60% B in 35min. The flow rate was 0.3 mL/min, and 1 µL of each solution was injected in duplicate for the LC analysis.

The UHPLC instrument was coupled to an Agilent 6120 quadrupole mass spectrometer with a dual ESI and APCI interface. Both ESI and APCI in positive and negative modes were evaluated, and ESI(+) in full scan mode from 300–1000 amu was selected for the analysis of TGs. The fragmentor voltage was optimized to 140 V. The drying gas flow was 12 L/min and the nebulizer pressure was 35 psi. The drying gas temperature and vaporizer temperature were set to 325 °C and 250 °C, respectively. The capillary voltage was 4000 V and the corona current was 4.0 µA.

### 3.4. *Determination of FAs Using GC/MS*

Stock solutions of palmitic acid, palmitoleric acid, stearic acid, oleic acid, vaccenic acid, linoleic acid and linolenic acid in the form of methyl ester were prepared in *n*-hexane to make the 10 mg/mL stock solutions. These solutions were diluted with the same solvent yielding 12 calibration working solutions between 5–1000 µg/mL. Undecanoic acid methyl ester was used as the internal standard and added to each calibration solution at a fixed concentration of 200 µg/mL.

A modified procedure based on the AOAC 996.06 method [28] and the method proposed by Ai, et al. [38] for the hydrolytic reaction and transesterification of TGs was employed. In brief, 70 µL of each oil sample was accurately weighed (57.2–67.7 mg) and mixed with 2 mL glyceryl triundecanoate (2.5 mg/mL) and 200 µL 2N potassium hydroxide, both in methanol. The resulting cloudy solution was vortexed for 2 min, and then sonicated for 45 min at 60 °C in a water bath until the solution became clear. Then, 2 mL *n*-hexane was added to the solution. After sonication for 30 min and centrifugation, the supernatants were taken and diluted (2, 4 and 10 times) to obtain three different concentrations for GC/MS analysis.

The FAMEs analysis was performed on an Agilent 7890B gas chromatography (Santa Clara, CA, USA) coupled with an Agilent 5977A mass spectrometer. An Agilent J&W HP-88 column (60 m × 0.25 mm × 0.20 µm) was used. Helium was used as the carrier gas at a constant flow rate of 1.2 mL/min. The oven temperature was first set at a 60 °C hold for 1 min, and the temperature was subsequently increased to 145 °C at a rate of 10 °C/min, then to 190 °C at a rate of 1 °C/min, and finally to 240 °C at a rate of 5 °C/min. The inlet temperature was 260 °C. The split ratio was 100:1 with 1 µL injection. Full scan data was acquired in the mass range of *m/z* 30–500 amu and the EI voltage at 70 V. The temperature of the transfer line was 260 °C. The temperatures of ion source and quadrupole were set to 230 °C and 150 °C, respectively.

### 3.5. *Statistical Analysis*

The raw data for both UHPLC/MS and GC/MS were pre-processed using MassHunter Profinder (version 8.0, Agilent Technologies, Santa Clara, CA, USA) for finding features. The extracted features were exported as a cef file and then imported to Mass Profiler Professional software package (version B.12.05, Agilent Technologies) and SIMCA-P software (Version 12.0, Umetrics, Umeå, Sweden) where the features were further aligned, normalized and statistically evaluated.

## 4. Conclusions

The current study endeavors to establish the comprehensive profiles and quality standards of avocado oil. The "solvent-free" SFE method used for avocado oil extraction is highly recommended in the food industry. Two independent and complementary analytical methods (LC and GC) were used to investigate different classes of compounds (TG and FA) in avocado oils. Characterization and quantitative analysis of 13 TGs in oils extracted from different parts of avocado fruit, viz. peel, pulp and seed, were conducted using UHPLC/ESI-MS. The complex TGs can be conclusively identified using the MS detection with the correct attribution of the acyl in the sn-2 position, as well as the calculation of ISYs for ions formed in the IS-CID. The efficiency and accuracy of the quantification were improved by the selection of one single reference standard from the same ECN group when reference standards are commercially unavailable. The FA compositions yielded from GC/MS method were in good agreement with UHPLC/MS. Although chemical profiles of avocado pulp and peel were very similar, a significant difference was observed for the avocado seed. It is concluded that the combination of TG and FA analysis using UHPLC/ESI-MS and GC/MS, as well as multivariate statistical analysis may provide comprehensive information for the characterization, standardization and authentication of avocado oils. The reported techniques might be useful for assessing the quality of other plant oils.

**Supplementary Materials:** The following are available online: Table S1: Information of commercial avocado (*P. americana*) oils.

**Author Contributions:** M.W. and P.Y.: chemical analysis, data analysis and manuscript preparation; A.G.C.: study design and manuscript preparation; D.C.: sample extraction and measurement of physicochemical properties; J.Z.: statistical analysis and manuscript editing; B.A. and Y.-H.W.: data interpretation and manuscript editing; I.A.K.: manuscript final version approval and guarantee the integrity of the entire study. All authors have read and agreed to the published version of the manuscript.

**Funding:** This research is supported in part by "Science Based Authentication of Dietary Supplements" funded by the Food and Drug Administration grant number 1U01FD004246-05, and the United States Department of Agriculture, Agricultural Research Service, Specific Cooperative Agreement No. 58-6060-6-015.

**Acknowledgments:** We would like to thank Jon F. Parcher and Iffat Parveen for manuscript editing, proof-reading and valuable suggestions.

**Conflicts of Interest:** The authors declare no conflict of interest.

## References

1. Costagli, G.; Betti, M. Avocado oil extraction processes: Method for cold-pressed high-quality edible oil production versus traditional production. *J. Agric. Eng.* **2015**, *46*, 115–122. [CrossRef]
2. Dabas, D.; Shegog, R.M.; Ziegler, G.R.; Lambert, J.D. Avocado (*Persea americana*) seed as a source of bioactive phytochemicals. *Curr. Pharm. Des.* **2013**, *19*, 6133–6140. [CrossRef] [PubMed]
3. dos Santos, M.A.Z.; Alicieo, T.V.R.; Pereira, C.M.P.; Ramis-Ramos, G.; Mendonca, C.R.B. Profile of bioactive compounds in avocado pulp oil: Influence of the drying processes and extraction methods. *J. Am. Oil Chem. Soc.* **2014**, *91*, 19–27. [CrossRef]
4. Barros, H.D.F.Q.; Coutinho, J.P.; Grimaldi, R.; Godoy, H.T.; Cabral, F.A. Simultaneous extraction of edible oil from avocado and capsanthin from red bell pepper using supercritical carbon dioxide as solvent. *J. Supercrit. Fluids* **2016**, *107*, 315–320. [CrossRef]
5. Woolf, A.; Wong, M.; Eyres, L.; McGhie, T.K. *Avocado Oil*; AOCS Press: Urbana, IL, USA, 2008.
6. Qin, X.; Zhong, J. A review of extraction techniques for avocado oil. *J. Oleo Sci.* **2016**, *65*, 881–888. [CrossRef] [PubMed]
7. Hurtado-Fernández, E.; Fernández-Gutiérrez, A.; Carrasco-Pancorbo, A. Avocado fruit—*Persea americana*. In *Exotic Fruits*; Rodrigues, S., de Oliveira Silva, E., de Brito, E.S., Eds.; Academic Press: London, UK, 2018.
8. Tan, C.X.; Chong, G.H.; Hamzah, H.; Ghazali, H.M. Effect of virgin avocado oil on diet-induced hypercholesterolemia in rats via $^1$H NMR-based metabolomics approach. *Phytother. Res.* **2018**, *32*, 2264–2274. [CrossRef] [PubMed]
9. Tan, C.X.; Chong, G.H.; Hamzah, H.; Ghazali, H.M. Hypocholesterolaemic and hepatoprotective effects of virgin avocado oil in diet-induced hypercholesterolaemia rats. *Int. J. Food Sci. Technol.* **2018**, *53*, 2706–2713. [CrossRef]
10. Marquez-Ramirez, C.A.; Hernandez de la Paz, J.L.; Ortiz-Avila, O.; Raya-Farias, A.; Gonzalez-Hernandez, J.C.; Rodriguez-Orozco, A.R.; Salgado-Garciglia, R.; Saavedra-Molina, A.; Godinez-Hernandez, D.; Cortes-Rojo, C. Comparative effects of avocado oil and losartan on blood pressure, renal vascular function, and mitochondrial oxidative stress in hypertensive rats. *Nutrition* **2018**, *54*, 60–67. [CrossRef]

11. Ortiz-Avila, O.; Gallegos-Corona, M.A.; Sanchez-Briones, L.A.; Calderon-Cortes, E.; Montoya-Perez, R.; Rodriguez-Orozco, A.R.; Campos-Garcia, J.; Saavedra-Molina, A.; Mejia-Zepeda, R.; Cortes-Rojo, C. Protective effects of dietary avocado oil on impaired electron transport chain function and exacerbated oxidative stress in liver mitochondria from diabetic rats. *J. Bioenerg. Biomembr.* **2015**, *47*, 337–353. [CrossRef]
12. Furlan, C.P.B.; Valle, S.C.; Ostman, E.; Marostica, M.R., Jr.; Tovar, J. Inclusion of Hass avocado-oil improves postprandial metabolic responses to a hypercaloric-hyperlipidic meal in overweight subjects. *J. Funct. Foods* **2017**, *38*, 349–354. [CrossRef]
13. Lu, Q.-Y.; Arteaga, J.R.; Zhang, Q.; Huerta, S.; Go, V.L.W.; Heber, D. Inhibition of prostate cancer cell growth by an avocado extract: Role of lipid-soluble bioactive substances. *J. Nutr. Biochem.* **2005**, *16*, 23–30. [CrossRef] [PubMed]
14. Dreher, M.L.; Davenport, A.J. Hass avocado composition and potential health effects. *Crit. Rev. Food Sci. Nutr.* **2013**, *53*, 738–750. [CrossRef] [PubMed]
15. FAO (Food and Agriculture Organisation). FAO-STAT. 2014. Available online: http://faostat3.fao.org/home/index.html (accessed on 25 January 2020).
16. FAO (ORGANIZAÇÃO DAS UNIDAS PARA ALIMENTAÇÃOE AGRICULTURA). Available online: http://www.fao.org (accessed on 16 January 2020).
17. Quinones-Islas, N.; Meza-Marquez, O.G.; Osorio-Revilla, G.; Gallardo-Velazquez, T. Detection of adulterants in avocado oil by Mid-FTIR spectroscopy and multivariate analysis. *Food Res. Int.* **2013**, *51*, 148–154. [CrossRef]
18. Jimenez-Sotelo, P.; Hernandez-Martinez, M.; Osorio-Revilla, G.; Meza-Marquez, O.G.; Garcia-Ochoa, F.; Gallardo-Velazquez, T. Use of ATR-FTIR spectroscopy coupled with chemometrics for the authentication of avocado oil in ternary mixtures with sunflower and soybean oils. *Food Addit. Contam. Part A* **2016**, *33*, 1105–1115. [CrossRef] [PubMed]
19. Merchak, N.; Rizk, T.; Silvestre, V.; Remaud, G.S.; Bejjani, J.; Akoka, S. Olive oil characterization and classification by $^{13}$C NMR with a polarization transfer technique: A comparison with gas chromatography and 1H NMR. *Food Chem.* **2018**, *245*, 717–723. [CrossRef] [PubMed]
20. Wei, F.; Ji, S.-X.; Hu, N.; Lv, X.; Dong, X.-Y.; Feng, Y.-Q.; Chen, H. Online profiling of triacylglycerols in plant oils by two-dimensional liquid chromatography using a single column coupled with atmospheric pressure chemical ionization mass spectrometry. *J. Chromatogr. A* **2013**, *1312*, 69–79. [CrossRef]
21. Hou, J.-J.; Cao, C.-M.; Xu, Y.-W.; Yao, S.; Cai, L.-Y.; Long, H.-L.; Bi, Q.-R.; Zhen, Y.-Y.; Wu, W.-Y.; Guo, D.-A. Exploring lipid markers of the quality of coix seeds with different geographical origins using supercritical fluid chromatography mass spectrometry and chemometrics. *Phytomedicine* **2018**, *45*, 1–7. [CrossRef]
22. Lisa, M.; Holcapek, M.; Rezanka, T.; Kabatova, N. High-performance liquid chromatography-atmospheric pressure chemical ionization mass spectrometry and gas chromatography-flame ionization detection characterization of Δ5-polyenoic fatty acids in triacylglycerols from conifer seed oils. *J. Chromatogr. A* **2007**, *1146*, 67–77. [CrossRef]
23. Lisa, M.; Netusilova, K.; Franek, L.; Dvorakova, H.; Vrkoslav, V.; Holcapek, M. Characterization of fatty acid and triacylglycerol composition in animal fats using silver-ion and non-aqueous reversed-phase high-performance liquid chromatography/mass spectrometry and gas chromatography/flame ionization detection. *J. Chromatogr. A* **2011**, *1218*, 7499–7510. [CrossRef]
24. Byrdwell, W.C.; Neff, W.E. Dual parallel electrospray ionization and atmospheric pressure chemical ionization mass spectrometry (MS), MS/MS and MS/MS/MS for the analysis of triacylglycerols and triacylglycerol oxidation products. *Rapid Commun. Mass Spectrom.* **2002**, *16*, 300–319. [CrossRef]
25. Yoshida, H.; Tomiyama, Y.; Yoshida, N.; Mizushina, Y. Characteristics of lipid components, fatty acid distributions and triacylglycerol molecular species of adzuki beans (*Vigna angularis*). *Food Chem.* **2009**, *115*, 1424–1429. [CrossRef]
26. Krumreich, F.D.; Borges, C.D.; Mendonca, C.R.B.; Jansen-Alves, C.; Zambiazi, R.C. Bioactive compounds and quality parameters of avocado oil obtained by different processes. *Food Chem.* **2018**, *257*, 376–381. [CrossRef] [PubMed]
27. *Official and Tentative Methods of the American Oils Chemists' Society*; American Oil Chemists' Society: Champaign, IL, USA, 1992.
28. AOAC. *Official Methods of Analysis of AOAC International*, 18th ed.; Horwitz, W., Latimer, G., Eds.; AOAC International: Maryland, MD, USA, 2007.

29. Wei, F.; Hu, N.; Lv, X.; Dong, X.-Y.; Chen, H. Quantitation of triacylglycerols in edible oils by off-line comprehensive two-dimensional liquid chromatography-atmospheric pressure chemical ionization mass spectrometry using a single column. *J. Chromatogr. A* **2015**, *1404*, 60–71. [CrossRef] [PubMed]
30. Lisa, M.; Holcapek, M. Triacylglycerols profiling in plant oils important in food industry, dietetics and cosmetics using high-performance liquid chromatography-atmospheric pressure chemical ionization mass spectrometry. *J. Chromatogr. A* **2008**, *1198–1199*, 115–130. [CrossRef]
31. Holcapek, M.; Lisa, M.; Jandera, P.; Kabatova, N. Quantitation of triacylglycerols in plant oils using HPLC with APCI-MS, evaporative light-scattering, and UV detection. *J. Sep. Sci.* **2005**, *28*, 1315–1333. [CrossRef]
32. Lee, K.W.Y.; Porter, C.J.H.; Boyd, B.J. A simple quantitative approach for the determination of long and medium chain lipids in bio-relevant matrices by high performance liquid chromatography with refractive index detection. *AAPS PharmSciTech* **2013**, *14*, 927–934. [CrossRef]
33. Holcapek, M.; Jandera, P.; Zderadicka, P.; Hruba, L. Characterization of triacylglycerol and diacylglycerol composition of plant oils using high-performance liquid chromatography-atmospheric pressure chemical ionization mass spectrometry. *J. Chromatogr. A* **2003**, *1010*, 195–215. [CrossRef]
34. Parcher, J.F.; Wang, M.; Chittiboyina, A.G.; Khan, I.A. In-source collision-induced dissociation (IS-CID): Applications, issues and structure elucidation with single-stage mass analyzers. *Drug Test. Anal.* **2018**, *10*, 28–36. [CrossRef]
35. Crellin, K.C.; Sible, E.; Van Antwerp, J. Quantification and confirmation of identity of analytes in various matrices with in-source collision-induced dissociation on a single quadrupole mass spectrometer. *Int. J. Mass Spectrom.* **2003**, *222*, 281–311. [CrossRef]
36. Hou, J.-J.; Guo, J.-L.; Cao, C.-M.; Yao, S.; Long, H.-L.; Cai, L.-Y.; Da, J.; Wu, W.-Y.; Guo, D.-A. Green quantification strategy combined with chemometric analysis for triglycerides in seeds used in traditional chinese medicine. *Planta Med.* **2018**, *84*, 457–464. [CrossRef]
37. Yang, Y.; Ferro, M.D.; Cavaco, I.; Liang, Y. Detection and Identification of Extra Virgin Olive Oil Adulteration by GC-MS Combined with Chemometrics. *J. Agric. Food Chem.* **2013**, *61*, 3693–3702. [CrossRef] [PubMed]
38. Ai, F.-F.; Bin, J.; Zhang, Z.-M.; Huang, J.-H.; Wang, J.-B.; Liang, Y.-Z.; Yu, L.; Yang, Z.-Y. Application of random forests to select premium quality vegetable oils by their fatty acid composition. *Food Chem.* **2014**, *143*, 472–478. [CrossRef] [PubMed]


**Sample Availability:** Samples of the compounds are available from the authors.

*Article*

# Co-Loaded Curcumin and Methotrexate Nanocapsules Enhance Cytotoxicity against Non-Small-Cell Lung Cancer Cells

**Loanda Aparecida Cabral Rudnik [1], Paulo Vitor Farago [1,2], Jane Manfron Budel [1,*], Amanda Lyra [1], Fernanda Malaquias Barboza [1], Traudi Klein [1], Carla Cristine Kanunfre [3], Jessica Mendes Nadal [1], Matheus Coelho Bandéca [4], Vijayasankar Raman [5], Andressa Novatski [1], Alessandro Dourado Loguércio [1] and Sandra Maria Warumby Zanin [2]**

1. Postgraduate Program in Pharmaceutical Sciences, Department of Pharmaceutical Sciences, State University of Ponta Grossa, 84030-900 Ponta Grossa, Brazil; loandacabral@hotmail.com (L.A.C.R.); pvfarago@gmail.com (P.V.F.); amandinhamlyra@gmail.com (A.L.); fer_barboza@hotmail.com (F.M.B.); traudiklein@gmail.com (T.K.); jessicabem@hotmail.com (J.M.N.); anovatski2@gmail.com (A.N.); aloguercio@hotmail.com (A.D.L.)
2. Postgraduate Program in Pharmaceutical Sciences, Department of Pharmacy, Federal University of Paraná, 81020-430 Curitiba, Brazil; sandrazanin@ufpr.br
3. Postgraduate Program in Biomedical Science, Department of General Biology, State University of Ponta Grossa, 84030-900 Ponta Grossa, Brazil; cckanunfre@gmail.com
4. Postgraduate Program in Dentistry, Ceuma University, 65065-470 São Luís, Brazil; mbandeca@gmail.com
5. National Center for Natural Products Research, School of Pharmacy, University of Mississippi, University, MS 38677, USA; vraman@olemiss.edu

* Correspondence: janemanfron@hotmail.com; Tel.: +55-42-3220-3124

Academic Editor: Maria Carla Marcotullio
Received: 6 March 2020; Accepted: 17 April 2020; Published: 21 April 2020

**Abstract:** *Background*: As part of the efforts to find natural alternatives for cancer treatment and to overcome the barriers of cellular resistance to chemotherapeutic agents, polymeric nanocapsules containing curcumin and/or methotrexate were prepared by an interfacial deposition of preformed polymer method. *Methods*: Physicochemical properties, drug release experiments and in vitro cytotoxicity of these nanocapsules were performed against the Calu-3 lung cancer cell line. *Results*: The colloidal suspensions of nanocapsules showed suitable size (287 to 325 nm), negative charge (−33 to −41 mV) and high encapsulation efficiency (82.4 to 99.4%). Spherical particles at nanoscale dimensions were observed by scanning electron microscopy. X-ray diffraction analysis indicated that nanocapsules exhibited a non-crystalline pattern with a remarkable decrease of crystalline peaks of the raw materials. Fourier-transform infrared spectra demonstrated no chemical bond between the drug(s) and polymers. Drug release experiments evidenced a controlled release pattern with no burst effect for nanocapsules containing curcumin and/or methotrexate. The nanoformulation containing curcumin and methotrexate (NCUR/MTX-2) statistically decreased the cell viability of Calu-3. The fluorescence and morphological analyses presented a predominance of early apoptosis and late apoptosis as the main death mechanisms for Calu-3. *Conclusions*: Curcumin and methotrexate co-loaded nanocapsules can be further used as a novel therapeutic strategy for treating non-small-cell lung cancer.

**Keywords:** Calu-3 cell line; cancer chemotherapy; drug resistance; poly(ε-caprolactone); poly(ethylene glycol)

---

## 1. Introduction

Lung cancer is the deadliest of all cancers in the world. Sudden changes in lifestyle, environmental pollution, and smoking are strongly related to the development of lung cancer [1,2]. Treatment for lung cancer is typically guided by stage, although individual factors, such as overall health and coexisting medical conditions, are also important. Chemotherapy is beneficial at most stages of the disease, although it and radiation therapy are curative in only a minority of patients. The failure of chemotherapy in lung cancer treatment is mainly due to resistance mechanisms against chemotherapeutic agents, which result in a lack of therapeutic response [3,4]. Resistance phenomenon can occur by efflux pumps as P-glycoprotein (P-gp), which is expressed in human cells and acts as a localized drug transport mechanism, actively exporting drugs out of the cell [3,5,6]. In that sense, the efflux mechanism shows great clinical importance on lung cancer treatment and several compounds with P-gp inhibitor properties have been studied [7] to modulate chemotherapy resistance and to provide a more pronounced effect to chemotherapeutic drugs.

Curcumin (CUR) is a polyphenol obtained mainly from turmeric, the rhizome of *Curcuma longa* L. (Zingiberaceae), which is widely used in cooking as a spice and color additive due to its characteristic taste and deep yellow coloration [8,9]. CUR has demonstrated a variety of biological properties, such as antioxidant, anti-inflammatory, antimicrobial, and antitumor properties [8,10,11]. CUR has also shown therapeutic feasibility in improving wound healing activities by the use of nanotechnology-based delivery systems [11]. Moreover, CUR is a non-competitive inhibitor of P-gp by blocking the ATP hydrolysis process in efflux pump, which also paves the way for its use against lung cancer cells [7,12]. Despite these interesting effects, CUR has some disadvantages, such as its low aqueous solubility, low photostability, limited absorption, low bioavailability, rapid metabolism and elimination [13].

Methotrexate (MTX) is a well-known cytotoxic agent, which competitively and irreversibly inhibits dihydrofolate reductase, an enzyme that participates in tetrahydrofolate synthesis [14,15]. It is currently used in lung cancer treatment alone or along with other chemotherapeutic agents by oral and parenteral routes. High doses of MTX may circumvent at least two known mechanisms of resistance to this drug, membrane transport and high levels of the target enzyme. However, MTX can cause substantial toxicity by its systemic administration and can lead to remarkable side effects, such as hepatotoxicity, bone marrow depression, leucopenia, among others [16,17]. In that sense, the traditional use of this drug has two main problems when used against lung cancer: (a) its efflux from cancer cells by P-gp and (b) its high toxicity at usual therapeutic doses [18,19]. Thereby, new strategies to overcome these limitations are required.

Some controlled release strategies have been reported in the literature concerning the co-delivery of CUR and MTX. Dey et al. [20] developed gold nanoparticles containing CUR and MTX and evaluated their cytotoxic effect on C6 glioma cells and MCF-7 breast cancer cells. Curcio et al. (2018) [21] obtained pH-responsive polymersomes by self-assembling of a carboxyl-terminated PEG amphiphile achieved via esterification of PEG diacid with PEG40stearate. A highly hemocompatible co-delivery system of CUR and MTX was obtained. Vakilinezhad et al. [22] prepared PLGA nanoparticles for the co-administration of MTX and CUR as a potential breast cancer therapeutic system. Even though these authors reported a burst release from such nanoparticles, higher cytotoxicity was demonstrated against SK-Br-3 breast adenocarcinoma cell line. Curcio et al. [23] effectively delivered MTX to breast cancer cells by the use of a nanocarrier system derived from the self-assembly of a dextran-CUR conjugate prepared via enzyme chemistry with immobilized laccase acting as a solid biocatalyst. However, to the best of our knowledge, no previous paper was devoted to the preparation of co-loaded CUR and MTX nanocapsules using poly(ε-caprolactone) (PCL) as biodegradable polymer wall and poly(ethylene glycol) (PEG) as coating polymer focused on treating lung cancer.

Polymeric nanocapsules (NCs) are attractive colloidal systems to develop formulations containing labile and toxic substances. By definition, NCs are vesicular systems composed of a core, generally oily, surrounded by a polymer wall [24]. These carriers can present several advantages such as enhancing the dissolution process, increasing the therapeutic index, providing controlled delivery and achieving protection from the photo and chemical degradation [25]. Therefore, NCs can circumvent limitations provided by both CUR and MTX since they allow drug protection against degradation, improve bioavailability, and reduce possible side effects. In particular, NCs can reach target tissues, certain affected organs, and tumors due to their superior features as small particle size, large surface area, Brownian motion, and surface functionality which could provide higher cytotoxic effect even at low doses of the chemotherapeutic agents [11,13]. Moreover, the two-drug combination into NCs can produce a higher objective response since CUR can potentiate MTX activity by delaying its efflux from the lung cancer cells.

Taking all these factors into consideration, this study was devoted to developing NCs for co-administration of CUR and MTX to provide a controlled release and a synergistic cytotoxic effect on non-small-cell lung cancer cell (Calu-3) growth. Moreover, in vitro studies were performed to evaluate the cytotoxic mechanism of these co-loaded NCs against Calu-3 cells.

## 2. Results and Discussion

### *2.1. Preparation and Characterization of Polymeric Nanocapsules (NCs) Containing Curcumin (CUR) and/or Methotrexate (MTX)*

Nanocapsules with or without CUR and/or MTX were successfully obtained by the interfacial deposition of the preformed polymer method. Formulations containing no CUR showed a liquid aspect with a slightly bluish-white opalescent coloring. However, CUR-loaded nanocapsules presented a liquid aspect and an intense yellow color.

#### 2.1.1. Determination of Mean Diameter, Polydispersity Index, and Zeta Potential

Results of particle size, polydispersity, and zeta potential are summarized in Table 1. CUR and/ or MTX-loaded and non-loaded NCs revealed mean sizes between 287.83 and 325.16 nm with a polydispersity index (PDI) varying from 0.290 to 0.351, which represent a certain system homogeneity.

**Table 1.** Mean diameter, polydispersity index (PDI), zeta potential, and encapsulation efficiency (EE) of the obtained CUR and/or MTX-loaded and non-loaded nanocapsules.

| Formulation | Mean Particle Size * (nm) | PDI * | Zeta Potential * (mV) | EE * (%) |
|---|---|---|---|---|
| NCUR/MTX-0 | 313.30 ± 48.20 | 0.351 ± 0.15 | −39.83 ± 2.46 | - |
| NCUR-1 | 307.26 ± 29.40 | 0.323 ± 0.07 | −38.80 ± 7.45 | 99.4 ± 0.51 |
| NCUR-2 | 315.96 ± 10.30 | 0.351 ± 0.02 | −41.20 ± 7.20 | 98.7 ± 0.66 |
| NMTX-1 | 287.83 ± 28.00 | 0.290 ± 0.11 | −40.60 ± 5.80 | 88.9 ± 0.47 |
| NMTX-2 | 298.60 ± 25.00 | 0.292 ± 0.01 | −38.96 ± 1.60 | 83.3 ± 0.42 |
| NCUR/MTX-1 | 312.03 ± 23.60 | 0.344 ± 0.05 | −39.43 ± 6.26 | 99.1 ± 0.63 (CUR)<br>84.4 ± 0.57 (MTX) |
| NCUR/MTX-2 | 325.16 ± 28.90 | 0.351 ± 0.03 | −33.40 ± 3.29 | 99.3 ± 0.54 (CUR)<br>82.4 ± 0.44 (MTX) |

* Values are depicted as mean ± standard deviation (SD).

In general, NCs obtained by the interfacial deposition of preformed polymer method have demonstrated mean diameters between 200 and 300 nm and PDI between 0.2 and 0.3, particularly for the works carried out using poly(lactic-co-glycolic acid) and PCL [26]. However, NCs of larger diameter may be related to the presence of PEG in their composition. PEG chains create a more viscous organic phase, which affects its dispersion into the aqueous phase during stirring and leads to higher particle sizes with broader polydispersity [27].

Also, other aspects may directly influence the particle diameter in nanosystems. One of them is the oil used for preparing nanocapsules that influences some of the core structure properties, such as viscosity, hydrophobicity, and surface tension [28]. PDI values close to 0 are considered monodisperse and greater than 0.5 indicate heterogeneous dispersion [29]. Taking all these into account, suitable nanometric-scaled size and adequate polydispersity were recorded for NCs with or without CUR and/or MTX.

Negative zeta potential values were observed for all the colloidal suspensions of nanocapsules (Table 1). Zeta potential analysis allows identifying the electrical charges that occur on the surface of the nanoparticles. Particles are regarded as stable when their zeta potential values are higher than ± 30 mV [30]. Moreover, negative values were achieved for NCs on account of the anionic nature of PCL due to the presence of carboxylic acid functional groups [28]. The statistical analysis showed that mean diameter, PDI and zeta potential were similar for the formulations with or without CUR and/or MTX ($p > 0.05$).

### 2.1.2. Encapsulation Efficiency

The encapsulation efficiency (EE) of CUR and/or MTX-loaded NCs are also depicted in Table 1. Mean EE values higher than 98.7% were obtained for formulations containing CUR alone or in combination with MTX. These values are attributed to the poor aqueous solubility of CUR (3.12 $\mu g.mL^{-1}$ at 25 °C) [31], which avoided its partitioning in the aqueous phase. For those nanocapsules containing MTX, mean values varying from 82.4% (NCUR/MTX-2) to 88.9% (NMTX-1). These values are higher than 70%, hence suitable for nanocapsules containing lipophilic compounds as MTX (aqueous solubility of 2.60 $mg{\cdot}mL^{-1}$ at 20 °C) [32] when emulsion–diffusion methods were used [33]. These EE values were further used for obtaining final concentrations in $\mu mol{\cdot}L^{-1}$ during in vitro cell culture-based assays. For formulations containing both CUR and MTX, the EE compensation was performed using MTX due to its lower content in co-loaded NCs.

### 2.1.3. Field Emission Scanning Electron Microscopy (FESEM)

Nanoscale dimensions were registered for CUR and/or MTX-loaded and non-loaded NCs when their images were assessed by FESEM (Figure 1). NCs were spherically-shaped and had a smooth surface. Their particle size and PDI were similar to those previously recorded [25] by photon correlation spectroscopy. Also, it was found that even after changing CUR and/or MTX concentrations, their morphologies were similar and no drug crystals were seen on their surfaces.

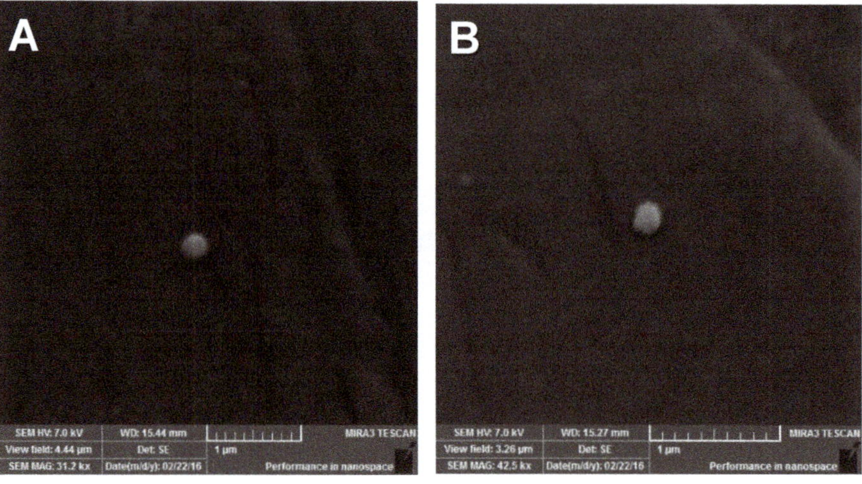


**Figure 1.** Photomicrographs of non-loaded and CUR and MTX-loaded NCs observed by FESEM: (**A**) NCUR/MTX-0 (31,200× magnification) and (**B**) NCUR/MTX-2 (42,500× magnification).

2.1.4. X-ray Diffraction (XRD)

The diffractograms obtained for pure drugs (CUR and MTX), polymers (PCL and PEG 6000), and nanocapsules NCUR-1, NCUR-2, NMTX-1, NMTX-2, NCUR/MTX-0, NCUR/MTX-1, and NCUR/MTX-2 are shown in Figure 2. CUR presented typical peaks at 2θ values of 8.85°, 14.25°, 17.04°, and 23.10° as previously reported [34,35], which confirmed its crystalline nature. MTX showed main crystalline peaks at 9.16°, 12.8°, 19.42°, and 26.74° as previously described [36]. Also, PCL revealed two crystalline peaks at 21.59° and 23.72°. XRD data for PEG 6000 showed 2θ values at 19.17° and 23.41°.

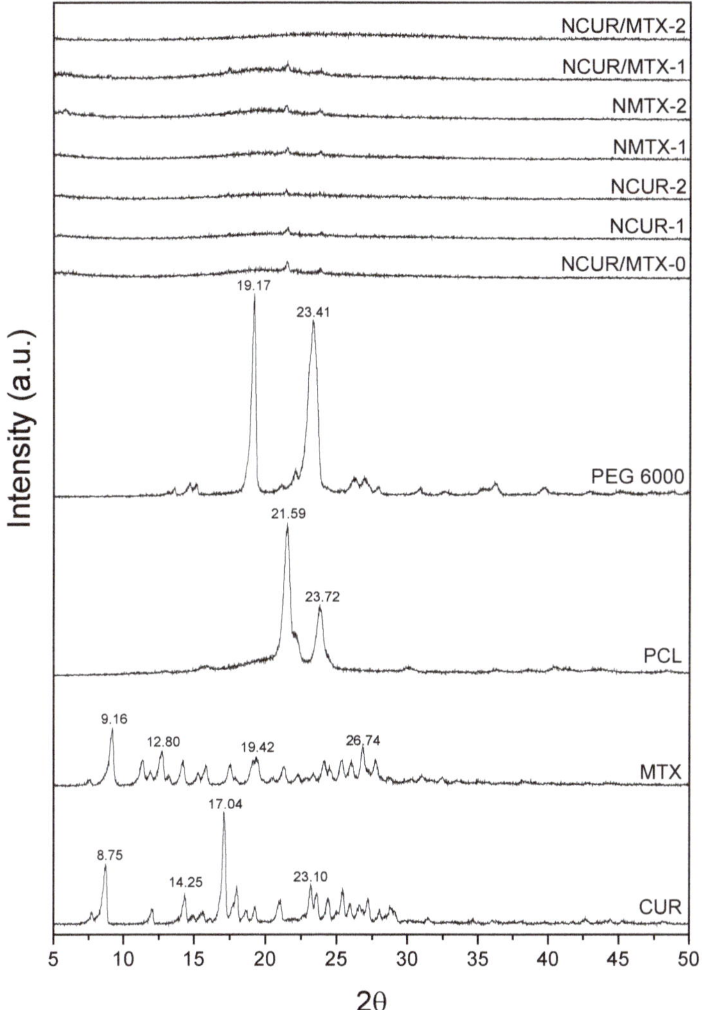


**Figure 2.** Diffractograms of CUR, MTX, PCL, PEG 6000, and CUR and/or MTX-loaded and non-loaded NCs obtained by XRD analysis.

CUR and/or MTX-loaded and non-loaded NCs exhibited similar non-crystalline pattern with a remarkable decrease of crystalline peaks from the drug(s) and polymers. In that sense, drug amorphization was demonstrated for all CUR and/or MTX-loaded formulations. This behavior is related to the molecular dispersion of such drugs when NCs were obtained using the interfacial deposition of preformed polymer method. Solid-state substances may have crystalline and/or amorphous characteristics. In general, the amorphous solids are more soluble than those in the crystalline state due to the free energies present in the dissolution process. In the amorphous state, molecules are randomly arranged and, therefore, lower energy is required to separate them, resulting in a faster dissolution when compared to the crystalline form [37]. Therefore, the amorphous pattern of CUR and/or MTX into NCs is considered to be advantageous because it allows the drug(s) dissolution and leads to drug(s) diffusion through the polymer wall, resulting in a sustained release of the encapsulated drug(s) [34,38].

### 2.1.5. Fourier-Transform Infrared Spectroscopy (FTIR)

The FTIR spectra recorded for CUR, MTX, PCL, PEG 6000, physical mixture, and CUR and/or MTX-loaded and non-loaded NCs are depicted in Figure 3.

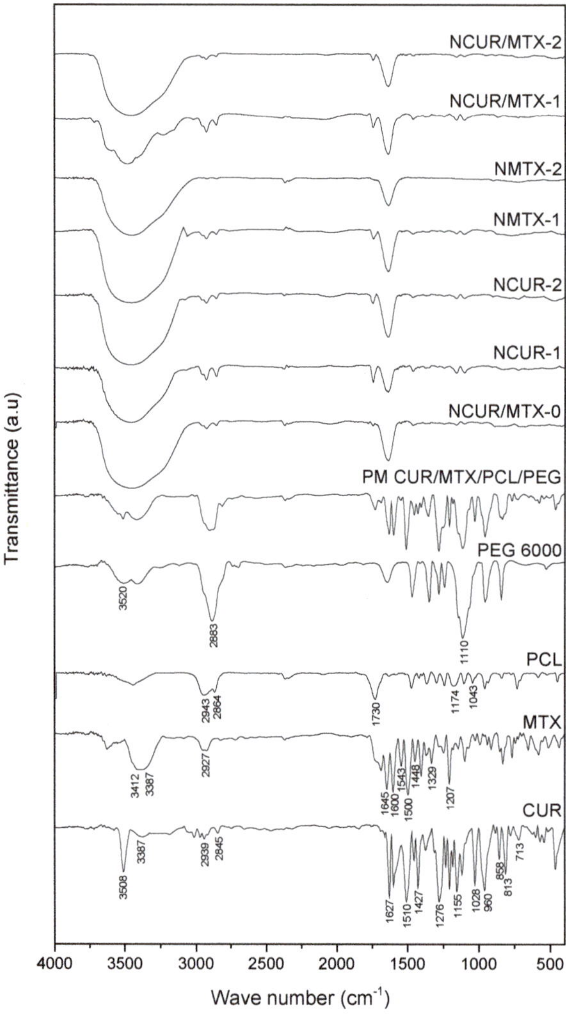


**Figure 3.** FTIR spectra of CUR, MTX, PCL, PEG 6000, physical mixture, and CUR and/or MTX-loaded and non-loaded NCs.

The FTIR spectrum of pure CUR presented the typical bands previously reported [35,39,40]. A sharp band at 3508 cm$^{-1}$ and a broad band at 3387 cm$^{-1}$ were assigned to the stretching of –OH group. Stretching vibrations of the carbonyl group and C–C bond of benzene ring were observed at 1627 cm$^{-1}$ and 1510 cm$^{-1}$, respectively. CUR showed a bending vibration at 1427 cm$^{-1}$ assigned to –CH group connected to the benzene rings. A band at 1276 cm$^{-1}$ was recorded for C–O stretching vibration. Two bands at 1155 cm$^{-1}$ and 1028 cm$^{-1}$ were assigned to stretching vibrations of C–O from ether group and C–O–C group, respectively. Bands at 960, 813, and 713 cm$^{-1}$ were related to bending vibrations of the C–H bond of alkene groups.

MTX presented bands at 3412 and 3387 cm$^{-1}$ assigned to the stretching vibration of amine groups. A band at 2927 cm$^{-1}$ was related to stretching of the O–H group from a carboxylic acid. The stretching of the C=O bond was observed at 1645 cm$^{-1}$ while the stretching of N-H bong from amide was recorded at 1600 cm$^{-1}$. Stretching bands at 1500, 1543, and 1448 cm$^{-1}$ were assigned to the aromatic ring. A band at 1207 cm$^{-1}$ was related to the stretching of the C–O bond from carboxylic acid [41].

PCL presented an intense band at 1730 cm$^{-1}$, corresponding to the stretching of the C=O group from aliphatic esters. Two bands at 2943 and 2864 cm$^{-1}$ were attributed to the asymmetric and symmetric stretching vibrations of –$CH_2$ group, respectively. Stretching bands at 1174 and 1043 cm$^{-1}$ were assigned to C–O bond.

The FTIR spectrum of PEG exhibited its typical broad band at 3520 cm$^{-1}$, corresponding to the stretching of –OH group. In addition, two bands at 2883 cm$^{-1}$ and 1110 cm$^{-1}$ were assigned to –CH group from an aliphatic chain and asymmetric stretching of C–O–C from dialkyl ethers, respectively.

Bands at the same wavenumber range of FTIR spectrum to those recorded for the physical mixture was observed for CUR and/or MTX-loaded NCs. Considering the mild nanoencapsulation conditions used, no covalent functionalization occurred among polymers and drugs when formulations were obtained.

### 2.2. In Vitro Drug Release Study

NCUR-2, NMTX-1, and NCUR/MTX-2 were initially chosen for in vitro drug release experiments due to their appropriate morphologies and sizes, suitable drug loadings and encapsulation efficiencies, and high amorphization of the drug(s) into NCs. The release profiles for free drugs and these NCs recorded by a dialysis method are summarized in Figure 4.

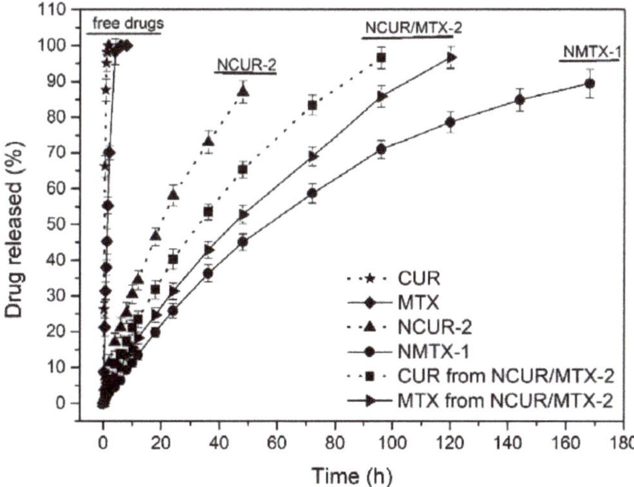


**Figure 4.** In vitro release profiles for free CUR, free MTX and nanocapsules (NCUR-2, NMTX-1, and NCUR/MTX-2).

Free CUR and free MTX showed drug release patterns with a fast diffusion in the dissolution medium used. Free CUR achieved a mean drug-released value of 80% at 41 min of the experiment. For free MTX, a mean drug release of 80% was obtained at 195 min of the experiment. NCUR-2 released 80% of CUR at 2,640 min (44 h). NMTX-1 released 80% of MTX at 8,160 min (136 h). NCUR/MTX-2 had a mean drug release of 80% at 4,140 min (69 h) and 5,400 min (90 h), respectively for CUR and MTX. Therefore, NCs demonstrated prolonged release profiles with no burst effect in comparison to free drugs. Vakilinezhad et al. [22] described PLGA nanoparticles containing CUR and MTX with an initial burst release that was attributed to the drug dispersion into the superficial layers of these particles. In the present study, a typical core-shell structure was strategically planned to avoid this effect and to achieve a prolonged drug release for circumventing the administration of multiple doses per day of chemotherapy [42].

Considering the release profiles obtained for NCs, the difference in the release rate could be explained by the concentration gradient between each formulation and the dissolution medium. At higher drug concentrations, this process was faster since this gradient was the driving force for drug dissolution. Moreover, at higher drug-loading concentration, CUR and/or MTX could be distributed near to the nanocapsules surface providing a more rapid in vitro release [43]. In that sense, NCUR-2, prepared at the higher theoretical drug concentration of 3.0 mg·mL$^{-1}$, presented a faster dissolution and reached a complete release in almost 2 days. On the other hand, NMTX-1, obtained at the lowest theoretical drug concentration of 0.1 mg·mL$^{-1}$, revealed a very slow dissolution rate in about 7 days. The co-loaded formulation NCUR/MTX-2 presented the best dissolution features for controlling drug release. The theoretical drug concentrations proposed for NCUR/MTX-2, 0.5 mg·mL$^{-1}$ of CUR and 0.3 mg·mL$^{-1}$ of MTX, achieved a suitable intermediate dissolution rate in about 3 days. In this formulation, the release behavior of these drugs was closer as depicted in Figure 4, when compared to the other formulations tested. This performance is desired as it could ensure a simultaneous effect in vivo to provide both chemotherapeutic activity and inhibition of P-gp.

### 2.3. *In Vitro Cell Culture-Based Assays*

#### 2.3.1. Cell Viability by MTT and SRB Tests

NCUR-2, NMTX-1, and NCUR/MTX-2 were used for cell culture-based assays due to their known drug release profiles. In order to explore whether or not these formulations had a cytotoxic effect on the Calu-3 cell line, MTT and SRB assays were first performed. The cell viability results for NCUR-2 at high drug concentrations and NMTX-1 at low drug concentrations are shown in Table 2.

**Table 2.** Cell viability of Calu-3 by MTT and SRB assays after its treatment with NCUR-2 and NMTX-1 at different concentrations for 72 h.

| Sample | Final Concentration (μmol·L$^{-1}$) | Cell Viability (%) | |
|---|---|---|---|
| | | MTT | SRB |
| NCUR-2 | 0 | 100.00 ± 1.60 | 99.98 ± 1.75 |
| | 25 | 91.48 ± 3.36 | 88.30 ± 4.01 |
| | 50 | 87.74 ± 5.62 | 60.68 ± 6.10 *** |
| | 100 | 49.91 ± 5.73 *** | 33.87 ± 3.07 *** |
| | 200 | 4.16 ± 0.67 *** | 3.32 ± 0.47 *** |
| NMTX-1 | 0 | 100.00 ± 0.95 | 99.79 ± 1.22 |
| | 0.5 | 100.37 ± 1.96 | 96.10 ± 2.57 |
| | 1 | 100.60 ± 2.15 | 93.01 ± 4.48 |
| | 2 | 99.53 ± 1.67 | 90.85 ± 3.80 |
| | 4 | 93.90 ± 4.60 | 90.78 ± 4.67 |

Results are expressed as mean ± standard error of the mean from 4 independent experiments. Asterisks denote significance levels compared to control: significantly different, *** $p < 0.0001$.

NCUR-2 significantly reduced Calu-3 viability at 200 and 100 μmol·L$^{-1}$ for both MTT and SRB assays. This formulation also provided a significant reduction of cell viability at 50 μmol·L$^{-1}$ for the SRB test. No statistically significant decrease was observed for other concentrations tested of NCUR-2 and for all concentrations investigated of NMTX-1.

In this study, CUR demonstrated cytotoxicity only at high concentrations. This effect may be related to the nanoencapsulation procedure since PCL/PEG NCs may have provided a controlled drug release during the 72 h assays, resulting in low concentrations of released CUR in contact within Calu-3 cells. Different results were observed when free CUR was used. Dhanasekaran et al. [44] investigated the effect of CUR on KG-1 cells using MTT reduction assay, where CUR achieved a dose-dependent reduction in cell viability at different exposure times. Cell viability was reduced to 57% at 50 μmol·L$^{-1}$ for 24 h and to 20% at 100 μmol.L$^{-1}$ for 48 h.

Although these data for pure CUR are more attractive, its free form is not a suitable option for its clinical use. CUR presents several problems related to its in vivo use due to its low aqueous solubility, low bioavailability, rapid hepatic metabolism, high decomposition rate at neutral or basic pH, and susceptibility to photochemical degradation. CUR generates inactive metabolites when administered orally, intraperitoneally, or intravenously, thus making it unfeasible to use in its free form [8].

In that sense, using CUR-loaded NCs is a viable and efficient therapeutic strategy in minimizing the negative features of CUR as well as to retain its antitumoral effect. Similar results have been reported for different tumor cell lines. Wang et al. [45] prepared solid lipid nanoparticles containing CUR and tested them on NCL-41299 and A549 cells. Yallapu et al. [46] obtained CUR-loaded polymeric nanoparticles and investigated their use against A2780CP and MDA-MB-231 lines.

Also, better results were obtained from the SRB assay instead of the MTT test. These findings were concerning to the high sensibility of SRB when cytotoxicity is investigated for cells growing adhered to dish bottom [47]. SRB evaluates cell viability through the ability of the dye to bind to protein components [47]. On the other hand, MTT examines the activity of mitochondrial dehydrogenase enzymes and their respective redox potential [48].

When cells were treated with NMTX-1, no significant reduction in Calu-3 viability was observed. This result is related to the very slow dissolution rate of MTX and the resistance phenomenon, in which MTX is a well-known substrate of P-gp [18,19]. Calu-3 is a tumor cell line that can express P-gp [49]. Therefore, lower intracellular accumulation of MTX was obtained since P-gp efflux pump was activated and led to the removal of the antitumor drug from within the cell.

To overcome this resistance mechanism, a co-loaded formulation (NCUR/MTX-2) was proposed since CUR even at low doses could block P-gp and could ensure the antitumor effect expected for MTX. The cell viability by MTT and SRB techniques of Calu-3, when treated with NCUR/MTX-2, is demonstrated in Figure 5.

NCUR/MTX-2 showed a significant reduction in Calu-3 cell viability by MTT procedure when 9.88:4 and 4.94:2 ratios of CUR:MTX were tested. For SRB assay, all concentrations (9.88:4; 4.94:2; and 2.47:1 ratio) provided a statistically significant decrease in tumor cell viability.

NMTX-1 at different drug concentrations of 1, 2, and 4 μmol·L$^{-1}$ did not reduce Calu-3 viability when tested alone. However, co-loaded formulations containing the same MTX concentrations and low CUR concentrations achieved a statistical reduction of cell viability. This positive result may be due to the suitable dissolution profiles and the P-gp inhibition. NCUR/MTX-2 released both CUR and MTX in almost 72 h that was indeed the time interval during the in vitro cell culture-based assay was performed. Also, it is possible to infer that low CUR concentration in NCUR/MTX-2 plays a remarkable role as P-gp inhibitor, which allows that MTX remains within the cell and performs its antimetabolic/antitumor function. The effect of CUR was attributed only to its capacity of blocking P-gp because this drug demonstrated cytotoxicity against Calu-3 at concentrations higher than 50 μmol.L$^{-1}$ as represented in Table 2. Dey et al. [20] studied alginate-stabilized gold nanoparticles containing CUR and MTX and investigated their effect on C6 glioma and MCF-7 cancer cells. Cell viability was reduced by the use of CUR:MTX ratios of 42:41 and 21:20.5 μmol·L$^{-1}$. In this study, a statistically significant decrease in tumor cell viability was observed by the use of NCs containing CUR even at lower concentrations than those reported in the literature [13].

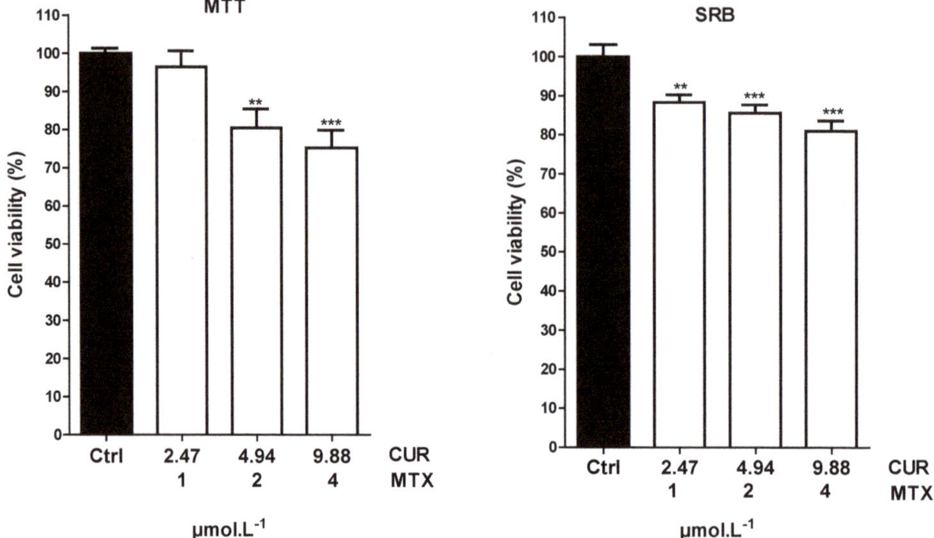


**Figure 5.** Cell viability of Calu-3 by MTT and SRB assays after its treatment with NCUR/MTX-2 at different (low) concentrations for 72 h. Results are expressed as mean ± standard error of the mean from 4 independent experiments. Asterisks denote significance levels compared to control: significantly different, ** $p < 0.001$ and *** $p < 0.0001$.

Another mechanism that may be responsible for the improved cytotoxicity of MTX by CUR is the increase of folate receptors (FRs) expression. FRs are transport systems that show a high affinity for folate and are expressed in large numbers of cancer cells. They are important for cell uptake of folic acid for its metabolism and cell division. However, MTX use can lead to cell resistance since this drug reduces FRs activity [50]. Furthermore, polysorbate 80, a pharmaceutical excipient used as a surfactant during NCs preparation, is a well-known P-gp inhibitor. Its molecules insert themselves between lipid tails of the lipid bilayer and fluidize the cell membrane. It may also interact with the bilayer's polar heads and change the hydrogen bond or ionic bond forces which may contribute to its inhibitory action against P-gp [51]. In that sense, NC composition may enhance the effect of efflux pump inhibition provided by CUR and may lead to improved cytotoxicity against non-small-cell lung cancer cell growth even at low doses of CUR and MTX.

Taking all these into account, the combination of low doses of CUR and MTX by the use of colloidal suspensions of NCs may be a promising strategy for the treatment of lung cancer, reducing the toxic effects associated with the use of high doses of MTX [44,50].

### 2.3.2. Combination Index

The next question addressed was whether NCUR/MTX-2 could provide a synergistic effect on Calu-3 lung cancer cells using MTT cytotoxicity assay. To reveal the synergistic activity, a combination index (CI) was performed considering the half-maximal inhibitory concentration ($IC_{50}$) data of separate and combined cytotoxic effects of NCs containing CUR and/or MTX. $IC_{50}$ (mean value ± SD) for CUR from NCUR-2 was 100.18 ± 3.74 μmol.L$^{-1}$. NMTX-1 presented an $IC_{50}$ of 23.06 ± 1.13 μmol·L$^{-1}$. $IC_{50}$ for CUR and MTX from NCUR/MTX-2 were 69.81 ± 3.07 and 1.89 ± 0.06 μmol·L$^{-1}$, respectively.

Thereby, the CI value achieved was 0.78. In brief, the CI equal to 1 indicates that the two drugs have additive effects, the CI lower than 1 suggests a synergism and the CI higher than 1 indicates antagonism [52]. Consequently, a synergistic activity was demonstrated for NCUR/MTX-2 due to the combination of these drugs induced greater cytotoxicity against Calu-3. This finding shows that cytotoxic chemotherapy using NCUR/MTX-2 appears to be a promising strategy for the treatment of non-small-cell lung cancer which merits further preclinical and clinical investigation since it has allowed a suitable effect even at low chemotherapeutic doses into this nanocarrier.

### 2.3.3. Cell Death Pattern through Acridine Orange/Ethidium Bromide (AO/EB) Test

Considering the results observed for NCUR/MTX-2 in reducing cell viability of Calu-3, the cell death pattern of this formulation was investigated after the 24 h treatment by AO/EB staining to verify the death mechanism.

The cytotoxic effect of NCUR/MTX-2 formulations can be observed in Figure 6. The control showed viable cells with normal and bright green nuclei (Figure 6A, white arrows). NCUR/MTX-2 resulted in early-stage apoptotic cells that were marked by crescent-shaped or granular yellow-green acridine orange nuclear staining (Figure 6B, yellow arrows), and late-stage apoptotic cells showing concentrated and asymmetrically located orange nuclear ethidium bromide staining (Figure 6B, blue arrow). In addition, the occurrence of apoptotic membrane blebbing (Figure 6B, red arrow) was verified since membrane blebbing is required for redistribution of fragmented DNA from the nuclear region into membrane blebs and apoptotic bodies [53]. No necrotic cell was observed after treating the Calu-3 cells with the co-loaded formulation.

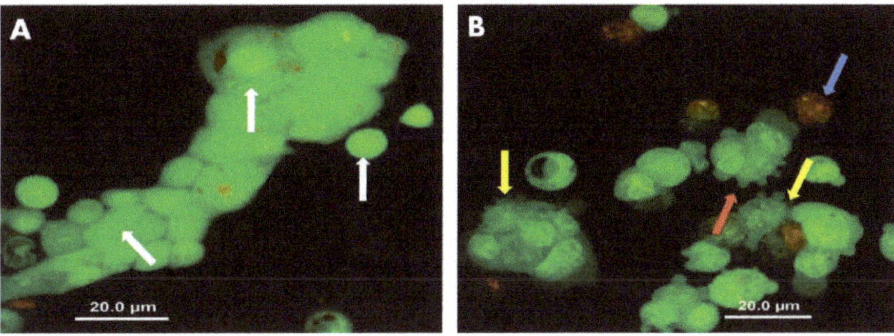


**Figure 6.** Effect of NCUR/MTX-2 on the morphology of Calu-3 cells stained with acridine orange/ ethidium bromide by light fluorescence microscopy at 400× magnification. The negative control (**A**, vehicle) presented viable cells with normal nucleus staining represented by the bright green chromatin (white arrows); NCUR/MTX-2 (**B**) showed early-stage apoptotic cells that were marked by crescent-shaped or granular yellow-green acridine orange nuclear staining (yellow arrows), late-stage apoptotic cells showing concentrated and asymmetrically localized orange nuclear ethidium bromide staining (blue arrow), and apoptotic membrane blebbing (red arrow). Images are representative of results obtained from three independent biological replicates ($n = 6$ slides per sample).

Apoptosis is typically represented by a series of intracellular events, which ultimately lead to DNA fragmentation and the internucleosomal degradation of genomic DNA due to the activation of endogenous endonucleases. This mechanism is less aggressive because apoptosis is controlled and energy-dependent and can affect individual cells or clusters of cells [54]. On the other side, necrotic cells swell and rupture, releasing the cytoplasmic material. This process is characterized by injury to a group of cells rather than by individual death. Necrosis occurs by disruption of plasma membranes and organelles, marked dilation of mitochondria with the emergence of large amorphous densities, probably representing denatured proteins. The nucleus may shrink, suffer fragmentation or even totally disappear by unspecific DNA fragmentation [55]. In that sense, the co-loaded formulation was able to provide a Calu-3 death pattern by a carefully regulated energy-dependent process, characterized by programmed cell death, which is a widespread component of both health and disease. Moreover, apoptosis is a superior mechanism for feasible therapeutic interventions on the pathophysiology of lung cancer.

## 3. Materials and Methods

### 3.1. Materials

Curcumin (CUR, ≥94% curcuminoid content, Sigma-Aldrich, St. Louis, MO, USA), methotrexate (MTX, >99% pure, Fermion, Espoo, Finland), poly(ε-caprolactone) (PCL, Mw 10,000–14,000 g.mol$^{-1}$, Sigma-Aldrich), poly(ethylene glycol) 6000 (PEG, Mw 5,400–6,600 g.mol$^{-1}$, Cromato Produtos Químicos, Diadema, Brazil), sorbitan monooleate (Span 80, Oxiteno, Mauá, Brazil), polysorbate 80 (Tween 80, Delaware, Porto Alegre, Brazil), medium chain triglycerides (MCT, 99% pure, Focus Química, São Paulo, Brazil), acridine orange base (AO, Sigma-Aldrich), ethidium bromide (EB, Sigma-Aldrich), methylthiazolyldiphenyltetrazolium bromide (MTT, Sigma-Aldrich), sulpho-rhodamine B (SRB, Sigma-Aldrich), penicillin-streptomycin (Sigma-Aldrich), and acetone (≥99.9% pure, Vetec Química, Rio de Janeiro, Brazil) were used as received. HPLC-grade methanol was purchased from Tedia (Rio de Janeiro, Brazil). RPMI 1640 medium and fetal bovine serum were obtained from Vitrocell (Campinas, Brazil). Water was purified in a Milli-Q Plus water purification system (Millipore, Bedford, MA, USA). All other solvents and reagents were analytical grade. The Calu-3 cell line was obtained from the Bank of Cells of Rio de Janeiro (BCRJ, Brazil) and was kindly provided by Dr. Katia Sabrina Paludo.

### 3.2. Preparation of Polymeric Nanocapsules (NCs) Containing Curcumin (CUR) and/or Methotrexate (MTX)

The interfacial deposition of the preformed polymer method was used for preparing NCs containing CUR and/or MTX [20]. Six different formulations (Table 3) were obtained depending on the amount of CUR and/or MTX in their composition. Briefly, PCL and PEG 6000 were solvated in the organic phase containing span 80, CUR and/or MTX, MCT, and acetone. This phase was carefully added to the aqueous phase containing tween 80 and purified water under vigorous magnetic stirring at 45 °C. The obtained colloidal emulsion was kept under magnetic stirring for 10 min. The organic solvent was then quickly removed by evaporation under reduced pressure at 40 °C. Non-loaded nanocapsules were also prepared as the negative control. All formulations were obtained in triplicate from three different batches. For providing a comparative analysis, a physical mixture of polymers and drugs (PM PCL/PEG/CUR/MTX) at the same molar ratio was prepared by mortar and pestle mixing before characterization procedures.

**Table 3.** Composition of polymeric nanocapsules containing curcumin (CUR) and/or methotrexate (MTX).

| Formulation | Composition | | | | | | | | |
|---|---|---|---|---|---|---|---|---|---|
| | CUR (g) | MTX (g) | PEG (g) | PCL (g) | Span 80 (g) | MCT (g) | Acetone (mL) | Tween 80 (g) | Water (mL) |
| NCUR/MTX-0 | - | - | 0.020 | 0.080 | 0.077 | 0.300 | 27 | 0.077 | 53 |
| NCUR-1 | 0.010 | - | 0.020 | 0.080 | 0.077 | 0.300 | 27 | 0.077 | 53 |
| NCUR-2 | 0.030 | - | 0.020 | 0.080 | 0.077 | 0.300 | 27 | 0.077 | 53 |
| NMTX-1 | - | 0.001 | 0.020 | 0.080 | 0.077 | 0.300 | 27 | 0.077 | 53 |
| NMTX-2 | - | 0.003 | 0.020 | 0.080 | 0.077 | 0.300 | 27 | 0.077 | 53 |
| NCUR/MTX-1 | 0.030 | 0.001 | 0.020 | 0.080 | 0.077 | 0.300 | 27 | 0.077 | 53 |
| NCUR/MTX-2 | 0.005 | 0.003 | 0.020 | 0.080 | 0.077 | 0.300 | 27 | 0.077 | 53 |

### 3.3. *Characterization of Polymeric Nanocapsules (NCs) Containing Curcumin (CUR) and/or Methotrexate (MTX)*

#### 3.3.1. Determination of Mean Diameter, Polydispersity Index, and Zeta Potential of NCs

Mean particle size, polydispersity, and zeta potential were measured at 25 °C using a Zetasizer Nanoseries ZS903600 apparatus (Malvern Instruments, Malvern, UK). Each sample was previously diluted in water (1:500) and each analysis was performed at a scattering angle of 90° and a temperature of 25 °C for mean particle size and polydispersity measurements. For zeta potential, each sample was placed into the electrophoretic cell where a potential of ± 150 mV was used.

#### 3.3.2. Encapsulation Efficiency

CUR and/or MTX content of NCs was determined by the indirect method. In brief, suspensions of nanocapsules were submitted to a combined ultrafiltration/centrifugation using centrifugal devices (Amicon®® 10.000 MW, Millipore, Bedford, MA, USA) at 2200 g during 30 min in triplicate. Free CUR and/or MTX were determined in ultrafiltrate using an HPLC method in a Merck-Hitachi Lachrom equipment (Tokyo, Japan), interface D-7000, UV detector module L-74000, equipped with pumps L-7100 and an integral degasser, controller software (Chromquest, Thermo Fisher Scientific, Incorporated, Pittsburgh, PA, USA), and manual injector (Rheodyne, Rohnert Park, CA, USA) equipped with a 20 µL injector loop and a 100 µL syringe (Microliter 710, Hamilton, Bonaduz, Switzerland). Chromatographic separation was accomplished using a Inertsil®® ODS3 (GL Sciences, Torrance, CA, USA) reversed-phase analytical column (150 mm × 4.6 mm, 5 µm) and a GL Sciences Inertsil®® ODS3 guard cartridge system (10 mm × 4 mm, 5 µm) at room temperature (20 ± 2 °C) using UV detection at 261 nm. Gradient elution was carried out using a mobile phase consisted of methanol:water acidified with 0.5% acetic acid at a flow rate of 1.0 mL.min$^{-1}$. The mobile phase gradient program was as follows: it was started at 44% MeOH for 3 min, increased to 90% MeOH in 5 min, held constant until 11 min, then returned to 44% MeOH in 12 min. The encapsulation efficiency (EE, %) was calculated using Equation (1):

$$EE(\%) = \frac{total\ drug\ content - free\ drug\ content}{total\ drug\ content} \times 100 \tag{1}$$

#### 3.3.3. Field Emission Scanning Electron Microscopy (FESEM)

The freeze-dried formulations were mounted on aluminum stubs and sputtered with gold (IC-50 Ion Coater, Shimadzu, Kyoto, Japan). Morphological analysis was performed and photomicrographs were prepared using a Mira3 LM FESEM (Tescan, Brno, Czech Republic) at an accelerating voltage of 5 kV.

#### 3.3.4. X-ray Diffraction (XRD)

Polymeric nanocapsules were previously deposited on a glass coverslip and dried at room temperature (25 ± 2 °C). XRD data were recorded in an Ultima IV diffractometer (Rigaku, Tokyo, Japan). The 2θ value was increased from 4° to 50° at a scan rate of 0.05°·min$^{-1}$ using a Cu-Kα source (λ = 1.5418 Å) at 30 kV and 40 mA.

#### 3.3.5. Fourier-Transform Infrared Spectroscopy (FTIR)

FTIR analyses of raw materials, freeze-dried polymeric nanocapsules, and physical mixture were carried out from 4000 to 400 cm$^{-1}$ using a Prestige-21 IR spectrophotometer (Shimadzu, Kyoto, Japan) in KBr pellets with 32 scans and a resolution of 4 cm$^{-1}$.

### 3.4. *In Vitro Drug Release Study*

The release experiments were performed for free CUR, free MTX, and formulations NCUR-2, NMTX-1, and NCUR/MTX-2 by dialysis diffusion procedure in phosphate-buffered saline (PBS,

20 mmol.L$^{-1}$) containing 0.1% (*w*/*v*) Tween®® 80 at pH 7.4 [21,22] by adding each sample in PBS (1.5 mL) into a dialysis bag (Spectra/Por®® molecular porous membrane tubing, MWCO 10,000, Spectrum Laboratories, Rancho Dominguez, CA, USA), and dialyzed against fresh PBS (13.5 mL) at 37 °C and under continuous magnetic stirring of 50 rpm. Aliquots of 500 µL were withdrawn at predetermined time intervals and replaced by the same volume of fresh medium. CUR and MTX concentrations were determined individually in each sample by the previously described HPLC method. The in vitro drug release assay was carried out in triplicate from three different batches.

### 3.5. *In Vitro Cell Culture-Based Assays*

#### 3.5.1. Cell Culture

Calu-3 cell line was cultured in RPMI 1640 medium at pH 7.4, containing 10% fetal bovine serum, supplemented with 24 mmol·L$^{-1}$ sodium bicarbonate, 2 mmol·L$^{-1}$ L-glutamine, 1 mmol·L$^{-1}$ sodium pyruvate, 10,000 U·L$^{-1}$ penicillin, and 10 mg·L$^{-1}$ streptomycin. The cultures were maintained in a humidified oven at 37 °C with 5% $CO_2$ atmosphere.

#### 3.5.2. Cell Treatment

Calu-3 cells were seeded in 96-well plates at a density of 1.5 × 10$^4$ cells.well$^{-1}$ and incubated for 24 h in the culture medium. For assessment of cell viability and cytotoxicity, Calu-3 cell line was incubated with the samples NCUR-2 (200.00–25.00 µmol·L$^{-1}$), NMTX-1 (100.00–12.50 µmol·L$^{-1}$ and 4.00–0.50 µmol·L$^{-1}$) and NCUR/MTX-2 (247.00–61.75 µmol·L$^{-1}$ and 9.88–2.47 µmol·L$^{-1}$ to CUR; 100.00–25.00 µmol·L$^{-1}$ and 4.00–1.00 µmol·L$^{-1}$ to MTX) for 72 h at 37 °C with 5% $CO_2$ atmosphere. For co-loaded formulation, final concentrations were standardized using MTX due to its lower content in NCs and EE compensation was also performed for CUR content, resulting in fractional values. Tests were obtained using serial dilution procedure and were performed as four independent experiments containing $n = 4$ samples per assay.

#### 3.5.3. Cell Viability by Methylthiazolyldiphenyl-tetrazolium Bromide (MTT) Test

After 72 h of the treatments, 200 µL of a solution of MTT at 0.5 mg·mL$^{-1}$ was added to the wells following a standard method [56]. The cultures were then incubated at 37 °C for 2 h, protected from light, until the presence of formazan crystals. The supernatant was then removed. For the solubilization of these crystals, 200 µL of dimethyl sulfoxide was added. The spectrophotometric absorbance reading was performed at a wavelength of 550 nm in a µQuant microplate reader (BioTek, Winooski, VT, USA). To calculate cell viability (%), Equation (2) was used. The concentration that inhibited 50% of cell growth ($IC_{50}$) was then calculated by Probit regression [57]:

$$Cell\ viability\ (\%) = \frac{absorbance\ of\ test}{absorbance\ of\ control} \times 100 \tag{2}$$

#### 3.5.4. Combination Index

To define the drug-drug interaction potential from the use of NCs containing CUR and MTX against Calu-3 cells, the combination index (CI) was calculated by Equation (3) [58] considering the results obtained by the MTT method:

$$Combination\ index\ (CI) = \frac{(D)1}{(Dx)1} + \frac{(D)2}{(Dx)2} \tag{3}$$

where (Dx)1 and (Dx)2 are the concentration of the tested substance 1 (CUR) and the tested substance 2 (MTX) used in the single treatment that was required to decrease the cell viability by 50% and (D)1 and (D)2 are the concentration of the tested substance 1 (CUR) in combination with the concentration of the tested substance 2 (MTX) that together decreased the cell viability by 50%.

#### 3.5.5. Cell Viability by Sulphorhodamine B (SRB) Test

For the analysis of SRB, the method described by Papazisis et al. [59] was used. In brief, Calu-3 cell line was treated with each sample for 72 h and the supernatant was then discarded. The cells were washed with 200 μL of sodium phosphate buffer at pH = 7.4. Then, 200 μL of 10% cold trichloroacetic acid was added and the plates were placed in the refrigerator for 30 min to fix the cells. Subsequently, the cells were washed three times with 200 μL of distilled water and maintained for 24 h at room temperature (20 ± 2 °C) to dryness. Later, 200 μL of 0.2% SRB solution was added and kept for 30 min. The plates were then washed five times with 200 μL of 1% acetic acid and again dried for 30 min. Finally, 150 μL of 10 mmol.$L^{-1}$ TrisBase was added and the spectrophotometric absorbance reading was performed at a wavelength of 432 nm in a microplate reader (μQuant, BioTek). Equation (2) was also used for calculating cell viability (%).

#### 3.5.6. Cell Death Pattern through Acridine Orange/Ethidium Bromide (AO/EB) Test

In 24-well plates, CALU-3 cells were seeded onto coverslips at the concentration of $1 \times 10^5$ cells.$well^{-1}$ using RPMI 1640 culture medium. After 24 h for cell adhesion, the medium was then discarded and NCUR/MTX-2 sample (CUR 9.88 μmol·$L^{-1}$ and MTX 4.00 μmol·$L^{-1}$) was added in a volume of 500 μL. Cells were then incubated at 37 °C in a wet atmosphere and 5% $CO_2$.

After 24 h treatment, the wells were washed with 200 μL of sodium phosphate buffer at pH = 7.4. Following this, 10 μL of acridine orange/ethidium bromide dye mixture (200 μg·$mL^{-1}$) was placed on a coverslip, which was then inverted on a slide. The staining was viewed under a BX41 fluorescence microscope (Olympus, Tokyo, Japan) with an excitation filter at 480/30 nm and emission at 535/40 nm. The typical fields were recorded with an attached Olympus DP71 camera.

Classification of cell type to AO/EB staining was performed based on the criteria proposed by Ribble et al. [60]. In summary, the viable cells present bright green nucleus, necrotic cells depict orange-red fluorescence, early-stage apoptotic cells are marked by crescent-shaped or granular yellow-green AO nuclear staining, and late-stage apoptotic cells show concentrated and asymmetrically sited orange nuclear EB staining.

### *3.6. Statistical Analysis*

Statistical analyses were performed using one-way analysis of variance (ANOVA), and when necessary, the post-hoc Tukey test was used. The results were expressed as a mean ± standard error of the mean (SEM) and mean ± standard deviation (SD). P values lower than 0.05 ($p < 0.05$) were considered significant. Calculations were carried out using the statistical software GraphPad Prism version 5.03 (GraphPad Inc., San Diego, CA, USA).

## 4. Conclusions

In conclusion, CUR and/or MTX co-loaded PCL/PEG6000 nanocapsules were successfully prepared by interfacial deposition of the pre-formed polymer. Nanometer-sized, monodisperse, amorphous/non-crystalline formulations with high drug-loading efficiencies were obtained. No changes in FTIR assignments were recorded after the nanoencapsulation procedure. In addition, NCUR/MTX-2 provided a statistically significant decrease of Calu-3 cell viability by MTT and SRB assays even at low concentrations of these chemotherapeutic agents. A synergistic effect was proven by the use of this co-loaded nanocarrier. The fluorescence-morphological analysis revealed a cell death pattern based on early and late apoptosis. No necrotic cell was observed in the Calu-3 cells after treating with the co-loaded formulation NCUR/MTX-2. In summary, the co-loaded nanocapsules method can be further used as a novel therapeutic strategy by intravenous or pulmonary route for treating non-small-cell lung cancer to hold the potential for achieving therapeutic outcomes while reducing the incidence of adverse drug effects.

**Author Contributions:** Preparation and Characterization of nanoparticles, L.A.C.R., P.V.F., J.M.N. and A.N.; Encapsulation Efficiency analyses, A.L., F.M.B. and T.K.; Microscopy analyses, Manuscript writing, J.M.B.; In vitro Drug Release study, P.V.F., J.M.N., M.C.B. and A.D.L.; In vitro Cell Culture-based assays, L.A.C.R., P.V.F., M.C.B. and C.C.K.; Statistical analysis, P.V.F. and S.M.W.Z.; Insights on Discussion topics, English editing, V.R.; Supervision of the Laboratory work, S.M.W.Z. All authors have read and agreed to the published version of the manuscript.

**Funding:** This research was funded by CNPq—Conselho Nacional de Pesquisa e Desenvolvimento, grant number 313704/2019-8.

**Acknowledgments:** The authors are grateful to CLABMU-UEPG for technical support.

**Conflicts of Interest:** The authors declare no conflict of interest.

## References

1. Waheed, A.; Gupta, A.; Patel, P. Targeted Drug Delivery Systems for Lung Cancer. *PharmaTutor* **2015**, *3*, 38–42.
2. Chen, L.; Nan, A.; Zhang, N.; Jia, Y.; Li, X.; Ling, Y.; Dai, J.; Zhang, S.; Yang, Q.; Yi, Y.; et al. Circular RNA 100146 functions as an oncogene through direct binding to miR361-3p and miR-615-5p in non-small cell lung cancer. *Mol. Cancer* **2019**, *18*, 1–8. [CrossRef]
3. Huber, P.C.; Maruiama, C.H.; Almeida, W.P. Glicoproteína-p, resistência a múltiplas drogas (mdr) e relação estrutura-atividade de moduladores. *Quim. Nova* **2010**, *33*, 2148–2154. [CrossRef]
4. Niu, S.; Williams, G.R.; Wu, J.J.; Zhang, X.; Zheng, H.; Li, S.; Zhu, L.M. A novel chitosan-based nanomedicine for multi-drug resistant breast cancer therapy. *Chem. Eng. J.* **2019**, *369*, 134–149. [CrossRef]
5. Ke, X.J.; Cheng, Y.F.; Yu, N.; Di, Q. Effects of carbamazepine on the P-gp and CYP3A expression correlated with PXR or NF-κB activity in the bEnd.3 cells. *Neurosci. Lett.* **2019**, *690*, 48–55. [CrossRef] [PubMed]
6. Albano, J.M.R.; Ribeiro, L.N.M.; Couto, V.M.; Messias, M.B.; Silva, G.H.R.; Breitkreitz, M.C.; Paula, E.; Pickholz, M. Rational design of polymer-lipid nanoparticles for docetaxel delivery. *Colloids Surf. B Biointerfaces* **2019**, *175*, 56–64. [CrossRef] [PubMed]
7. Chearwae, W.; Anuchapreeda, S.; Nandigama, K.; Ambudkar, S.V.; Limtrakul, P. Biochemical mechanism of modulation of human P-glycoprotein (ABCB1) by curcumin I, II and III purified from Turmeric powder. *Biochem. Pharmacol.* **2004**, *68*, 2043–2052. [CrossRef]
8. Santiago, V.S.; Silva, G.P.M.; Ricardo, D.D.; Lima, M.E.F. Curcumina, o Pó Dourado do Açafrão-Da-Terra: Introspecções Sobre Química e Atividades Biológicas. *Quím. Nova* **2015**, *38*, 538–552.
9. Mishra, H.; Kesharwani, R.K.; Singh, D.B.; Tripathi, S.; Dubey, S.K.; Misra, K. Computational simulation of inhibitory effects of curcumin, retinoic acid and their conjugates on GSK-3 beta. *New Model Anal. Health Inform. Bioinform* **2019**, *8*, 1–3. [CrossRef]
10. Sharma, R.A.; Gescher, A.J.; Steaward, W.P. Curcumin: The story so far. *Eur. J. Cancer* **2005**, *41*, 1955–1968. [CrossRef]
11. Huaasin, Z.; Thu, H.E.; Ng, S.F.; Khan, S.; Katas, H. Nanoencapsulation, an efficient and promising approach to maximize wound healing efficacy of curcumin: A review of new trends and state-of-the-art. *Colloids Surf. B Biointerfaces* **2017**, *150*, 223–241. [CrossRef] [PubMed]
12. Teng, Y.N.; Hsieh, Y.W.; Hung, C.C.; Lin, H.Y. Demethoxycurcumin Modulates Human P-Glycoprotein Function via Uncompetitive Inhibition of ATPase Hydrolysis Activity. *J. Agric. Food Chem.* **2015**, *63*, 847–855. [CrossRef] [PubMed]
13. Khalil, N.M.; Nascimento, T.C.F.; Casa, D.M.; Dalmolin, L.F.; Mattos, A.C.; Hoss, I.; Romano, M.A.; Mainardes, R.M. Pharmacokinetics of curcumin-loaded PLGA and PLGA-PEG blend nanoparticles after oral administration in rats. *Colloids Surf. B Biointerfaces* **2013**, *101*, 353–360. [CrossRef] [PubMed]
14. Brandt, J.V.; Piazza, R.D.; Santos, C.C.; Chacón, J.V.; Amantéa, B.E.; Pinto, G.C.; Magnani, M.; Piva, H.L.; Tedesco, A.C.; Primo, F.L.; et al. Synthesis and colloidal characterization of folic acid-modified PEG-b-PCL Micelles for methotrexate delivery. *Colloids Surf. B Biointerfaces* **2019**, *177*, 228–234. [CrossRef] [PubMed]
15. Katiyar, S.S.; Kushwah, V.; Dora, C.P.; Jain, S. Lipid and TPGS based novel core-shell type nanocapsular sustained release system of methotrexate for intravenous application. *Colloids Surf. B Biointerfaces* **2019**, *174*, 501–510. [CrossRef]
16. Rubino, F.M.J. Separation methods for methotrexate, its structural analogues and metabolites. *Chromatogr B* **2001**, *764*, 217–254. [CrossRef]

17. Prasad, R.; Koul, V.; Anand, S.; Khar, R.K. Effect of DC/mDC iontophoresis and terpenes on transdermal permeation of methotrexate: In vitro study. *Int. J. Pharm.* **2007**, *333*, 70–78. [CrossRef]
18. Choi, C.H. ABC transporters as multidrug resistance mechanisms and the development of chemosensitizers for their reversal. *Cancer Cell Int.* **2005**, *5*, 1–13. [CrossRef]
19. Zhu, Y.; Meng, Q.; Wang, C.; Liu, Q.; Huo, X.; Zhang, A.; Sun, P.; Sun, H.; Li, H.; Liu, K. Methotrexate-bestatin interaction: Involvement of P-glycoprotein and organic anion transporters in rats. *Int. J. Pharm.* **2014**, *465*, 368–377. [CrossRef]
20. Dey, S.; Sherly, M.C.D.; Rekha, M.R.; Sreenivasan, K. Alginate stabilized gold nanoparticle as multidrug Carrier: Evaluation of cellular interactions and hemolytic potential. *Carbohydr. Polym.* **2016**, *136*, 71–80. [CrossRef]
21. Curcio, M.; Mauro, L.; Naimo, G.D.; Amantea, D.; Cirillo, G.; Tavano, L.; Casaburi, I.; Nicoletta, F.P.; Alvarez-Lorenzo, C.; Iemma, F. Facile synthesis of pH-responsive polymersomes based on lipidized PEG for intracellular co-delivery of curcumin and methotrexate. *Colloids Surf. B Biointerfaces.* **2018**, *1*, 568–576. [CrossRef]
22. Vakilinezhad, M.A.; Amini, A.; Dara, T.; Alipour, S. Methotrexate and Curcumin co-encapsulated PLGA nanoparticles as a potential breast cancer therapeutic system: In vitro and In vivo evaluation. *Colloids Surf. B Biointerfaces.* **2019**, *184*, 1–10. [CrossRef]
23. Curcio, M.; Cirillo, G.; Tucci, P.; Farfalla, A.; Bevacqua, E.; Vittorio, O.; Iemma, F.; Nicoletta, F.P. Dextran-Curcumin Nanoparticles as a Methotrexate Delivery Vehicle: A Step Forward in Breast Cancer Combination Therapy. *Pharmaceuticals* **2020**, *2*, 2. [CrossRef]
24. Gomes, M.L.S.; Nascimento, N.S.; Borsato, D.M.; Pretes, A.P.; Nadal, J.M.; Novatski, A.; Gomes, R.Z.; Fernandes, D.; Farago, P.V.; Zanin, S.M.W. Long-lasting anti-platelet activity of cilostazol from poly(ε-caprolactone) poly(ethylene glycol) blend nanocapsules. *Mater Sci Eng C.* **2019**, *94*, 694–702.
25. Chassot, J.M.; Ribas, D.; Silveira, E.F.; Grünspan, L.D.; Pires, C.C.; Farago, P.V.; Braganhol, E.; Tasso, L.; Cruz, L. Beclomethasone Dipropionate-Loaded Polymeric Nanocapsules: Development, In Vitro Cytotoxicity, and In Vivo Evaluation of Acute Lung Injury. *J. Nanosci. Nanotechnol.* **2014**, *14*, 1–10. [CrossRef]
26. Ferreira, L.M.; Cervi, V.F.; Sari, M.H.M.; Barbieria, A.V.; Ramos, A.P.; Copetti, P.M.; Brum, G.F.; Nascimento, K.; Nadal, J.M.; Farago, P.V.; et al. Diphenyl diselenide loaded poly(ε-caprolactone) nanocapsules with selective antimelanoma activity: Development and cytotoxic evaluation. *Mater. Sci. Eng. C.* **2018**, *91*, 1–9. [CrossRef]
27. Aditya, N.; Ravi, P.R.; Avula, U.S.R.; Vats, R. Poly (e-caprolactone) nanocapsules for oral delivery of raloxifene: Process optimization by hybrid design approach, in vitro and in vivo evaluation. *J Microencapsul.* **2014**, *31*, 508–518. [CrossRef]
28. Schaffazick, S.R.; Guterres, S.S. Caracterização e Estabilidade Físico-Química de Sistemas Poliméricos Nanoparticulados para Administração de Fármacos. *Quím. Nova* **2003**, *26*, 726–737. [CrossRef]
29. Avadi, M.R.; Sadeghi, A.M.M.; Mohammadpour, N.; Abedin, S.; Atyabi, F.; Dinarvand, R.; Tehrani, M.R. Preparation and characterization of insulin nanoparticles using chitosan and Arabic gum with ionic gelation method. *Nanomedicine* **2010**, *6*, 58–63. [CrossRef]
30. Neves, A.R.; Lúcio, M.; Martins, S.; Lima, J.; Reis, S. Novel resveratrol nanodelivery systems based on lipid nanoparticles to enhance its oral bioavailability. *Int. J. Nanomed.* **2013**, *8*, 177–187.
31. O'Neil, M.J.; Heckelman, P.E.; Dobbelaar, P.H.; Roman, K.J.; Kenney, C.M.; Karaffa, L.S. *The Merck Index: An Encyclopedia of Chemicals, Drugs, and Biologicals*; Maryadele, J.O.N., Patricia, E.H., Cherie, B.K., Kristin, J.R., Eds.; Royal Society of Chemistry: Cambridge, UK, 2013; p. 474.
32. Michael, D.L.; Richard, J.L., Sr.; Robert, A.L. *Hawley's Condensed Chemical Dictionary*, 13rd ed.; John Wiley & Sons: New York, NY, USA, 1997; p. 722.
33. Mora-Huertas, C.E.; Fessi, H.; Elaissari, A. Polymer-based nanocapsules for drug delivery. *Int. J. Pharm.* **2010**, *385*, 113–142. [CrossRef]
34. Zhang, Q.; Polyakov, N.E.; Chistyachenko, Y.S.; Khvostov, M.V.; Frolova, T.S.; Tolstikova, T.G.; Dushkin, A.V.; Su, W. Preparation of curcumin self-micelle solid dispersion with enhanced bioavailability and cytotoxic activity by mechanochemistry. *Drug Deliv.* **2018**, *25*, 198–209. [CrossRef]
35. Teng, Z.; Luo, Y.; Wang, Q. Nanoparticles Synthesized from Soy Protein: Preparation, Characterization, and Application for Nutraceutical Encapsulation. *J. Agric. Food Chem.* **2012**, *60*, 2712–2720. [CrossRef]

36. Oliveira, A.R.; Molina, E.F.; Mesquita, P.C.; Fonseca, J.L.C.; Rossanezi, G.; Pedrosa, M.F.F.; Oliveira, A.G.; Júnior, A.A.S. Structural and thermal properties of spray-dried methotrexate-loaded biodegradable microparticles. *J. Therm. Anal. Calorim.* **2013**, *112*, 555–565. [CrossRef]
37. Mendes, J.B.E.; Riekes, M.K.; Oliveira, V.M.; Michel, M.D.; Stulzer, H.K.; Khalil, N.M.; Zawadzki, S.F.; Mainardes, R.M.; Farago, P.V. PHBV/PCL Microparticles for Controlled Release of Resveratrol: Physicochemical Characterization, Antioxidant Potential, and Effect on Hemolysis of Human Erythrocytes. *Sci. World J.* **2012**, *2012*, 1–13. [CrossRef]
38. Song, Z.; Zhu, W.; Liu, N.; Yang, F.; Feng, R. Linolenic acid-modified PEG-PCL micelles for curcumin delivery. *Int. J. Pharm.* **2014**, *471*, 312–321. [CrossRef]
39. Hu, L.; Kong, D.; Hu, Q.; Gao, N.; Pang, S. Evaluation of High-Performance Curcumin Nanocrystals for Pulmonary Drug Delivery Both In Vitro and In Vivo. *Nanoscale Res. Lett.* **2015**, *10*, 1–9. [CrossRef]
40. Bich, V.T.; Thuy, N.T.; Binh, N.T.; Huong, N.T.M.; Yen, P.N.D.; Luong, T.T. Structural and Spectral Properties of Curcumin and Metal-Curcumin Complex Derived from Turmeric (Curcuma longa). In *Physics and Engineering of New Materials, Heidelberg, Germany, 15 January 2009, D.T.*; Pucci, A., Wandelt, K., Eds.; Springer: Berlin, Germany, 2009; pp. 271–278.
41. Ajmal, M.; Yunus, U.; Matin, A.; Haq, N.U. Synthesis, characterization and in vitro evaluation of methotrexate conjugated fluorescent carbon nanoparticles as drug delivery system for human lung cancer targeting. *m Photobiol. B* **2015**, *153*, 111–120. [CrossRef]
42. Yurgel, V.; Collares, T.; Seixas, F. Developments in the use of nanocapsules in oncology. *Braz. J. Med. Biol. Res.* **2013**, *46*, 486–501. [CrossRef]
43. Mirza, M.A.; Rahman, M.A.; Talegaonkar, S.; Iqbal, Z. In vitro/in vivo performance of different complexes of itraconazole used in the treatment of vaginal candidiasis. *Braz. J. Pharm. Sci.* **2012**, *48*, 759–772. [CrossRef]
44. Dhanasekaran, S.; Biswal, B.K.; Sumantran, V.N.; Verma, R.S. Augmented sensitivity to methotrexate by curcumin induced overexpression of folate receptor in KG-1 cells. *Biochimie* **2013**, *95*, 1567–1573. [CrossRef]
45. Wang, P.; Zhang, L.; Peng, H.; Li, Y.; Xiong, J.; Xu, Z. The formulation and delivery of curcumin with lipid nanoparticles for the treatment of non-small cell lung cancer both in vitro and in vivo. *Mater. Sci. Eng. C.* **2013**, *33*, 4802–4808.
46. Yallapu, M.M.; Gupta, B.K.; Jaggi, M.; Chauhan, S.C.J. Fabrication of curcumin encapsulated PLGA nanoparticles for improved therapeutic effects in metastatic cancer cells. *Colloid Interface Sci.* **2010**, *351*, 19–29. [CrossRef]
47. Houghton, P.; Fang, R.; Techatanawat, I.; Steventon, G.; Hylands, P.J.; Lee, C.C. The sulphorhodamine (SRB) assay and other approaches to testing plant extracts and derived compounds for activities related to reputed anticancer activity. *Methods* **2007**, *42*, 377–387. [CrossRef]
48. Nadal, J.M.; Gomes, M.L.S.; Borsato, D.M.; Almeida, M.A.; Barboza, F.M.; Zawadzki, S.F.; Kanunfre, C.C.; Farago, P.V.; Zanin, S.M.W. Spray-dried Eudragit®® L100 microparticles containing ferulic acid: Formulation, in vitro cytoprotection and in vivo anti-platelet effect. *Mater. Sci. Eng. C* **2016**, *64*, 318–328.
49. Hamilton, K.O.; Backstrom, G.; Yazdanian, M.A.; Audus, K.L. P-Glycoprotein Efflux Pump Expression and Activity in Calu-3 Cells. *J. Pharm. Sci.* **2001**, *90*, 647–658. [CrossRef]
50. Srivalli, K.M.R.; Lakshmi, P.K. Overview of P-glycoprotein inhibitors: A rational outlook. *Braz. J. Pharm. Sci.* **2012**, *48*, 353–367. [CrossRef]
51. Coleman, M.L.; Sahai, E.A.; Yeo, M.; Bosch, M.; Dewar, A.; Olson, M.F. Membrane blebbing during apoptosis results from caspase-mediated activation of ROCK I. *Nat. Cell. Biol.* **2001**, *3*, 339–345. [CrossRef]
52. Wang, F.; Dai, W.; Wang, Y.; Shen, M.; Chen, K.; Cheng, P.; Zhang, Y.; Wang, C.; Li, J.; Zheng, Y.; et al. The Synergistic In Vitro and In Vivo Antitumor Effect of Combination Therapy with Salinomycin and 5-Fluorouracil against Hepatocellular Carcinoma. *PLoS ONE* **2014**, *9*, 1–10. [CrossRef]
53. Kroemer, G.; Dallaporta, B.; Resche-Rigon, M. The mitochondrial death/life regulator in apoptosis and necrosis. *Annu. Ver. Physiol.* **1998**, *60*, 619–642. [CrossRef]
54. Zakeri, Z.; Lockshin, R.A. Cell death during development. *J. Immunol. Methods* **2002**, *265*, 3–20. [CrossRef]
55. Fessi, H.; Puisieux, F.; Devissaguet, J.P.; Ammoury, N.; Benita, S. Nanocapsule formation by interfacial polymer deposition following solvent displacement. *Int. J. Pharm.* **1989**, *55*, R1–R4. [CrossRef]
56. Mosmann, T. Rapid colorimetric assay for cellular growth and survival: Application to proliferation and cytotoxicity assays. *J. Immunol. Methods* **1983**, *65*, 55–63. [CrossRef]

57. D'oll-Boscardin, P.M.; Sartoratto, A.; Maia, B.H.L.N.S.; de Paula, J.P.; Nakashima, T.; Farago, P.V.; Kanunfre, C.C. In vitro cytotoxic potential of essential oils of *Eucalyptus benthamii* and its related terpenes on tumor cell lines. *Evid. Based Complement Altern. Med.* **2012**, *2012*, 1–8. [CrossRef]
58. Kifer, D.; Jakši'c, D.; Klari´c, M.Š. Assessing the Effect of Mycotoxin Combinations: Which Mathematical Model Is (the Most) Appropriate? *Toxins* **2020**, *12*, 153. [CrossRef]
59. Papazisis, K.T.; Geromichalos, G.D.; Dimitriadis, K.A.; Kortsaris, A.H. Optimization of the sulforhodamine B colorimetric assay. *J. Immunol. Methods* **1997**, *208*, 151–158. [CrossRef]
60. Ribble, D.; Goldstein, N.B.; Norris, D.A.; Shellman, Y.G. A simple technique for quantifying apoptosis in 96-well plates. *BMC Biotechnol.* **2005**, *5*, 1–12. [CrossRef]


**Sample Availability:** Samples of the compounds curcumin and methotrexate, and the polymeric nanocapsules NCUR-2, NMTX-1, and NCUR/MTX-2 are available from the authors.

*Communication*

# *Dalbergia ecastaphyllum* (L.) Taub. and *Symphonia globulifera* L.f.: The Botanical Sources of Isoflavonoids and Benzophenones in Brazilian Red Propolis

**Gari Vidal Ccana-Ccapatinta [1], Jennyfer Andrea Aldana Mejía [1], Matheus Hikaru Tanimoto [1], Milton Groppo [2], Jean Carlos Andrade Sarmento de Carvalho [3] and Jairo Kenupp Bastos [1,*]**

1 Laboratory of Pharmacognosy, School of Pharmaceutical Sciences of Ribeirão Preto, University of São Paulo (USP), Av. do Café s/n, Ribeirão Preto 14040-903, SP, Brazil; ccana.ccapatinta@usp.br (G.V.C.-C.); j.aldana@usp.br (J.A.A.M.); tanimoto@usp.br (M.H.T.)

2 Laboratory of Plant Systematics, Department of Biology, Faculty of Philosophy, Sciences and Letters at Ribeirão Preto, USP, Av. dos Bandeirantes 3900, Ribeirão Preto 14040-901, SP, Brazil; groppo@ffclrp.usp.br

3 Cooperativa de Apicultores de Canavieiras (COAPER), Av. Burundanga 1900, Canavieiras 45860-000, BA, Brazil; polemj@hotmail.com

* Correspondence: jkbastos@fcfrp.usp.br; Tel.: +55-16-3315-4230

Received: 27 March 2020; Accepted: 25 April 2020; Published: 28 April 2020

**Abstract:** The Brazilian red propolis (BRP) constitutes an important commercial asset for northeast Brazilian beekeepers. The role of *Dalbergia ecastaphyllum* (L.) Taub. (Fabaceae) as the main botanical source of this propolis has been previously confirmed. However, in addition to isoflavonoids and other phenolics, which are present in the resin of *D. ecastaphyllum*, samples of BRP are reported to contain substantial amounts of polyprenylated benzophenones, whose botanical source was unknown. Therefore, field surveys, phytochemical and chromatographic analyses were undertaken to confirm the botanical sources of the red propolis produced in apiaries located in Canavieiras, Bahia, Brazil. The results confirmed *D. ecastaphyllum* as the botanical source of liquiritigenin (**1**), isoliquiritigenin (**2**), formononetin (**3**), vestitol (**4**), neovestitol (**5**), medicarpin (**6**), and 7-*O*-neovestitol (**7**), while *Symphonia globulifera* L.f. (Clusiaceae) is herein reported for the first time as the botanical source of polyprenylated benzophenones, mainly guttiferone E (**8**) and oblongifolin B (**9**), as well as the triterpenoids $\beta$-amyrin (**10**) and glutinol (**11**). The chemotaxonomic and economic significance of the occurrence of polyprenylated benzophenones in red propolis is discussed.

**Keywords:** isoflavonoids; polyisoprenylated benzophenones; propolis; botanical sources

## 1. Introduction

The red propolis is the second most produced and traded type of propolis in Brazil and constitutes an important commercial asset for northeast Brazilian beekeepers. Apiaries devoted to Brazilian red propolis (BRP) production are frequently located around native populations of the fabaceous species *Dalbergia ecastaphyllum* (L.) Taub., which produces a red resin that is collected by bees to produce propolis in the beehive [1]. The studies by Daugsch et al. (2008) and Silva et al. (2008) were the first ones to describe, simultaneously in 2008, *D. ecastaphyllum* as the main botanical source of the BRP [2,3]. These initial studies together with the report of Piccinelli et al. (2011) confirmed the presence of a rich variety of phenolic compounds, in both the propolis and the plant resin, such as chalcones (e.g., isoliquiritigenin), flavonoids (e.g., luteolin, liquiritigenin), isoflavones (e.g., formononetin, biochanin A), isoflavans (e.g., vestitol, neovestitol, 7-O-methylvestitol), pterocarpanes (e.g., medicarpin, homopterocarpin, vesticarpan), and C30 isoflavans (retusapurpurins A and B) [4].

Subsequent studies have shown the great qualitative and quantitative variability of the constituents present in red propolis, which are influenced by regional and seasonal factors [5]. Nevertheless, reports on the presence of substantial amounts of polyprenylated benzophenones, such as gutifferone E and oblongifolin A in BRP, have appeared in the literature more frequently [6,7]. Even though the botanical source of these latter constituents was unknown, Piccinelli et al. (2011) inferred from a chemotaxonomic point of view that these compounds must derive from a member of the Clusiaceae family [4]. Important biological activities of BRP extracts, such as antimicrobial, fungicidal, and cytotoxic properties, have been correlated with the occurrence of polyprenylated benzophenones [6,8,9]. Thus, establishing its botanical source may help to increase red propolis production and to attain a higher degree of chemical standardization. Therefore, field surveys, phytochemical and chromatographic analyses were undertaken in order to identify the botanical source of polyprenylated benzophenones present in the red propolis produced by apiaries of the Beekeepers Association of Canavieiras (Cooperativa de Apicultores de Canavieiras—COAPER), Bahia, Brazil.

## 2. Results

A filed survey was carried out in March 2019 in order to collect local red propolis samples, the reddish resin of *D. ecastaphyllum*, and to identify other resin-producing plant species, especially of the Clusiaceae family, located in the surrounding flora of apiaries from the COAPER beekeepers association. The presence of individuals of *Symphonia globulifera* L.f. was evidenced and its resin was collected, transported to the laboratory, lyophilized, and kept under refrigeration.

### *2.1. Isolation of Polyprenylated Benzophenones from Symphonia globulifera and BRP*

Samples of both BRP and *S. globulifera* resin were extracted with aqueous ethanol 70% and submitted to partition with solvents of increasing polarities. The hexane fractions were separately submitted to chromatographic procedures that resulted in the isolation of two polyprenylated benzophenones (**8** and **9**) and two pentacyclic triterpenoids (**10** and **11**) from both fractions. The structures of the isolated compounds were established based on 1D and 2D-NMR spectroscopy, high-resolution MS, and comparison with the literature data for the polyprenylated benzophenones guttiferone E (**8**) [10], in a mixture with its double bond position isomer xanthochymol [11], oblongifolin B (**9**) [12], and the triterpenoids $\beta$-amyrin (**10**) [13] and glutinol (**11**) [14]. The NMR spectra of **8** and **9** in $CDCl_3$ displayed abnormalities that disappeared when spectra were recorded in $CD_3OD$ + 0.1% trifluoroacetic acid (TFA). The chemical structure of those isolated constituents (**8–11**), as well as other phenolic compounds (**1–7**) identified in BRP are displayed in Figure 1.

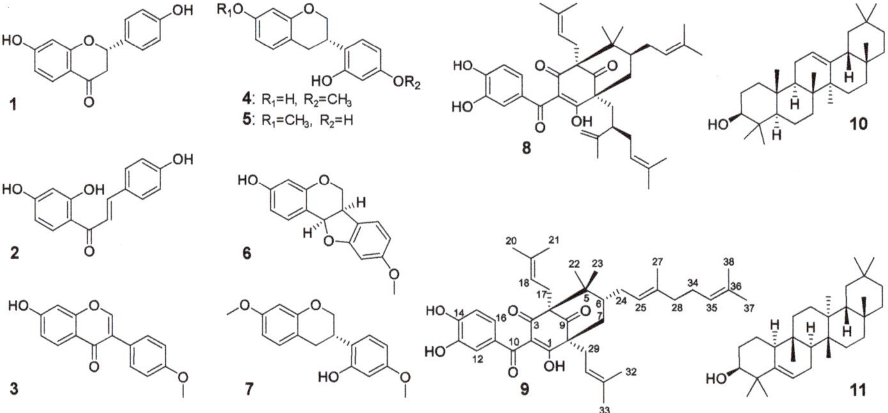


**Figure 1.** Chemical constituents of Brazilian red propolis.

The oblongifolins were first isolated from *Garcinia oblongifolia* by Hamed et al. (2006) [12]. The occurrence of oblongifolin A in samples of BRP was initially reported by Piccinelli et al. (2011), while further publications replicated this finding by LC-MS analyses. However, mass spectrometry of oblongifolins is not enough to identify them, since they bear the same molecular weight, while none of those previous reports have provided NMR data to confirm the occurrence of oblongifolin A in BRP [4,6–8]. Therefore, the NMR spectroscopic data of the isolated compound **9** in comparison with published data of oblongifolins A and B is presented in Table S1 (Supplementary Material). The $^{1}$H and $^{13}$C NMR chemical shifts of these two stereoisomers are almost identical. Nonetheless, significant differences are found in the $^{13}$C NMR spectra of these two oblongifolins at $\delta_{C\text{-}6}$ (47.8 for A and 44.2 for B), $\delta_{C\text{-}7}$ (40.8 for A and 43.3 for B), and $\delta_{C\text{-}8}$ (61.8 for A and 64.3 for B). Similarly, the $^{1}$H NMR spectra of oblongifolin B displays a characteristic axial coupling constant for H-7 of 13.0 Hz. Therefore, the isolated compound **9** corresponds to oblongifolin B and its isolation from BRP is here reported for the first time.

### *2.2. The Botanical Sources of Brazilian Red Propolis*

The chromatographic profiles of a typical red propolis sample and the resins of *D. ecastophyllum* and *S. globulifera* were recorded by a HPLC analysis method (Figure 2); the chromatographic parameters are described in the material and methods section.

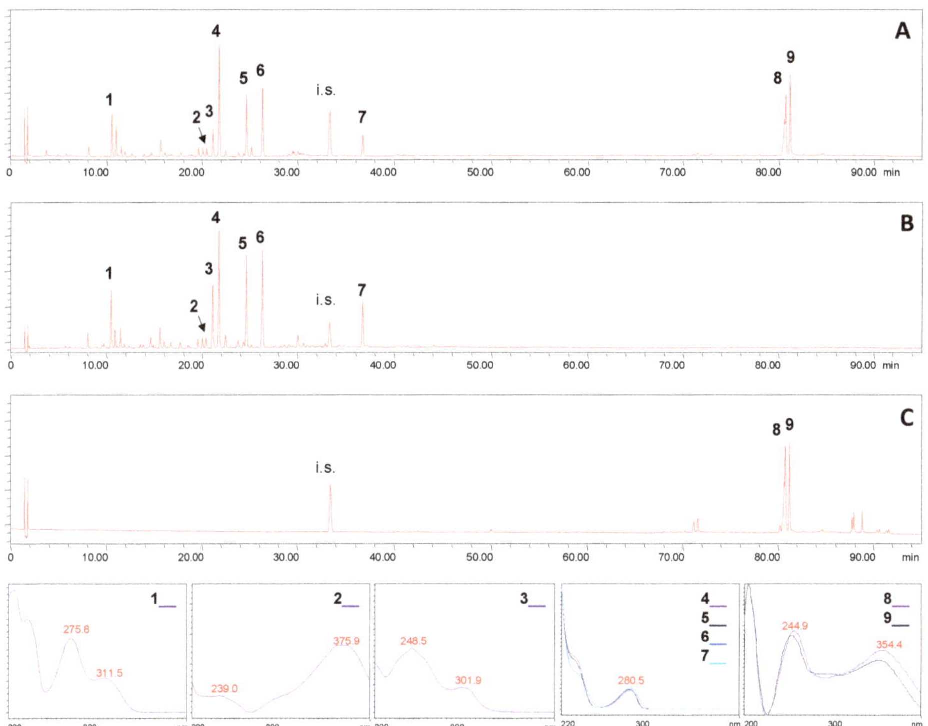


**Figure 2.** HPLC chromatographic profiles (275 nm) of Brazilian red propolis (**A**) in comparison with the resins of *Dalbergia ecastaphyllum* (**B**) and *Symphonia globulifera* (**C**). Numbers correspond to liquiritigenin (**1**), isoliquiritigenin (**2**), formononetin (**3**), vestitol (**4**), neovestitol (**5**), medicarpin (**6**), 7-*O*-neovestitol (**7**), guttiferone E (**8**), and oblongifolin B (**9**). UV spectra of compounds **1–9** are displayed at the bottom of the figure.

Comparison of both the HPLC chromatographic retention times and its corresponding UV spectra with the ones of the standard compounds **1–9** were used for assigning the identity of peaks in the chromatographic profiles of the analyzed samples. The chromatographic prolife of a typical sample of BRP (Figure 2A) displayed at least nine main chromatographic peaks corresponding to phenolic compounds (**1** and **2**), isoflavonoids (**3–5**, **7**), a pterocarpane (**6**) and benzophenones (**8–9**). The resin of *D. ecastaphyllum* (Figure 2B) displayed the presence of chromatographic peaks mainly in the region of the HPLC chromatogram from 10 to 40 min, confirming *D. ecastaphyllum* as the botanical source of liquiritigenin (**1**), isoliquiritigenin (**2**), formononetin (**3**), vestitol (**4**), neovestitol (**5**), medicarpin (**6**), and 7-*O*-neovestitol (**7**). In the lipophilic region of the chromatogram (Figure 2C), the resin of *S. globulifera* displayed two main chromatographic peaks, which were assigned to guttiferone E (**8**) and oblongifolin B (**9**), confirming *S. globulifera* as the botanical source of polyprenylated benzophenones.

## 3. Discussion

The occurrence of the flavonoid liquiritigenin (**1**); the chalcone isoliquiritigenin (**2**); the isoflavone formononetin (**3**); the isoflavans vestitol (**4**), neovestitol (**5**), and 7-*O*-methylvestitol (**7**); as well as the pterocarpane medicarpin (**6**) have been previously reported both in BRP and in the resin of its botanical source, *D. ecastaphyllum* [4]. From a chemotaxonomic point of view, these constituents, especially the isoflavonoids, are characteristic chemotaxonomic markers for Papilionoideae (Fabaceae) subfamily members, to which the *Dalbergia* genus belongs [15]. The results reported here confirm that the main botanical source of the red propolis produced in the COPAER beekeepers association of Canavieiras is *D. ecastaphyllum*. This species, which occurs in coastal sand dune and mangrove ecosystems, is present in large populations in the surrounding areas of the apiaries not only in Bahia state but also in the northeast Brazilian localities where red propolis is produced.

On the other hand, the presence of the prenylated benzophenones guttiferone E (**8**) and oblongifolin A in commercial samples of BRP was previously reported [6–8] and proposed as a characteristic that could differentiate Brazilian samples from other red propolis, e.g., Cuban propolis [4]. However, the botanical source of polyprenylated benzophenones present in BRP samples was unknown, but it was inferred from a chemotaxonomic point of view that these compounds would be collected by bees from a resin-producing plant belonging to the Clusiaceae family [4]. The isolation of the polyprenylated benzophenones guttiferone E (**8**) and oblongifolin B (**9**) reported here from samples of BRP and the resin of *S. globulifera* confirms that the botanical source of these compounds is a member of the Clusiaceae family, in agreement with the chemotaxonomical inference of Piccinelli et al. (2011). The occurrence of compounds **8** and **9** in both BRP and *S. globulifera* was also confirmed here by HPLC analysis. Therefore, the present report identifies, for the first time, *S. globulifera* as the botanical source of polyprenylated benzophenones of BRP produced in the locality of Canavieiras, Bahia. Unlike the data previously reported, describing the occurrence of oblongifolin A in BRP, we identified, instead, its stereoisomer oblongifolin B (**9**), the occurrence of which in BRP and *S. globulifera* is here reported for the first time. Although *S. globulifera* is a well-known source of polyprenylated benzophenones, this is the first report on the occurrence of guttiferone E (**8**) in this species. Additionally, this report demonstrates the value of chemotaxonomy for establishing the botanical sources of propolis.

The knowledge of propolis plant sources can help to increase propolis production and to attain a higher degree of chemical standardization [16]. Unlike the Brazilian green propolis, the most produced propolis in Brazil, the market value of red propolis raw material is double the price of green propolis. Therefore, commercial and academic interest has been awakened to extend the area of bee pasture to increase the production of red propolis. The findings reported here demonstrate that BRP does not have only a single botanical source but at least two, *D. ecastaphyllum* and *S. globulifera*, which contribute different chemical constituents, principally isoflavonoids and polyprenylated benzophenones, which should be taken into account when installing new beehives for red propolis production in the northeast of Brazil.

## 4. Material and Methods

### 4.1. Field Survey

A field survey was carried out from 24 March to 5 April 2019 aiming to collect local red propolis samples, the resin of *D. ecastaphyllum*, and to identify other resin-producing species, especially of the Clusiaceae family, located in the surrounding flora of apiaries from the beekeepers association of Canavieiras (*Cooperativa de Apicultores de Canavieiras*—COAPER), Bahia, Brazil. The presence of individuals of *Symphonia globulifera* was evidenced, and their resins were collected, transported to the laboratory, lyophilized, and kept under refrigeration. The plant vouchers were identified and deposited as *Symphonia globulifera* L.f., SPFR 17770; *Dalbergia ecastaphyllum* (L.) Taub., SPFR 17,771 at the Herbarium SPFR of the Department of Biology, Faculty of Philosophy, Sciences and Letters at Ribeirão Preto, FFCLRP, University of São Paulo, USP.

### 4.2. Extraction and Isolation of Triterpenoids and Polyprenylated Benzophenones

Two hundred grams of BRP were extracted with aqueous ethanol 70% (Vetec Química, Rio de Janeiro, Brazil) for three times, 1:10 sample/solvent ratio, and g/mL furnishing 140 g of crude extract. The extract was then mixed with 200 g of microcrystalline cellulose (Sigma Aldrich, St. Louis, MO, USA) and submitted to solid-liquid partition with hexanes (Vetec Química, Rio de Janeiro, Brazil), three times of 500 mL, furnishing 23.8 g of hexane crude fraction. The hexane fraction was then submitted to vacuum liquid chromatography (VLC) by using 150 g of silica gel (Sigma Aldrich, St. Louis, MO, USA), 40–63 µm particle size, as the stationary phase and mixtures of increasing polarity of hexanes:ethyl acetate (100:0→30:70) as the mobile phase. The eluted fractions were monitored by thin layer chromatography (TLC) and pooled by similarity, generating four fractions: F1 (6 g), F2 (9.26 g), F3 (2.27), and F4 (1.1 g). TLC was carried out on precoated glass TLC silica gel $60F_{254}$ plates (Merck, Darmstadt, Germany), with detection accomplished by visualization with a UV lamp at 254 and 360 nm, followed by spraying with a 1% solution of 2-aminoethyl diphenylborinate in methanol (w/v). A portion of fraction F2 (500 mg) was submitted to centrifugal thin-layer chromatography on a Chromatotron device (Harrison Research, USA) by using a 1-mm disk of silica gel (10–40 µm particle size) as the stationary phase, and mixtures of increasing polarity of hexanes:ethyl acetate (100:0→30:70) as the mobile phase, affording compounds **8** (70 mg) and **9** (30 mg). A portion of fraction F1 (500 mg) was submitted to flash chromatography on an Isolera One equipment (Biotage, Sweden) by using a 10-g cartridge of silica gel (40 µm particle size) and mixtures of increasing polarity of hexanes:ethyl acetate (100:0→50:50) as the mobile phase, affording compounds **10** (30 mg) and **11** (40 mg). A portion of the *S. globulifera* resin (100 g) was also submitted to the procedures as describe above furnishing the same compounds **8–11**.

One-dimensional and 2-D NMR spectra of compounds **8–11** were acquired in a Brucker DRX500 NMR spectrometer (Brucker, Santa Barbara, CA, USA) operating at a frequency of 500 MHz for $^1H$ and 125 MHz for $^{13}C$ by using $CDCl_3$ and $CD_3OD$ + 0.1% TFA as deuterated solvents (Sigma Aldrich, St. Louis, MO, USA). High-resolution mass spectrometry data of compounds **8** and **9** were obtained by direct infusion in negative ionization mode on an orbitrap mass spectrometer (Thermo Scientific, San Jose, CA, USA).

*Guttiferone E* (**8**): yellow amorphous powder; UV (HPLC-online) $\lambda_{max}$: 249.6, 354.4 nm; $^1H$ and $^{13}C$ NMR data as reported by Gustafson et al., 1992 [10]. HRESIMS negative mode *m/z* 601.3546 $[M - H]^-$ (calcd. for $C_{38}H_{50}O_6$, 601.3529), 525.3232, 183.0116, 109.0284.

*Oblongifolin B* (**9**): yellow amorphous powder; UV (HPLC-online) $\lambda_{max}$: 244.9, 350.9 nm; $^1H$ and $^{13}C$ NMR data as reported by Hamed et al., 2006 [12]. HRESIMS negative mode *m/z* 601.3546 $[M - H]^-$ (calcd. for $C_{38}H_{50}O_6$, 601.3529), 525.3232, 333.1349, 183.0116, 109.0284.

*β-Amyrin* (**10**): white needles; $^1H$ and $^{13}C$ NMR data as reported by Lima et al., 2004 [13].

*Glutinol* (**11**): white needles; $^{1}$H and $^{13}$C NMR data as reported by Mahato et al., 1981 [14]

### 4.3. HPLC Analyses of Botanical Sources and BRP

The chromatographic profiles of BRP samples, and the resins of *D. ecastaphyllum* and *S. globulifera* were obtained by using an HPLC-DAD method on a Waters 1500-series setup (Milford, CT, USA) with an Ascentis Express C18 column (2.7 μm, 150 × 4.60 mm) as the stationary phase and mixtures of water with 0.1% formic acid (B) and acetonitrile (B) as the mobile phase. The flow rate was set at 1.0 mL/min in a gradient elution mode as follows: 20→50% B in 40 min, 50→100% B in 90 min, 100% B (isocratic) until 95 min, 100→20% B until 100 min, 20% B (isocratic) up to 105 min. Benzophenone (20 μg/mL) was used as the internal standard (i.s.). One milligram of each resin sample or propolis extract was dissolved in methanol and subjected to chromatographic analysis. Standard compounds liquiritigenin (**1**), isoliquiritigenin (**2**), formononetin (**3**), vestitol (**4**), neovestitol (**5**), medicarpin (**6**), and 7-O-neovestitol (**7**) were available in the isolated compound library of the laboratory. The standard compounds **1–7** and the isolated compounds guttiferone E (**8**) and oblongifolin B (**9**) were dissolved in methanol (20 μg/mL) for HPLC analyses.

**Supplementary Materials:** The $^{1}$H and $^{13}$C NMR spectra of isolate compounds **8** and **9** are available online.

**Author Contributions:** Conceptualization, G.V.C.-C., J.A.A.M., M.G., J.C.A.S.d.C. and J.K.B.; methodology, G.V.C.-C., J.A.A.M. and J.K.B.; isolation procedures, G.V.C.-C., J.A.A.M. and M.H.T.; resources, J.K.B.; plant identification, M.G.; field research, G.V.C.-C., J.C.A.S.d.C. and J.K.B.; manuscript preparation, G.V.C.-C., J.A.A.M. and J.K.B.; funding acquisition, J.K.B. All authors have read and agreed with the published version of the manuscript.

**Funding:** This research was funded by São Paulo Research Foundation (FAPESP, grants #2017/04138-8, 2017/26252-7, #2018/13700-4 and #2019/01697-1), Coordination for the Improvement of Higher Education Personnel (CAPES) and the National Council for Scientific and Technological Development (CNPq).

**Acknowledgments:** Jairo Kenupp Bastos would like to register that it was an honor to be a student of James D. McChesney, an outstanding scientist and human being.

**Conflicts of Interest:** The authors declare no conflict of interest.

## References

1. Salatino, A.; Salatino, M.L.F. Brazilian red propolis: Legitimate name of the plant resin source. *MOJ Food Process. Technol.* **2018**, *6*, 21–22. [CrossRef]
2. Daugsch, A.; Moraes, C.S.; Fort, P.; Park, Y.K. Brazilian red propolis—Chemical composition and botanical origin. *Evidence-Based Complement. Altern. Med.* **2008**, *5*, 435–441. [CrossRef] [PubMed]
3. Silva, B.B.; Rosalen, P.L.; Cury, J.A.; Ikegaki, M.; Souza, V.C.; Esteves, A.; Alencar, S.M. Chemical composition and botanical origin of red propolis, a new type of brazilian propolis. *Evidence-Based Complement. Altern. Med.* **2008**, *5*, 313–316. [CrossRef] [PubMed]
4. Piccinelli, A.L.; Lotti, C.; Campone, L.; Cuesta-Rubio, O.; Campo Fernandez, M.; Rastrelli, L. Cuban and Brazilian red propolis: Botanical origin and comparative analysis by high-performance liquid chromatography–photodiode array detection/electrospray ionization tandem mass spectrometry. *J. Agric. Food Chem.* **2011**, *59*, 6484–6491. [CrossRef] [PubMed]
5. do Nascimento, T.G.; dos Santos Arruda, R.E.; da Cruz Almeida, E.T.; dos Santos Oliveira, J.M.; Basílio-Júnior, I.D.; Celerino de Moraes Porto, I.C.; Rodrigues Sabino, A.; Tonholo, J.; Gray, A.; Ebel, R.A.E.; et al. Comprehensive multivariate correlations between climatic effect, metabolite-profile, antioxidant capacity and antibacterial activity of Brazilian red propolis metabolites during seasonal study. *Sci. Rep.* **2019**, *9*, 1–16. [CrossRef] [PubMed]
6. Rufatto, L.C.; dos Santos, D.A.; Marinho, F.; Henriques, J.A.P.; Roesch Ely, M.; Moura, S. Red propolis: Chemical composition and pharmacological activity. *Asian Pac. J. Trop. Biomed.* **2017**, *7*, 591–598. [CrossRef]
7. Fasolo, D.; Bergold, A.M.; von Poser, G.; Teixeira, H.F. Determination of benzophenones in lipophilic extract of Brazilian red propolis, nanotechnology-based product and porcine skin and mucosa: Analytical and bioanalytical assays. *J. Pharm. Biomed. Anal.* **2016**, *124*, 57–66. [CrossRef] [PubMed]

8. Trusheva, B.; Popova, M.; Bankova, V.; Simova, S.; Marcucci, M.C.; Laguna Miorin, P.; Da, F.; Pasin, R.; Tsvetkova, I. Bioactive constituents of brazilian red propolis. *Evidence-Based Complement. Altern. Med.* **2006**, *3*, 249–254. [CrossRef] [PubMed]
9. Dantas Silva, R.P.; Machado, B.A.S.; de Barreto, G.A.; Costa, S.S.; Andrade, L.N.; Amaral, R.G.; Carvalho, A.A.; Padilha, F.F.; Barbosa, J.D.V.; Umsza-Guez, M.A. Antioxidant, antimicrobial, antiparasitic, and cytotoxic properties of various Brazilian propolis extracts. *PLoS ONE* **2017**, *12*, e0172585. [CrossRef] [PubMed]
10. Gustafson, K.R.; Blunt, J.W.; Munro, M.H.G.; Fuller, R.W.; McKee, T.C.; Cardellina, J.H.; McMahon, J.B.; Cragg, G.M.; Boyd, M.R. The guttiferones, HIV-inhibitory benzophenones from *Symphonia globulifera*, *Garcinia livingstonei*, *Garcinia ovalifolia* and *Clusia rosea*. *Tetrahedron* **1992**, *48*, 10093–10102. [CrossRef]
11. Li, J.; Gao, R.; Zhao, D.; Huang, X.; Chen, Y.; Gan, F.; Liu, H.; Yang, G. Separation and preparation of xanthochymol and guttiferone E by high performance liquid chromatography and high speed counter-current chromatography combined with silver nitrate coordination reaction. *J. Chromatogr. A* **2017**, *1511*, 143–148. [CrossRef] [PubMed]
12. Hamed, W.; Brajeul, S.; Mahuteau-Betzer, F.; Thoison, O.; Mons, S.; Delpech, B.; Van Hung, N.; Sévenet, T.; Marazano, C.; Oblongifolins, A.-D. Polyprenylated benzoylphloroglucinol derivatives from *Garcinia oblongifolia*. *J. Nat.* **2006**, *69*, 774–777.
13. Da Paz Lima, M.; De Campos Braga, P.A.; Macedo, M.L.; Da Silva, M.F.D.G.F.; Ferreira, A.G.; Fernandes, J.B.; Vieira, P.C. Phytochemistry of *Trattinnickia burserifolia*, *T. rhoifolia*, and *Dacryodes hopkinsii*: Chemosystematic implications. *J. Braz. Chem. Soc.* **2004**, *15*, 385–394. [CrossRef]
14. Mahato, S.B.; Das, M.C.; Sahu, N.P. Triterpenoids of *Scoparia dulcis*. *Phytochemistry* **1981**, *20*, 171–173. [CrossRef]
15. Wink, M. Evolution of secondary metabolites in legumes (Fabaceae). *South African J. Bot.* **2013**, *89*, 164–175. [CrossRef]
16. Bankova, V.; Popova, M.; Trusheva, B. The phytochemistry of the honeybee. *Phytochemistry* **2018**, *155*, 1–11. [CrossRef] [PubMed]


**Sample Availability:** Samples of the compounds **1–11** are available from the authors.

*Communication*

# Oxyresveratrol Possesses DNA Damaging Activity

**Sarayut Radapong [1,2,*], Satyajit D. Sarker [2] and Kenneth J. Ritchie [2]**

[1] Medicinal Plant Research Institute, Department of Medical Sciences, Ministry of Public Health, Nonthaburi 11000, Thailand

[2] Centre for Natural Products Discovery, School of Pharmacy and Biomolecular Sciences, Liverpool John Moores University, Byrom Street, Liverpool L3 3AF, UK; S.Sarker@ljmu.ac.uk (S.D.S.); k.j.ritchie@ljmu.ac.uk (K.J.R.)

* Correspondence: Sarayutradapong@gmail.com; Tel.: +44-737-873-6587

Academic Editor: Muhammad Ilias
Received: 30 April 2020; Accepted: 29 May 2020; Published: 1 June 2020

**Abstract:** *Artocarpus lakoocha* Wall. ex Roxb. (family: *Moraceae*) has been used as a traditional Thai medicine for the treatment of various parasitic diseases. This species has been reported to be the source of phytochemicals, which show potent biological activities. The objective of this study was to investigate the phytochemical profile of the extracts of the heartwood of *A. lakoocha* and their pro-oxidant activity in vitro. The heartwood was ground, extracted, and then chromatographic and spectroscopic analyses were carried out; oxyresveratrol was identified as the major component in the extracts. The pro-oxidant activity was investigated using DNA-nick, reactive oxygen species and reducing assays. The results showed that oxyresveratrol induced DNA damage dose-dependently in the presence of copper (II) ions. It was also found to generate reactive oxygen species (ROS) in a dose-dependent manner and reduce copper (II) to copper (I). It is concluded that oxyresveratrol is the most abundant stilbenoid in *A. lakoocha* heartwood. The compound exhibited pro-oxidant activity in the presence of copper (II) ions, which may be associated with its ability to act as an anticancer compound.

**Keywords:** oxyresveratrol; *Artocarpus lakoocha*; DNA damaging; pro-oxidant

---

## 1. Introduction

Oxyresveratrol belongs to the group of phytochemicals known as hydroxystilbenoids and has a molecular structure similar to the well-known phytochemical resveratrol. They are monomeric stilbene comprising two aromatic rings joined by an ethylene bridge, of which, the *trans* isomer is the most common configuration (Figure 1) [1]. The natural sources of these stilbenes are abundant in grapes, itadori, peanuts, and *A. lakoocha* [2–4]. The compounds demonstrate similar biological activities and are used in the treatment of atherosclerosis, inflammatory, pigmentation, and carcinogenesis [5]. Interestingly, however, several biological activities are unique to oxyresveratrol (antivirus and antihelminthics) [6,7]. Further, several reports suggest that oxyresveratrol or the heartwood extract of *A. lakoocha* might cause DNA damage and cause oxidative stress [8–10]. Previous studies report that resveratrol and its derivatives are capable of double-stranded DNA cleavage specifically in the presence of copper ions and oxygen [10]. Piceatannol, a resveratrol derivative, fragmented DNA in human peripheral blood [9]. Some polyphenolic compounds are already well-known to cause DNA damage and are being investigated as useful leads in the development of chemotherapeutic drugs [11]. In this study, we consequently investigated the ability of oxyresveratrol from the heartwood of *A. lakoocha* to induce DNA damage and the mechanism by which the damage occurs.

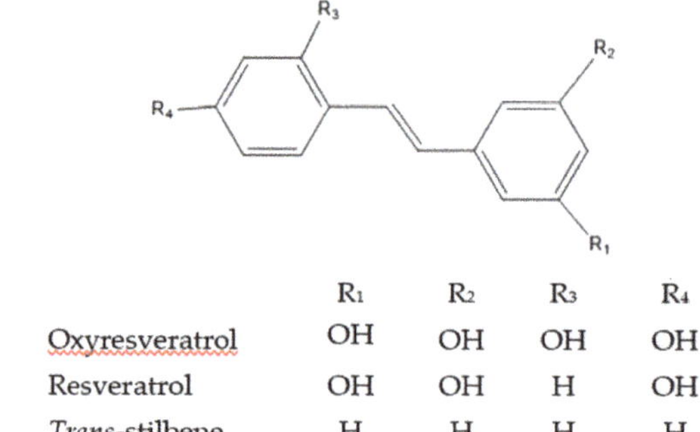


| | | $R_1$ | $R_2$ | $R_3$ | $R_4$ |
|---|---|---|---|---|---|
| 1 | Oxyresveratrol | OH | OH | OH | OH |
| 2 | Resveratrol | OH | OH | H | OH |
| 3 | *Trans*-stilbene | H | H | H | H |

**Figure 1.** Chemical structure of oxyresveratrol, resveratrol and *trans*-stilbene.

## 2. Results and Discussion

### 2.1. HPLC Analysis

Phytochemical profiling of *A. lackoocha heartwood* was carried out by HPLC. Each extract was investigated (water, ethanol and ethyl acetate) and found to contain one single dominant peak with a retention time consistent with that of the chromatogramic standard of oxyresveratrol (Figure 2A). The chromatographic standard of resveratrol was also used allowing accurate calculation of the amounts of each compound in each extract (Figure 2B). Further authentication of the predominant peak was carried out using NMR with the $^{13}$C-NMR and $^{1}$H-NMR; the chemical shifts and structure elucidated in Figure 3. were in good agreement with the standard. Mass spectra of the main peak detected at RT 14.0 supposed to be oxyresveratrol showed the spectra patterns were the same as the standard (showed in the Supplementary Materials). Its molecular formula $C_{14}H_{12}O_4$ was deduced from the ESI-MS spectrum in positive ion mode by the hydriated molecular ion peak at $m/z$ 245 $[M + H]^+$. The amount of oxyresveratrol in the plant was consistent with the range reported by Maneechai et al. [3], while the amount of resveratrol was lower compared to the report from Borah et al. [4].

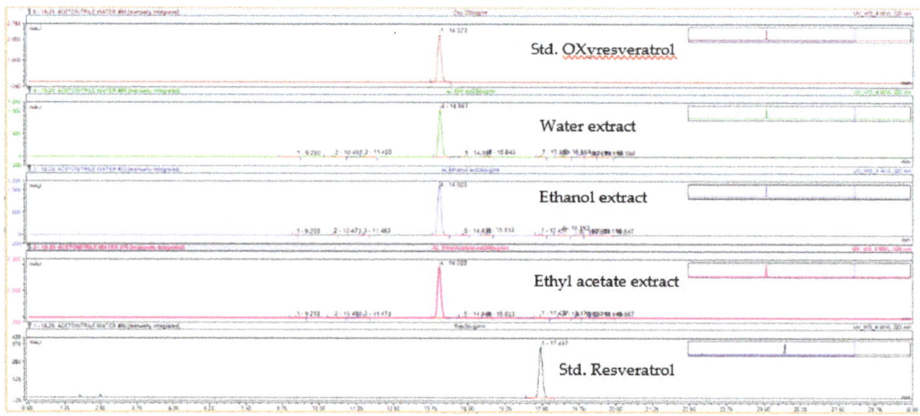


**(A)**

**Figure 2.** *Cont.*

| | Oxyresveratrol (μg/mL) | Yield (%) | Resveratrol (μg/mL) | Yield (%) |
|---|---|---|---|---|
| Water extract | 335.45±0.83 | 33.55±0.08 | 3.73±0.00 | 0.37±0.00 |
| Ethanol extract | 378.14±9.03 | 37.81±0.90 | 4.46±0.01 | 0.45±0.00 |
| Ethyl acetate extract | 642.26±0.28 | 64.23±0.03 | 6.54±0.14 | 0.65±0.01 |

**(B)**

**Figure 2.** Chemical profiles and contents of oxyresveratrol and resveratrol in *A. lakoocha* heartwood extracts. (**A**) Chemical profiles of the compounds in the extracts at 250 μg/mL were analysed by HPLC at a wavelength of 320 nm; water extract, ethanol extract, ethyl acetate extract, and standard oxyresveratrol and resveratrol at 10 μg/mL. (**B**) Amount of oxyresveratrol and resveratrol detected in *A. lakoocha* heartwood extracts.

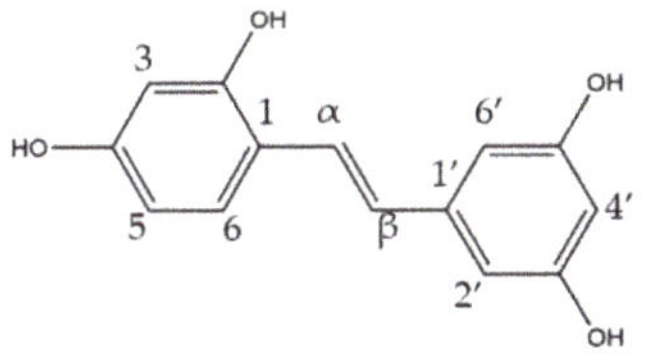

| Position | $\delta_C$ | $\delta_H$ m (*J* in Hz) | Position | $\delta_C$ | $\delta_H$ m (*J* in Hz) |
|---|---|---|---|---|---|
| 1 | 116.4 | - | 1′ | 140.7 | - |
| 2 | 156.0 | - | 2′ | 104.6 | 6.53 dd (2.1, 2.5) |
| 3 | 102.7 | 6.44 d (2.4) | 3′ | 158.6 | - |
| 4 | 158.2 | - | 4′ | 101.4 | 6.24 t (2.1) |
| 5 | 107.5 | 6.39 dd (2.4, 8.4) | 5′ | 158.6 | - |
| 6 | 127.3 | 7.41 d (8.4) | 6′ | 104.6 | 6.53 dd (2.1, 2.5) |
| α | 123.4 | 7.41 d (16.4) | 2-OH | | 8.48 brs |
| β | 125.4 | 6.89 d (16.4) | 4-OH | | 8.30 brs |
| | | | 3′-OH | | 8.09 brs |
| | | | 5′-OH | | 8.09 brs |

**Figure 3.** $^{13}$C-NMR (300 MHz, Acetone-$d_6$) and $^{1}$H-NMR (300 MHz, Acetone-$d_6$) data of oxyresveratrol.

### 2.2. DNA Nicking Assay

The ability of oxyresveratrol to cause DNA damage in the presence of 50 μM copper(II) was investigated (Figure 4A). Oxyresveratrol was consequently found to cause double-strand breaks in supercoiled plasmid DNA in a dose-dependent manner (lane 5–10); 200 μM was the most effective at inducing this DNA damage, while no DNA damage was found at low oxyresveratrol concentrations (0.4–2 μM). The DNA damaging activity of oxyresveratrol was found to be dependent on the presence of Cu(II) ions, and not Fe(II) or Zn(II)(data not shown). Comparison of the DNA-damaging capacity of equivalent concentrations (50 μM) of stilbene derivatives revealed only DNA damage being induced by oxyresveratrol and resveratrol (Figure 4B), of which oxyresveratrol was the most damaging. The DNA-damaging capacity of a series of hydrostilbenoids was previously investigated and in contrast to the data reported here resveratrol was found to be the most active [6]. The DNA-damaging capacity was specifically induced by copper ions, one of the most abundant metal ions in biological systems [12]. This raises the possibility that the DNA damage observed may be due to ROS generation via the Fenton reaction consistent previous observations made by Moran et al. [13] who observed similar findings with a number of other phenolic compounds.

### *2.3. Pro-oxidant Activity of ROS Generation and Copper Reduction*

Oxyresveratrol was found to generate ROS in the presence of copper(II) in a dose-dependent manner and in a higher amount compared to resveratrol and trans-stilbene (Figure 4C,D). The three extracts also exhibited ROS formation consistent with oxyresveratrol content (data not shown). Investigating the ability of oxyresveratrol to reduce $Cu^{2+}$ ions, we report the amount of $Cu^{+}$ produced by oxyresveratrol was dose-dependent (Figure 4E). Oxyresveratrol was also found to produce significantly higher amounts of copper (I) than resveratrol (Figure 4F). Trans-stilbene showed no copper-reducing activity; the ability of oxyresveratrol to generate ROS is consequently reported to be specifically induced by copper ions. Speculation exists as to the mechanism responsible although it has been previously noted that copper–oxo complex or copper–peroxide complex have the capacity to cause DNA damage following the reactions shown in Scheme 1 [6,14,15]. However, the exact structural feature of the complexes that caused damage to the DNA is still unclear. Copper to some certain extent is well-known to be one of the transition metals playing an important role in biological functions. However, It has been reported that significantly elevated levels of copper have been found in both serum and tissue of cancer patients [16]. Moreover, elevated copper levels have been documented in breast, cervical, ovarian, lung, prostate, stomach cancers, and leukemia. Serum copper concentration has also been found to correlate with tumour incidence and burden; malignant progression in Hodgkin's lymphoma; and leukemic, sarcoma, brain, breast, cervical, liver, and lung cancer [17]. The generation of ROS by stilbenoids in the presence of copper may be one of the key mechanisms by which they (stilbenoids) may be useful in the treatment of cancer.

$$\text{Oxyresveratrol} + Cu^{2+} \rightarrow Cu^{1+} + \text{Oxyresveratrol radical} + H^{+}$$

$$Cu^{1+} + O_2 \rightarrow Cu^{2+} + \bullet O_2^{-}$$

$$\bullet 2O_2^{-} + 4H^{+} \rightarrow 2H_2O_2 + O_2$$

$$Cu^{1+} + 2H_2O_2 \rightarrow \text{Copper–peroxide complex or}$$

$$Cu^{1+} + 2H_2O_2 \rightarrow \text{Copper -Oxo-species complex}$$

**Scheme 1.** The assumed mechanism of oxyresveratrol for ROS production. The compound reduces cupric (II) into cuprous (I), then the cuprous reacting with $O_2$ generates superoxide radical ($\bullet O_2^{-}$); the radical causes direct oxidative stress or reacts to hydrogen ions presenting hydrogen peroxide ($H_2O_2$). Lastly, the cuprous ion and hydrogen peroxide in the reaction generates a copper–peroxide complex or copper oxo-species complex.

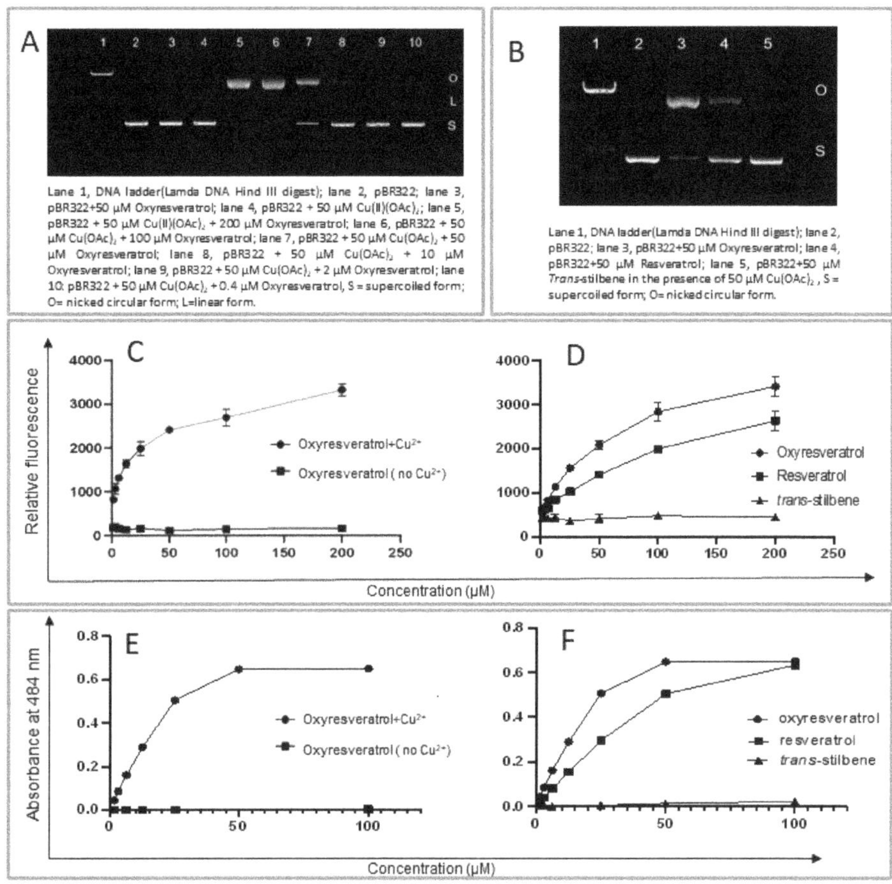


**Figure 4.** Effect of oxyresveratrol on DNA, ROS generation and copper reduction in cell-free assays. (**A**) DNA damage induced by oxyresveratrol at six concentrations of $Cu(OAc)_2$. (**B**) Comparison of DNA damage induced by oxyresveratrol, resveratrol, and *trans*-stilbene. (**C**) Generation of ROS by oxyresveratrol in the presence or absence of Cu(II). (**D**) Comparison of generation of ROS by oxyresveratrol, resveratrol, and *trans*-stilbene. (**E**) Reduction of Cu(II) by oxyresveratrol. (**F**) Comparison of copper-reducing ability of the three stilbenoids. Values express mean ± SD ($n$ = 3).

## 3. Materials and Methods

### 3.1. Standard

Oxyresveratrol (>97%), resveratrol, *trans*-stilbene, and copper acetate ($Cu(OAc)_2$) (99.99%) were purchased from Sigma-Aldrich (Buchs, Switzerland); pBR322 Supercoiled plasmid DNA was purchased from Thermo Fisher Scientific (Vilnius, Lithuania); deionized water (<18 MΩ cm resistivity) was obtained from the Milli-Q Element water purification system Millipore S.A.S. (Molcheim, France).

### 3.2. Plant Material

The heartwood of *A. lakoocha* was collected from Chanthaburi province, Thailand, in June 2018. The voucher specimen (DMSC 5237) was deposited at the DMSC International Herbarium, Department of Medical Sciences, Ministry of Public Health.

### *3.3. Extraction*

The heartwood was ground into powder and dried until constant weight. The dried powder was refluxed with distilled water or ethanol (250 g of the powder: 3 L of the solvents) following the procedure modified from Borah et al. [4]. The other fraction was extracted by ethyl acetate using Soxhlet apparatus (800 g of the powder: 4.5 L of the solvents). All solvents were then filtered with Whatman™ filter paper. Finally, the three extracts were concentrated and dried by rotary evaporation or lyophilizaion.

### *3.4. Chemical Analysis*

Analytical HPLC was performed on a Dionex UPLC 3000 (Thermoscientific, UK) HPLC coupled with a photo-diode-array (PDA) detector (Thermoscientific). Extracts were diluted in methanol and analysed on a Phenomenex C18 column (4.6 mm × 15 cm, 5 μm, Phenomenex, Torrance, CA, USA), flow rate 1 mL/min, mobile phase gradient of water (A) and acetonitrile (B) both containing 0.1% TFA: 10–50% B, 0–30 min; 100% B, 30–32 min; 10–100% B, 33–35 min, monitored at variable UV–vis wavelengths (210, 254, 280 and 320 nm). The column temperature was set at 25 °C. The NMR spectroscopic analysis was performed in acetone-$d_6$ solution on a Bruker AMX300 NMR spectrometer (300 MHz for $^{13}C$ and $^{1}H$).

### *3.5. DNA Nicking Assay*

The DNA break assay was modified from Subramanian et al. [10] using the reaction mixture of pBR322 plasmid DNA (200 ng) in the presence of $Cu(OAc)_2$ (50 μM) or 50 μM $FeCl_3$ or 50 μM $ZnCl_2$ in 10 mM HEPES buffer pH 7.2. The stock solutions of oxyresveratrol, resveratrol and *trans*-stilbenoid were prepared at the concentration of 1000 μM (1mM) in 1% DMSO/Milli-Q water. Whereas, *A. lakoocha* heartwood extracts were prepared at the concentration of 1.00 mg/mL in 1% DMSO/Milli-Q water. The reactions were initiated by adding 5 μL of each compound into 20 μL of the reaction mixture and incubated at 37 °C for 15 min. After the incubation, the products were loaded into 0.80% agarose gel subjected to electrophoresis at 75 V for 1.50 h. The image of the gel was quantified by Bio-Rad documentation imaging system (Hercules, CA, USA).

### *3.6. Reactive Oxygen Species Assay*

2′,7′-Dichlorodihydrofluorescein diacetate ($H_2$DCFDA) was used to measure ROS generated in the reactions between oxyresveratrol in the presence or absence of copper (II) ions. The reaction was initiated by adding 50 μL of 200 μM Cu(II) or HEPES buffer into each reaction of 200 μL that contained oxyresveratrol or other derivatives. After the addition, the tubes were incubated at 37 °C for 15 min. Dichlorofluorescein (DCF), the fluorescence product of the reaction, was quantitated spectrofluorometric.

### *3.7. Copper Reducing Assay*

Oxyresveratrol was colorimetrically determined using the Cu(I)-stabilizing reagent bathocuproine disulfonic acid. The reaction mixture (total volume for each tested concentration 200 μL) contained 50 μL of 2,000 mM (2M) bathocuproine disulfonic acid and 100 μL of oxyresveratrol in 10 mM HEPES buffer. The reaction was initiated by adding 50 μL of 200 mM $Cu(OAc)_2$ in the sample tubes and the absorbance of the mixture at 484 nm was read at the end of 5 min.

## 4. Conclusions

Oxyresveratrol was the major stilbenoid found in all three extracts tested (water, ethanol and ethyl acetate). The compound exhibited both dose-dependent DNA damage in the presence of copper ions and ROS generation. In consideration of these findings, it can be concluded that oxyresveratrol has the potential to be developed as a promising anticancer drug candidate.

**Supplementary Materials:** The following are available online, Figure S1: $^1$H-NMR Oxyresveratrol, Figure S2: $^{13}$C-NMR Oxyresveratrol, Figure S3: ESI-MS spectrum of standard oxyresveratrol, Figure S4: ESI-MS spectrum of the peak RT 14.0 in the water extract, Figure S5: ESI-MS spectrum of the peak RT 14.0 in the ethanol extract, Figure S6: ESI-MS spectrum of the peak RT 14.0 in the ethyl acetate extract.

**Author Contributions:** Conceptualization, S.R., S.D.S. and K.J.R.; methodology, S.R. and K.J.R.; software, K.J.R.; validation, S.R., S.D.S. and K.J.R.; formal analysis, S.R.; investigation, S.R. and K.J.R.; resources, S.R., S.D.S. and K.J.R.; data curation, S.R.; writing—original draft preparation, S.R.; writing—review and editing, S.D.S. and K.J.R.; visualization, S.R.; supervision, S.D.S. and K.J.R.; project administration, K.J.R.; funding acquisition, S.D.S. and K.J.R. All authors have read and agreed to the published version of the manuscript.

**Funding:** This research was funded by The Royal Thai government, Thailand; in the PhD project no. R253184.

**Acknowledgments:** The author would like to express profound gratitude to The Medicinal Plant Research Institute, Department of Medical Sciences, Thailand for providing the plant materials and identifying the plant species and the excellent support from Liverpool John Moores University for the laboratory facilities.

**Conflicts of Interest:** The authors declare no conflict of interest.

## References

1. Riviere, C.; Pawlus, A.D.; Merillon, J.-M. Natural stilbenoids: Distribution in the plant kingdom and chemotaxonomic interest in Vitaceae. *Nat. Prod. Rep.* **2012**, *29*, 1317–1333. [CrossRef] [PubMed]
2. Burns, J.; Yokota, T.; Ashihara, H.; Lean, M.E.J.; Crozier, A. Plant Foods and Herbal Sources of Resveratrol. *J. Agric. Food Chem.* **2002**, *50*, 3337–3340. [CrossRef] [PubMed]
3. Maneechai, S.; Likhitwitayawuid, K.; Sritularak, B.; Palanuvej, C.; Ruangrungsi, N.; Sirisa-Ard, P. Quantitative analysis of oxyresveratrol content in Artocarpus lakoocha and 'Puag-Haad'. *Med Princ Pr.* **2009**, *18*, 223–227. [CrossRef] [PubMed]
4. Borah, H.J.; Singhal, R.; Hazarika, S. Artocarpus lakoocha roxb.: An untapped bioresource of resveratrol from North East India, its extractive separation and antioxidant activity. *Ind. Crop. Prod.* **2017**, *95*, 75–82. [CrossRef]
5. Akinwumi, B.C.; Bordun, K.-A.M.; Anderson, H.D. Biological activities of stilbenoids. *Int. J. Mol. Sci.* **2018**, *19*, 792. [CrossRef] [PubMed]
6. Galindo, I.; Hernaez, B.; Berna, J.; Fenoll, J.; Cenis, J.L.; Escribano, J.M.; Alonso, C. Comparative inhibitory activity of the stilbenes resveratrol and oxyresveratrol on African swine fever virus replication. *Antivir. Res.* **2011**, *91*, 57–63. [CrossRef] [PubMed]
7. Preyavichyapugdee, N.; Sangfuang, M.; Chaiyapum, S.; Sriburin, S.; Pootaeng-on, Y.; Chusongsang, P.; Jiraungkoorskul, W.; Preyavichyapugdee, M.; Sobhon, P. Schistosomicidal Activity of the Crude Extract of Artocarpus Lakoocha. *Southeast Asian J. Trop. Med. Public Health* **2016**, *47*, 1–15. [PubMed]
8. Singhatong, S.; Leelarungrayub, D.; Chaiyasut, C. Antioxidant and toxicity activities of Artocarpus lakoocha Roxb. heartwood extract. *J. Med. Plants Res.* **2010**, *4*, 947–953.
9. Azmi, A.S.; Bhat, S.H.; Hadi, S.M. Resveratrol-Cu(II) induced DNA breakage in human peripheral lymphocytes: Implications for anticancer properties. *Febs Lett.* **2005**, *579*, 3131–3135. [CrossRef] [PubMed]
10. Subramanian, M.; Shadakshari, U.; Chattopadhyay, S. A mechanistic study on the nuclease activities of some hydroxystilbenes. *Bioorg. Med. Chem.* **2004**, *12*, 1231–1237. [CrossRef] [PubMed]
11. Yamada, K.; Shirahata, S.; Murakami, H.; Nishiyama, K.; Shinohara, K.; Omura, H. DNA breakage by phenyl compounds. *Agric. Biol. Chem.* **1985**, *49*, 1423–1428.
12. Kalinowski, D.S.; Stefani, C.; Toyokuni, S.; Ganz, T.; Anderson, G.J.; Subramaniam, N.V.; Trinder, D.; Olynyk, J.K.; Chua, A.; Jansson, P.J.; et al. Redox cycling metals: Pedaling their roles in metabolism and their use in the development of novel therapeutics. *Biochim. Biophys. ActaMol. Cell Res.* **2016**, *1863*, 727–748. [CrossRef] [PubMed]
13. Moran, J.F.; Klucas, R.V.; Grayer, R.J.; Abian, J.; Becana, M. Complexes of iron with phenolic compounds from soybean nodules and other legume tissues: Prooxidant and antioxidant properties. *Free Radic. Biol. Med.* **1997**, *22*, 861–870. [CrossRef]
14. Alarcon de la lastra, C.; Villegas, I. Resveratrol as an antioxidant and pro-oxidant agent: Mechanisms and clinical implications. *Biochem. Soc. Trans.* **2007**, *35*, 1156–1160.
15. Valko, M.; Izakovic, M.; Mazur, M.; Rhodes, C.J.; Telser, J. Role of oxygen radicals in DNA damage and cancer incidence. *Mol. Cell. Biochem.* **2004**, *266*, 37–56. [CrossRef] [PubMed]

16. Kuo, H.W.; Chen, S.F.; Wu, C.C.; Chen, D.R.; Lee, J.H. Serum and tissue trace elements in patients with breast cancer in Taiwan. *Biol. Trace Elem. Res.* **2002**, *89*, 1–11. [CrossRef]
17. Gupte, A.; Mumper, R.J. Elevated copper and oxidative stress in cancer cells as a target for cancer treatment. *Cancer Treat. Rev.* **2009**, *35*, 32–46. [CrossRef]

**Sample Availability:** Samples of the compounds are not available from the authors.

Article

# Teratopyrones A–C, Dimeric Naphtho-γ-Pyrones and Other Metabolites from *Teratosphaeria* sp. AK1128, a Fungal Endophyte of *Equisetum arvense*

**Ya-Ming Xu [1], A. Elizabeth Arnold [2], Jana M. U'Ren [2], Li-Jiang Xuan [3], Wen-Qiong Wang [3] and A. A. Leslie Gunatilaka [1,*]**

1 Southwest Center for Natural Products Research, School of Natural Resources and the Environment, College of Agriculture and Life Sciences, University of Arizona, 250 E, Valencia Road, Tucson, AZ 85706, USA; yamingx@arizona.edu

2 School of Plant Sciences, College of Agriculture and Life Sciences, University of Arizona, Tucson, AZ 85721, USA; arnold@ag.arizona.edu (A.E.A.); juren@email.arizona.edu (J.M.U.)

3 State Key Laboratory of Drug Research, Shanghai Institute of Materia Medica, Chinese Academy of Sciences, 501 Haike Road, Zhangjiang Hi-Tech Park, Shanghai 201203, China; ljxuan@simm.ac.cn (L.-J.X.); wenqiong1019@126.com (W.-Q.W.)

* Correspondence: leslieg@cals.arizona.edu; Tel.: +520-621-9932

Academic Editors: Muhammad Ilias and Dhammika Nanayakkara
Received: 12 October 2020; Accepted: 26 October 2020; Published: 30 October 2020

**Abstract:** Bioassay-guided fractionation of a cytotoxic extract derived from a solid potato dextrose agar (PDA) culture of *Teratosphaeria* sp. AK1128, a fungal endophyte of *Equisetum arvense*, afforded three new naphtho-γ-pyrone dimers, teratopyrones A–C (**1–3**), together with five known naphtho-γ-pyrones, aurasperone B (**4**), aurasperone C (**5**), aurasperone F (**6**), nigerasperone A (**7**), and fonsecin B (**8**), and two known diketopiperazines, asperazine (**9**) and isorugulosuvine (**10**). The structures of **1–3** were determined on the basis of their spectroscopic data. Cytotoxicity assay revealed that nigerasperone A (**7**) was moderately active against the cancer cell lines PC-3M (human metastatic prostate cancer), NCI-H460 (human non-small cell lung cancer), SF-268 (human CNS glioma), and MCF-7 (human breast cancer), with $IC_{50}$s ranging from 2.37 to 4.12 μM while other metabolites exhibited no cytotoxic activity up to a concentration of 5.0 μM.

**Keywords:** endophytic fungus; *Teratosphaeria*; teratopyrones; naphtho-γ-pyrones; nigerasperone A

## 1. Introduction

Fungal endophytes constitute an abundant and underexplored group of fungi that inhabit plants in diverse natural and human-managed ecosystems [1,2]. These symbiotic fungi often produce bioactive metabolites, some of which may improve the growth or resiliency of the host plant [3]. Recent studies have demonstrated that fungal endophytes are rich sources of small-molecule natural products with novel structures and biomedical potential [4,5]. In our continuing search for bioactive and/or novel metabolites from endosymbiotic microorganisms [6], we have investigated a cytotoxic extract of the fungal endophyte, *Teratosphaeria* sp. AK1128, isolated from a photosynthetic stem of *Equisetum arvense* (field horsetail, Equisetaceae). Herein, we report cytotoxicity assay-guided fractionation of this extract, resulting in isolation and characterization of ten metabolites, including three new naphtho-γ-pyrone dimers, teratopyrones A–C (**1–3**), and the constituent responsible for cytotoxic activity, nigerasperone A (**7**). *Teratosphaeria* is a genus within the newly established fungal family Teratosphaeriaceae (Dothideomycetes, Ascomycota), which has been distinguished recently from *Mycosphaerella* (Mycosphaerellaceae) [7]. To the best of our knowledge, this constitutes the second report on metabolites of a fungal strain of the family Teratosphaeriaceae [8].

## 2. Results and Discussion

The EtOAc extract of a PDA (potato dextrose agar) culture of *Teratosphaeria* sp. AK1128 exhibiting cytotoxic activity, on bioassy-guided fractionation involving solvent-solvent partitioning, Sephadex LH-20 size-exclusion and silica gel chromatography followed by HPLC purification, afforded metabolites **1–10** (Figure 1). Of these, **4–10** were previously known and were identified as naphtho-γ-pyrones, aurasperone B (**4**) [9], aurasperone C (**5**) [10], aurasperone F (**6**) [11], nigerasperone A (**7**) [12], and fonsecin B (**8**) [9], and diketopiperazines, asperazine (**9**) [13] and isorugulosuvine (**10**) [14], by comparison of their spectroscopic ($^{1}$H NMR, $^{13}$C NMR, and LR-MS) data with those reported.

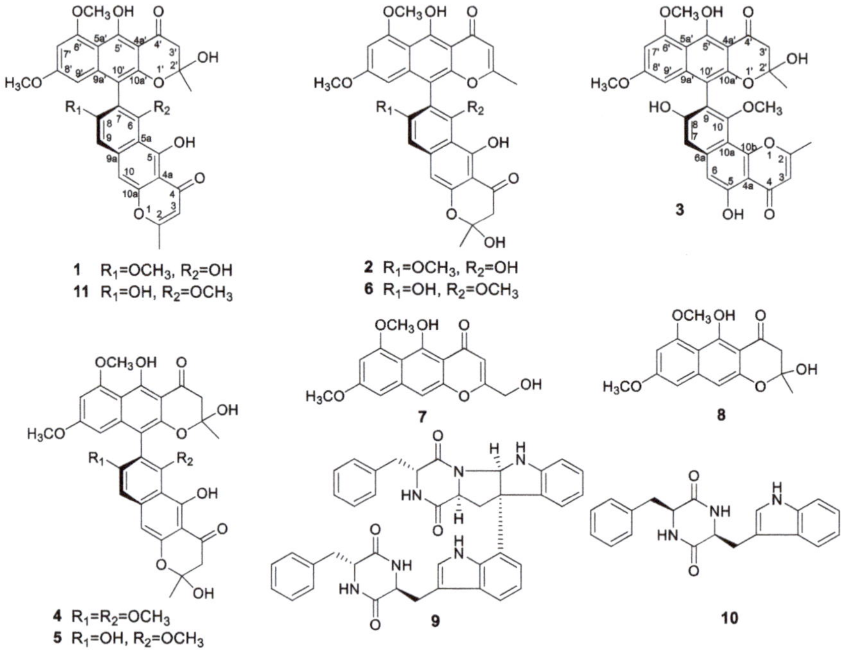

**Figure 1.** Structures of metabolites isolated from *Teratosphaeria* sp. AK1128.

Spectroscopic ($^{1}$H and $^{13}$C NMR, HRESIMS, and UV) data of teratopyrones A–C (**1–3**), together with their common molecular formula, $C_{31}H_{26}O_{11}$, suggested that they are dimeric naphtho-γ-pyrones [11]. In their $^{1}$H and $^{13}$C NMR spectra, two sets of signals were observed in different intensity ratios and this was suspected to be due to the atropisomerism around the $C_7$–$C_{10'}$ axis [10] and/or the presence of C-2 and C-2′ hemi-ketal stereoisomeric mixtures as a result of non-enzymatic formation of this moiety during the biosynthesis of naphtho-γ-pyrones [15]. Although several recent publications on dimeric naphtho-γ-pyrones report only one set of the NMR data, careful examination of the spectra of these dimeric naphtho-γ-pyrones provided in the Supporting Information of these papers indicated the presence of two sets of signals, corresponding to two possible tautomers [16–18]. The atropisomerization around the $C_7$–$C_{10'}$ axis is known to be restricted under mild conditions [10], and the atropisomer ratio for a given dimeric naphtho-γ-pyrone may depend on its producer fungus [19]. Although it is not possible to obtain the atropisomer ratio directly, the use circular dichroism (CD) data for the assignment of stereochemistry of the binaphthyl moiety of the major atropisomer of dimeric naphtho-γ-pyrones by the exciton chirality method has been reported [20,21].

In its $^{1}$H NMR spectrum, teratopyrone A (**1**) exhibited signals due to two methyls ($\delta_H$ 1.43 (s) and 2.36 (s)), three methoxyls ($\delta_H$ 3.39 (s), 3.60 (s), and 3.90 (s)), five aromatic protons ($\delta_H$ 5.99 (s), 6.12 (br. s),

6.26 (br. s), 7.06 (s), and 7.07 (s)), and two hydrogen-bonded phenolic hydroxy groups ($\delta_H$ 14.48 (s) and 14.97 (s)). The $^{13}$C NMR spectrum of **1** (Table 1) contained 31 carbon signals due to its major tautomer and these were assigned with the help of HSQC and HMBC spectra. The two carbonyl signals at $\delta_C$ 197.9 and 184.6 revealed that **1** was made up of two naphtho-γ-pyrone monomers, of which one was hydrated at $C_{2'}$–$C_{3'}$. The HMBC data for **1** (Figure 2) revealed that the linkage of two monomers of **1** was the same as that of nigerasperone C (**11**; Figure 1)) in which the hydrated monomer consisted of the upper unit [12]. The $^{13}$C NMR data of **1** closely resembled those of **11** [12], except for the signals in the vicinity of C-5–C-8. However, some differences were observed for C-5 ($\delta_C$ 162.5 for **1**; 166.1 for **11**), C-6 ($\delta_C$ 107.1 for **1**; 111.8 for **11**), and C-8 ($\delta_C$ 157.3 for **1**; 163.1 for **11**), suggesting that **1** differed from **11** in the attachment of hydroxy/methoxy groups at C-6/C-8. The attachment of the methoxy group to C-8 in **1** was confirmed by the HMBC correlations of 8-$OCH_3$ ($\delta_H$ 3.38) and H-9 ($\delta_H$ 7.06) to C-8 ($\delta_C$ 157.3). The ECD spectrum of **1** (Figure 3) showed Davydov-split Cotton effects as negative ([θ] $-1.46 \times 10^5$, 289.5 nm) first and positive ([θ] $+1.78 \times 10^5$, 272 nm) second, suggesting negative chirality [21] and M-configuration of the 7′–10′ bond [20]. Thus, the structure of teratopyrone A was determined as (10′*S*)-2′,5,5′′,6-tetrahydroxy-6′,8,8′-trimethoxy-2,2′-dimethyl-2′,3′-dihydro-4*H*,4′*H*-[7,10′-bibenzo[g] chromene]-4,4′-dione (**1**) (Figure 1).

**Table 1.** $^{13}$C NMR data (100 MHz, δ) for teratopyrones A–C (**1–3**).

| **Position** | **1** | **2** | **3** |
|---|---|---|---|
| | **$CDCl_3$** | **DMSO-$d_6$** | **DMSO-$d_6$** |
| 2 | 168.1 s | 100.0 s | 167.9/167.8 s |
| 3 | 107.1 d | 47.8 t | 109.7 d |
| 4 | 184.6 s | 198.4 s | 182.4/182.3 s |
| 4a | 104.2 s | 103.0 s | 107.9 s |
| 5 | 162.5 s | 163.0 4/163.98 s | 155.2/155.0 s |
| 5a (6) * | 107.1 s | 108.6 s | 104.3 d |
| 6 (6a) * | 160.9 s | 159.2 s | 139.8 s |
| 7 | 117.6 s | 116.3 s | 104.9 d |
| 8 | 157.3 s | 159.3 s | 158.9 s |
| 9 | 105.7 d | 105.4 d | 117.9 s |
| 9a (10) * | 142.3 s | 142.2 s | 156.9 s |
| 10 (10a) * | 101.2 d | 101.4 d | 106.6 s |
| 10a (10b) * | 152.8 s | 153.3 s | 154.8 s |
| 2-Me | 20.7 q | 27.6 q | 19.9 q |
| 6-OMe | | | 60.8 q |
| 8-OMe | 62.0 q | 56.4 q | |
| 2′ | 100.2 s | 167.9/167.8 s | 100.3 s |
| 3′ | 46.7 t | 109.8 d | 48.4 t |
| 4′ | 197.9 s | 182.3 s | 198.4 s |
| 4a′ | 103.5 s | 107.6 s | 102.7 s |
| 5′ | 162.5 s | 153.7 s | 163.9 5 |
| 5a′ | 107.1 s | 109.7 s | 106.2/106.4 s |
| 6′ | 161.8 s | 159.0 s | 160.7/161.5 s |
| 7 | 96.1 d | 96.4/96.3 d | 94.9/95.1 d |
| 8′ | 162.5 s | 161.0 s | 161.3/161.7 s |
| 9′ | 96.5 d | 97.0/96.8 d | 96.9/97.6 d |
| 9a′ | 140.2 s | 140.2 s | 142.3 s |
| 10′ | 103.1 s | 104.1 s | 105.4/105.5 s |
| 10a′ | 153.3 s | 155.0 s | 154.7 s |
| 2′-Me | 27.8 q | 20.1 q | 27.6/26.8 q |
| 6′-OMe | 56.0 q | 55.2/55.1 q | 56.1 q |
| 8′-OMe | 55.2 q | 61.2/61.4 q | 55.0/54.9 q |

* The position number in parentheses is for teratopyrone C (**3**).

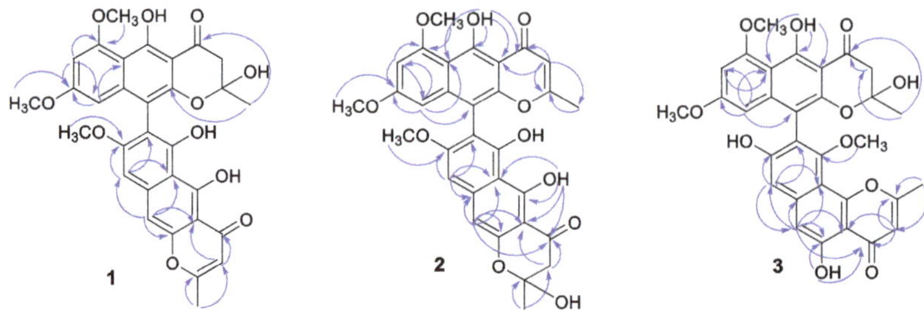

**Figure 2.** Selected HMBC correlations of teratopyrones A–C (**1–3**).

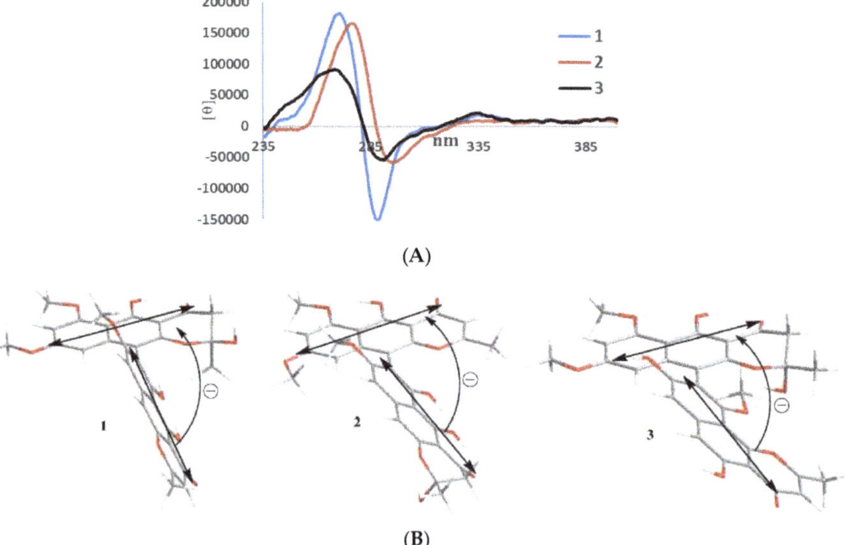


**Figure 3.** ECD spectra (**A**) and negative exciton chirality (**B**) of teratopyrones A–C (**1–3**).

Teratopyrone B (**2**) also exhibited UV adsorption bands at 202, 238, 281.5, and 386 nm typical for naphtho-γ-pyrones [11]. Similar to **1**, the $^{1}$H NMR spectrum of **2** in DMSO-d$_6$ solution showed two sets of signals due to the presence of two tautomeric forms in almost equal amounts (10:8 ratio) and these consisted of two methyl singlets ($\delta_H$ 1.65 (1.64 for the other conformer) and 2.55), three methoxy singlets ($\delta_H$ 3.41 (3.40), 3.60 (3.59), and 4.00), five aromatic signals ($\delta_H$ 6.19 (6.17), 6.55, 6.61 (2H), and 6.90), and two singlets due to hydrogen-bonded phenolic hydroxy groups ($\delta_H$ 14.27 (14.25), and 13.19 (13.18)). The $^{13}$C NMR spectrum of **2** (Table 1) resembled closely that of aurasperone F (**6**). The $^{13}$C NMR signals belonging to each of the conformers of **2** were recognized and assigned with the help of HSQC and HMBC data. The presence of two carbonyl signals at $\delta_C$ 198.4 and 182.3 indicated that **2** is a dimer consisting of a naphtho-γ-pyrone monomer and its hydrated form. The positions of attachment of $OCH_3$ groups in **2** were determined by the analysis of its HMBC spectrum (Figure 2). The presence of 8-$OCH_3$ in **2** was confirmed by the HMBC correlations of H-9 ($\delta_H$ 6.90) to C-8 ($\delta_C$ 159.3) and 8-$OCH_3$ ($\delta_H$ 4.00) to C-8 ($\delta_C$ 159.3). The ECD spectrum of **2** (Figure 3) showed Davydov-split Cotton effects as negative ([θ] $-5.70 \times 10^4$, 297 nm) first and positive ([θ] $+1.63 \times 10^5$, 278 nm) second, suggesting the negative chirality [21] and M-configuration of the 7–10′ bond [20]. Therefore, the structure teratopyrone B was determined as (10′*S*)-2,5,5′,6-tetrahydroxy-6′,8,8′-trimethoxy-2,2′-dimethyl-2,3-dihydro-4*H*,4′*H*-[7,10′-bibenzo[g] chromene]-4,4′-dione (2) (Figure 1), suggesting that it is the isomer of aurasperone F (6) with different methyl ether positions at C-6 and C-8.

The $^{1}$H NMR spectrum of teratopyrone C (3) also exhibited two sets of signals belonging to two tautomers in the ratio 3:2 as a result of the hemi-ketal tautomerism and consisted of two methyl singlets ($\delta_H$ 1.40 (1.78 for another tautomer) and 2.47), three methoxy singlets ($\delta_H$ 3.46 (3.53), 3.55, and 3.91 (3.89)), five aromatic proton singlets ($\delta_H$ 5.99 (5.95), 6.44 (6.39), 6.50, 6.93, and 7.00 (7.02)), and two hydrogen-bonded phenolic hydroxy singlets ($\delta_H$ 14.23 and 12.89 (12.88)). The hydrogen-bonded phenolic hydroxy singlet at $\delta_H$ 12.89 (12.88) suggested the existence of an angular naphtho-γ-pyrone moiety [9]. The $^{13}$C NMR spectrum of 3 (Table 1) contained signals due to each tautomer that were assigned separately with the help of HSQC and HMBC data. The two carbonyl signals at $\delta_C$ 198.4 and 182.4 (182.3) indicated that 3 consisted of one each of naphtho-γ-pyrone and hydrated naphtho-γ-pyrone monomers. The positions of attachment of $OCH_3$ groups in 3 were also determined by the analysis of its HMBC spectrum (Figure 2). The HBMC correlation of 5-OH ($\delta_H$ 12.89) to C-6 ($\delta_C$ 104.3), further supporting the existence of angular naphtho-γ-pyrone moiety, and the correlations of H-7 ($\delta_H$ 7.01) to C-8 ($\delta_C$ 158.9), H-7 ($\delta_H$ 7.01) to C-9 ($\delta_C$ 117.9), and 10-$OCH_3$ ($\delta_H$ 3.53 (3.46)) to C-8 ($\delta_C$ 156.8) established the $C_9$–$C_{10'}$ linkage and 8-methoxyl group. The ECD spectrum of 3 (Figure 3) exhibited Davydov-split Cotton effects as negative ([θ] $-5.14 \times 10^4$, 292 nm) first and positive ([θ] $+8.82 \times 10^4$, 278 nm) second, suggesting the negative chirality [21] and M-configuration of the 9–10′ bond [20]. Therefore, the structure of teratopyrone C was determined as 9-((10*R*)-2,5-dihydroxy-6,8-dimethoxy-2-methyl-4-oxo-3,4-dihydro-2*H*-benzo[g]chromen-10-yl)-5,8-dihydroxy-10-methoxy-2-methyl-4*H*-benzo[h]chromen-4-one (3) (Figure 1).

Previous studies have shown that some naphtha-γ-pyrone dimers such as chaetochromins exhibit cytotoxic activity [22]. Thus, teratopyrones A–C and other metabolites encountered were evaluated for their cytotoxic activity, employing a panel of five cancer cell lines and normal cells. Among these, only nigerasperone A (7) showed cytotoxic activity at a concentration <5.0 μM. The $IC_{50}$s of 7 for these cancer cell lines were determined as 4.12 ± 0.32 μM (PC-3M (human metastatic prostate cancer)), 3.01 ± 0.11 μM (NCI-H460 (human non-small lung cancer)), 2.37 ± 0.15 μM (SF-268 (human central nervous system glioma)), 3.90 ± 0.33 μM (MCF-7 (human breast cancer)), and >5.0 μM for MDA-MB-231 (human metastatic breast cancer) and WI-38 (normal human lung fibroblast cells). Interestingly, nigerasperone A (7) has been reported previously to be inactive against A549 (human alveolar adenocarcinoma) and SMMC-7721 (human hepatocellular carcinoma) cancer cell lines [12]. These and our data suggest that nigerasperone A (7) may be selectively cytotoxic against some cancer cell lines.

## 3. Materials and Methods

### 3.1. General Procedures

Optical rotations were measured with a Jasco Dip-370 polarimeter (JASCO Inc., Easton, MD, USA.) using $CHCl_3$ as solvent. ECD spectra were measured with JASCO J-810 (JASCO Inc., Easton, MD, USA). First, 1D and 2D NMR spectra were recorded in $CDCl_3$ or DMSO-$d_6$ with a Bruker Avance III 400 instrument (Bruker BioSpin Corporation, San Jose, CA, USA) at 400 MHz for $^{1}$H NMR and 100 MHz for $^{13}$C NMR using residual $CHCl_3$ as the internal standard. Chemical shift values (δ) are given in parts per million (ppm) and the coupling constants are in Hz. Low-resolution and high-resolution MS were recorded on Shimadzu LCMS-DQ8000α (Shimadzu Scientific Instruments, Inc., Columbia, MD, USA) and Agilent G6224A TOF mass spectrometers (Agilent Technologies Co. Ltd., Beijing, China), respectively. HPLC (Waters Corporation, Milford, MA, USA) was carried out on a 10 × 250 mm Phenomenex Luna 5μ C18 (2) column with Waters Delta Prep system consisting of a PDA 996 detector.

### 3.2. Fungal Isolation and Identification

In June 2008, a healthy individual of *Equisetum arvense* was collected from Beringian tundra in the Seward Peninsula of Western Alaska (64°30′04″ N, 165°24′23″ W; 6 m.a.s.l.) [23]. The photosynthetic stem was washed in tap water and cut into ca. 2 $mm^2$ segments that were surface-sterilized by agitating sequentially in 95% EtOH for 30 s, 0.5% NaOCl for 2 min, and 70% EtOH for 2 min [23]. Forty-eight tissue segments were surface-dried under sterile conditions and then placed individually onto 2% malt extract agar (MEA) in sterile 1.5 mL micro centrifuge tubes. Tubes were sealed with Parafilm™ and incubated under ambient light/dark conditions at room temperature (ca. 21.5 °C) for up to one year. Emergent fungi were isolated into pure culture on 2% MEA, vouchered in sterile water, and deposited as living vouchers at the Robert L. Gilbertson Mycological Herbarium at the University of Arizona. One fungus of interest, isolate AK1128, was used for the present study. This fungus has been deposited at the University of Arizona Robert L. Gilbertson Mycological Herbarium (accession number AK1128).

Total genomic DNA was isolated from fresh mycelium of the isolate AK1128 and the nuclear ribosomal internal transcribed spacers and 5.8s gene (ITS rDNA; ca. 600 base pairs (bp)) and an adjacent portion of the nuclear ribosomal large subunit (LSU rDNA; ca. 500 bp) was amplified as a single fragment by PCR. Positive amplicons were sequenced bidirectionally as described previously [23]. A consensus sequence was assembled and basecalls were made by *phred* [24] and *phrap* [25] with orchestration by Mesquite [26], followed by manual editing in Sequencher (Gene Codes Corp.). The resulting sequence was deposited in GenBank (accession JQ759476).

Because the isolate did not produce diagnostic fruiting structures in culture, two methods were used to tentatively identify isolate AK1128 via molecular sequence data. First, the LSU rDNA portion of the sequence was evaluated using the naïve Bayesian classifier for fungi [27] available through the Ribosomal Database Project (http://rdp.cme.msu.edu/). The Bayesian classifier estimated placement within the Capnodiales (Dothideomycetes) with high support, but placement at finer taxonomic levels was not possible. Therefore, the entire sequence was compared against the GenBank database using BLAST [28]. The top ten BLAST matches were to unidentified dothideomycetous endophytes or uncultured ascomycetous clones, except for two strains of *Colletogloeopsis dimorpha* (strains CBS 120085 and CBS 120086). The matches to *Colletogloeopsis dimorpha* had 97% coverage and a maximum identity of only 92%, and similar levels of match precision to other taxa restricted our taxonomic inference.

Therefore, to clarify the phylogenetic placement and taxonomic assignment of AK1128, two phylogenetic analyses were conducted. First, the top 99 BLAST matches were downloaded from GenBank, four problematic sequences for which quality was suspect were removed, and AK1128 and the resulting dataset was aligned automatically via MUSCLE (http://www.ebi.ac.uk/Tools/msa/muscle/) with default parameters. The alignment consisted of dothideomycetous endophytes as well as described species of Dothideomycetes that mostly comprised taxa affiliated with some lineages recognized within *Mycosphaerella* (e.g., *Teratosphaeria, Colletogloeopsis, Catulenostroma*, etc.). The alignment was trimmed so that starting and ending points were generally consistent with the sequence length for AK1128 and adjusted manually in Mesquite [26] prior to analysis. The final dataset consisted of 96 sequences and 1084 characters. This dataset was analyzed by maximum likelihood in GARLI (Zwickl, D. J. Genetic algorithm approaches for the phylogenetic analysis of large biological sequence datasets under the maximum likelihood criterion (Ph.D. Dissertation, The University of Texas at Austin, Austin, TX, USA, 2008) using the GTR + I + G model of evolution as determined by ModelTest [29]. The resulting topology indicated that AK1128 had an affinity for *Teratosphaeria, Catenulostroma*, and relatives but was not phylogenetically affiliated with *Colletogloeopsis* (data not shown). Because this analysis included many unknown taxa (endophytes) and we could not root the tree with certainty, the taxon sampling was found to be insufficient to infer with confidence the placement of the AK1128.

A second phylogenetic analysis was therefore conducted using taxa affiliated with *Teratosphaeria, Catulenostroma,* and relatives, as analyzed previously [30]. The alignment of ITS rDNA sequences from [30] was obtained from TreeBase and pruned to include only those taxa of interest based on our analysis above (i.e., the lower half of Figure 1 in [30] in a preliminary analysis, and then only those taxa most closely related within *Teratosphaeria* in the final analysis). The sequence for AK1128 was integrated into the pruned dataset and the data were realigned, adjusted, and analyzed as described above.

The resulting dataset consisted of 79 sequences and 582 characters in the preliminary analysis. A bootstrap analysis with 1000 replicates was conducted to assess topological support and the topology was rooted with *Devriesia strelitziae* [30]. The resulting tree suggested that AK1128 is affiliated with *Teratosphaeria* species (data not shown).

Therefore, the dataset was pruned further, focusing only on those *Teratosphaeria* species suggested to be close relatives of AK1128. The final analysis included 42 terminal taxa and 554 characters and was rooted with *Catulenostroma macowanii* based on the topology of the preliminary analysis. The final tree with maximum likelihood bootstrap values is shown in Supplemental Figure S13. The final analysis placed AK1128 with certainty within the genus *Teratosphaeria* (Teratosphaeriaceae), likely in affiliation with *T. associata (*known from *Eucalyptus* [31] and *Protea* [30] from Australia). Given the distinctive geographical origin and host of AK1128, and the sister relationship of AK1128 to known strains of *T. associata* (Supplemental Figure S13), we designate the strain as *Teratosphaeria* sp. AK1128, affiliated with but distinct from known variants of *T. associata*.

### 3.3. Cultivation and Isolation of Metabolites

A culture of *Teratosphaeria* sp. AK1128 grown on PDA for two weeks was used for extraction. The PDA cultures from 20 T-flasks were combined and extracted with MeOH (5.0 L) in an ultrasonic bath for 1.0 h. After filtration, the MeOH solution was concentrated to around one-third of its volume in vacuo, and the resulting solution was extracted with EtOAc (3 × 700.0 mL). The EtOAc solution was concentrated in vacuo to afford the crude extract (1.223 g). The crude extract, which showed activity in cytotoxicity assay, was fractionated by solvent–solvent partitioning using 80% aq. MeOH (100.0 mL) and hexanes (3 × 100.0 mL), followed by 50% aq. MeOH (obtained from 80% aq. MeOH layer by adding calculated volume of water) and $CHCl_3$ (3 × 100.0 mL) to afford hexanes, $CHCl_3$, and 50% aq. MeOH fractions of which only the $CHCl_3$ fraction was found to be cytotoxic. Therefore, the $CHCl_3$ fraction (1.094 g) was subjected to gel-permeation chromatography on Sephadex LH-20 (50.0 g). The column was eluted with 250.0 mL each of 1:4 hexanes-$CH_2Cl_2$, 3:2 $CH_2Cl_2$-acetone, 1:4 $CH_2Cl_2$-acetone, and MeOH to give four fractions. The cytotoxic 3:2 $CH_2Cl_2$-acetone fraction (778.6 mg) was further fractionated by $SiO_2$ gel (100.0 g) column chromatography using 95:5 $CHCl_3$-MeOH as eluting solvent. Eight combined fractions (fractions 1–8) were obtained by combining the fractions based on their TLC profiles. Fraction 1 (12.5 mg) was separated by reversed-phase HPLC (70% aq. MeOH as eluant) to afford **1** (7.4 mg, $t_R$ 16.5 min) and **8** (2.7 mg, $t_R$ 10.5 min). Reversed-phase HPLC separation of a portion (55.6 mg) of fraction 2 (135.8 mg) using 65% aq. MeOH gave additional amounts of **8** (3.7 mg, $t_R$ 14.4 min) and **4** (29.6 mg, $t_R$ 25.6 min). Compounds **3** (16.9 mg, $t_R$ 26.2 min), **4** (10.8 mg, $t_R$ 18.1 min), and **7** (4.2 mg, $t_R$ 15.5 min) were obtained from fraction 3 (70.8 mg) by reversed-phase HPLC using 65% aq. MeOH as the eluent. Fraction 4 (29.1 mg) was further purified by reversed-phase HPLC (65% aq. MeOH) to afford **2** (8.3 mg, $t_R$ 27.7 min) and **6** (9.8 mg, $t_R$ 21.6 min). A portion (78.5 mg) of fraction 5 (299.7 mg) was subjected to reversed-phase HPLC. Elution with 65% aq. MeOH yielded metabolites **5** (54.5 mg, $t_R$ 15.3 min) and **6** (10.3 mg). Reversed-phase HPLC separation of fraction 6 (106.0 mg) using 60% aq. MeOH as eluent gave two crude fractions, A ($t_R$ 10.8 min) and B ($t_R$ 27.7 min). Further purification of fractions A and B by normal-phase $SiO_2$ HPLC (eluent: 97.5:2.5 $CHCl_3$-MeOH) afforded **10** (2.3 mg, $t_R$ 15.0 min) and **9** (26.0 mg, $t_R$ 17.5 min), respectively.

Teratopyrone A (**1**). Yellow amorphous solid; $[\alpha]_D^{25}$ −54.4 (c 0.1, $CHCl_3$); UV (MeOH) $\lambda_{max}$ (log $\varepsilon$) 203 (4.45), 229 (4.63), 281.5 (4.84), 403 (4.09); ECD (MeOH) $[\theta]$ +1.78 × $10^5$ (272 nm), −1.46 × $10^5$ (289.5 nm); $^1$H NMR (400 MHz, $CDCl_3$) $\delta$ 14.97 (s, 1H, 5-OH), 14.48 (s, 1H, 5′-OH), 7.08/7.03 (br s, 1H, H-10), 7.07/7.04 (br s, 1H, H-9), 6.26/6.28 (br s, 1H, H-7′), 6.12 (br s, 1H, H-9′), 5.98 (s, 1H, H-3′), 3.90/3.94 (s, 3H, 6′-OMe), 3.60/3.59 (s, 3H, 8′-OMe), 3.39/3.55 (s, 3H, 8-OMe), 3.35 (m, 1H, H-3′), 2.80/2.81 (m, 1H, H-3′), 2.35 (s, 3H, 2-Me), 1.42/1.47 (s, 3H, 2′-Me); $^{13}$C NMR data, see Table 1; HRESIMS *m/z* 575.1546 $[M + H]^+$ (calcd. for $C_{31}H_{27}O_{11}$, 575.1548).

Teratopyrone B (**2**). Yellow amorphous solid; $[\alpha]_D^{25}$ +39.7 (*c* 0.105, $CHCl_3$); UV (MeOH) $\lambda_{max}$ (log $\varepsilon$) 202 (4.36), 238 (4.66), 281.5 (4.73), 386 (3.95); ECD (MeOH) $[\theta]$ + 1.63×$10^5$ (278 nm), −5.70 × $10^4$

(297 nm); $^1$H NMR (400 MHz, DMSO-$d_6$) $\delta$ 14.27/14.25 (s, 1H, 5-OH), 13.19/13.18 (s, 1H, 5′-OH), 6.90/6.85 (br s, 1H, H-9), 6.61 (br s, 2H, H-10, H-7′), 6.54 (s, 1H, H-3′), 6.19/6.17 (d, 1H, $J$ = 2.0, H-9′), 4.00 (s, 3H, 8-OMe), 3.60/3.59 (s, 3H, 8′-OMe), 3.41/3.40 (s, 3H, 6′-OMe), 3.25/3.23 (d, 1H, $J$ = 12.6, H-3′), 2.78/2.79 (d, 1H, $J$ = 12.6, H-3′), 2.55 (s, 3H, 2′-Me), 1.65/1.64/1.69 (s, 3H, 2-Me); $^{13}$C NMR data, see Table 1; HRESIMS *m/z* 575.1543 [M + H]$^+$ (calcd. for $C_{31}H_{27}O_{11}$, 575.1548).

Teratopyrone C (**3**). Yellow amorphous powder; $[\alpha]_D^{25}$ +19.3 ($c$ 0.07, $CHCl_3$); UV (MeOH) $\lambda_{max}$ (log $\varepsilon$) 204 (4.52), 237.5 (4.80), 281 (4.80), 395 (4.03); ECD (MeOH) $[\theta]$ +8.82 × $10^4$ (278 nm), −5.14 × $10^4$ (292 nm); $^1$H NMR (400 MHz, DMSO-$d_6$) $\delta$ 14.42 (s, 1H, 5′-OH), 12.89/12.88 (s, 1H, 5-OH), 7.00/7.02 (br s, 1H, H-7), 6.94 (br s, 1H, H-6), 6.50 (s, 1H, H-3), 6.44/6.39 (br s, 1H, H-7′), 5.99/5.95 (br s, 1H, H-9′) 3.91/3.89 (s, 3H, 6′-OMe), 3.55 (s, 3H, 8′-OMe), 3.46/3.53 (s, 3H, 10-OMe), 3.09 (m, 1H, H-3′), 2.86 (m, 1H, H-3′), 2.47 (s, 3H, 2-Me), 1.40/1.78 (s, 3H, 2′-Me); $^{13}$C NMR data, see Table 1; HRESIMS *m/z* 575.1551 [M + H]$^+$ (calcd for $C_{31}H_{27}O_{11}$, 575.1548).

### 3.4. Cytotoxicity Assay

A resazurin-based colorimetric (alamarBlue) assay was used for evaluating the cytotoxic activity of metabolites **1–10** against androgen-sensitive human prostate adenocarcinoma (LNCaP), metastatic human prostate adenocarcinoma (PC-3M), human breast (MCF-7), human non-small cell lung (NCI-H460), and human CNS glioma (SF-268) cancer cell lines and normal human lung fibroblast (WI-38) cells as described previously [32]. Doxorubicin and DMSO were used as positive and negative controls, respectively.

## 4. Conclusions

Ten metabolites, including three new dimeric naphtha-γ-pyrones, teratopyrones A–C (**1–3**), were isolated from a PDA culture of the endophytic fungus, *Teratosphaeria* sp. AK1128. This constitutes the second report of the occurrence of secondary metabolites in a fungus of the family Teratosphaeriaceae. The ECD spectra of teratopyrones A–C suggested that all three of them have negative exciton chirality. Cytotoxicity data for nigerasperone A (**7**) from this and previous studies suggest that it has selective activity for certain cancer cell lines.

**Supplementary Materials:** The following are available online, Figure S1: $^1$H NMR Spectrum (400 MHz) of Teratopyrone A (1) in $CDCl_3$, Figure S2. $^{13}$C NMR Spectrum (100 MHz) of Teratopyrone A (1) in $CDCl_3$, Figure S3. HSQC Spectra (400 MHz) of Teratopyrone A (1) in $CDCl_3$, Figure S4. HMBC Spectra (400 MHz) of Teratopyrone A (1) in $CDCl_3$, Figure S5. $^1$H NMR Spectrum (400 MHz) of Teratopyrone B (2) in DMSO-$d_6$, Figure S6. $^{13}$C NMR Spectrum (100 MHz) of Teratopyrone B (2) in DMSO-$d_6$, Figure S7. HSQC Spectrum (400 MHz) of Teratopyrone B (2) in DMSO-$d_6$, Figure S8. HMBC Spectrum (400 MHz) of Teratopyrone B (2) in DMSO-$d_6$, Figure S9. $^1$H NMR Spectrum (400 MHz) of Teratopyrone C (3) in DMSO-$d_6$, Figure S10. $^{13}$C NMR Spectrum (100 MHz) of Teratopyrone C (3) in DMSO-$d_6$, Figure S11. HSQC Spectrum (400 MHz) of Teratopyrone C (3) in DMSO-$d_6$, Figure S12. HMBC Spectrum (400 MHz) of Teratopyrone C (3) in DMSO-$d_6$, Figure S13. Results of maximum likelihood analysis placing strain AK1128 within *Teratosphaeria* with high support.

**Author Contributions:** Conceptualization, A.A.L.G. and A.E.A.; methodology, Y.-M.X., A.E.A., and J.M.U.; investigation, Y.-M.X., A.E.A., and J.M.U.; writing—original draft preparation, Y.-M.X.; writing—review and editing, A.A.L.G. and A.E.A.; supervision, A.A.L.G. and A.E.A.; funding acquisition, A.A.L.G. and A.E.A.; resources, A.A.L.G., A.E.A, L.-J.X. and W.-Q.W. All authors have read and agreed to the published version of the manuscript.

**Funding:** This research was funded by National Institutes of Health, grant numbers R01 CA090625 and P41 GM094060.

**Acknowledgments:** We thank Patricia Espinosa-Artiles for fungal cultivation and Manping Liu for her help with the cytotoxicity assays.

**Conflicts of Interest:** A.A.L.G. has disclosed financial interests in Sun Pharma Advanced Research Co. Ltd., India and Regulonix, LLC. USA that are unrelated to the subject of the research presented here. All other authors declare no conflict of interest. The funders had no role in the design of the study; in the collection, analyses, or interpretation of data; in the writing of the manuscript, or in the decision to publish the results.

## References


1. Rodriguez, R.J.; White, J.F., Jr.; Arnold, A.E.; Redman, R.S. Fungal endophytes: Diversity and functional roles: Tansley review. *New Phytol.* **2009**, *182*, 314–330. [CrossRef] [PubMed]
2. Higginbotham, S.J.; Arnold, A.E.; Ibañez, A.; Spadafora, C.; Coley, P.D.; Kursar, T.A. Bioactivity of fungal endophytes as a function of endophyte taxonomy and the taxonomy and distribution of their host plants. *PLoS ONE* **2013**, *8*, e73192. [CrossRef] [PubMed]
3. Kusari, S.; Pandey, S.V.; Spiteller, M. Untapped mutualistic paradigms linking host plant and endophytic fungal production of similar bioactive secondary metabolites. *Phytochemistry* **2013**, *91*, 81–87. [CrossRef] [PubMed]
4. Gunatilaka, A.A.L. Natural Products from Plant-associated Microorganisms: Distribution, Structural Diversity, Bioactivity, and Implications of Their Occurrence. *J. Nat. Prod.* **2006**, *69*, 509–526. [CrossRef]
5. Zhang, H.W.; Song, Y.C.; Tan, R.X. Biology and chemistry of endophytes. *Nat. Prod. Rep.* **2006**, *23*, 753–771. [CrossRef]
6. Wijeratne, E.M.K.; Gunaherath, G.M.K.B.; Chapla, V.M.; Tillotson, J.; de la Cruz, F.; Kang, M.-J.; U'Ren, J.M.; Araujo, A.R.; Arnold, A.E.; Chapman, E.; et al. Oxaspirol B with p97 inhibitory activity and other oxaspirols from *Lecythophora* sp. FL1375 and FL1031, endolichenic fungi inhabiting *Parmotrema tinctorum* and *Cladonia evansii*. *J. Nat. Prod.* **2016**, *79*, 340–352.
7. Crous, P.W.; Braun, U.; Groenewald, J.Z. Mycosphaerella is polyphyletic. *Stud. Mycol.* **2007**, *58*, 1–32. [CrossRef]
8. Padumadasa, C.; Xu, Y.M.; Wijeratne, E.M.K.; Espinosa-Artiles, P.; U'Ren, J.M.; Arnold, A.E.; Gunatilaka, A.A.L. Cytotoxic and noncytotoxic metabolites from *Teratosphaeria* sp. FL2137, a fungus associated with *Pinus clausa*. *J. Nat. Prod.* **2018**, *81*, 616–624. [CrossRef]
9. Priestap, H.A. New naphthopyrones from *Aspergillus fonsecaeus*. *Tetrahedron* **1984**, *40*, 3617–3624. [CrossRef]
10. Tanaka, H.; Wang, P.-L.; Namiki, M. Yellow pigments of *Aspergillus niger* and *A. awamori*. III. Structure of aurasperone C. *Agric. Biol. Chem.* **1972**, *36*, 2511–2517. [CrossRef]
11. Bouras, N.; Mathieu, F.; Coppel, Y.; Lebrihi, A. Aurasperone F–a new member of the naphtho-gamma-pyrone class isolated from a cultured microfungus, *Aspergillus niger* C-433. *Nat. Prod. Res.* **2005**, *19*, 653–659. [CrossRef] [PubMed]
12. Zhang, Y.; Li, X.M.; Wang, B.G. Nigerasperones A–C, new monomeric and dimeric naphtho-γ-pyrones from a marine alga-derived endophytic fungus *Aspergillus niger* EN-13. *J. Antibiot.* **2007**, *60*, 204–210. [CrossRef] [PubMed]
13. Varoglu, M.; Corbett, T.H.; Valeriote, F.A.; Crews, P. Asperazine, a selective cytotoxic alkaloid from a sponge-derived culture of *Aspergillus niger*. *J. Org. Chem.* **1997**, *62*, 7078–7079. [CrossRef]
14. Tullberg, M.; Grotli, M.; Luthman, K. Efficient synthesis of 2,5-diketopiperazines using microwave assisted heating. *Tetrahedron* **2006**, *62*, 7484–7491. [CrossRef]
15. Fujii, I.; Watanabe, A.; Sankawa, U.; Ebizuka, Y. Identification of Claisen cyclase domain in fungal polyketide synthase WA, a naphthopyrone synthase of *Aspergillus nidulans*. *Chem. Biol.* **2001**, *8*, 189–197. [CrossRef]
16. Shaaban, M.; Shaaban, K.A.; Abdel-Aziz, M.S. Seven naphtho-γ-pyrones from the marine-derived fungus *Alternaria alternata*: Structure elucidation and biological properties. *Org. Med. Chem. Lett.* **2012**, *2*, 6. [CrossRef]
17. Siriwardane, A.; Kumar, N.S.; Jayasingh, L.; Fujimoto, Y. Chemical investigation of metabolites produced by an endophytic *Aspergillus* sp. isolated from *Limonia acidissima*. *Nat. Prod. Res.* **2015**, *29*, 1384–1387. [CrossRef]
18. He, Y.; Tian, J.; Chen, X.; Sun, W.; Zhu, H.; Li, Q.; Lei, L.; Yao, G.; Xue, Y.; Wang, J.; et al. Fungal naphtho-γ-pyrones: Potent antibiotics for drug-resistant microbial pathogens. *Sci. Rep.* **2016**, *6*, 24291. [CrossRef]
19. Akiyama, K.; Teraguchi, S.; Hamasaki, Y.; Mori, M.; Tatsumi, K.; Ohnishi, K.; Hayashi, H. New dimeric naphthopyrones from *Aspergillus niger*. *J. Nat. Prod.* **2003**, *66*, 136–139. [CrossRef]
20. Koyama, K.; Natori, S.; Iitaka, Y. Absolute configurations of chaetochromin A and related bis(naphtho-γ-pyrone) mold metabolites. *Chem. Pharm. Bull.* **1987**, *35*, 4049–4055. [CrossRef]
21. Harada, N.; Nakanishi, K. The exciton chirality method and its application to configurational and conformational studies of natural products. *Acc. Chem. Res.* **1972**, *5*, 257–263. [CrossRef]

22. Koyama, K.; Ominato, K.; Natori, S.; Tashiro, T.; Tsuruo, T. Cytotoxicity and antitumor activities of fungal bis(naphtho-γ-pyrone) derivatives. *J. Pharmacobiodyn.* **1988**, *11*, 630–635. [CrossRef]
23. U'Ren, J.M.; Lutzoni, F.; Miadlikowska, J.; Laetsch, A.D.; Arnold, A.E. Host and geographic structure of endophytic and endolichenic fungi at a continental scale. *Am. J. Bot.* **2012**, *99*, 898–914. [CrossRef] [PubMed]
24. Ewing, B.; Green, P. Base-calling of automated sequencer traces using phred. II. Error probabilities. *Genome Res.* **1998**, *8*, 186–194. [CrossRef]
25. Ewing, B.; Hillier, L.; Wendl, M.C.; Green, P. Base-calling of automated sequencer traces using phred. I. Accuracy assessment. *Genome Res.* **1998**, *8*, 175–185. [CrossRef]
26. Maddison, W.P.; Maddison, D.R. Mesquite: A Modular System for Evolutionary Analysis, Version 2.6. 2009. Available online: http://mesquiteproject.org (accessed on 26 October 2020).
27. Liu, K.L.; Porras-Alfaro, A.; Kuske, C.R.; Eichorst, S.A.; Xie, G. Accurate, rapid taxonomic classification of fungal large-subunit rRNA genes. *Appl. Environ. Microbiol.* **2011**, *78*, 1523–1533. [CrossRef] [PubMed]
28. Altschul, S.F.; Gish, W.; Miller, W.; Myers, E.W.; Lipman, D.J. Basic local alignment search tool. *J. Mol. Biol.* **1990**, *215*, 403–410. [CrossRef]
29. Posada, D.; Crandall, K.A. MODELTEST: Testing the model of DNA substitution. *Bioinformatics* **1998**, *14*, 817–818. [CrossRef] [PubMed]
30. Crous, P.W.; Summerell, B.A.; Mostert, L.; Groenewald, J.Z. Host specificity and speciation of *Mycosphaerella* and *Teratosphaeria* species associated with leaf spots of Proteaceae. *Persoonia* **2008**, *20*, 59–86. [CrossRef]
31. Crous, P.W.; Summerell, B.A.; Carnegie, A.J.; Mohammed, C.; Himaman, W.; Groenewald, J.Z. Foliicolous *Mycosphaerella* spp. and their anamorphs on *Corymbia* and *Eucalyptus*. *Fungal Divers.* **2007**, *26*, 143–185.
32. Wijeratne, E.M.K.; Bashyal, B.P.; Liu, M.X.; Rocha, D.D.; Gunaherath, G.M.K.B.; U'Ren, J.M.; Gunatilaka, M.K.; Arnold, A.E.; Whitesell, L.; Gunatilaka, A.A.L. Geopyxins A-E, ent-kaurane diterpenoids from endolichenic fungal strains *Geopyxis aff. majalis* and *Geopyxis* sp. AZ0066: Structure–activity relationships of geopyxins and their analogues. *J. Nat. Prod.* **2012**, *75*, 361–369. [PubMed]

**Sample Availability:** Samples of the compounds **1–3** are available from the authors.

Article

# Selective Interactions of *O*-Methylated Flavonoid Natural Products with Human Monoamine Oxidase-A and -B

**Narayan D. Chaurasiya [1,2], Jacob Midiwo [3], Pankaj Pandey [2,4], Regina N. Bwire [5], Robert J. Doerksen [4], Ilias Muhammad [2,*] and Babu L. Tekwani [1,2,*]**

1. Department of Infectious Diseases, Division of Drug Discovery, Southern Research, Birmingham, AL 35205, USA; nchaurasiya@southernresearch.org
2. National Center for Natural Products Research, Research Institute of Pharmaceutical Sciences, School of Pharmacy, University of Mississippi, Oxford, MS 38677, USA; ppandey@olemiss.edu
3. Department of Chemistry, University of Nairobi, Nairobi P.O. Box 30197-00100, Kenya; jmidiwo@uonbi.ac.ke
4. Department of BioMolecular Sciences, Division of Medicinal Chemistry, Research Institute of Pharmaceutical Sciences, School of Pharmacy, University of Mississippi, Oxford, MS 38677, USA; rjd@olemiss.edu
5. Department of pure and applied Chemistry, Masinde Muliro University of Science and Technology, Kakamega P.O. Box 190-50100, Kenya; rbwire@mmust.ac.ke

* Correspondence: milias@olemiss.edu (I.M.); btekwani@southernresearch.org (B.L.T.); Tel.: +1-662-915-1051 (I.M.); +1-205-581-2205 (B.L.T.)

Academic Editor: Valeria Patricia Sülsen
Received: 17 September 2020; Accepted: 9 November 2020; Published: 17 November 2020

**Abstract:** A set of structurally related *O*-methylated flavonoid natural products isolated from *Senecio roseiflorus* (**1**), *Polygonum senegalense* (**2** and **3**), *Bhaphia macrocalyx* (**4**), *Gardenia ternifolia* (**5**), and *Psiadia punctulata* (**6**) plant species were characterized for their interaction with human monoamine oxidases (MAO-A and -B) in vitro. Compounds **1**, **2**, and **5** showed selective inhibition of MAO-A, while **4** and **6** showed selective inhibition of MAO-B. Compound **3** showed ~2-fold selectivity towards inhibition of MAO-A. Binding of compounds **1–3** and **5** with MAO-A, and compounds **3** and **6** with MAO-B was reversible and not time-independent. The analysis of enzyme-inhibition kinetics suggested a reversible-competitive mechanism for inhibition of MAO-A by **1** and **3**, while a partially-reversible mixed-type inhibition by **5**. Similarly, enzyme inhibition-kinetics analysis with compounds **3**, **4**, and **6**, suggested a competitive reversible inhibition of MAO-B. The molecular docking study suggested that **1** selectively interacts with the active-site of human MAO-A near N5 of FAD. The calculated binding free energies of the *O*-methylated flavonoids (**1** and **4–6**) and chalcones (**2** and **3**) to MAO-A matched closely with the trend in the experimental $IC_{50}$'s. Analysis of the binding free-energies suggested better interaction of **4** and **6** with MAO-B than with MAO-A. The natural *O*-methylated flavonoid (**1**) with highly potent inhibition ($IC_{50}$ 33 nM; Ki 37.9 nM) and >292 fold selectivity against human MAO-A (vs. MAO-B) provides a new drug lead for the treatment of neurological disorders.

**Keywords:** recombinant monoamine oxidase-A; monoamine oxidase-B; neurological disorder; enzyme kinetics; molecular docking; inhibition activity; flavonoid

## 1. Introduction

Monoamine oxidases (EC.1.4.3.4; MAO-A and -B) are FAD-dependent enzymes that are responsible for the metabolism of neurotransmitters such as dopamine, adrenaline, serotonin, and noradrenaline, and also for the inactivation of exogenous arylalkyl amines [1–3]. Due to their vital role in neurotransmitter metabolism, these enzymes signify attractive drug targets for the pharmacological therapy of neurodegenerative diseases and neurological disorders [4–7]. Recent efforts toward the development of MAO inhibitors have been focused on selective, reversible MAO-A or MAO-B inhibitors. The identification of MAO inhibitors could be helpful for numerous aspects of new drug discovery. Selective MAO-A inhibitors are effective in the treatment of depression and anxiety [8–11]. By contrast, MAO-B inhibitors are suitable for the treatment of the neurodegenerative diseases Alzheimer's disease and Parkinson's disease [5,7,8,11,12]. Historically, monoamine oxidase inhibitors (MAOI's) have been used to treat neurological disorders including depression [10]. Presently, MAO-A inhibitors play an important role in the control of neurological disorders, anxiety, and depression, while MAO-B inhibitors could potentially be used as therapeutic agents for Parkinson's and Alzheimer's diseases [13]. Pharmacotherapeutic limitations and adverse effects of the currently available MAO inhibitor drugs require the discovery of new MAO inhibitors with selective inhibition profiles and multi-target neuropharmacological profiles [12]. Natural products provide useful sources for MAO inhibitors combined with neuroprotective actions [3,14,15]. Several plant extracts and bioactive natural product metabolites with catecholaminergic neuropharmacological properties [16] have shown promising utility for the treatment of Alzheimer's disease and Parkinson's disease [17].

Recent studies from our lab have reported selective inhibition of MAO with flavonoid natural products [18–20]. In continuation of these studies on selected classes of flavonoids as well as other related published reports on various flavonoids [21,22], we have selected a series of methoxylated flavones and chalcones from our repository for further evaluation to explore their activities. The structure of chalcone is different from flavone/flavonol/flavanone, but they are biogenetically correlated. However, from our previous studies on flavone/flavanone [18,19], and those reported for chalcones [23] as valid MAO inhibitors, the difference in structures did not explain the structure–activity relationship for the inhibition of MAO A/B. Therefore, we chose to study the MAO inhibitory activity of these two types of flavonoids (flavone/flavonol and chalcone), including their docking studies. Therefore, these studies were further extended to test a set of related *O*-methylated flavonoids isolated from different plants, namely, *Senecio roseiflorus* (3,4′-di-*O*-methylkaempferol; **1**) [24], *Polygonum senegalense* (2′-hydroxy-4′,6′-dimethoxy-chalcone; **2** and 2′,4′-dihydroxy-6′-methoxy-chalcone; **3**) [25], *Bhaphia macrocalyx* (8-demethylsideroxylin; **4**) [26] *Gardenia ternifolia* (4′-*O*-methylkaempferol; **5**) [27], and *Psiadia punctulata* (5,7-dihydroxy-2,3,4,5-tetramethoxyflavone; **6**), [28] for experimental activities against MAO-A and -B. The isolated compounds are all non-polar chalcones (di-*O*-methylated, **2**, and **3**) and flavones exhibiting mono-*O*-methylation (**4** and **5**) or di-*O*-methylation (**1**) or tetra-*O*-methylation (**6**). There was one (**4**) which even showed ring A C-methylation (**4**) (Figure 1). This study was also extended to determine enzyme kinetics and the mechanism of inhibition of the compounds which showed the best $IC_{50}$ values in the recombinant human monoamine oxidases assays (MAO-A and -B). Furthermore, molecular docking simulations were performed to understand the putative binding modes of the best compounds to MAO-A and -B.

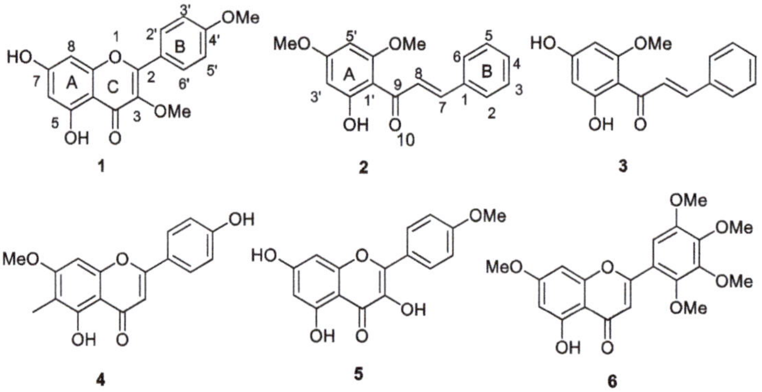

**Figure 1.** Chemical structures of *O*-methylated flavonoids **1–6**.

## 2. Results

### 2.1. Isolation, Purification, and Characterization of O-Methylated Flavonoids

The *O*-methylated flavonoids reported in this paper were isolated from various plants, using general methods reported earlier [25,27]. Aerial parts (leaves and branches) were dipped in a non-polar solvent for short periods to wash off the exudates into the solvent (without affecting cell vacuole compounds). The solvents used were normally medium polarity solvents such as acetone or ethyl acetate. The solvent was removed using a rotary evaporator and the remaining solid materials were subjected to column chromatography using silica gel as a stationary phase and eluting with hexane/dichloromethane in a gradient fashion continuously increasing polarity followed by dichloromethane/methanol. Compounds **1–6** were isolated from different plants, namely, *S. roseiflorus* (**1**) [24], *P. senegalense* (**2** and **3**) [25], *G. ternifolia* (**5**) [27], *P. punctulata* (**6**) [28], and *B. macrocalyx* (**4**). Compound **4** is the first report from the genus *Baphia*. The isolation of compound **4** was not reported in the literature by us, therefore, its structure was determined using 1D and 2D NMR spectral data and TOF-MS (see Material and Methods section for details).

### 2.2. Enzyme Inhibition and Kinetics Mechanism of MAO-A and -B with Compounds 1–6

The inhibition ($IC_{50}$) of the MAO-A and -B enzymes by compounds **1–6** are shown in Table 1. Compounds **1**, **2**, and **5** showed selective potent inhibition of MAO-A compared to compound **3**, which was potent at MAO-A but only slightly selective for MAO-A over -B. Compounds **4** and **6** were more potent than **3** at MAO-B and were selective for MAO-B over -A.

**Table 1.** Inhibition of recombinant human MAO-A and MAO-B by a selected set of *O*-methylated flavonoids **1–6**. The results with significantly potent inhibition are presented in bold.

| Compounds | MAO-A $IC_{50}$ (μM) [a] | MAO-B $IC_{50}$ (μM) [a] | SI [b] | SI [c] |
|---|---|---|---|---|
| 1 | **0.033 ± 0.042** | 9.667 ± 2.309 | 0.0034 | **292.93** |
| 2 | **0.407 ± 0.075** | 5.933 ± 0.833 | 0.082 | **14.57** |
| 3 | **1.167 ± 0.513** | 2.700 ± 0.794 | 0.432 | 2.31 |
| 4 | 5.167 ± 1.106 | **0.800 ± 0.180** | **6.451** | 0.154 |
| 5 | **1.350 ± 0.198** | >100 | - | >74.07 |
| 6 | 87.501 ± 3.536 | **0.875 ± 0.035** | **100.0** | 0.009 |
| Clorgyline [b] | 0.0065 ± 0.001 | - | - | - |
| Deprenyl [c] | - | 0.043 ± 0.011 | - | - |

[a] The $IC_{50}$ values computed from the dose–response inhibition curves are mean ± S.D. of at least triplicate observations. [b] SI Selectivity index: MAO-A $IC_{50}$/MAO-B $IC_{50}$. [c] Selectivity Index: MAO-A $IC_{50}$ / MAO-B $IC_{50}$. [b]Clorgyline and [c]L-deprenyl were used as positive controls for MAO-A and -B, respectively.

Furthermore, the MAO-A inhibition mechanisms of compounds **1–3** and **5** were studied, using varying concentrations of kynuramine, a nonselective substrate, to investigate the nature of inhibition of the enzymes. Based on dose–response inhibition results, at least two concentrations of **1–3** and **5** were selected for the inhibition kinetics assay—one below and another above the $IC_{50}$ value. Three sets of assays were completed at varying concentrations of the substrate for each experiment, one control without inhibitor and the others with two different concentrations of the inhibitor. The data were evaluated by double reciprocal Lineweaver-Burk plots for determination of the Ki (i.e., inhibition/binding affinity) values. Binding of compounds **1–3** and **5**, with human MAO-A, yielded the Km value (i.e., the affinity of the substrate for the enzyme) as well as Vmax (maximum enzyme activity) (Figure 2A–D). Ki values were computed from the double reciprocal plots (Table 2). Binding of compounds **3**, **4**, and **6** to human MAO-B yielded the Km value (i.e., the affinity of the substrate for the enzyme) as well as Vmax (maximum enzyme activity) (Figure 3A–C). Ki values were computed from the double reciprocal plots (Table 2). Compounds **3**, **4**, and **6** showed inhibitory activity of MAO-B with substantially high affinity (Ki = 1.242, 0.809, and 0.874 μM, respectively) (Table 2).

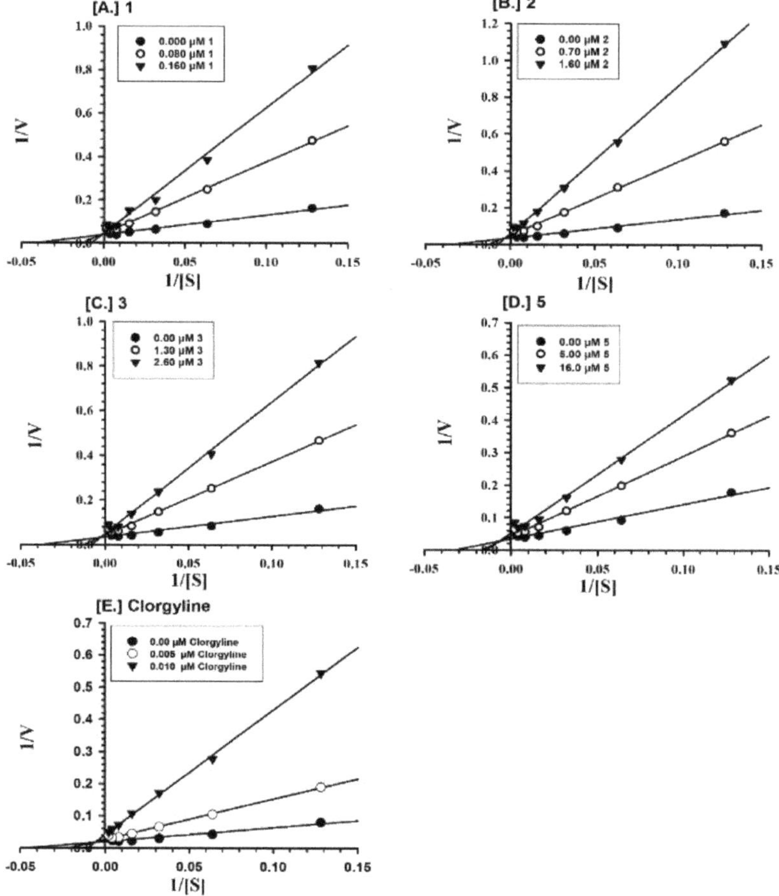


**Figure 2.** Kinetics characteristics of the inhibition mechanism of recombinant human MAO-A by compounds (**A**) **1**; (**B**) **2**; (**C**) **3**; (**D**) **5**; and (E) Clorgyline. Each point shows the mean value of three observations.

**Table 2.** Inhibition/binding affinity constants (Ki) for inhibition of recombinant human MAO-A by compounds **1–3** and **5** and of MAO-B by compounds **3**, **4**, and **6**.

| Compounds | Monoamine Oxidase-A | | Monoamine Oxidase-B | |
|---|---|---|---|---|
| | Ki (μM) [a] | Type of Inhibition | Ki (μM) [a] | Type of Inhibition |
| 1 | 0.0379 ± 0.0008 | Competitive/Reversible | - | - |
| 2 | 0.339 ± 0.219 | Competitive/Reversible | - | - |
| 3 | 0.633 ± 0.107 | Competitive/Reversible | 1.242 ± 0.600 | Competitive/Reversible |
| 4 | - | - | 0.809 ± 0.093 | Mixed/Reversible |
| 5 | 3.531 ± 0.265 | Mixed/Partially Reversible | - | - |
| 6 | - | - | 0.874 ± 0.069 | Competitive/Reversible |
| Clorgyline [b] | 0.0018 ± 0.0003 | Mixed/Irreversible | - | - |
| Deprenyl [b] | - | - | 0.0101 ± 0.0034 | Mixed/Irreversible |

[a] The results are presented as the mean ± SD of three observations; [b] Clorgyline and L-deprenyl were used as positive controls for MAO-A and -B, respectively.

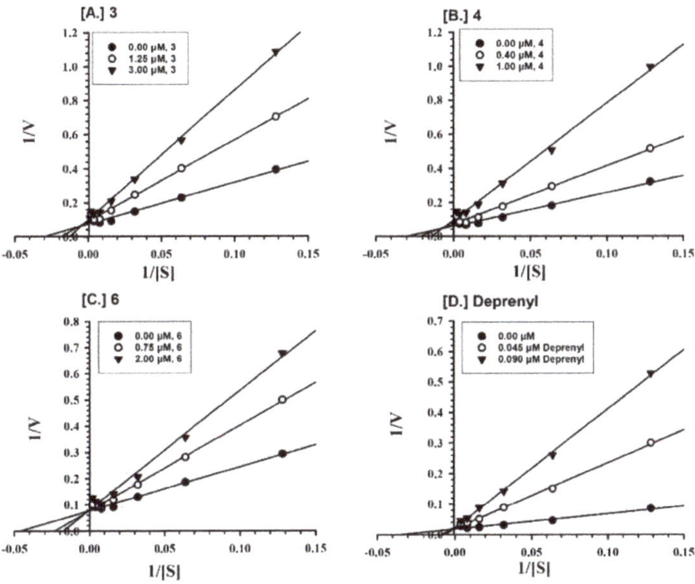


**Figure 3.** Kinetics characteristics of the inhibition mechanism of recombinant human MAO-B by compounds (**A**) **3**; (**B**) **4**; (**C**) **6**; and (**D**) L-Deprenyl. Each point shows the mean value of three observations

## *2.3. Binding and Time-Dependent Assays of MAO-A and -B with Compounds **1–6***

The characteristics of binding of compounds **1–3** and **5** with MAO-A were investigated by the equilibrium-dialysis assay. High concentrations of the compounds **1–3** and **5** (10.0, 25.0, 25.0, and 100.0 μM, respectively) were incubated with the MAO-A enzyme for 20 min and the resulting enzyme–inhibitor–complex preparation was dialyzed overnight against the 0.025 M phosphate buffer (pH 7.4). The activities of the enzyme were analyzed before and after the dialysis (Figure 4). The binding of compounds **1–3** with MAO-A was reversible and compound **5** showed partial reversibility (Table 2). Incubation of MAO-B with compounds **3**, **4**, and **6** (50.0, 50.0, and 50.0 μM, respectively) produced more than 60% inhibition of activity, and 80% of the activity of the enzyme was recovered after dialysis (Figure 5). Thus, the binding of compounds **3**, **4**, and **6** with MAO-B was reversible (Table 2). The selective MAO-B inhibitor deprenyl was confirmed to bind irreversibly with the enzyme (Table 2).

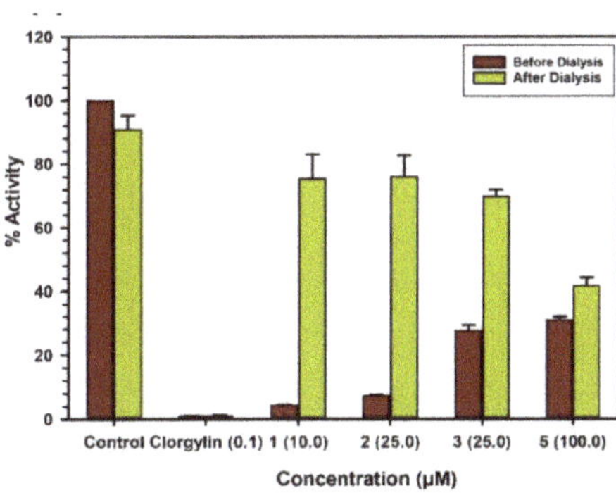


**Figure 4.** Binding modes of compounds **1** (10.0 μM), **2** (25.0 μM), **3** (25.0 μM), **5** (100.0 μM) and clorgyline (0.100 μM) with MAO-A.

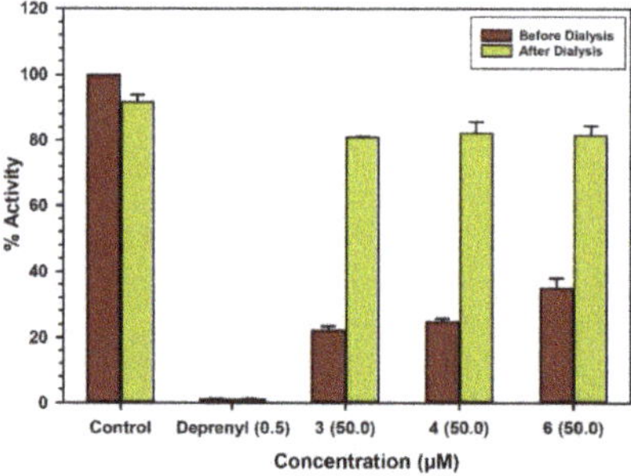


**Figure 5.** Binding modes of compounds **3** (50.0 μM), **4** (50.0 μM), **6** (50.0 μM) and L-deprenyl (0.500 μM) with MAO-B.

Further investigation of the time dependence of the assay showed that the inhibition of MAO-A by compounds **1–3** and **5** was not time-dependent (Figure 6A). The compounds **3**, **4**, and **6** also did not show time-dependent inhibition of MAO-B (Figure 6B). For validation, we have run MAO-A and -B standards simultaneously for the time-dependent assay.

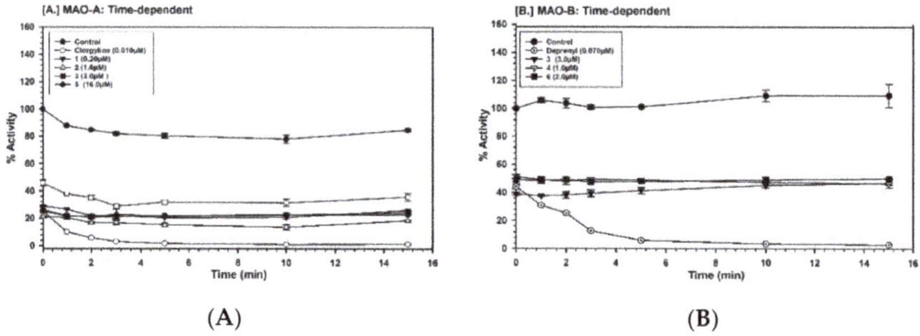


**Figure 6.** (**A**) Time-dependent inhibition of recombinant human MAO-A by compounds **1** (0.20 μM), **2** (1.6 μM), **3** (3.0 μM), **5** (16.0 μM) and L-deprenyl (0.010 μM). The remaining activity was expressed as % of initial activity. Each point represents the mean ± S.D. of triplicate values. (**B**) Time-dependent inhibition of recombinant human MAO-B by compounds **3** (3.0 μM), **4** (1.0 μM), **6** (2.0 μM) and L-deprenyl (0.070 μM). The remaining activity was expressed as % of the initial activity. Each point represents the mean ± S.D. of triplicate values.

### *2.4. Computational Analysis of Enzyme–Inhibitor Interactions*

A molecular docking study was performed to understand the binding pose and interaction profiles of compounds **1–6** to MAO-A and -B. Schrödinger's Induced-Fit docking protocol was adopted to consider the optimal geometry of the protein–ligand complex after conformational changes induced by the bound ligand. The GlideScores and binding free-energies of compounds **1–6** in the active sites of the *h*MAO-A and *h*MAO-B X-ray crystal structures are presented in Table 3. The docking protocol used in this study was validated by self- or native-docking. The native ligands, harmine and pioglitazone, were extracted from the X-ray structures of MAO-A and -B, respectively, and docked into their corresponding protein models. The calculated RMSD between the docked and experimental poses were found to be identical <0.6 Å, which verified the suitability of the docking method for the current study. The putative binding mode and interactions of the best compounds with the X-ray crystal structures of MAO-A and -B are presented in Figure 7. The calculated binding free energies vary between 28–76 kcal/mol against MAO-A and -B. Since some of the measured Ki values are in the micromolar range, the binding affinities should be somewhere around 6–10 kcal/mol. This is a known limitation of the employed computations, which are useful not on an absolute scale but in relative terms among structurally similar ligands, which is the focus here Compound **1** exhibited a strong binding affinity to the MAO-A receptor in terms of GlideScore and binding free energy (ΔG = −57.522 kcal/mol) compared to the other compounds. The *p*-methoxy phenyl at the C-2 position (Ring-B) of **1** showed π–π stacking with Phe208 and was surrounded by an array of hydrophobic residues, including Leu97, Phe108, Ala111, Ile180, and Ile325. The hydroxyl at C-5 of ring A formed H-bonding with N5 and C=O of FAD and the hydroxyl at C-7 of ring A exhibited water-mediated H-bonding with Tyr444. In addition, Ring A was surrounded by strong hydrophobic residues Tyr69, Tyr197, Tyr407, and Tyr444. The best GlideScore and binding free energy matched well with the experimental binding affinity of **1**. Interestingly, **1** and **5** had the difference of $-OCH_3$ and –OH, respectively, at the C-3 position of ring C but had significant differences in their MAO-A binding affinity. Our docking results also predicted relatively poor GlideScores and binding free energy for compound **5** compared to compound **1**. After careful observation, we have found that the methoxy group at C-3 of **1** exhibited strong hydrophobic interactions with Ile335 and Leu337 compared to the hydroxyl group at C-3 of **5**. Compounds **1** and **5** have very similar poses; however, compound **1** slightly shifted towards FAD (~1.5 Å) compared to **5**. In addition, the hydroxyl at Ring A of **5** did not show any H-bonding with FAD, further helping to explain the poorer binding affinity of compound **5** for MAO-A. Interestingly, compound **1** showed better GlideScore and binding free energy for MAO-B than for MAO-A (see Table 3); however, its best-ranked docking pose left it 15 Å away from the N5 of FAD, which apparently is an unrealistic docking prediction. In the search for an alternative pose for **1** to MAO-B, we found a pose in which **1** fit into the active site of MAO-B (near N5 of FAD) with a GlideScore of −9.705 kcal/mol, and that is the one reported in Table 3. The substitutions of acetyl ($-CH_3CO$) and methylsulfone ($-SO_2CH_3$) at the C-3 and C-4′ of Ring B of **1** are predicted to enhance the affinity towards MAO-A. The poly-substituted methoxy group at Ring B caused a loss of binding affinity towards MAO-A but submicromolar activity towards MAO-B.

**Table 3.** GlideScores and binding free energies of compounds **1–6** to MAO-A and -B.

| Compounds | Experimental $IC_{50}$ (μM) [a] | | GlideScore (kcal/mol) | | Binding Free-Energy (kcal/mol) | |
|---|---|---|---|---|---|---|
| | MAO-A | MAO-B | MAO-A | MAO-B | MAO-A | MAO-B |
| 1 | 0.033 ± 0.04 | 9.667 ± 2.309 | −11.667 | −10.028 [b] | −57.522 | −76.353 [b] |
| 2 | 0.407 ± 0.075 | 5.933 ± 0.833 | −10.686 | −9.999 | −47.724 | −28.119 |
| 3 | 1.167 ± 0.513 | 2.700 ± 0.794 | −9.951 | −10.579 | −37.683 | −51.309 |
| 4 | 5.167 ± 1.106 | 0.800 ± 0.180 | −9.664 | −10.225 | −35.043 | −53.574 |
| 5 | 1.350 ± 0.198 | >100 | −10.567 | ND [c] | −47.035 | ND [c] |
| 6 | 87.501 ± 3.536 | 0.875 ± 0.035 | −7.239 | −11.191 | −34.651 | −68.053 |

[a] The data are mean ± SD of three observations.; [b] The best pose is 15 Å away from the N5 of FAD (substrate active site); the numbers given here are for an alternate pose (see text); [c] ND = Not determined.

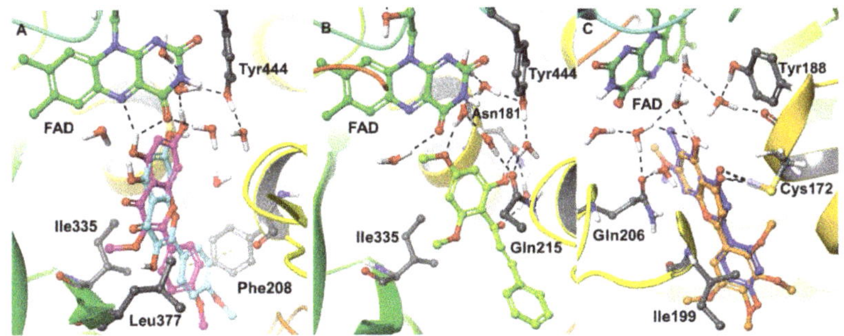


**Figure 7.** Three-dimensional (3D) representation of the protein–ligand interactions of **1–6** with the X-ray crystal structures of MAO-A and -B. (**A**) **1** (C magenta, stick model) and **5** (C cyan, stick model) with MAO-A, (**B**) **2** (C light green, stick model) with MAO-A, (**C**) **4** (C blue, stick model) and **6** (C orange, stick model) with MAO-B. Some crystallographic waters (O red, H white, stick model), FAD (C dark green), and the important residues of MAO-A and MAO-B (C gray) are also shown. The black dashed lines represent H-bonding.

The next structural category, chalcone, represented by compounds **2** and **3**, was analyzed. Compound **2** showed the more negative binding free energy (ΔG = −47.724 kcal/mol) compared to compound **3** (ΔG = −37.683 kcal/mol) for binding to MAO-A, and these data match closely with the experimental binding affinities (cf. Table 1). The only structural difference between compounds **2** and **3** involves C-4′ carrying methoxy and hydroxyl moieties, respectively. The docked pose of **2** in the MAO-A receptor showed H-bonding of its hydroxyl moiety at C-6′ and its C-1 carbonyl (water-mediated H-bonding) with Asn181. In addition, the oxygen of the methoxy at C-4′ exhibited water-mediated H-bonding with Gln215 and Tyr444. The major difference of binding free energy between **2** and **3** was because of an additional strong hydrophobic interaction (CH ... C, C ... C, and CH ... π) of the C-4′ methoxy group of **2** with Tyr69, Phe352, and Tyr407, respectively.

The GlideScores and binding free energies of the flavonoids **4** and **6** showed a better binding affinity for interaction with MAO-B (**4**: GlideScore = −10.225 kcal/mol, ΔG = −53.574 kcal/mol and **6**: GlideScore = −11.191 kcal/mol, ΔG = −68.053 kcal/mol) than with MAO-A. Compounds **4** and **6** docked in a very similar orientation and showed H-bonding interactions between their C-4 carbonyl and Cys172. In addition, the C-5 hydroxyl of **4** and **6** showed water-mediated H-bonding with Tyr188, and Gln206 and the C-4 hydroxyl had direct hydrogen-bonding with the backbone carbonyl of Cys172. Ring A had an orientation towards the isoalloxazine ring of FAD and was surrounded by an array of hydrophobic residues, including Tyr60, Phe343, Tyr398, and Tyr435. Furthermore, Ring B was surrounded by hydrophobic residues Leu164, Leu167, Phe168, Ile199, Ile316, and Tyr326 (including π–π stacking for Tyr326). Overall, the docking results of compounds **1**–**6** were in good agreement with the experimental binding data for MAO-A and -B.

## 3. Discussion

The molecules with reversible selective inhibition of MAO-A or MAO-B have therapeutic potential for the treatment of neurological and psychiatric disorders, especially caused due to depletion of neurotransmitter biogenic amines [9,29,30]. Previous studies from our lab have reported selective inhibition of human MAO-B with flavonoid natural products [18–20]. A recent study has also reported MAO-A and MAO-B inhibition activity by acacetin 7-*O*-(6-*O*-malonylglucoside), a derivative of acacetin isolated from *Agastache rugosa* plant leaves [31]. The follow-up studies presented here with a select set of *O*-methylated flavonoids (**1**–**6**) identified MAO inhibitors selective against both MAO-A (**1**–**3** and **5**) and MAO-B (**4** and **6**). Compounds **1**–**3** interact with MAO-A through reversible binding as assessed by the enzyme–inhibitor complex equilibrium dialysis assay, while the binding of compound **5** with MAO-A was partially reversible. The inhibition of MAO-A by compounds **1**–**3** and **5** was not time-dependent. A recent paper reported MAO-A and -B inhibition activity by natural constituent acacetin 7-*O*-(6-*O*-malonylglucoside) that was isolated and purified from *Agastache rugosa* plant leaves [32]. Compounds **3**, **4**, and **6** also interact with MAO-B reversibly, as assessed by the enzyme–inhibitor complex equilibrium dialysis assay, and the inhibition was not time-dependent.

Computational analysis of the binding of **1**–**6** with human MAO-A and -B revealed the putative binding mode and interaction profiles of the compounds with MAO-A and -B. Among all the *O*-methylated flavonoids, **1** showed the strongest computed interaction with MAO-A and exhibited H-bonding with N5 and C=O of FAD through the hydroxyl at C-5 of Ring A. In addition, the hydroxyl at C-7 of Ring A also exhibited water-mediated H-bonding with Tyr444. **1** also showed π–π interactions with Phe208 and was surrounded by an array of hydrophobic residues. On the other hand, **4** and **6** showed strong interactions with MAO-B and shared an identical binding mode with MAO-B. This study suggests that it would be worthwhile to perform further evaluation of compounds **1**–**6** including considering the effects of their MAO-A and -B inhibitory actions in experimental animal models of neurological and/or neurodegenerative disorders.

The *O*-methylated flavonoids are predominant bioactive secondary metabolites present in several plants [32]. The *O*-methylated flavonoids are generated in plants through the action of specific *O*-methyltransferase (OMT) enzymes [33]. *O*-methylation changes the solubility of flavonoids and improves bioactive properties compared to their non-methylated counterparts [34]. The natural product O-methylated flavonoid 3,4′-di-O-methylkaempferol (**1**) isolated from *S. roseiflorus* was identified as a highly potent inhibitor of human MAO-A with $IC_{50}$ and Ki values of 33 nM and 37.8 nM, respectively. The metabolite **1** was more than 292-fold selective for MAO-A over MAO-B. The compound formed a reversible enzyme–inhibitor complex and was had a very low Ki for MAO-A. With its highly potent MAO-A inhibition and extraordinary selectivity for human MAO-A over MAO-B, **1** is worth optimizing further as a new-drug lead and merits advancement to preclinical evaluations regarding utility for the treatment of neurological and psychiatric disorders.

## 4. Materials and Methods

### 4.1. Enzymes

Recombinant human monoamine oxidase (rhMAO-A and -B) enzymes were purchased from BD Biosciences (Bedford, MA, USA). Kynuramine, clorgyline, deprenyl, and DMSO were obtained from Sigma Chemical (St Louis, MO, USA).

### 4.2. Isolation and Identification of Compounds 1–6

Compounds **1–3**, **5**, and **6** (Figure 1) were isolated and characterized by $^{1}H$ and $^{13}C$-NMR spectra from *Senecio roseiflorus* (**1**), *Polygonum senegalense* (**2** and **3**), *Gardenia ternifolia* (**5**), and *Psiadia punctulata* (**6**) plant species collected from Kenya [24,25,27,28] (*vide supra*). In addition, compound **4** was isolated from the leaves of *B. macrocalyx*, collected from Southern Tanzania (Mikindaniya Leo village). A voucher specimen (FMM 3579) was deposited at the Herbarium of the Department of Botany, University of Dar es Salaam, Tanzania. The dried pulverized plant material (5 kg) underwent sequential cold solvent extraction with hexane, EtOAc, and finally MeOH, each soaked in each solvent for 24 h, then the solvents were evaporated with a rotavap yielding hexane, EtOAc, and MeOH extracts (100, 300 and 200 g, respectively). Column chromatography of EtOAc extract (70 g) on silica gel (Merck silica gel 60; 0.40–0.63 mm) eluting with a hexane-DCM gradient, from 100% hexane to 100% DCM and finally in methanol afforded 10 fractions. The fractions eluting with hexane: DCM (3:7) afforded compounds **4** (8-desmethylsideroxylin, 13.2 mg); obtained as yellow needles, mp 275–277 °C (Lit. 275–277 °C; ESI MS (TOF +ve, Finnigan Mat SSQ 7000): *m/z* 299.1 ($[M+H]^+$, $C_{17}H_{14}O_5$+H) $^{1}H$-NMR (600 MHz, Avance Bruker): 6.82, s, 1H (H-3), 6.82, s, 1H (H-8), 7.95, *dd*, 2H, J = 2.4, 9.0 Hz (H-2′/6′), 6.93, *dd*, 2H, J = 2.4, 9.0 Hz (H-3′/5′), 1.98, *s*, 3H (C-6-Me), 3.90, *s*, 3H (C-7-OMe), 13.08, *s*, 1H (C-5-OH), $^{1}H$-NMR spectra was in agreement with those reported in the literature); $^{13}C$-NMR: 161.3 (C-2), 103.1 (C-3), 182.0 (C-4), 104.4 (C-4a), 157.5 (C-5), 107.6 (C-6), 163.1 (C-7), 90.4 (C-8), 155.5 (C-8a), 121.2 (C-1′), 128.6 (C-2′/6′), 116.2 (C-3′/5′),163.9 (C-4′), 7.4 ($CH_3$ at C-6), 56.3 ($OCH_3$ at C-7). All NMR spectra of **1–6** are provided in Supplementary Information.

### 4.3. MAO Inhibition Assay

In this study, we have investigated the effect of the isolated constituents (**1–6**) on human recombinant MAO-A and -B. The kynuramine oxidation deamination assay was performed in 96-well plates as previously reported, with modification [18,35]. A fixed concentration of kynuramine substrate and varying concentrations of test compounds or inhibitor were used to determine the $IC_{50}$ values. Kynuramine concentrations for MAO-A and -B were 80 μM and 50 μM, respectively. The concentrations of compounds **1–6** varied from 0.001 μM to 100 μM for the rhMAO-A and -B enzyme activity inhibition. The test compounds **1–6** were dissolved in DMSO, diluted in the buffer solution just before the assay, and pre-incubated with the enzyme for 10 min at 37 °C. The final concentration of DMSO in the enzyme-assay reaction mixtures did not exceed 1%. The enzymatic reactions were initiated by the addition of MAO-A (5 μg/mL) or -B (12.5 μg/mL), and incubated for 20 min at 37 °C. The enzyme reactions were terminated by the addition of 78 μL of 2N NaOH. The formation of 4-hydroxyquinoline (the enzyme reaction end product) was recorded fluorometrically on a SpectraMax M5 fluorescence plate reader (Molecular Devices, Sunnyvale, CA, USA) with an excitation (320 nm) and emission (380 nm) wavelength, using the Soft MaxPro-6 program. The inhibition effects of enzyme activity were calculated as the percent of product formation compared to the corresponding controls (enzyme–substrate reaction) without inhibitors. The assay controls, to define the interference of the test compounds with the fluorescence measurements, were set up simultaneously, and the enzyme or the substrate was added after stopping the reaction.

### 4.4. Determination of $IC_{50}$ Values

The enzyme assays were performed at a fixed concentration of the substrate kynuramine (80 µM for MAO-A and 50 µM for MAO-B) and different concentrations of the test compounds (**1–6**). The dose–response enzyme-inhibition curves were generated using Microsoft® Excel and the $IC_{50}$ values were computed with XLfit®.

### 4.5. Enzyme Kinetics and Mechanism Studies

For determination of the binding affinity of the inhibitor (Ki) to MAO-A and -B, the enzyme assays were carried out at different concentrations of kynuramine substrate (1.90 µM to 500 µM) and varying concentrations of the inhibitors/compound. The flavonoids (**1–6**) were tested at 0.030–0.100 µM for MAO-A and 0.100–0.500 µM for -B. The controls without inhibitor were also run simultaneously. The results were analyzed by SigmaPlot version 10 using standard double reciprocal Lineweaver-Burk plots for computing Km and Vmax values, which were further analyzed to determine the Ki values [18,36,37].

### 4.6. Analysis of Binding of Inhibitor with The Enzymes

Enzyme-inhibitors mostly produce inhibition of the target enzyme through the formation of an enzyme–inhibitor complex. The formation of the enzyme–inhibitor complex may be accelerated in the presence of a high concentration of the test inhibitor. The property of binding of test compounds to MAO-A or -B was determined by the formation of the enzyme–inhibitor complex by incubation of the enzyme with a high concentration of the test compound. This was followed by extensive equilibrium dialysis of the enzyme–inhibitor complex. Recovery of catalytic activity of MAO-A and -B was determined before and after the dialysis. The MAO-A enzyme (0.2 mg/mL protein) was incubated with each test compound: **1** (10.0 µM), **2** (25.0 µM), **3** (25.0 µM), **5** (100.0 µM) and clorgyline (0.100 µM), in 1 mL of potassium phosphate buffer (100 mM, pH 7.4). After 20 min incubation at 37 °C, the reaction was stopped by chilling the tubes in an ice bath. Similarly, the MAO-B enzyme (0.2 mg/mL protein) was incubated with each test compound: **3** (50.0 µM), **4** (50.0 µM), **6** (50.0 µM), and deprenyl (0.500 µM), in 1.0 mL potassium phosphate buffer (100 mM, pH 7.4). After 20 min incubation at 37 °C, the reaction was stopped by chilling the tubes in an ice bath. All the samples with enzyme–inhibitor complex were individually dialyzed against potassium phosphate buffer (25 mM; pH 7.4) at 4 °C for 16–18 h (including three buffer changes). The control enzyme (without inhibitor) was also run through the same procedure and the activity of the enzyme was determined before and after the dialysis [36].

### 4.7. Time-Dependent Inhibition of the Enzyme

To investigate if the binding of the inhibitor with MAO-A and -B followed time-dependent inhibition kinetics, the enzyme was pre-incubated with the inhibitor for different time periods (0–15 min). The compound concentrations used to test time-dependent inhibition were: **1** (0.20 µM), **2** (1.6 µM), **3** (3.0 µM) and **4** (16.0 µM) and clorgyline (0.010 µM), with MAO-A (5.0 µg/mL). The inhibitor concentrations used to test time-dependent inhibition were: **3** (3.0 µM), **4** (1.0 µM), **6** (2.0 µM), and deprenyl (0.070 µM), with MAO-B (12.5 µg/mL). The controls without inhibitors were also run simultaneously. The activities of the MAO-A and -B enzymes were determined as described above and the percentage of enzyme activity remaining was plotted against the pre-incubation time to determine time-dependent inhibition.

### 4.8. Computational Analysis of The Interaction of Test Compounds with MAO-A or -B

The X-ray crystal structures of MAO-A (PDB ID: 2Z5X) and MAO-B (PDB ID: 4A79 [38] were directly imported from the Protein Data Bank website (https://www.rcsb.org) to Maestro [39] using the Protein Preparation Wizard module of the Schrödinger software (Cambridge, MA, USA) [40]. We followed a similar method and protocol for docking as previously described [19]. The protein structures of MAO-A and -B were each used as monomers in the docking study. In brief, these proteins were prepared by adding hydrogens, adjusting bond orders, adding missing side chains, setting the proper ionization states at pH 7.4, refining overlapping atoms, and making H-bond assignments using PROPKA at pH 7.0. The water molecules beyond 5 Å from the co-crystallized ligands were deleted and the protonation states of the co-crystallized ligands were generated using Epik at pH 7.4. During the refinement process, water molecules with fewer than two H-bonds to non-waters were also removed and, finally, restrained minimization of hydrogens only was performed using the Optimized Potentials for Liquid Simulations 3 (OPLS3) force field [41]. The cofactor FAD was not removed during protein preparation and docking. The 2D structures of compounds **1–6** were sketched in the 2D sketcher module of Maestro, prepared, and energy-minimized at a physiological pH of 7.4 using the LigPrep module [42] of the Schrödinger software. The compounds were docked as neutral molecules. The OPLS3 force field was used for protein and ligand preparation, and docking. The active sites of the MAO-A and -B proteins were generated using the centroid of the co-crystallized ligands of 2Z5X and 4A79, respectively. The Induced Fit docking [43] protocol was used for the docking of compounds **1–6.** The standard precision (SP) docking method was applied during the initial docking stages. In the initial Glide docking, the receptor and the ligand were "softened" by van der Waals radii scaling. The scaling factor was chosen to be 0.50 for both the ligand and the receptor to permit enough flexibility for the ligand to dock in the best poses. The "trim-side chains" option was not used in this study. The maximum number of poses was chosen to be 20. In the next step, residues that are within 5 Å of the active site (ligand) were refined using the "Prime Refinement" Table In the final step, a threshold of 30 kcal/mol was used to redock the best structure, for eliminating high-energy structures from the Prime refinement step. The top 20 poses were kept for analysis and the best poses were selected based on IFD scores and visual inspection of protein–ligand interactions. The best docking poses were subjected to binding free-energy calculations using the Prime MM-GBSA module of the Schrödinger software allowing protein flexibility within 5 Å of the ligand. Only protein side-chains were minimized during the calculations. Finally, the Maestro Version 11.5 molecular graphics system was used to create all the computationally derived figures.

## 5. Conclusions

Screening of a selected set of *O*-methylated flavonoid constituents isolated from *Senecio roseiflorus*, *Bhaphia macrocalyx*, *Polygonum sengalense*, *Psiadia punctulata*, and *Gardenia ternifolia* identified compounds **1–3** and **5** as potent and selective inhibitors of human MAO-A, relative to MAO-B, and compounds **4** and **6** as selective inhibitors of human MAO-B. Further investigations suggested compounds **1–3** as reversible and competitive inhibitors and compound **5** as a partially reversible mixed-type inhibitor of MAO-A and compounds **3**, **4**, and **6** as reversible and competitive inhibitors of MAO-B. The computational results for compounds **1–6** were in good agreement with the experimental binding data for MAO-A and -B. The compounds 1 and 6 with high potency and selectivity of inhibition against MAO-A and MAO-B, respectively, may be promising new drug leads for further development as therapeutic treatment of neurological disorders, depression, Alzheimer's disease, and Parkinson's disease. It is important to mention that the flavonoid scaffold possesses promiscuous biological activity that may be due to inherent structural features. For this reason, they should be treated with caution as lead compounds for drug development.

**Supplementary Materials:** All NMR spectra of compounds **1–6** are provided as Supplementary Information.

**Author Contributions:** B.L.T. and I.M. conceptualized the study. B.L.T. and N.D.C. planned the experiments for enzyme-inhibition assays. J.M. and I.M. authenticated and provided the natural products, R.N.B. carried out the chemistry work. N.D.C. performed the experiments on enzyme assays. B.L.T. and N.D.C. analyzed the results of enzyme inhibition. P.P. and R.J. planned the computational studies and analyzed the results. P.P. carried out computational studies. N.D.C., B.L.T., P.P., R.J.D., and I.M. prepared the original draft of the manuscript. All the authors contributed to the writing, review, and editing of the manuscript. All the authors have read and agreed to the published version of the manuscript.

**Funding:** The International Foundation of Uppsala University, Sweden is acknowledged for supporting the KEN; 02 project at the University of Nairobi, Kenya, where the compounds were isolated. This research was funded in part by grant number P20GM104932 from the National Institute of General Medical Sciences (NIGMS), a component of the National Institutes of Health (NIH), and was conducted in part in a facility constructed with support from the Research Facilities Improvements Program (C06RR14503) from the NIH National Center for Research Resources; its contents are solely the responsibility of the authors and do not necessarily represent the official view of University of Mississippi NIGMS, NIH or Southern Research.

**Acknowledgments:** The authors are thankful to the National Center for Natural Products Research (NCNPR), the University of Mississippi for the facilities to support this work.

**Conflicts of Interest:** The authors declare no conflict of interest.

## References

1. Shih, J.C.; Chen, K.; Ridd, M.J. Monoamine oxidase: From genes to behavior. *Annu. Rev. Neurosci.* **1999**, *22*, 197–217. [CrossRef] [PubMed]
2. Abell, C.W.; Kwan, S.W. Molecular characterization of monoamine oxidases A and B. *Prog. Nucl. Acid Res. Mol. Biol.* **2001**, *65*, 129–156.
3. Carradori, S.; D'Ascenzio, M.; Chimenti, P.; Secci, D.; Bolasco, A. Selective MAO-B inhibitors: A lesson from natural products. *Mol. Divers.* **2014**, *18*, 219–243. [CrossRef]
4. Cesura, A.M.; Pletscher, A. The new generation of monoamine oxidase inhibitors. *Prog. Drug Res.* **1992**, *38*, 171–297. [PubMed]
5. Yamada, M.; Yasuhara, H. Clinical pharmacology of MAO inhibitors: Safety and future. *Neurotoxicology* **2004**, *25*, 215–221. [CrossRef]
6. Youdim, M.B.; Edmondson, D.; Tipton, K.F. The therapeutic potential of monoamine oxidase inhibitors. *Nat. Rev. Neurosci.* **2006**, *7*, 295–309. [CrossRef]
7. Youdim, M.B.; Fridkin, M.; Zheng, H. Novel bifunctional drugs targeting monoamine oxidase inhibition and iron chelation as an approach to neuroprotection in Parkinson's disease and other neurodegenerative diseases. *J. Neural Transm.* **2004**, *111*, 1455–1471. [CrossRef]
8. Youdim, M.B.; Bhakle, Y.S. Monoamine oxidase: Isoforms and inhibitors in Parkinson's disease and depressive illness. *Br. J. Pharmacol. Chemother.* **2006**, *147*, S287–S296. [CrossRef]
9. Finberg, J.P.M.; Rabey, J.M. Inhibitors of MAO-A and MAO-B in Psychiatry and Neurology. *Front. Pharmacol.* **2016**, *7*, 340–355. [CrossRef]
10. Riederer, P.; Laux, G. MAO-inhibitors in Parkinson's Disease. *Exp. Neurobiol.* **2011**, *20*, 1–17. [CrossRef]
11. Cai, Z. Monoamine oxidase inhibitors: Promising therapeutic agents for Alzheimer's disease. *Mol. Med. Rep.* **2014**, *9*, 1533–1541. [CrossRef] [PubMed]
12. Riederer, P.; Muller, T. Monoamine oxidase-B inhibitors in the treatment of Parkinson's disease: Clinical-pharmacological aspects. *J. Neural Transm.* **2018**, *125*, 1751–1757. [CrossRef] [PubMed]
13. Guglielmi, P.; Carradorri, S.; Ammazzalorso, A.; Secci, D. Novel approaches to the discovery of selective human monoamine oxidase-B inhibitors: Is there room for improvement. *Expert Opin. Drug Discov.* **2019**, *14*, 995–1035. [CrossRef] [PubMed]
14. Orhan, I.E. Potential of Natural Products of Herbal Origin as Monoamine Oxidase Inhibitors. *Curr. Pharm. Des.* **2016**, *22*, 268–276. [CrossRef] [PubMed]
15. Mathew, B.; Suresh, J.; Mathew, G.E.; Parasuraman, R.; Abdulla, N. Plant secondary metabolites- potent inhibitors of monoamine oxidase isoforms. *Cent. Nerv. Syst. Agents. Med. Chem.* **2014**, *14*, 28–33. [CrossRef] [PubMed]
16. Bhattacharjee, M.; Perumal, E. Potential plant-derived catecholaminergic activity enhancers for neuropharmacological approaches: A review. *Phytomedicine* **2019**, *55*, 148–164. [CrossRef] [PubMed]

17. Sharifi-Rad, M.; Lankatillake, C.; Dias, D.A.; Docea, A.O.; Mahomoodally, M.F.; Lobine, D.; Chazot, P.L.; Kurt, B.; Tumer, T.B.; Moreira, A.C.; et al. Impact of Natural Compounds on Neurodegenerative Disorders: From Preclinical to Pharmacotherapeutics. *J. Clin. Med.* **2020**, *9*, 1061. [CrossRef]
18. Chaurasiya, N.D.; Ibrahim, M.A.; Muhammad, I.; Walker, L.A.; Tekwani, B.L. Monoamine oxidase inhibitory constituents of propolis: Kinetics and mechanism of inhibition of recombinant human MAO-A and MAO-B. *Molecules* **2014**, *19*, 18936–18952. [CrossRef]
19. Chaurasiya, N.D.; Zhao, J.; Pandey, P.; Doerksen, R.J.; Muhammad, I.; Tekwani, B.L. Selective Inhibition of Human Monoamine Oxidase B by Acacetin 7-Methyl Ether Isolated from Turnera diffusa (Damiana). *Molecules* **2019**, *24*, 810. [CrossRef]
20. Chaurasiya, N.D.; Gogineni, V.; Elokely, K.M.; León, J.F.; Núñez, M.J.; Klein, M.L.; Walker, L.A.; Cutler, S.J.; Tekwani, B.L. Isolation of Acacetin from Calea urticifolia with Inhibitory Properties against Human Monoamine Oxidase-A and -B. *J. Nat. Prod.* **2016**, *79*, 2538–2544. [CrossRef]
21. Jalili-Baleh, L.; Babaei, E.; Abdpour, S.; Nasir Abbas Bukhari, S.; Foroumadi, A.; Ramazani, A.; Sharifzadeh, M.; Abdollahi, M.; Khoobi, M. A review on flavonoid-based scaffolds as multi-target-directed ligands (MTDLs) for Alzheimer's disease. *Eur. J. Med. Chem.* **2018**, *152*, 570–589. [CrossRef] [PubMed]
22. Larit, F.; Elokely, K.M.; Chaurasiya, N.D.; Benyahia, S.; Nael, M.A.; León, F.; Abu-Darwish, M.S.; Efferth, T.; Wang, Y.H.; Belouahem-Abed, D.; et al. Inhibition of human monoamine oxidase A and B by flavonoids isolated from two Algerian medicinal plants. *Phytomedicine* **2018**, *40*, 27–36. [CrossRef] [PubMed]
23. Chimenti, F.; Fioravanti, R.; Bolasco, A.; Chimenti, P.; Secci, D.; Rossi, F.; Yáñez, M.; Orallo, F.; Ortuso, F.; Alcaro, S. Chalcones: A valid scaffold for monoamine oxidases inhibitors. *J. Med. Chem.* **2009**, *52*, 2818–2824. [CrossRef] [PubMed]
24. Kerubo, L.; Midiwo, J.O.; Derese, S.; Langat, M.K.; Akala, H.M.; Waters, N.C.; Peter, M.; Heydenreich, M. Antiplasmodial activity of compounds from surface exudates of Senecio roseiflorus. *Nat. Prod. Commun.* **2013**, *8*, 175–176. [CrossRef] [PubMed]
25. Midiwo, J.O.; Matasi, J.J.; Wanjau, O.M.; Mwangi, R.W.; Waterman, P.G.; Wollenweber, E. Anti-feedant effects of surface accumulated flavonoids of Polygonum senegalense. *Bull. Chem. Soc. Ethiop.* **1990**, *4*, 123.
26. Cardona, M.; Seoan, E. Flavonoid and xanthonolignoids of Hypericum ericoides. *Phytochemistry* **1982**, *21*, 2759–2760. [CrossRef]
27. Awas, L.K.; Omosa, J.O.; Midiwo, A.; Ndakala, J. Anti-oxidant Activities of flavonoid aglycones from Kenyan Gardenia ternifolia Schum and Thonn. *J. Pharm. Biol. Sci.* **2016**, *11*, 136–141.
28. Bernard, F.; Yenesev, A.J.; Midiwo, J.O.; Waterman, P.G. Flavones and phenypropanoids in exudate of Psiadia punctulata. *Phytochemistry* **2001**, *57*, 571–574.
29. Tripathi, R.K.P.; Ayyannam, S.R. Monoamine oxidase-B inhibitors as potential neurotherapeutic agents: An overview and update. *Med. Res. Rev.* **2019**, *39*, 1603–1706. [CrossRef]
30. Carradori, S.; Secci, D.; Petzer, J.P. MAO inhibitors and their wider applications: A patent review. *Expert. Opin. Ther. Pat.* **2018**, *28*, 211–226. [CrossRef]
31. Lee, H.W.; Ryu, H.W.; Baek, S.C.; Kang, M.G.; Park, O.D.; Han, H.Y.; An, J.H.; Oh, S.R.; Kim, H. Potent inhibitions of monoamine oxidase A and B by acacetin and its 7-O-(6-O-malonylglucoside) derivative from Agastache rugosa. *Int. Biol. Macromol.* **2017**, *104*, 547–553. [CrossRef] [PubMed]
32. Koirala, N.; Thuan, N.H.; Ghimire, G.P.; Van Thang, D.; Sohng, J.K. Methylation of flavonoids: Chemical structures, bioactivities, progress and perspectives for biotechnological production. *Enzym. Microb. Technol.* **2016**, *86*, 103–116. [CrossRef] [PubMed]
33. Schroder, G.; Wehinger, E.; Lukacin, R.; Wellmann, F.; Seefelder, W.; Schwab, W.; Schröder, J. Flavonoid methylation: A novel 4′-O-methyltransferase from Catharanthus roseus, and evidence that partially methylated flavanones are substrates of four different flavonoid dioxygenases. *Phytochemistry* **2004**, *65*, 1085–1094. [CrossRef] [PubMed]
34. Quideau, S.; Deffieux, D.; Douat-Casassus, C.; Pouységu, L. Plant polyphenols: Chemical properties, biological activities, and synthesis. *Angew. Chem. Int. Ed. Engl.* **2011**, *50*, 586–621. [CrossRef] [PubMed]
35. Haider, S.; Alhusban, M.; Chaurasiya, N.D.; Tekwani, B.L.; Chittiboyina, A.G.; Khan, I.A. Isoform selectivity of harmine-conjugated 1,2,3-triazoles against human monoamine oxidase. *Future Med. Chem.* **2018**, *10*, 1435–1448. [CrossRef]

36. Chaurasiya, N.D.; Leon, F.; Ding, Y.; Gomez-Betancur, I.; Benjumea, D.; Walker, L.A.; Cutler, S.J.; Tekwani, B.L. Interactions of Desmethoxyyangonin, a Secondary Metabolite from Renealmia alpinia, with Human Monoamine Oxidase-A and Oxidase-B. *Evid. Complem. Altern. Med. eCAM* **2017**, *2017*, 1–10. [CrossRef]
37. Ramsay, R.R.; Albreht, A. Kinetics, mechanism, and inhibition of monoamine oxidase. *J. Neural Transm.* **2018**, *125*, 1659–1683. [CrossRef]
38. Binda, C.; Aldeco, M.; Geldenhuys, W.J.; Tortorici, M.; Mattevi, A.; Edmondson, D.E. Molecular insights into human monoamine oxidase B inhibition by the glitazone antidiabetes drugs. *ACS Med. Chem. Lett.* **2011**, *3*, 39–42. [CrossRef]
39. *Maestro, Version 10.6*; Schrödinger; LLC: New York, NY, USA, 2016.
40. *Schrödinger, Version 2016-2*; Schrödinger; LLC: New York, NY, USA, 2016.
41. Harder, E.; Damm, W.; Maple, J.; Wu, C.; Reboul, M.; Xiang, J.Y.; Wang, L.; Lupyan, D.; Dahlgren, M.K.; Knight, J.L.; et al. OPLS3: A force field providing broad coverage of drug-like small molecules and proteins. *J. Chem. Theory Comput.* **2016**, *12*, 1281–1296. [CrossRef]
42. *LigPrep, Version 3.8*; Schrödinger; LLC: New York, NY, USA, 2016.
43. Sherman, W.; Day, T.; Jacobson, M.P.; Friesner, R.A.; Farid, R. Novel procedure for modeling ligand/receptor induced fit effects. *J. Med. Chem.* **2006**, *49*, 534–553. [CrossRef]


**Sample Availability:** Samples of the compounds are available from the authors after the execution of Intellectual Property and Material Transfer Agreements.

*Article*

# Newly Generated Atractylon Derivatives in Processed Rhizomes of *Atractylodes macrocephala* Koidz

**Chunmei Zhai [1,2], Jianping Zhao [2], Amar G. Chittiboyina [2], Yonghai Meng [1,2], Mei Wang [3] and Ikhlas A. Khan [1,2,4,*]**

1 School of Pharmacy, Heilongjiang University of Chinese Medicine, Harbin 150040, China; zhaicm163@163.com (C.Z.); myhdx163@163.com (Y.M.)
2 National Center for Natural Products Research, School of Pharmacy, University of Mississippi, Oxford, MS 38677, USA; jianping@olemiss.edu (J.Z.); amar@olemiss.edu (A.G.C.)
3 Natural Products Utilization Research Unit, Agricultural Research Service, Department of Agriculture, University of Mississippi, Oxford, MS 38677, USA; mei.wang@usda.gov
4 Division of Pharmacognosy, Department of BioMolecular Sciences, School of Pharmacy, University of Mississippi, Oxford, MS 38677, USA
* Correspondence: ikhan@olemiss.edu; Tel.: +1-662-915-7821

Academic Editors: Luisella Verotta and Derek J. McPhee
Received: 13 November 2020; Accepted: 8 December 2020; Published: 13 December 2020

**Abstract:** Thermally processed rhizomes of *Atractylodes macrocephala* (RAM) have a long history of use in traditional Chinese medicine (TCM) for treating various disorders, and have been an integral part of various traditional drugs and healthcare products. In TCM, herbal medicines are, in most cases, uniquely processed. Although it is thought that processing can alter the properties of herbal medicines so as to achieve desired functions, increase potency, and/or reduce side effects, the underlying chemical changes remain unclear for most thermally processed Chinese herbal medicines. In an attempt to shed some light on the scientific rationale behind the processes involved in traditional medicine, the RAM processed by stir-frying with wheat bran was investigated for the change of chemical composition. As a result, for the first time, five new chemical entities, along with ten known compounds, were isolated. Their chemical structures were determined by spectroscopic and spectrometric analyses. The possible synthetic pathway for the generation of such thermally-induced chemical entities was also proposed. Furthermore, biological activity evaluation showed that none of the compounds possessed cytotoxic effects against the tested mammalian cancer and noncancer cell lines. In addition, all compounds were ineffective at inhibiting the growth of the pathogenic microorganisms.

**Keywords:** *Atractylodes macrocephala*; wheat bran; processing; NMR; atractylon derivatives; isolation and identification

## 1. Introduction

The rhizome of *Atractylodes macrocephala* Koidz (RAM)—called Baizhu in traditional Chinese medicine (TCM)—is a well-known herbal medicine. It was documented in the earliest existing book on TCM—"Shen Nong's Materia Medica" written during the Han Dynasty (A.D. 25-220). RAM has been traditionally used for the treatment of various disorders, such as loss of appetite, diarrhea, limb weakness, gastrointestinal dysfunction, immune dysfunction, diabetes, and some chronic inflammatory diseases [1]. It was reported that RAM is used in more than 835 TCM preparations, as well as an integral part of more than 4340 classic prescriptions for treating chronic diseases [1]. Indeed, RAM is considered a functional food, a tonic, and a constituent of various health products purported for promoting digestion, alleviating fatigue, improving sleeping, enhancing immunity, and treating alimentary anemia [2]. RAM is traditionally used in its processed form which is commonly achieved by stir-frying raw RAM with wheat bran [3,4]. This type of processing technique, known as Pao-Zhi in TCM, has been widely used for the preparation of Chinese Materia Medica [5] with a long history. It is believed that after processing, property and function of remedies can be changed, medical potency can be increased, and/or toxicity and side effect can be reduced. The practice of processing RAM can date back to the Tang dynasty (A.D. 618–907) [6], and the earliest processing protocol for RAM by stir-frying with wheat bran was recorded in the famous Sheng Ji Zong Lu, the TCM prescription book written in A.D. 1117. This processing method is also currently documented in the Chinese Pharmacopeia [7]. Although processed RAM is commonly used as an ingredient of various Chinese medicines and health products, so far, the chemical compositional change as a result of processing has not been clearly delineated.

Extensive phytochemical and pharmacological studies have been conducted on raw RAM, revealing the presence of volatile organics, sesquiterpenoids, triterpenoids, polyacetylenes, coumarins and phenylpropanoids, flavonoids and their glycosides, steroids, benzoquinones, and polysaccharides [1]. Among them, sesquiterpene lactones (i.e., atractylenolids I, II, and III) are considered to be the characteristic and bioactive constituents of RAM [8,9]. Efforts have been previously undertaken to investigate the changes occurring in the chemical components of RAM before and after processing by utilizing HPLC, GC/MS, and other analytical techniques. Substantial changes in the concentration of constituents, including those in essential oil, as well as the eudesmane-type sesquiterpenoids (such as atractylon, atractylnolides I, II, and III, etc.), were observed [3,10–13]. These changes can be explained as the result of evaporation, conversion, and/or degradation by heating, due to their physicochemical properties such as volatility and instability [13,14]. Nevertheless, in continuation of the quest to probe the overall integrity of botanical preparations by using NMR-based metabolomics to investigate the change of chemical composition which occurred during processing, we observed that some new characteristic signals appeared in the NMR spectra of the processed RAM, which suggested that some unknown compounds could be generated during processing. Herein, the discovery of five new compounds isolated from the RAM processed by stir-frying with wheat bran is reported. Their chemical structures were determined by both spectroscopic and spectrometric analyses. In addition, a plausible mechanism for the generation of these new compounds was proposed. Furthermore, their cytotoxic, antimicrobial, and antileishmanial activities were evaluated.

## 2. Results and Discussion

### 2.1. Structure Elucidation

A previous study using a NMR-based metabolomics method to investigate the effects of processing on the Chinese herbal medicine *Flos Lonicerae* revealed that the NMR approach can provide not only a holistic view on the change of chemical composition during processing, but also the information on identity of individual components [5]. In the present study, with the aid of information obtained from the NMR-based metabolomic analyses of RAM before and after processing, we could narrow the scope of isolation targets to the hexane partition fraction and its sub-fractions based on new signals of interest observed in their NMR spectra.

Compound **1** was obtained as a white amorphous powder. Its molecular formula was established as $C_{31}H_{40}O_2$ based on its high-resolution atmospheric pressure chemical ionization mass data (HR-APCI-MS, *m/z* = 445.3134 $[M + H]^+$). From its $^{13}$C-NMR spectrum, only 16 resonance signals were observed, suggesting **1** might contain two symmetrical sub-structures with 15 carbons for each structure. Referring to the multiplicity-edited HSQC spectrum, the signals in the $^{13}$C spectrum were assigned to two methyls, six methylenes, one exocyclic methylidene, one aliphatic methine, one aliphatic quaternary, and five olefinic carbons (Table 1). The $^{1}$H-NMR spectrum showed signals of methyl groups at $\delta_H$ 1.87 (s) and 0.76 (s), exocyclic methylidene at $\delta_H$ 4.68 and 4.84, aliphatic methine at $\delta_H$ 2.10 (m), five methylenes in the range of $\delta_H$ 1.40–2.50, and one methylene at $\delta_H$ 3.82 (s). The integral value of the peak of methylene at $\delta_H$ 3.82 was found to be only 1/3 of the value for the singlet peak of methyl at $\delta_H$ 0.76 or 1.87, indicating that this methylene is the group bridging the two symmetrical moieties together with each moiety containing 15 carbons (Figure 1). The HMBC spectrum of **1** reveals the cross-peaks of the methyl signal at $\delta_H$ 0.76 (H-14,14′) with the carbon signals at $\delta_C$ 42.0 (C-1,1′), $\delta_C$ 45.7 (C-5,5′), $\delta_C$ 39.2 (C-9,9′) and $\delta_C$ 36.7 (C-10,10′). Additionally, the cross-peaks of the methyl signal at $\delta_H$ 1.87 (H-13,13′) with the carbon signals at $\delta_C$ 116.6 (C-7,7′), $\delta_C$ 114.3 (C-11,11′), and $\delta_C$ 145.0 (C-12,12′), and the cross-peaks of the exocyclic methylidene at $\delta_H$ 4.68 and 4.84 (H-15,15′) with the carbon signals at $\delta_C$ 37.3 (C-3,3′), $\delta_C$ 150.0 (C-4,4′), and $\delta_C$ 45.7 (C-5,5′) were observed in the HMBC spectrum. These observations suggested that the 15-carbon moiety had a furanoeudesmane carbon skeleton in the structure as in atractylon, which is a known furanosesquiterpene isolated from RAM in significant quantity [15]. The analysis of the $^1$H-$^1$H COSY spectrum further confirmed the assignment. The nuclear Overhauser effect (NOE) interactions between H-14 (14′) and H-2β (2′β)/H-6β (6′β), but no NOE interaction between H-14 (14′) and H-5 (5′), were observed in the NOESY spectrum, indicating that the *trans*-fused A/B ring junction and α,β-orientation of H-5 (5′) and Me-14 (14′) in the furanoeudesmane skeleton was the same as atractylon. Observations of HMBC correlations of the methylene signal at $\delta_H$ 3.82 (H-16) to C-11 (11′) at $\delta_C$ 114.3 and C-12 (12′) at $\delta_C$ 145.0 in the HMBC spectrum indicated that the two symmetrical moieties were linked together through the methylene group attaching to C-11 and C-11′. Accordingly, **1** was identified to be bis(3,8α-dimethyl-5-methylene-4,4α,5,6,7,8,8α,9-octahydronaphtho-[2,3-β]-furan-2-yl)methane, trivially named methylene-biatractylon.

Compound **2** was obtained as a white amorphous powder. As determined from the [M + H]$^+$ peak at *m/z* = 459.3267 in HR-ACPI-MS, **2** has a molecular formula of $C_{32}H_{42}O_2$ with 12 degrees of unsaturation (DOU). Compound **2** has the same number of DOU as compound **1**, but has an additional $CH_2$ in molecular formula. The $^1$H NMR and HSQC spectra of **2** exhibit similar signal patterns as those of **1**, respectively, i.e., the signals of methyl groups at $\delta_H$ 1.82/1.76 (s) and 0.75/0.77 (s), exocyclic methylidene at $\delta_H$ 4.68 and 4.84, aliphatic methine at $\delta_H$ 2.10 (m), six methylenes in the range of $\delta_H$ 1.40–2.50. A significant difference lies in that one methylene at $\delta_H$ 3.82 (s) was observed for **1**, but one aliphatic methine at $\delta_H$ 4.17 (q) and one methyl at $\delta_H$ 1.57 (d) were observed for **2** (Table 1). The $^{13}$C-NMR spectrum of **2** also displays a similar signal pattern as that of **1**. A significant difference between **2** and **1** is the presence of one methine carbon at $\delta_C$ 30.1 and one methyl carbon at $\delta_C$ 18.1 in **2**, corresponding, respectively, to the proton signals at $\delta_H$ 4.17 and $\delta_H$ 1.57. The analyses of the COSY and HSQC spectra confirmed the presence of an ethylidene unit in **2**, instead of the presence of a methylene in **1**. Comparing **2** to **1**, another difference in their $^{13}$C spectra is that slight splitting of carbon signals for the pair groups of C-1/1′, C-5/5′, C-9/9′, C-11/11′, C-13/13′, and C-14/14′ were observed for **2** but not for **1** (Table 1). Analyses of the HMBC and COSY spectra of **2** revealed that these pair of signals were attributed to the two atractylon moieties presented in **2**. Furthermore, the *trans*-fused A/B ring junction in the two atractylon moieties was confirmed by the NOE observations from the NOESY spectrum. The HMBC correlations of the methyl protons (Me-17) in the ethylidene unit at $\delta_H$ 1.75 with the carbons C-16 at $\delta_C$ 30.1, C-11 (11′) at $\delta_C$ 112.9/113.0, and C-12 (12′) at $\delta_C$ 149.1, and the methine proton (H-16) in the ethylidene unit at $\delta_H$ 4.17 with the carbons C-17 at $\delta_C$ 18.1, C-11 (11′) at $\delta_C$ 112.9/113.0, and C-12 (12′) at $\delta_C$ 149.1 indicated that the two atractylon moieties were linked together through the methine in the ethylidene unit connecting at C-11 and C-11′. On the basis of the aforementioned evidence, the structure of **2** was established to be 2,2′-(ethane-1,1-diyl)-bis(3,8α-dimethyl-5-methylene-4,4α,5,6,7,8,8α,9-octahydronaphtho-[2,3-β]-furan), trivially named ethylidene-biatractylon.

**Table 1.** $^{1}$H-(500 MHz) and $^{13}$C-(125 MHz) NMR data of compounds **1–5** (recorded in $CDCl_3$, ppm).

| No. * | 1 | | 2 | | 3 | | 4 | | 5 | |
|---|---|---|---|---|---|---|---|---|---|---|
| | $\delta_C$ | $\delta_H$ | $\delta_C$ | $\delta_H$ | $\delta_C$ | $\delta_H$ | $\delta_C$ | $\delta_H$ | $\delta_C$ | $\delta_H$ |
| 1, 1′, (1′′) | 42.0 (t) | 1.47, 1.68 (m, 4H) | 41.9/42.0 (t) | 1.49, 1.67 (m, 4H) | 42.0 (t) | 1.47, 1.66 (m, 4H) | 41.9 (t) | 1.48, 1.65 (m, 4H) | 42.0 (t) | 1.47, 1.67 (m, 6H) |
| 2, 2′, (2′′) | 23.6 (t) | 1.57, 1.63 (m, 4H) | 23.6 (t) | 1.56, 1.63 (m, 4H) | 23.6 (t) | 1.55, 1.64 (m, 4H) | 23.6 (t) | 1.53, 1.63 (m, 4H) | 23.6 (t) | 1.54, 1.64 (m, 6H) |
| 3, 3′, (3′′) | 37.3 (t) | 2.02, 2.37 (m, 4H) | 37.4 (t) | 2.02, 2.37 (m, 4H) | 37.3 (t) | 2.02, 2.38 (m, 4H) | 37.3 (t) | 2.02, 2.38 (m, 4H) | 37.3/36.7 (t) | 2.02, 2.37 (m, 6H) |
| 4, 4′, (4′′) | 150.0 (s) | | 150.0 (s) | | 149.9 (s) | | 149.9 (s) | | 149.9/150.0 (s) | |
| 5, 5′, (5′′) | 45.7 (d) | 2.10 (m, 2H) | 45.7/45.8 (d) | 2.10 (m, 2H) | 45.7 (d) | 2.10 (m, 2H) | 45.6/45.7 (d) | 2.09 (m, 2H) | 45.6/45.7 (d) | 2.10 (m, 3H) |
| 6, 6′, (6′′) | 21.2 (t) | 2.25, 2.32 (m, 4H) | 21.1 (t) | 2.25, 2.32 (m, 4H) | 21.1 (t) | 2.27, 2.34 (m, 4H) | 21.1 (t) | 2.25, 2.32 (m, 4H) | 21.1 (t) | 2.25, 2.32 (m, 6H) |
| 7, 7′, (7′′) | 116.6 (s) | | 116.6 (s) | | 116.8 (s) | | 116.8 (s) | | 116.8/116.7 (s) | |
| 8, 8′, (8′′) | 146.9 (s) | | 146.6 (s) | | 147.6 (s) | | 147.6 (s) | | 147.3/147.0 (s) | |
| 9, 9′, (9′′) | 39.2 (t) | 2.41 (m, 4H) | 39.2/39.3 (t) | 2.42 (m, 4H) | 39.2 (t) | 2.42 (m, 4H) | 39.2 (t) | 2.40 (m, 4H) | 39.2/39.1 (t) | 2.40/2.39 (m, 6H) |
| 10, 10′, (10′′) | 36.7 (s) | | 36.7 (s) | | 36.7 (s) | | 36.7 (s) | | 36.7/36.4 (s) | |
| 11, 11′, (11′′) | 114.3 (s) | | 113.0 (s) | | 115.2 (s) | | 115.2 (s) | | 115.1/114.9 (s) | |
| 12, 12′, (12′′) | 150.0 (s) | | 149.1 (s) | | 144.7 (s) | | 144.5 (s) | | 144.8/144.3 (s) | |
| 13, 13′, (13′′) | 8.0 (q) | 1.87 (s, 6H) | 7.8/8.0 (q) | 1.76/1.82 (s, 6H) | 7.8/7.9 (q) | 1.80/1.81 (s, 6H) | 7.8/7.9 (q) | 1.79/1.80 (s, 6H) | 7.8/7.9/8.0 (q) | 1.74/1.76/1.82 (s, 9H) |
| 14, 14′, (14′′) | 17.6 (q) | 0.76 (s, 6H) | 17.6/17.7 (q) | 0.75/0.77 (s, 6H) | 17.6/17.7 (q) | 0.75/0.76 (s, 6H) | 17.6/17.7 (q) | 0.75/0.76 (s, 6H) | 17.6/17.7 (q) | 0.75/0.76 (s, 9H) |
| 15, 15′, (15′′) | 107.1 (t) | 4.68, 4.84 (d, 4H) | 107.1 (t) | 4.68, 4.84 (d, 4H) | 107.1 (t) | 4.68, 4.84 (d, 4H) | 107.2 (t) | 4.67, 4.83 (d, 4H) | 107.1/107.2 (t) | 4.67, 4.83 (d, 6H) |
| 16 | 23.9 (t) | 3.82 (s, 2H) | 30.1 (d) | 4.17 (q, 1H) | 36.1 (d) | 5.47 (s, 1H) | 36.2 (d) | 5.44 (s, 1H) | 36.2 (d) | 5.42 (s, 1H) |
| 17 | | | 18.1 (q) | 1.57 (d, 3H) | 153.1 (s) | | 153.2 (s) | | 151.5 (s) | |
| 18 | | | | | 107.2 (d) | 6.07 (d, 1H) | 108.1 (d) | 6.02 (d, 1H) | 107.9 (d) | 5.93 (d, 1H) |
| 19 | | | | | 110.2 (d) | 6.30 (dd, 1H) | 108.7 (d) | 6.20 (d, 1H) | 106.5 (d) | 5.89 (d, 1H) |
| 20 | | | | | 141.4 (d) | 7.35 (d, 1H) | 153.0 (s) | | 151.4 (s) | |
| 21 | | | | | | | 57.6 (t) | 4.55 (s, 2H) | 25.8 (t) | 3.86 (s, 2H) |

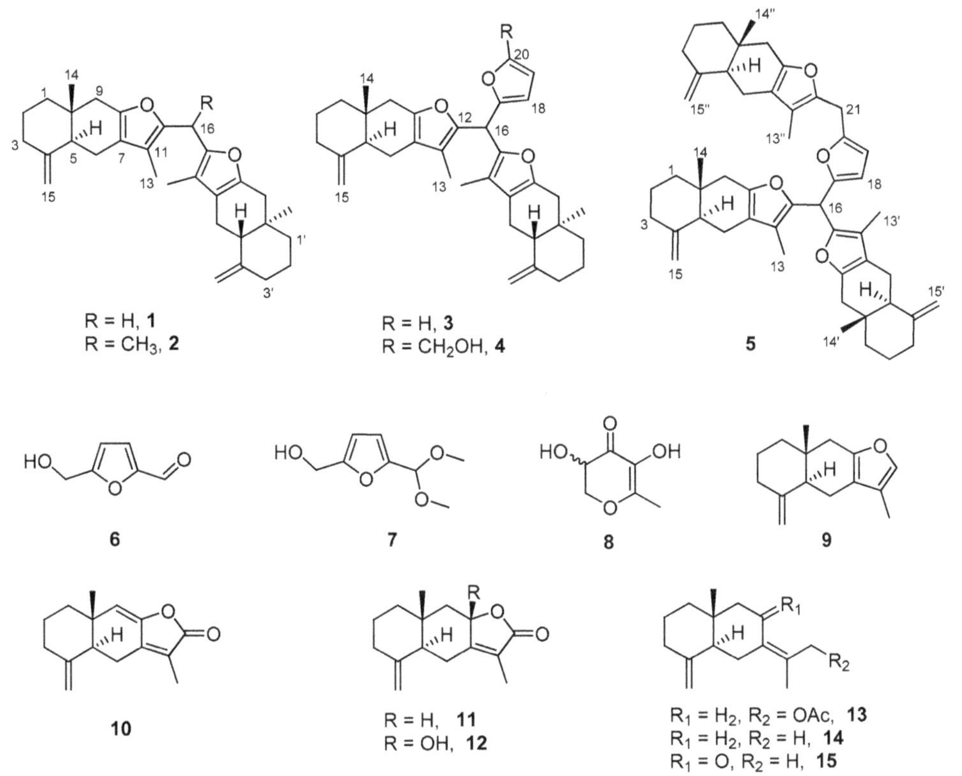

**Figure 1.** Structures of compounds **1–15**.

Compound **3** was isolated as a white amorphous powder. Its molecular formula was determined as $C_{35}H_{42}O_3$ based on the HR-ACPI-MS data for the $[M + H]^+$ peak at $m/z = 511.3221$, accounting for 15 degrees of unsaturation. Both the $^1$H and $^{13}$C signals of **3** in the upfield NMR chemical shift range showed close similarities to those of **1** and **2**, however, differences were observed in the downfield range. Four additional carbon signals at $\delta_C$ 153.1, 141.4, 110.2, and 107.2 in the $^{13}$C spectrum, and three additional proton signals at $\delta_H$ 7.35 (d), 6.30 (dd), and 6.07 (d) in the $^1$H spectrum were observed for **3**, as compared to **1** or **2**. These signals were assigned to a furan ring (3 degrees of unsaturation) by the analyses of COSY and HMBC spectra. Apart from these differences, the other signals displayed almost the same features as in **1** or **2**. The analyses of the 2D NMR spectra confirmed that **3** possessed a similar biatractylon skeleton in its structure as was found in **1** and **2**. Clear HMBC correlations of the methine proton (CH-16, with proton and carbon signals at $\delta_H$ 5.47 (s) and $\delta_C$ 36.1, respectively) to C-17 at $\delta_C$ 153.1, C-18 at $\delta_C$ 107.2, C-12 at $\delta_C$ 144.7, and C-11 $\delta_C$ 115.2 were observed, indicating that the furan ring and the two atractylon moieties were linked together through this methine. Hence, the structure of **3** was determined to be 2,2′-(furan-2-yl-methylene)-bis(3,8α-dimethyl-5-methylene-4,4α,5,6,7,8,8α,9-octahydronaphtho-[2,3-β] -furan), trivially named furan-2-methanetriyl-biatractylon.

Compound **4** was obtained as a white amorphous powder. The $[M + H]^+$ ion at $m/z = 541.3299$ in the HR-ACPI-MS spectrum revealed its molecular formula as $C_{36}H_{44}O_4$, with the same 15 degrees of unsaturation but with one more carbon, two more protons, and one more oxygen when compared to **3**. The $^1$H-NMR spectrum of **4** showed almost the same features as that of **3**, except for having one more singlet signal at $\delta_H$ 4.55. The $^{13}$C-NMR spectrum of **4** also displayed close similarity to that of **3**, except for the addition of an oxygenated methylene signal at $\delta_C$ 57.6 (corresponding to the above signal at $\delta_H$ 4.55 in the HSQC spectrum) and the significant downfield shift of the olefinic carbon signal from $\delta_C$ 114.4 to 153.0. In the HMBC spectrum of **4**, the protons of this oxygenated methylene demonstrated clear correlations to the downfield-shifted carbons C-20 at $\delta_c$ 153.0 and C-19 at $\delta_c$ 108.7, indicating its connection at C-20. No other significant differences were observed when comparing the NOESY spectra of **4** and **3**. Accordingly, the structure of **4** was established to be (5-(bis(3,8α-dimethyl-5-methylene-4,4α,5,6,7,8,8α,9-octahydronaphtho-[2,3-β]-furan-2-yl)methyl) furan-2-yl)methanol, trivially named 5-furanmethanol-2-methanetriyl-biatractylon.

Compound **5** gave the $[M + H]^+$ ion at $m/z$ = 739.4719 in its HR-APCI-MS spectrum, which was consistent with a molecular formula $C_{51}H_{62}O_4$ possessing 21 degrees of unsaturation. As compared with **4**, compound **5** has six more degrees of unsaturation, 15 more carbons, and 18 more hydrogens in the molecular formula. Both its $^1H$ and $^{13}C$-NMR spectra exhibit similar signal patterns to those of **4**, respectively, except for the occurrence of additional peaks which were attributed to atractylon moieties. Given that an atractylon moiety possesses 6 degrees of unsaturation, 15 carbons, and 19 hydrogens; it is implied that **5** should contain three atractylon moieties in its structure. Further analyses of the 1D and 2D NMR spectra confirmed the assignment, with the supporting data listed in Table 1. The methine proton (H-16) at $\delta_H$ 5.42 displayed clear HMBC correlations to the carbon pairs C-11/11′ at $\delta_C$ 115.1/115.1, C-12/12′ at $\delta_C$ 144.8/144.8, as well as C-17 at $\delta_C$ 151.5 and C-18 at $\delta_C$ 107.9, indicating two atractylon moieties were linked through this methine to the furan unit. The methylene protons (H-21) at $\delta_H$ 3.86 showed clear HMBC correlations to the carbon C-11″ at $\delta_C$ 114.9, C-12″ at $\delta_C$ 144.3, C-19 at $\delta_C$ 106.5, and C-20 at $\delta_C$ 151.4, revealing that the third atractylon moiety is linked through this methylene to the furan unit. On the basis of the above information, the structure of **5** was assigned as 2,2′-((5-((3,8α-dimethyl-5-methylene-4,4α,5,6,7,8,8α,9-octahydronaphtho-[2,3-β]-furan-2-yl)methyl) furan-2-yl)methylene)-bis(3,8α-dimethyl-5-methylene-4,4α,5,6,7,8,8α,9-octahydronaphtho-[2,3-β]-furan), trivially named furan-5-methanediyl-2-methanetriyl-triatractylon.

In addition to the aforementioned five new compounds, ten known compounds (**6**–**15**) were also isolated from the processed RAM in the present study (Figure 1). They were identified as 5-(hydroxymethyl)furfural (**6**) [16], 5-(Hydroxymethyl)-2-(dimethoxymethyl)furan (**7**) [17], 2,3-dihydro-3,5-dihydroxy-6-methyl-4-pyranone (**8**) [18], atractylon (**9**) [19], atractylenolide I (**10**) [20], atractylenolide II (**11**) [20], atractylenolide III (**12**) [20], atractylenolide VII (**13**) [21], γ-selinene (**14**) [22], and selina-4(14),7(11)-dien-8-one (**15**) [23] with comparing their spectroscopic data with those in the literature. Compound **7** is likely an artifact from **6** since methanol was used in the isolation process. Compound **8** was isolated from wheat [18], indicating that it might come from the wheat which was used during the process.

### 2.2. *Proposed Mechanism for the Formation of Compounds* **1–5**

To the best of our knowledge, this is the first report to delineate the possible role that processing plays in producing new chemical entities and impacting the chemical composition of traditional medicines. Five new compounds (**1**–**5**) were generated as a result of the processing of RAM by stir-frying with wheat bran. A plausible rationale for the formation of new atractylon derivatives (**1**–**5**) is outlined below. Both RAM and wheat bran contain fiber, polysaccharides, cellulose, resistant starch, inulin, lignins, and oligosaccharides [1,24]. Thermal processing of cellulose, hemicellulose, and other polysaccharides are known to produce significant amounts of diverse carbonyl compounds such as formaldehyde [25], acetaldehyde, furfural, 5-hydroxymethylfurfural, etc. [26]. Isolation of major quantities of both 5-(hydroxymethyl)furfural (**6**) and 5-(hydroxymethyl)-2-(dimethoxymethyl)furan (**7**) from the processed RAM in this study further supports the formation of carbonyl compounds. On the other hand, atractylon, the major sesquiterpene of *A. macrocephala* [15,27], can undergo electrophilic reactions with the pyrolytic aldehyde products. For example, as shown in the proposed plausible mechanism (Figure 2), atractylon would undergo electrophilic addition with RCHO (R can be hydrogen, methyl, 2-furanyl, or 5-hydroxymethylfuranyl) to yield a carbinol intermediate (**A**). Under pyrolytic conditions, the resulting carbinol would yield electrophilic species (**B**), which can be added to another molecule of atractylon to form an adduct (**C**). Further deprotonation would yield thermally stable adducts **1**–**4**. Moreover, compound **4** can undergo the same set of reactions (generation of electrophilic species, and addition to another molecule of atractylon and dehydration) to form compound **5**.

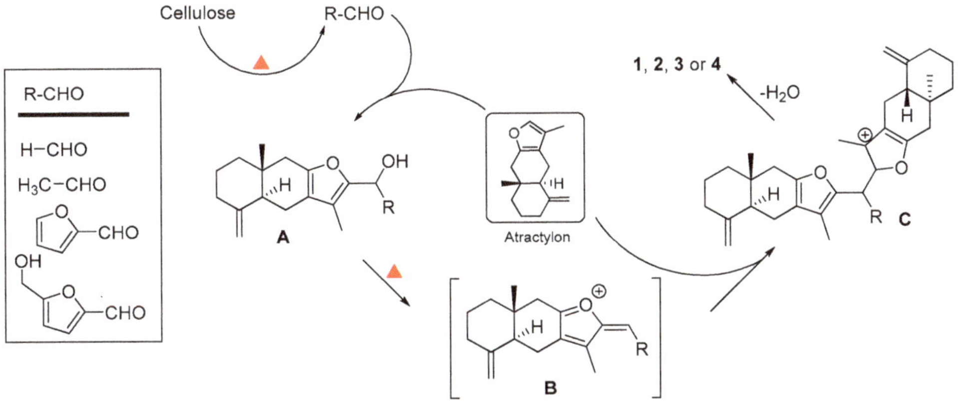


R= H, $CH_3$, 2-furanyl, 5-hydroxymethl-2-furanyl

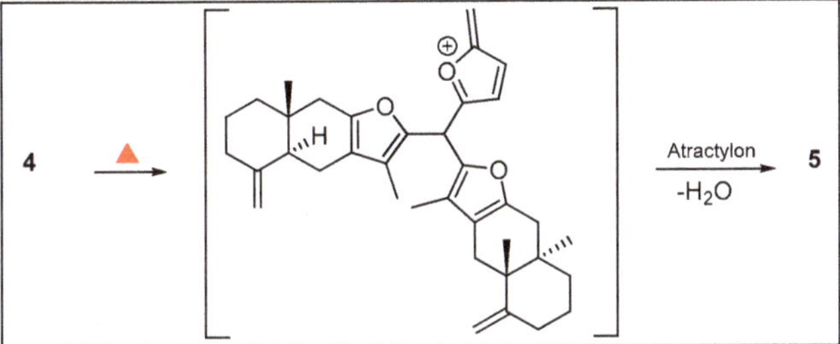


**Figure 2.** Proposed mechanism for the generation of compounds **1**–**5**.

### 2.3. Biological Activities of Compounds 1–5

The new compounds (**1**–**5**) were evaluated for their cytotoxicity to a panel of selected mammalian cancer and noncancer cell lines (SK-MEL, KB, BT-549, SK-OV-3, LLC-PK1, and Vero cells). None of the compounds exhibited cytotoxic effects. All compounds were ineffective at inhibiting the growth of the pathogenic microorganisms including three fungi (*C. albicans*, *C. neoformans*, and *A. fumigates*) and five bacteria (*S. aureus*, methicillin-resistant *S. aureus*, *E. coli*, *P. aeruginosa*, and *M. intracellulare*). In addition, the in vitro antileishmanial activity testing revealed that the compounds were not effective against *L. donavani*. Even though the preliminary biological data of these interesting class of compounds are not encouraging, further biological studies on this class of compounds concerning other biological targets, as well as SAR and medicinal properties need to be further investigated. In addition to biological activities, probing the comparative physicochemical properties of these polymeric compounds against monomeric atractylon might provide additional scientific rationale behind the traditional processes associated with polyherbal formulations.

## 3. Materials and Methods

### 3.1. General Experimental Procedures

UV and IR spectra were obtained on a HP 8452A UV-Vis spectrometer (Hewlett-Packard, Palo Alto, CA, USA) and an Cary 630 FTIR spectrometer (Agilent Technologies, Santa Clara, CA, USA), respectively.

High-resolution mass spectra were obtained on an Agilent TOF LC/MS spectrometer equipped with an atmospheric pressure chemical ionization (APCI) source and operated with the Analyst QS 1.1 software for data acquisition and processing. NMR spectra were acquired on an Agilent DD2-500 NMR spectrometer with a OneNMR probe at 500 MHz for $^1$H and 125 MHz for $^{13}$C using the pulse programs provided by the Agilent Vnmrj 4.0 software. Silica gel (J. T. Baker, 40 μm for flash chromatography) and Sephadex LH-20 were purchased from Fisher Scientific Co. (Waltham, MA, USA), and used for column chromatographic separations. Biotage Isolera$^{TM}$ Prime flash chromatography system (Biotage Co, Charlotte, NC, USA) was used for further separation and purification. TLC was performed on silica gel 60 GF 254 plates (Millipore Sigma, Burlington, MA, USA) with the TLC spots being observed at 254 nm, followed by spraying with 1% vanillin-$H_2SO_4$ derivatization reagent.

### 3.2. Plant Material and Preparation

The rhizomes of *A. macrocephala* were purchased from the Harbin Medicine Group Shiyitang Chinese Herbal Medicine Pharmaceutical Co. (Harbin City, Heilongjiang, China), and authenticated by Prof. Zhenyue Wang at the Heilongjiang University of Chinese Medicine. A voucher specimen (No. 20150013) was deposited in the Natural Products Laboratory, the School of Pharmacy of Heilongjiang University of Chinese Medicine. The wheat bran, produced in Heilongjiang, was bought from the Harbin Medicinal Materials market.

Processing of the rhizomes of *A. macrocephala* (2.5 kg) was conducted in the Processing Lab of the Experimental Training Center in the Heilongjiang University of Chinese Medicine. Following the processing protocol in the Chinese Pharmacopeia, the rhizomes were cut into slices and air-dried to reach a constant weight. The wheat bran was first stir-fried in a cauldron with a temperature around 170 °C until smoke appeared. The rhizome slices were then added into the cauldron with a ratio of the rhizome to wheat bran of 1:4 (*w/w*), the mixture was stir-fried approximately 26 mins until the slices turned to a yellow-brown color. Next, the processed rhizome slices were separated from the wheat bran by a sifter.

### 3.3. Extraction and Isolation

The processed slices of rhizomes (2 kg) of *A. macrocephala* were powdered and extracted with methanol (10 L × 3 times) using ultrasonic extraction for 1 hour for each cycle. After the solvent was evaporated under reduced pressure, a total of 206.8 g of crude extract was obtained. The obtained extract was suspended in water (3 L) and successively partitioned with hexane (3 L × 3 times) and EtOAc (3 L × 3 times) to yield 44.4 g and 44.0 g partition fractions, respectively, after removal of solvents in vacuo. The hexane fraction (44.0 g) was subjected to a silica gel column (CC) (820 g, 10 × 80 cm) for chromatographic fractionation, eluting with a hexane-$CHCl_3$ solvent system by stepwise increasing the polarity. The fractions were collected and combined based on TLC analysis to afford 18 fractions (Fr.1–18). Fr.5 (77.4 mg) was chromatographed on a Sephadex LH-20 column (45 g, 2 × 80 cm), eluted with $CHCl_3$-MeOH (2.5:1) to yield compound **1** (12.1 mg). Fr.4 (77.1 mg) was first separated on a Biotage chromatographic system (SNAP KP-Sil 100 g, eluting with hexane-$CH_2Cl_2$ solvent system in increasing $CH_2Cl_2$ concentration from 15% to 80%), then purified by Sephadex LH-20 (45 g, 2 × 80 cm, eluting with $CHCl_3$-MeOH 2:1) and silica gel column chromatography (CC) (9.25 g, 1.5× 40 cm, stepwise eluting with hexane-$CH_2Cl_2$) to give **2** (6.2 mg). Fr.8 (239.8 mg) was chromatographed on a silica gel column (120 g, 4 × 80 cm) eluted with MeOH-$CHCl_3$ (1:1), the targeted fraction was further purified using a Sephadex LH-20 column (45 g, 2 × 80 cm), eluting with $CHCl_3$-MeOH while gradually increasing the polarity, then further purified on a silica gel column (8.9 g, 2 × 40 cm) eluting with hexane-$CH_2Cl_2$ to yield **3** (13.5 mg). Fr.15 (4.20 g) was first chromatographed on a silica gel column (120g, 4× 80 cm) eluting with hexane-EtOAc in increasing polarity, and then further purified on a Sephadex LH-20 column (45 g, 2 × 80 cm) eluting with $CHCl_3$-MeOH (2:1) to yield **4** (7.4 mg). Fr.10 (297.1 mg) was treated similarly to that of Fr. 4 to obtain **5** (9.6 mg).

### 3.4. Characterization of Compounds **1–5**

*Bis(3,8α-dimethyl-5-methylene-4,4α,5,6,7,8,8α,9-octahydronaphtho-[2,3-β]-furan-2-yl)methane* (**1**), (methylene-biatractylon), white amorphous powder. UV (MeOH): $\lambda_{max}$ (logε) = 231 (3.95) nm. IR (ART): $V_{max}$ = 2926.0, 2849.5, 1641.9, 1440.6, 1378.1, 1231.9, 1250.5, 1136.8, 889.0, 758.5 $cm^{-1}$. $^1$H-NMR and $^{13}$C-NMR data see Table 1. HR-APCI-MS: *m/z* = 445.3134 $[M + H]^+$ (calc'd. for $C_{31}H_{41}O_2$ = 445.3106).

*2,2′-(Ethane-1,1-diyl)-bis(3,8α-dimethyl-5-methylene-4,4α,5,6,7,8,8α,9-octahydronaphtho-[2,3-β]-furan)* (**2**), (ethylidene-biatractylon), white amorphous powder. UV (MeOH): $\lambda_{max}$ (logε) = 228 (4.02) nm. IR (ART): $V_{max}$ = 2926.0, 2847.7, 1641.9, 1440.6, 1377.3, 1138.7, 889.0, 758.5 $cm^{-1}$. $^1$H-NMR and $^{13}$C-NMR data see Table 1. HR-APCI-MS: *m/z* = 459.3267 $[M + H]^+$ (calc'd. for $C_{32}H_{43}O_2$ = 459.3262).

*2,2′-(furan-2-ylmethylene)-bis(3,8α-dimethyl-5-methylene-4,4α,5,6,7,8,8α,9-octahydronaphtho-[2,3-β]-furan)* (**3**) (furan-2-methanetriyl–biatractylon), white amorphous powder. UV (MeOH): $\lambda_{max}$ (logε) = 229 (4.12) nm. IR (ART): $V_{max}$ = 2924.1, 2849.5, 1641.9, 1377.3, 1440.6, 1340.0, 1250.5, 1231.9, 1181.6, 1136.8, 899.0, 762.2, 730.6 $cm^{-1}$. $^1$H-NMR and $^{13}$C-NMR data see Table 1. HR-APCI-MS: *m/z* = 511.3221 $[M + H]^+$ (calc'd. for $C_{35}H_{43}O_3$ = 511.3212).

*(5-(bis(3,8α-dimethyl-5-methylene-4,4α,5,6,7,8,8α,9-octahydronaphtho-[2,3-β]-furan-2-yl)methyl)furan-2-yl)methanol* (**4**) (5-furanmethanol-2-methanetriyl-biatractylon), white amorphous powder. UV (MeOH): $\lambda_{max}$ (logε) = 230 (4.17) nm. IR (ART): $V_{max}$ = 2924.1, 2845.8, 1641.9, 1440.6, 1377.3, 1250.5, 1233.7, 1136.8, 1013.8, 889.0, 786.5, 758.5 $cm^{-1}$. $^1$H-NMR and $^{13}$C-NMR data see Table 1. HR-APCI-MS: *m/z* = 541.3299 $[M + H]^+$ (calc'd. for $C_{36}H_{45}O4$ = 541.3318).

*2,2′-((5-((3,8α-dimethyl-5-methylene-4,4α,5,6,7,8,8α,9-octahydronaphtho-[2,3-β]-furan-2-yl)methyl)furan-2-yl)methylene)bis(3,8α-dimethyl-5-methylene-4,4α,5,6,7,8,8α,9-octahydronaphtho-[2,3-β]-furan)* (**5**) (furan-5-methanediyl-2-methanetriyl-triatractylon), white amorphous powder. UV (MeOH): $\lambda_{max}$ (logε) 230 (4.30) nm. IR (ART): $V_{max}$ = 2953.9, 2924.1, 2853.3, 1459.3, 1377.3, 1272.9, 1136.8, 1123.8, 1073.5, 1015.7, 889.0 $cm^{-1}$. $^1$H-NMR and $^{13}$C-NMR data see Table 1. HR-APCI-MS: *m/z* = 739.4719 $[M + H]^+$ (calc'd. for $C_{51}H_{63}O_4$ = 739.4726).

The NMR, MS, UV, and IR spectra of above five compounds are presented in the Supplementary Materials, which are available free of charge via the internet at https://www.mdpi.com/journal/molecules.

### 3.5. Biological Activity Assays

#### 3.5.1. Cytotoxicity Assay

Cytotoxic activity was determined against four human cancer cell lines (SK-MEL, KB, BT-549, and SKOV-3) and two noncancerous kidney cell lines (LLC-PK1 and Vero) as described earlier [28]. All cell lines were obtained from the American Type Culture Collection (ATCC, Rockville, MD, USA). Each assay was performed in 96-well tissue culture-treated microplates. Cells were seeded at a density of 25,000 cells/well and incubated for 24 h. Samples at different concentrations were added, and cells were again incubated for 48 h. At the end of incubation, the cell viability was measured using a tetrazolium dye (WST-8) which was converted to a water-soluble formazan product. The absorbance was measured at 450 nm and the percent viability of sample treated cells was calculated in comparison to the vehicle-treated cells. Doxorubicin was used as a positive control, while DMSO was used as the negative (vehicle) control.

#### 3.5.2. Antimicrobial Assays

Isolates were tested against a panel of 8 pathogenic organisms including three fungi (*Candida albicans* ATCC 90028, *Cryptococcus neoformans* ATCC 90113, and *Aspergillus fumigates* ATCC 204305) and five bacteria (*Staphylococcus aureus* ATCC 29213, methicillin-resistant *S. aureus* ATCC 33591, *Escherichia coli* ATCC 35218, *Pseudomonas aeruginosa* ATCC 27853, and *Mycobacterium intracellulare* ATCC 23068). Microorganisms were obtained from the American Type Culture Collection. The assays were performed at the National Center for Natural Products Research (NCNPR), at the University of Mississippi as a part of the antimicrobial screening program following a previously reported method [29]. Drug controls, ciprofloxacin for bacteria and amphotericin B for fungi, were included in each assay.

3.5.3. Antileishmanial Assay

Compounds were tested in vitro for their ability to inhibit *Leishmania donovani*, employing the assay described by Jain et al. [30]. Amphotericin B was included as the drug control for *L. donovani*.

## 4. Conclusions

Pao-Zhi (frying and cooking of herbs) is an ancient pharmaceutic technique in TCM to facilitate the use of herbal medicines for specific clinic needs [31]. Traditionally, most Chinese herbal medicines undergo elaborate processing in order to become ingredients that are prescribed or utilized in the manufacturing of TCM proprietary drugs [32]. Although the practice of processing has a long history, the underlying mechanisms largely remain unclear for most Chinese herbal medicines. In the present study, through the characterization of chemical profiles coupled with NMR metabolomics approach, the chemical changes resulting from the traditional processing protocol associated with RAM preparation were investigated. For the first time, five new chemical adducts, which were formed during processing *A. macrocephala* with wheat bran, were isolated and their structures were identified. The findings allowed us to gain valuable insights into the chemical reactions which occur during the processing procedures. Processed RAM is widely used in various formulations of TCM drugs and health care products. Stir-frying with wheat bran is one of the most widely used traditional processing methods for RAM in TCM [12]. Although the processed RAM is listed as an item in the Chinese Pharmacopoeia, there are currently no modern standardized processing protocols or quality control standards for the processed RAM products. The findings of this study may provide useful information for developing such standards with a scientific basis. Furthermore, as the change of chemical profile will inevitably influence the associated pharmacological properties of processed RAM, further investigations of the bio-activities of the newly generated compounds are needed.

**Supplementary Materials:** The supplementary information is available online. Figure S1.1: $^{1}$H-NMR spectrum of compound **1**, Figure S1.2: $^{13}$C-NMR spectrum of compound **1**, Figure S1.3: HSQC spectrum of compound **1**, Figure S1.4: COSY spectrum of compound **1**, Figure S1.5: HMBC spectrum of compound **1**, Figure S1.6: HR-APCI-MS spectrum of compound **1**, Figure S1.7: IR spectrum of compound **1**, Figure S1.8: UV spectrum of compound **1**, Figure S2.1: $^{1}$H-NMR spectrum of compound **2**, Figure S2.2: $^{13}$C-NMR spectrum of compound **2**, Figure S2.3: HSQC spectrum of compound **2**, Figure S2.4: COSY spectrum of compound **2**, Figure S2.5: HMBC spectrum of compound **2**, Figure S2.6: HR-APCI-MS spectrum of compound **2**, Figure S2.7: IR spectrum of compound **2**, Figure S2.8: UV spectrum of compound **2**, Figure S3.1: $^{1}$H-NMR spectrum of compound **3**, Figure S3.2: $^{13}$C-NMR spectrum of compound **3**, Figure S3.3: HSQC spectrum of compound **3**, Figure S3.4: COSY spectrum of compound **3**, Figure S3.5: HMBC spectrum of compound **3**, Figure S3.6: HR-APCI-MS spectrum of compound **3**, Figure S3.7: IR spectrum of compound **3**, Figure S3.8: UV spectrum of compound **3**, Figure S4.1: $^{1}$H-NMR spectrum of compound **4**, Figure S4.2: $^{13}$C-NMR spectrum of compound **4**, Figure S4.3: HSQC spectrum of compound **4**, Figure S4.4: COSY spectrum of compound **4**, Figure S4.5: HMBC spectrum of compound **4**, Figure S4.6: HR-APCI-MS spectrum of compound **4**, Figure S4.7: IR spectrum of compound **4**, Figure S4.8: UV spectrum of compound **4**, Figure S5.1: $^{1}$H-NMR spectrum of compound **5**, Figure S5.2: $^{13}$C-NMR spectrum of compound **5**, Figure S5.3: HSQC spectrum of compound **5**, Figure S5.4: COSY spectrum of compound **5**, Figure S5.5: HMBC spectrum of compound **5**, Figure S5.6: HR-APCI-MS spectrum of compound **5**, Figure S5.7: IR spectrum of compound **5**, Figure S5.8: UV spectrum of compound **5**.

**Author Contributions:** Conceptualization, C.Z. and J.Z.; methodology, J.Z.; extraction, isolation, purification and identification, C.Z. and Y.M.; formal analysis, M.W.; writing—original draft preparation, J.Z., Y.M., and A.G.C.; writing—review and editing, M.W. and A.G.C.; project administration, I.A.K. All authors have read and agreed to the published version of the manuscript.

**Funding:** This work was supported in part by "Discovery & Development of Natural Products for Pharmaceutical & Agricultural Applications" funded by the United States Department of Agriculture, Agricultural Research Service, Specific Cooperative Agreement No. 58-6060-6-015.

**Acknowledgments:** The authors sincerely thank Shabana I. Khan and Bharathi Avula, at the NCNPR, University of Mississippi, for bioactivity testing and HR-APCI mass spectral recording, respectively.

**Conflicts of Interest:** The authors declare no conflict of interest. The funders had no role in the design of the study; in the collection, analyses, or interpretation of data; in the writing of the manuscript, or in the decision to publish the results.

## References

1. Zhu, B.; Zhang, Q.-L.; Hua, J.-W.; Cheng, W.-L.; Qin, L.-P. The traditional uses, phytochemistry, and pharmacology of Atractylodes macrocephala Koidz.: A review. *J. Ethnopharmacol.* **2018**, *226*, 143–167. [CrossRef] [PubMed]
2. Liu, D.; Li, Y.; Xu, Y.; Zhu, Y. Anti-aging traditional Chinese medicine: Potential mechanisms involving AMPK pathway and calorie restriction based on "medicine-food homology" theory. *China J. Chin. Mater. Med.* **2016**, *41*, 1144–1151. [CrossRef]
3. Li, Y.; Li, W.; Wang, L.; Shao, C.; Zhao, J.; Liu, J. Summary of research on bran fried atractylodes. *Shandong Zhong Yi Yao Da Xue* **2018**, *42*, 182–185.
4. Wang, C.S. The processing of Atractylodes macrocephala by stir-frying with wheat bran. *Zhong Yao Tong Bao (Beijing, China: 1981)* **1983**, *8*, 18–19.
5. Zhao, J.; Wang, M.; Avula, B.; Zhong, L.; Song, Z.; Xu, Q.; Li, S.; Ibrahim, M.A. Effect of Processing on the Traditional Chinese Herbal Medicine Flos Lonicerae: An NMR-based Chemometric Approach. *Planta Med.* **2015**, *81*, 754–764. [CrossRef] [PubMed]
6. Ye, D.; Yuan, S. *Dictionary of Chinese Herbal Processing Science*; Shanghai Science and Technology Press: Shanghai, China, 2005; p. 102.
7. State Pharmacopoeia Commission of the PRC. *Pharmacopoeia of the People's Republic of China. 1 (2015)*; China Medical Science Press: Beijing, China, 2015.
8. Hoang, L.S.; Tran, M.H.; Lee, J.-S.; Ngo, Q.M.T.; Woo, M.H.; Min, B.S. Inflammatory Inhibitory Activity of Sesquiterpenoids from Atractylodes macrocephala Rhizomes. *Chem. Pharm. Bull.* **2016**, *64*, 507–511. [CrossRef] [PubMed]
9. Li, Y.; Zhang, Y.; Wang, Z.; Zhu, J.; Tian, Y.; Chen, B. Quantitative analysis of atractylenolide I in rat plasma by LC–MS/MS method and its application to pharmacokinetic study. *J. Pharm. Biomed. Anal.* **2012**, *58*, 172–176. [CrossRef]
10. Cai, B.; Cai, H.; Xu, Z.; Luo, S.; Zhang, W.; Cao, G.; Liu, X.; Lou, Y.; Ma, X.; Qin, K. Study on chemical fingerprinting of crude and processed Atractylodes macrocephala from different locations in Zhejiang province by reversed-phase high-performance liquid chromatography coupled with hierarchical cluster analysis. *Pharmacogn. Mag.* **2012**, *8*, 300–307. [CrossRef]
11. Cao, G.; Xu, Z.; Wu, X.; Li, Q.; Chen, X. Capture and identification of the volatile components in crude and processed herbal medicines through on-line purge and trap technique coupled with GC × GC-TOF MS. *Nat. Prod. Res.* **2014**, *28*, 1607–1612. [CrossRef]
12. Gu, S.; Li, L.; Huang, H.; Wang, B.; Zhang, T. Antitumor, Antiviral, and Anti-Inflammatory Efficacy of Essential Oils from Atractylodes macrocephala Koidz. Produced with Different Processing Methods. *Molecules* **2019**, *24*, 2956. [CrossRef]
13. Shan, G.-S.; Zhang, L.; Zhao, Q.-M.; Xiao, H.-B.; Zhuo, R.-J.; Xu, G.; Jiang, H.; You, X.-M.; Jia, T.-Z. Metabolomic study of raw and processed Atractylodes macrocephala Koidz by LC–MS. *J. Pharm. Biomed. Anal.* **2014**, *98*, 74–84. [CrossRef] [PubMed]
14. Cui, X.; Shan, C.; Wen, H.; Li, W.; Wu, H. UFLC/Q-TOF-MS based analysis on material base of *Atractylodes macrocephalae* rhizoma stir-fried with wheat bran. *Zhongguo Zhongyao Za Zhi* **2013**, *38*, 1929–1933.
15. Zhao, C.-X.; He, C. Preparative isolation and purification of atractylon and atractylenolide III from the Chinese medicinal plant atractylodes macrocephala by high-speed counter-current chromatography. *J. Sep. Sci.* **2006**, *29*, 1630–1636. [CrossRef] [PubMed]
16. Wang, X.; Li, L.; Ran, X.; Dou, D.; Li, B.; Yang, B.; Li, W.; Koike, K.; Kuang, H. What caused the changes in the usage of Atractylodis Macrocephalae Rhizoma from ancient to current times? *J. Nat. Med.* **2015**, *70*, 36–44. [CrossRef] [PubMed]

17. Peng, X.; Wu, Q.; Gao, C.; Gai, C.; Yuan, D.; Fu, H. Chemical constituents from the fruit calyx of *Physalis alkekengi var. francheti*. *J. Chin. Pharm. Sci.* **2015**, *24*, 600–606. [CrossRef]
18. Jiang, D.; Peterson, D.G. Identification of bitter compounds in whole wheat bread. *Food Chem.* **2013**, *141*, 1345–1353. [CrossRef]
19. Hasada, K.; Yoshida, T.; Yamazaki, T.; Sugimoto, N.; Nishimura, T.; Nagatsu, A.; Mizukami, H. Quantitative determination of atractylon in Atractylodis Rhizoma and Atractylodis Lanceae Rhizoma by 1H-NMR spectroscopy. *J. Nat. Med.* **2010**, *64*, 161–166. [CrossRef]
20. Zhang, N.; Liu, C.; Sun, T.-M.; Ran, X.-K.; Kang, T.-G.; Dou, D. Two new compounds from Atractylodes macrocephala with neuroprotective activity. *J. Asian Nat. Prod. Res.* **2016**, *19*, 35–41. [CrossRef]
21. Ding, H.-Y.; Liu, M.-Y.; Chang, W.-L.; Lin, H.-C. New sesquiterpenoids from the rhizomes of *Atractylodes macrocephala*. *Chin. Pharm. J.* **2005**, *57*, 37–42.
22. Gupta, J.; Ali, M.; Pillai, K.K.; Velasco-Negueruela, A.; Pérez-Alonso, M.J.; Contreras, F.Ó. The Occurrence of Ishwarane and Ishwarone in the Roof Oil of Corallocarpus epigaeus Benth. ex Hook. f. *J. Essent. Oil Res.* **1997**, *9*, 667–672. [CrossRef]
23. Bohlmann, F.; Suwita, A. New sesquiterpenes from Peteravenia schultzii. *Phytochemistry* **1978**, *17*, 567–568. [CrossRef]
24. Stevenson, L.; Phillips, F.; O'Sullivan, K.; Walton, J. Wheat bran: Its composition and benefits to health, a European perspective. *Int. J. Food Sci. Nutr.* **2012**, *63*, 1001–1013. [CrossRef] [PubMed]
25. Li, S.; Lyons-Hart, J.; Banyasz, J.; Shafer, K. Real-time evolved gas analysis by FTIR method: An experimental study of cellulose pyrolysis. *Fuel* **2001**, *80*, 1809–1817. [CrossRef]
26. Wang, S.; Guo, X.; Liang, T.; Zhou, Y.; Luo, Z. Mechanism research on cellulose pyrolysis by Py-GC/MS and subsequent density functional theory studies. *Bioresour. Technol.* **2012**, *104*, 722–728. [CrossRef]
27. Yin, H.; Wang, Z.; Wang, L.; Zhou, A.-Z.; Li, Q.-L.; Cheng, Z. Simultaneous determination of atractylone, atractylenolide I, II, III in *Atractylodes macrocephala* by HPLC-wavelength switching method. *J. Tradit. Chin. Med.* **2013**, *1*, 235–238.
28. Yuan, H.; Zhao, J.; Wang, M.; Khan, S.I.; Zhai, C.; Xu, Q.; Huang, J.; Peng, C.; Xiong, G.; Wang, W.; et al. Benzophenone glycosides from the flower buds of Aquilaria sinensis. *Fitoterapia* **2017**, *121*, 170–174. [CrossRef]
29. Auker, K.M.; Coleman, C.M.; Wang, M.; Avula, B.; Bonnet, S.L.; Kimble, L.; Mathison, B.D.; Chew, B.P.; Ferreira, D. Structural Characterization of Cranberry Arabinoxyloglucan Oligosaccharides. *J. Nat. Prod.* **2019**, *82*, 606–620. [CrossRef]
30. Jain, S.K.; Sahu, R.; Walker, L.A.; Tekwani, B. A Parasite Rescue and Transformation Assay for Antileishmanial Screening Against Intracellular Leishmania donovani Amastigotes in THP1 Human Acute Monocytic Leukemia Cell Line. *J. Vis. Exp.* **2012**, *70*, e4054. [CrossRef]
31. Sheridan, H.; Kopp, B.; Krenn, L.; Guo, D.; Sendker, J. Traditional Chinese herbal medicine preparation: Invoking the butterfly effect. *Science* **2015**, *350*, S64–S66.
32. Zhao, Z.; Liang, Z.; Chan, K.; Lu, G.; Lee, E.L.M.; Chen, H.; Li, L. A Unique Issue in the Standardization of Chinese Materia Medica: Processing. *Planta Med.* **2010**, *76*, 1975–1986. [CrossRef]


**Sample Availability:** Samples of compounds **1–5** are available from the authors.

*Article*

# Antimicrobial Constituents from *Machaerium* Pers.: Inhibitory Activities and Synergism of Machaeriols and Machaeridiols against Methicillin-Resistant *Staphylococcus aureus*, Vancomycin-Resistant *Enterococcus faecium*, and Permeabilized Gram-Negative Pathogens

**Ilias Muhammad [1,*], Melissa R. Jacob [1], Mohamed A. Ibrahim [1], Vijayasankar Raman [1], Mallika Kumarihamy [1], Mei Wang [1,2], Taha Al-Adhami [3], Charlotte Hind [4], Melanie Clifford [4], Bethany Martin [4], Jianping Zhao [1], J. Mark Sutton [4] and Khondaker Miraz Rahman [3,*]**

1. National Center for Natural Products Research, Research Institute of Pharmaceutical Sciences, School of Pharmacy, University of Mississippi, Oxford, MS 38677, USA; mellie2880@gmail.com (M.R.J.); mmibrahi@olemiss.edu (M.A.I.); vraman@olemiss.edu (V.R.); mkumarih@olemiss.edu (M.K.); meiwang@olemiss.edu (M.W.); jianping@olemiss.edu (J.Z.)
2. Natural Products Utilization Research Unit, Agricultural Research Service, U.S. Department of Agriculture, University of Mississippi, Oxford, MS 38677, USA
3. Institute of Pharmaceutical Science, School of Cancer and Pharmaceutical Sciences, King's College London, Franklin-Wilkins Building, 150 Stamford Street, London SE1 9NH, UK; taha.al-adhami@kcl.ac.uk
4. National Infection Service, Public Health England, Manor Farm Road, Salisbury SP4 0JG, UK; Charlotte.Hind@phe.gov.uk (C.H.); Melanie.Clifford@phe.gov.uk (M.C.); Bethany.martin@phe.gov.uk (B.M.); mark.sutton@phe.gov.uk (J.M.S.)

* Correspondence: milias@olemiss.edu (I.M.); Miraz.rahman@kcl.ac.uk (K.M.R.); Tel.: +1-662-915-1051 (I.M.); +44-20-7848-1891 (K.M.R.)

Academic Editor: Daniela Rigano
Received: 18 November 2020; Accepted: 16 December 2020; Published: 18 December 2020

**Abstract:** Two new epimeric bibenzylated monoterpenes machaerifurogerol (**1a**) and 5-*epi*-machaerifurogerol (**1b**), and four known isoflavonoids (+)-vestitol (**2**), 7-*O*-methylvestitol (**3**), (+)-medicarpin (**4**), and 3,8-dihydroxy-9-methoxypterocarpan (**5**) were isolated from *Machaerium* Pers. This plant was previously assigned as *Machaerium multiflorum* Spruce, from which machaeriols A-D (**6–9**) and machaeridiols A-C (**10–12**) were reported, and all were then re-isolated, except the minor compound **9**, for a comprehensive antimicrobial activity evaluation. Structures of the isolated compounds were determined by full NMR and mass spectroscopic data. Among the isolated compounds, the mixture **10** + **11** was the most active with an MIC value of 1.25 μg/mL against methicillin-resistant *Staphylococcus aureus* (MRSA) strains BAA 1696, −1708, −1717, −33591, and vancomycin-resistant *Enterococcus faecium* (VRE 700221) and *E. faecalis* (VRE 51299) and vancomycin-sensitive *E. faecalis* (VSE 29212). Compounds **6–8** and **10–12** were found to be more potent against MRSA 1708, and **6**, **11**, and **12** against VRE 700221, than the drug control ciprofloxacin and vancomycin. A combination study using an in vitro Checkerboard method was carried out for machaeriols (**7** or **8**) and machaeridiols (**11** or **12**), which exhibited a strong synergistic activity of **12** + **8** (MIC 0.156 and 0.625 μg/mL), with >32- and >8-fold reduction of MIC's, compared to **12**, against MRSA 1708 and −1717, respectively. In the presence of sub-inhibitory concentrations on polymyxin B nonapeptide (PMBN), compounds **10** + **11**, **11**, **12**, and **8** showed activity in the range of 0.5–8 μg/mL for two strains of *Acinetobacter baumannii*, 2–16 μg/mL against *Pseudomonas aeruginosa* PAO1, and 2 μg/mL against *Escherichia coli* NCTC 12923, but were inactive (MIC > 64 μg/mL) against the two isolates of *Klebsiella pneumoniae*.

**Keywords:** *Machaerium* Rimachi 12161; machaerifurogerol; 5-*epi*-machaerifurogerol; machaeriol A–C; machaeridiol A–C; isoflavonoid; MRSA; VRE; gram negative bacteria

## 1. Introduction

The genus *Machaerium* Pers. (Fabaceae) consists of approximately 130 species, which are primarily distributed in the tropical Americas [1]. It is a genus of shrubs or lianas and small to medium-sized trees occurring throughout Southern Mexico to Brazil and Northern Argentina and Peru. These species are indigenous to all climatic regions ranging from equatorial rainforests to the verges of dry and cold deserts [2–4]. Several species of this genus are used in traditional medicines are considered to have multiple medicinal properties. Generally, various plant parts of *Machaerium* are used as an antitussive, and the sap is used to cure aphthous ulcers of the mouth [4]. *M. floribundum* is used to treat diarrhea and menstrual cramps [4]. The presence of a wide array of secondary metabolites from *Machaerium*, including flavonoids, terpenoids, and oxygenated phenolic compounds, together with their bioactivities, was recently reviewed by Amen et al. (2015) [2].

Earlier studies on one of the *Machaerium* species (Manuel Rimachi, Y-12161), named *M. multiflorum* Spruce, yielded four unique (+)-*trans*-hexahydrodibenzopyrans (HHDBP), machaeriols A-D, and three 5,6-*seco*-HHDBPs, machaeridiols A-C [5,6]. An unprecedented structural similarity for the HHDBP nucleus was observed in machaeriol and hexahydrocannabinol, and the 5,6-*seco*-HHDBP nucleus in machaeridiol and dihydrocannabidiol. Since these are the first reports of novel phytocannabinoids from a higher plant other than *Cannabis*, a recollection of the plant material was necessary from the original source. Unfortunately, there is an absence of documentary evidence for the existence of the species *M. multiflorum*. This species name was not included in the regional Floras as well as in major online databases (i.e., the International Plant Names Index (http://www.ipni.org/index.html) and The Plant List (http://www.theplantlist.org). Therefore, it was assumed that the plant sample was misidentified and was given the name combination *M. multiflorum* Spruce in error. An investigation was carried out on the identity of the plant, and a re-examination of the voucher specimen (Rimachi # 12161) at the Missouri Botanical Garden (MBG) concluded that this species should be treated only as an unidentified species of *Machaerium* Pers., as determined by the collection information (Manuel Rimachi, Y. 12161) [7].

The significance of the chemistry and biological activity of these aralkyl class of phytocannabinoid-type compounds led to the re-examination of the *n*-hexane and DCM fractions of the stem bark EtOH extract of the original plant material [5,6], as well as previously unexamined root and leaf extracts, which showed significant enhancement of antimicrobial activity against the various strains of methicillin-resistant *Staphylococcus aureus* (MRSA) and vancomycin resistant *Enterococci* (VRE). MRSA and VRE represent two potential threats to human health. According to the Centers for Disease Control and Prevention (CDC), MRSA can cause serious health problems, such as bloodstream infections and pneumonia. CA-MRSA occurs with a higher incidence rate in the United States and in particular amongst people who are in close physical contact, such as football athletes and childcare workers [8]. A recent national estimate for invasive MRSA incidence rates showed one in three people carry *S. aureus* in their nose and two in 100 people carry MRSA. *Enterococci* bacteria have the ability to survive for months in humans and animals. Similar to MRSA, VRE infections are commonly acquired by hospitalized patients. Enterococcal infections can be lethal, particularly those caused by VRE. According to the CDC, the number of nosocomial VRE isolates increased in the United States 20-fold, between 1989 and 1993. VRE is now the second to third most common cause of nosocomial infections in the USA [9].

In order to acquire substantial quantities of machaeriol A-D (**6–9**) and machaeridiol A-C (**10–12**) for comprehensive antimicrobial evaluations against MRSA and VRE, a reinvestigation was conducted on stem bark, leaves, and roots of the original plant material. During the course of this work, the novel epimeric mixture of bibenzylated furanoid monoterpenes, machaerifurogerol (**1a**), and 5-*epi*-machaerifurogerol (**1b**), together with the known isoflavons (+)-vestitol (**2**) and 7-*O*-methylvestitol (**3**), and pterocarpans (+)-medicarpin (**4**) and 3,8-dihydroxy-9-methoxypterocarpan (**5**), as well as previously isolated [5,6] machaeriol A-C (**6–8**) and machaeridiol A-C (**10–12**), were isolated. In this study, we report the correction of the previously reported botanical identity of the plant *M. multiflorum*, the structure elucidation of compounds **1–5**, and comprehensive antimicrobial activities of compounds **1–8** and **10–12**.

## 2. Results and Discussion

### 2.1. Botanical Identity of Machaerium sp. (Rimachi 12161)

The regional floras and other relevant publications were consulted for possible botanical identification of the species [7,10]. The voucher specimen (Rimachi 12161) was confirmed to belong to the genus *Machaerium*, based on the morphological features and available field information such as habitat, leaf, inflorescence, and fruit characters. The leaves showed partial similarities with those of *M. leiophyllum* var. *leiophyllum* and *M. glabrum*. However, complete identification of the specimen was not possible due to a lack of information on necessary diagnostic features, such as presence or absence of spines, features of stipules, and floral characters. It is possible that the specimen could represent an un-described taxon. The description (vide infra) is based on a single herbarium specimen and the associated collection information available from the original collection.

The *Machaerium* sp. (12161) plant was found to grow on sandy soils in open forests in Maynas, near Loreto, Peru, at an altitude of about 140–160 m. It is a woody liana with cylindrical stems, imparipinnate leaves with 17–21 leaflets, green flowers in axillary panicles, and 1-seeded samaroid fruits with a terminal wing showing reticulate venations.

### 2.2. Phytochemical Constituents

The dried EtOH extract of the stem bark was fractionated with *n*-hexane, followed by dichloromethane (DCM), and resulted in the isolation of compounds **1–5** (Figure 1 and Figure S1) (see the Experimental Section).

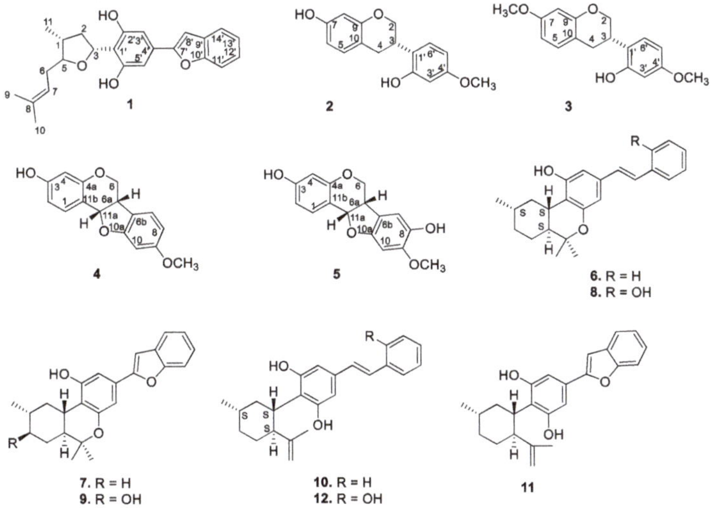

**Figure 1.** Structures of the isolated compounds.

Chromatographic separation of the DCM fraction 20–24 that led to the isolation of compound **1**, showed a single peak upon LC–MS analysis, which showed a protonated molecular ion peak at m/z 379.1906 $[M + H]^+$ in its ESI–HRMS, suggesting the molecular formula $C_{28}H_{29}O_3$. A careful analysis of the $^1H$ and $^{13}C$ NMR spectra (Table 1), and 2D NMR COSY, HMQC, HMBC, and NOESY spectra (Figures S2–S11) suggested that the compound was a mixture of C-5 epimers **1a** and **1b** (Figure 2).

**Table 1.** $^1H$ and $^{13}C$ NMR (in $CDCl_3$) data for epimeric compounds **1a** and **1b**.

| Position | 1a (Major Epimer) | | 1b (Minor Epimer) | | HMBC |
|---|---|---|---|---|---|
| | $\delta_C$ [a] (J in Hz) | $\delta_H$ [b] (J in Hz) | $\delta_C$ [a] (J in Hz) | $\delta_H$ [b] (J in Hz) | |
| 1 | 35.2 CH | 2.4 (m) | 35.4 CH | 2.4 (m) | 5, 2, 11 |
| 2 | 41.8 $CH_2$ | 2.72 (m), 1.59 (m) | 42.0 $CH_2$ | 2.39 (m), 1.58 (m) | 5, 11 |
| 3 | 77.7 CH | 5.32 (dd, 12, 7.6) | 77.4 CH | 5.59 (dd, 12, 8.0) | 2, 1′, 2′, 6′ |
| 5 | 83.1 CH | 4.00 (m) | 83.3 CH | 4.21 (m) | 11, 7 |
| 6 | 29.9 $CH_2$ | 1.24 (br s), 2.4 (m) | 29.4 $CH_2$ | 1.24 (br s), 2.4 (m) | 7, 8, 5 |
| 7 | 120.0 CH | 5.18 (t, 7.2) | 120.0 CH | 5.18 (t, 7.2) | 9, 10 |
| 8 | 134.5 C | | 134.4 CH | | |
| 9 | 18.2 $CH_3$ | 1.66 (s) | 18.3 $CH_3$ | 1.66 (s) | 7, 8, 10 |
| 10 | 26.1 $CH_3$ | 1.72 (s) | 26.1 $CH_3$ | 1.72 (s) | 7, 8, 9 |
| 11 | 15.5 $CH_3$ | 1.02 (d, 7.2) | 13.9 $CH_3$ | 1.03 (d, 7.2) | 1, 2, 5 |
| 1′ | 113.3 C | | 114.0 C | | |
| 2′ | 155.3 C | | 155.3 C | | |
| 3′ | 104.9 CH | 6.83 (br s) | 104.9 CH | 6.83 (br s) | 5′, 4′, 1′ |
| 4′ | 130.6 C | | 130.8 C | | |
| 5′ | 104.9 CH | 6.83 (br s) | 104.9 CH | 6.83 (br s) | 3′, 4′, 1′, 7′ |
| 6′ | 155.4 C | | 155.4 C | | |
| 7′ | 154.9 C | | 154.9 C | | |
| 8′ | 101.7 CH | 6.89 (br s) | 101.6 CH | 6.89 (br s) | 4′, 7′ |
| 9′ | 129.5 C | | 129.4 C | | |
| 10′ | 155.5 C | | 155.5 C | | |
| 11′ | 111.3 CH | 7.45 (d, 7.6) | 111.3 CH | 7.45 (d, 7.6) | 12′, 13′, 10′ |
| 12′ | 124.4 CH | 7.22 (m) | 124.4 CH | 7.22 (m) | 13′, 11′, 10′ |
| 13′ | 123.1 CH | 7.18 (m) | 123.1 CH | 7.18 (m) | 11′, 9′ |
| 14′ | 121.1 CH | 7.52 (d, 7.2) | 121.1 CH | 7.52 (d, 7.2) | 8′, 10′, 13′ |

[a] $^1H$ Spectra recorded at 400 MHz, and [b] $^{13}C$ spectra recorded at 100 MHz.

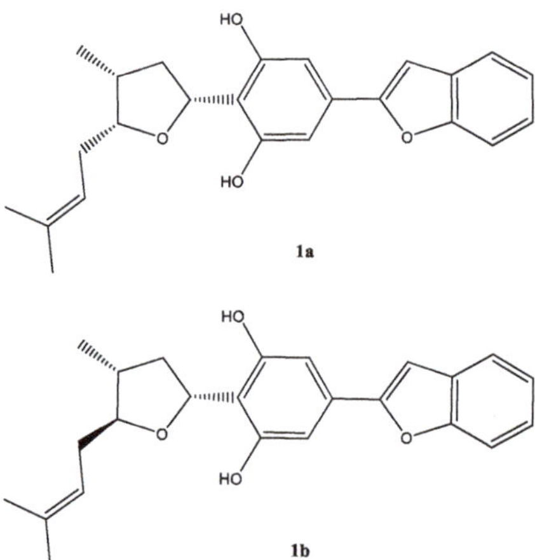


**Figure 2.** Structure of the epimeric compound (**1a**) (major) and (**1b**) (minor) in mixture.

Moreover, the NMR data were found to be partially comparable with the machaeridiol C (**11**) [6], suggesting the presence of a benzofuran side chain ($\delta_H$ 7.18–7.52) (4H) attached to the substituted resorcinol moiety. The HMBC spectrum (**1a**; Figure 2) showed $^2J$- and $^3J$ correlations between the H-8′

at $\delta_H$ 6.89 (brs) and the two sp$^2$-hybridized carbons at C-4′ and C-7′ ($\delta_C$ 130.6 and 154.9, respectively), supporting the attachment of C-4′ of the resorcinol unit to the C-7′ of benzofuran ring. The $^1$H NMR spectrum also showed signals at $\delta_H$ 6.83 (2H) for two identical protons (H-3′ and H-5′), suggesting the presence of C-1′,2′,4′,6′-tetra substituted resorcinol ring with two oxygenated carbons at C-2′ and C-6′ ($\delta_C$ 155.3 and 155.4). In addition, the $^1$H NMR spectrum (**1a**) showed signals at $\delta_H$ 4.0 (1H, m, H-5) and 5.32 (1H, dd, J = 1.2, 7.6 Hz, H-3) for the tetrahydrofuran ring, and at $\delta_H$ 1.02 (3H, d, J = 7.2 Hz, H-11) for a Me-group. The HMBC (Figure 3) spectrum showed correlations between the H-3 ($\delta_H$ 5.32) and the three sp$^2$ hybridized carbons at C-1′, C-2′, and C-6′ ($\delta_C$ 113.3, 155.3, and 155.4, respectively), supporting the attachment of the tetrahydrofuran ring at C-3 ($\delta_C$ 77.7) to C-1′ position. The HMBC spectrum also showed cross peaks between the methyl protons H-11 ($\delta_H$ 1.02) and C-1, C-2, and C-5 ($\delta_C$ 35.2, 41.8, and 83.1), supporting the attachment of the methyl (C-11, $\delta_C$ 15.5) to the tetrahydrofuran at C-1 ($\delta_C$ 35.2). The $^1$H, $^{13}$C, and 2D NMR spectra supported the presence of the 2-methylbut-2-ene unit. This was confirmed by the HMBC spectrum, which showed correlations between the H-6 at ($\delta_H$ 2.4) and the three carbons at C-5, C-7, and C-8 ($\delta_C$ 83.1, 120.0, and 134.5, respectively), confirming the attachment of the methylbut-2-ene unit to the tetrahydrofuran moiety at C-5. The relative configurations at C-1, C-3, and C-5 of the tetrahydrofuran were assigned via careful analysis of NOESY correlations for **1a**. In the NOESY spectrum (assigned for **1a**), H-5 ($\delta_H$ 4.0) showed correlation with H-1 ($\delta_H$ 2.4) and H-3 ($\delta_H$ 5.32), indicating the cofacial ($\beta$)-orientation of the three groups. Additionally, NOESY showed cross peaks between H-1 ($\delta_H$ 2.4.) and H-3 ($\delta_H$ 5.32), which supported their presence in the same plane of the molecule like H-5. On the other hand, such nOe signals were not evident in the NOESY of the minor compound **1b**. Moreover, in its $^1$H NMR spectrum, H-5 and H-3 were deshielded at $\delta_H$ 4.21 (1H, m) and 5.59 (1H, dd, J = 12.0, 8.0 Hz), respectively, for the tetrahydrofuran ring, and at $\delta_H$ 1.02 (3H, d, J = 7.2 Hz, H-11) for an Me-group. Moreover, a CD analysis for this compound revealed a weak spectrum in the range of 250–500 nm, which is reflective of the epimeric nature of the compound. Based on the foregoing discussion and comparing the NMR data with compound **11** [6], the structure **1a** and **1b** were determined for machaerifurogerol and 5-*epi*-machaerifurogerol, respectively.

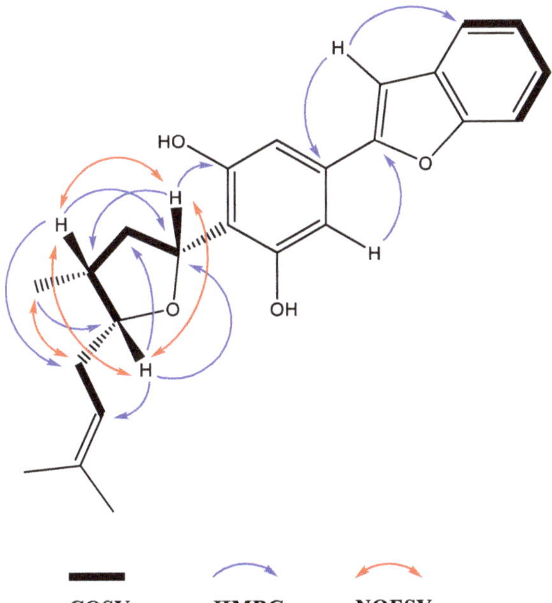


**Figure 3.** Key 2D NMR COSY, HMBC, and NOESY correlations of compound **1a**.

During the course of isolation, four isoflavonoid derivatives (**2–5**) and previously reported machaeriols (**6–8**) and machaeridiols (**10–12**) were isolated from the DCM partition of the stem bark and leaves extracts. However, the minor compound **9** could not be isolated due to a paucity of material. Compounds **2** and **3** were identified as known isoflavons (+)-vestitol and 7-*O*-methylvestitol, while **4** and **5** were identified as known peterocarpans. (+)-medicarpin and 3,8-dihydroxy-9-methoxy-pterocarpan, respectively, previously reported from *Machaerium vestitum* and Cuban propolis [10,11]. Compounds **4** and **5** were also reported from Cuban propolis [11]. The $^{1}H$ and $^{13}C$ NMR spectroscopic data (see Table S1) of compounds **2–5** were in agreement with those reported [11,12]. In addition, examination of the leaves of *Machaerium* sp. also yielded compounds **6–8** and **10–12**, as well as their presence in the root extract. The identities of compounds **6–12** were established by NMR spectra and by direct comparison with authentic samples (TLC, HPLC/ LC–MS).

### *2.3. Antimicrobial Activity against Gram-Positive Species and Fungi*

The availability of machaeriols A-C (**6–8**) and machaeridiols (**10–12**) [5,6] offered the opportunity to carry out a comprehensive investigation of antimicrobial activity. Among the tested fractions, DCM-25-32 (enriched with compounds **8** and **10–12**) and DCM-10-15 (enriched with compound **12**) were the most active against bacteria *S. aureus*, and methicillin-resistant *S. aureus* (MRSA), and the fungi *Candida glabrata*, *C. krusei*, and *Cryptococcus neoformans*, with $IC_{50}$ values of <0.8, <0.8, <0.8, 5.35, <0.8 μg/mL, and <0.8, 1.95, 3.0, 6.07, 12.58 μg/mL, respectively (Table 2). Antibacterial activities of **6–8**, **10–12**, and a mixture **10 + 11** (1:1) were evaluated against methicillin-resistant *S. aureus* (ATCC 1708, 1696, and 1717), the ex vivo MRSA XEN31 strain, and vancomycin-resistant *Enterococci* (VRE; *Enterococcus faecium* ATCC 700221), low-level VRE (*E. faecalis* ATCC 51299), and the vancomycin-sensitive strain (VSA; *E. faecalis* ATCC 29212) (Tables 3 and 4). Compound **11** and mixture **10 + 11** (1:1) showed the most potent activity against MRSA BAA 1696, BAA 1708, BAA 1717, and BBA 33591 with $IC_{50}$/MIC/MBC values of 0.43/1.25/5 μg/mL, 0.38/1.25/1.25 μg/mL, 0.38/1.25/2.5 μg/mL, 0.71/1.25/1.25 μg/mL; and 0.41/1.25/10 μg/mL, 0.34/1.25/1.25 μg/mL, 0.39/1.25/1.25 μg/mL and 0.61/1.25/10 μg/mL, respectively. On the other hand, compound **8** and mixture **10 + 11** were found to be the most potent against *E. faecium* ATCC 700221 and *E. faecalis* ATCC 51299, (VRE) and *E. faecalis* ATCC 29212 (VSE) with $IC_{50}$/MIC/MBC of 0.48/1.25/2.5 μg/mL, 1.02/1.25/5 μg/mL and 1.16/2.5/2.5 μg/mL; and 0.49/1.25/2.5 μg/mL, 0.70/1.25/5 μg/mL, and 0.72/1.25/5 μg/mL, respectively (Tables 3 and 4). The activities of compounds **6–8** and **10–12** were found to be more potent than ciprofloxacin and vancomycin against the MRSA BBA 1708 strain, while **6**, **8**, and **10 + 11** were more active against VRE 700221 than the positive controls.

**Table 2.** Antimicrobial activity ($IC_{50}$ in μg/mL) of *Machaerium* DCM fractions [a].

| | *C. glabrata* | *C. krusei* | *C. neoformans* | *S. aureus* | *MRSA* | *VRE* 29212 [b] | *VRE* 51299 [c] | *VRE* 700221 [d] |
|---|---|---|---|---|---|---|---|---|
| DCM-1-8 | >20 | >20 | >20 | 6.09 | 5.64 | - | - | - |
| DCM-10-15 | >20 | >20 | 4.26 | <0.8 | 0.81 | 11.41 | 1.89 | 3.47 |
| DCM-12 | 6.07 | 12.58 | <0.8 | 1.95 | 3 | 5.68 | 1.65 | 2.95 |
| DCM-25-32 | <0.8 | 5.35 | <0.8 | <0.8 | <0.8 | 3.70 | 1.34 | <0.8 |
| DCM-56 | >20 | >20 | >20 | >20 | >20 | - | - | 2.94 |
| DCM-1-8-A-31 | >20 | >20 | >20 | 9.91 | >20 | - | - | - |
| DCM-1-8-A-41 | >20 | >20 | 4.41 | 4.51 | 10.62 | 5.41 | 8.52 | 4.52 |
| DCM-1-8-A-63 | >20 | >20 | >20 | 4.29 | 4.55 | 11.39 | 14.49 | 9.97 |
| DCM-1-8-A-167 | >20 | >20 | >20 | 4.5 | 4.41 | - | - | - |

[a] $IC_{50}$ is the concentration causing 50% growth inhibition; MIC (minimum inhibitory concentration) is the lowest concentration that allows no growth; MFC (minimum fungicidal concentration) or MBC (minimum bactericidal concentration) is the lowest concentration at kills the test organism; [b] Vancomycin sensitive, [c] Low-level vancomycin resistant. [d] Vancomycin resistant strain; -: not active at the highest test concentration of 20 μg/mL.

**Table 3.** Anti-MRSA activities (in µg/mL) of compounds **6–8** and **10–12**.

| | **MRSA BAA-1708** | **MRSA BAA-1717** | **MRSA 33591** | **MRSA BAA-1696** | **MRSA XEN31** |
|---|---|---|---|---|---|
| Compound | $IC_{50}$/MIC/MBC | $IC_{50}$/MIC/MBC | $IC_{50}$/MIC/MBC | $IC_{50}$/MIC/MBC | IC50/MIC/MBC |
| **6** | 10.61/-/- | I1.93/-/- | _ | 11.87/-/- | 3.50/-/- |
| **7** | 5.36/-/- | 3.29/-/- | NT | 14.64/-/- | 4.95/20/20 |
| **8** | 0.69/1.25/1.25 | 0.72/1.25/10 | NT | 0.71/2.5/2.5 | 0.46/1.25/2.5 |
| **10** | 1.03/2.5/5 | 1.03/2.5/5 | NT | 1.07/2.5/2.5 | 1.01/2.5/5 |
| **11** | 0.38/1.25/1.25 | 0.38/1.25/2.5 | 0.71/1.25/1.25 | 0.43/1.25/5 | 0.33/0.63/1.25 |
| **12** | 1.52/5/5 | 0.38/2.5/5 | 1.57/2.5/5 | 1.40/5/10 | 1.64/5/10 |
| **10** + **11** | 0.34/1.25/1.25 | 0.39/1.25/1.25 | 0.61/1.25/10 | 0.41/1.25/10 | 0.32/0.63/1.25 |
| Ciprofloxacin | -/-/- | 0.14/0.63/1.25 | 0.04/0.16/0.31 | 6.17/-/- | 0.10/0.31/0.63 |
| Vancomycin | >20/>20/>20 | 0.73/1.25/>20 | 0.47/1.25/5.0 | 0.37/0.62/>20 | NT |
| Methicillin | 2.2/50/50 | - | - | 2.54/50/50 | 0.38/1.56/3.13 |
| Cefotaxime | 0.35/0.63/0.63 | 0.35/0.63/0.63 | NT | 2.47/12.5/25 | 0.29/0.78/3.13 |

NT not tested.

**Table 4.** Anti-VRE activities (in µg/mL) of compounds **6–8** and **10–12**.

| | ***Enterococcus faecalis*** **ATCC 29212** [a] | ***Enterococcus faecalis*** **ATCC 51299** [b] | ***Enterococcus faecium*** **ATCC 700221** [c] |
|---|---|---|---|
| Compound | $IC_{50}$/MIC/MBC | $IC_{50}$/MIC/MBC | $IC_{50}$/MIC/MBC |
| **6** | 3.51/-/- | 18.79/-/- | 0.55/1.25/10 |
| **8** | 1.16/2.5/2.5 | 1.02/1.25/5 | 0.48/1.25/2.5 |
| **12** | 2.99/5/10 | 2.96/5/10 | 1.91/2.5/5 |
| **10** + **11** | 0.72/1.25/5 | 0.70/1.25/5 | 0.49/1.25/2.5 |
| Ciprofloxacin | 0.25/0.78/6.25 | 0.22/0.39/6.25 | >20/>20/>20 |
| Vancomycin | 0.73/1.25/>20 | 3.8/10/>20 | >20/>20/>20 |
| Methicillin | 15.3/25/50 | 14.2/50.0/50.0 | >20/>20/>20 |

[a] vancomycin sensitive; [b] low-level vancomycin resistant; [c] vancomycin resistant.

## 2.4. Antimicrobial Combination Studies

In light of the strong antimicrobial activity of the DCM fraction 25–32, which is enriched with compounds **8**, **10–12** (Table 3), a combination study using an in vitro Checkerboard method [13,14] was carried out for machaeriol (**7** or **8**) and machaeridiol (**11** or **12**), to evaluate the synergy of combination treatment against the strains of MRSA and *Enterococcus* (VRE) (Table 4). Among these compounds, a combination of machaeridiol B (**12**; at MIC 5 µg/mL) and machaeriol C (**8**; at $\frac{1}{2}$ MIC 1.25 µg/mL) exhibited a potent activity, with the MIC values of 0.156 and 0.625 µg/mL exhibiting a >32- and >8-fold reduction of MICs, compared to those observed for **12**, against MRSA 1708 and MRSA 1717 strains. When these two compounds were tested with an inverse concentration (i.e., MIC of **8**; 2.5 µg/mL + $\frac{1}{2}$ MIC of **12**; 2.5 µg/mL), a strong synergism was also observed, but to a lesser extent. When tested against VRE (*E. faecium* 700221), this combination showed synergism with the MIC values of 1.25 µg/mL, a >4-fold reduction of MIC compared to **12** (Table 5).

Isobologram showing synergistic activity of the combination of compounds **8** and **12** in MRSA 1708 (red) and 1717 (blue) are presented in Figure 4. The green series represents the additivity line of compounds **8** and **12** (green dots represent the MIC of each compound alone; the green line represents all possible additive combinations). The red (MRSA 1708) and blue (1717) dots represent the combination of compounds **8** and **12**, and show that they fell below the additivity line (the combination of the compounds produces a synergistic effect beyond additivity). This synergism between machaeriol (HHDBP) and machaeridiol (seco-HHDBP) could be due to different molecular targets affected by these two molecules. A combination study of compounds **8** and **12** with antibiotics, either methicillin or ciprofloxacin, did not show any additive or synergistic effects.

**Table 5.** Combination study (MIC in μg/mL) [a] of compounds **7**, **8**, **11**, and **12** by Checkerboard assay against MRSA and VRE.

| Compound | MRSA 1708 | MRSA 1717 | MRSA 33591 | MRSA 1696 | MRS XEN31 | Ef 29212 [b] | Ef 51299 [c] | Ef 700221 [d] |
|---|---|---|---|---|---|---|---|---|
| **7** | 20 | >10 | NT | 10 | 20 | NT | NT | NT |
| **8** | 2.5 | 2.5 | 2.5 | 2.5 | 1.25 | 2.5 | 2.5 | 1.25 |
| **12** | 5 | 5 | 5 | 5 | 5 | 5 | >5 | 5 |
| **11** | 20 | 20 | NT | 10 | 10 | NT | NT | NT |
| **8** + **12** (+2.5 μg/mL) | 0.625 (↓4X) | 0.625 (↓4X) | 1.25 (↓2X) | 0.625 (↓4X) | NT | 1.25 (↓2X) | 1.25 (↓2X) | 0.625 (↓2X) |
| **12** + **8** (+1.25 μg/mL) | 0.156 (↓32X) | 0.625 (↓8X) | 1.25 (↓4X) | NT | NT | 2.5 (↓2X) | 2.5 (↓2X) | NT |
| **12** + **8** (+0.625 μg/mL) | NT | NT | NT | 2.5 (↓2X) | NT | NT | NT | 1.25 (↓4X) |
| **11** + **7** (+2.5 μg/mL) | 5.0 (↓4X) | 10.0 (↓2X) | NT | 5.0 (↓2X) | 2.5 (↓4X) | NT | NT | NT |
| DAPG [e] | >1.25 | NT | NT | NT | NT | NT | NT | >1.25 |
| **8** + DAPG (+0.63 μg/mL) | 2.5 (=) | NT | NT | NT | NT | NT | NT | 1.25 (=) |
| **12** + DAPG (+0.63 μg/mL) | >5 (=) | NT | NT | NT | NT | NT | NT | 5 (=) |
| Methicillin | 50 | - | - | 50 | 1.56 | 25 | 50.0 | >20 |
| Vancomycin | >20 | 1.25 | 1.25 | 0.62 | NT | 1.25 | 10 | >20 |

[a] In general, when the MIC of each compound decreased ≥4X in the presence of the other, it is considered synergistic; reduction of MIC in parentheses; [b] Vancomycin-sensitive *Enterococcus faecalis*; [c] Low-level vancomycin-resistant *Enterococcus faecium*; [d] Vancomycin-resistant *Enterococcus faecium*; [e] Diacetylphloroglucinol; NT: Not tested.

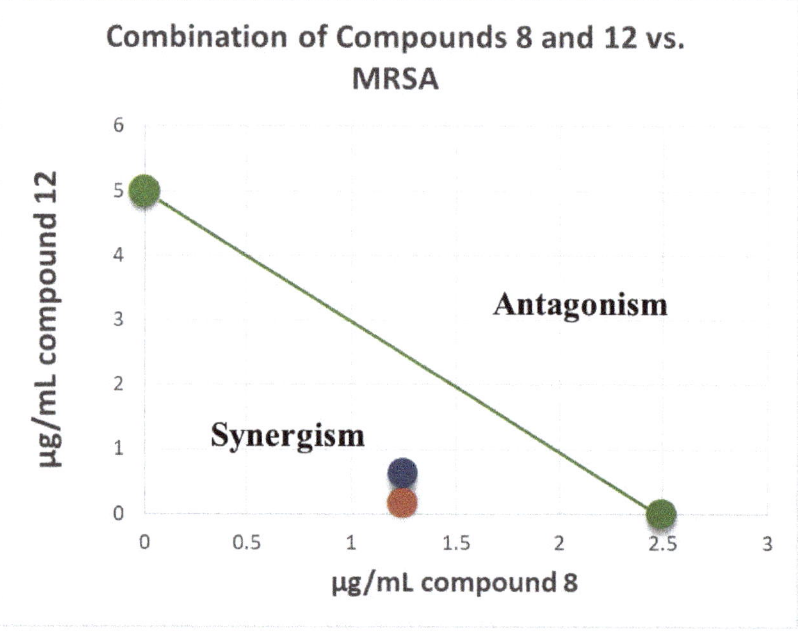


**Figure 4.** Isobologram of combination of compounds **8** and **12** in MRSA 1708 (red) and 1717 (blue). The green series represents the additivity line of compounds **8** and **12** green dots represent the MIC of each compound alone; the green line represents all possible additive combinations). The red (MRSA 1708) and blue (1717) dots represent the combination of compounds **8** and **12**, and show that they fell below the additivity line (the combination of compounds produces a synergistic effect beyond additivity).

### *2.5. Antimicrobial Activity against Gram-Negative Species*

The activity of the compounds was determined in the Gram-negative species of the ESKAPEE panel, *Klebsiella pneumoniae*, *Acinetobacter baumannii*, *Pseudomonas aeruginosa*, and *Escherichia coli* (Table 6).

None of the tested compounds displayed any antimicrobial activity in the Gram-negative species up to a concentration of 128 µg/mL. However, when the outer membrane was permeabilized using the membrane permeabilizer polymyxin-B-nonapeptide (PMBN), MICs as low as 0.5 µg/mL were observed, suggesting an intracellular target. With the exception of **6** + **7**, all compounds tested displayed a good activity in *A. baumannii* and *E. coli* strains and in *P. aeruginosa* strain PAO1. The *P. aeruginosa* strain NCTC 13437 is a near-pan drug resistant strain, and this was reflected in the MICs to compounds **6** + **7**, **10** + **11**, **11**, and **12**, although **8** had an MIC of 8 µg/mL in the presence of PMBN. None of the compounds displayed activity in the two *K. pneumoniae* strains, even in the presence of PMBN, suggesting the target/pathway might be missing or modified in this species.

**Table 6.** Gram-negative MIC Values (in μg/mL) of compounds **6–8** and **10–12** in the presence of the membrane permeabilizer polymyxin-B-nonapeptide at 30 μg/mL.

| | ***Klebsiella pneumoniae*** **NCTC13368** | ***Klebsiella pneumoniae*** **M6** | ***Acinetobacter baumannii*** **AYE** | ***Acinetobacter baumannii*** **ATCC17978** | ***Pseudomonas aeruginosa*** **PAO1** | ***Pseudomonas aeruginosa*** **NCTC13437** | ***Escherichia coli*** **NCTC12923** |
|---|---|---|---|---|---|---|---|
| Compound | MIC | MIC | MIC | MIC | MIC | MIC | MIC |
| **6** + **7** +PMBN | >64 | >64 | >64 | >64 | 32 | 64 | >64 |
| **10** + **11** +PMBN | >64 | >64 | 8 | 0.5 | 2 | 64 | 2 |
| **11** + PMBN | >64 | >64 | 2 | 1 | 4 | 64 | 2 |
| **12** + PMBN | >64 | >64 | 4 | 2 | 16 | 64 | 2 |
| **8** + PMBN | >64 | >64 | 8 | 2 | 2 | 8 | 2 |
| Ciprofloxacin [a] | 0.5 | ≤0.125 | 128 | 0.5 | 0.5 | 64 | ≤0.125 |
| Ceftazidime [a] | >128 | 0.25 | >128 | <0.5 | 1 | >128 | 0.5 |
| Gentamicin [a] | 4 | 0.25 | >512 | 0.5 | 4 | 256 | 1 |

[a] MICs in the absence of PMBN.

## 3. Materials and Methods

### 3.1. General Experimental Procedures

Optical rotations were recorded at ambient temperature using a Rudolph Research Analytical Autopol IV automatic polarimeter. IR spectra were obtained using a Bruker Tensor 27 instrument. NMR spectra were acquired on a Varian Mercury 400 MHz NMR spectrometer at 400 ($^1$H) and 100 MHz ($^{13}$C) in $CDCl_3$, using the residual solvent as an internal standard. Multiplicity determinations (DEPT) and 2D NMR spectra (HMQC, HMBC, NOESY) were obtained using the standard Bruker pulse programs. ESI-HRMS were acquired by direct injection using a Water Xevo G2-S TOF with electrospray ionization (ESI). TLC was carried out on pre-coated silica gel 60 F254 (EMD Chemicals Inc, Darmstadt, Germany) using toluene-EtOAc (9:1) and *n*-hexane-EtOAc (7.5:2.5) as solvents. Centrifugal preparative TLC (CPTLC, using a Chromatotron, Harrison Research Inc. model 8924, tagged with a fraction collector) was carried out on 6 mm custom-made RP $C_{18}$ silica gel [15], and silica gel $P_{254}$ (Analtech) 1, 2, and 4 mm rotors, using $H_2O$-MeOH, EtOAc-:*n*-hexane, and $CHCl_3$ as eluents. SPE cartridges $C_{18}$ (Supelco Inc., Bellefonte, PA, USA) were used in the fractionation work. Purifications were performed on prep-HPLC (silica gel-100 A 250 × 15.00, 5 μM; Phenomenex Luna, Torrance, CA, USA) using an HPLC Delta Prep 4000 equipped with a dual wavelength detector Model 2487 adjusted at 210 and 254 nm (Waters Corporation, Milford, MA, USA), Preparative HPLC was carried out on Waters LC module I plus, using Phenomenex C18, 22 mm, λ 254, flow 15 mL/min, 0–2 min [90% $H_2O$; 10% MeCN], 2–45 min, 10% MeCN→ 100% MeN, 45–50 min 100% MeCN]. Samples were dried using a Savant Speed Vac Plus SC210A concentrator. The compounds were visualized by spraying the TLC plates with 1% vanillin-$H_2SO_4$ spray reagent.

### 3.2. Plant Material

The stem bark, leaves, and roots of *Machaerium* Pers. (Manuel Rimachi, Y.-12161), previously identified as *M. multiflorum* Spruce [5] by (Late) Professor Sydney T. McDaniel, was collected in November, 1997, from open sandy forest near Loreto (Maynas), Peru. The voucher specimen (Manuel Rimachi Y. 12161) is deposited at the Missouri Botanical Garden (http://www.tropicos.org/Specimen/100326687).

### 3.3. Extraction and Isolation of Compounds from Stem Bark and Leaves

The powdered stem bark (0.5 kg) was extracted by percolation with 95% EtOH (3 × 2 L) and the combined extracts were evaporated under reduced pressure (yield 17.7 g). A portion of the dried EtOH extract (15 g) was percolated with *n*-hexane, followed by DCM, and finally the residual extract was washed with MeOH (each 200 mL × 3). The *n*-hexane, DCM, and MeOH fractions were separately filtered and dried, which afforded 3.8, 8.9, and 4.5 g, respectively. The antimicrobial activity was detected in the DCM fraction ($IC_{50}$ < 20 μg/mL against *S. aureus* and MRSA). A portion of the dried DCM fraction (1.65 g) was fractionated by CPTLC with a Chromatotron®instrument, using a 4 mm custom-made $C_{18}$ RP silica gel ChromatoRotor™ [15], eluting with a gradient of 60% to 100% MeCN-$H_2O$ to afford 30 fractions. The fractions were pooled by TLC analyses.

Fractions 1–8 (475 mg) were combined and further subjected to CPTLC, using a 4 mm silica gel $P_{254}$ disc, and gradient elution with MeCN:DCM. Elution with 2% MeCN:DCM afforded medicarpin (**4**; 4.5 mg), followed by elution with 4% MeCN:DCM, which gave 3,8-dihydroxy-9-methoxy-pterocarpan (**5**; 7.5 mg), and finally elution with 5% MeCN:DCM yielded vestitol (**2**; 9.8 mg). The combined fractions 10–15 (70 mg) was subjected to preparative $C_{18}$ RP-HPLC, using 90% MeCN:$H_2O$ as solvent, which afforded machaeridiol B (**12**), followed by machaeridiol A (**10**) and machaeridiol C (**11**). Similarly, combined fractions 25–32 (100 mg) was also separated by preparative $C_{18}$ RP-HPLC, which afforded additional quantities of **10–12** [total yields: **10** (10 mg), **11**; (18 mg), **12** (21 mg)] and machaeriol C (**8**; 34.6), however, the minor compound machaeriol D (**9**) could not be re-isolated due to a paucity of material. Further elution with 75% MeCN:$H_2O$ afforded 13 fractions, which contained the mixture of two compounds **6** + **7**, (50 mg). The mixture was then separated by preparative $C_{18}$ RP-HPLC (column: ODS prodigy 10μ, 250 × 10 mm; detector: UV-254 nm), using 95% MeCN:$H_2O$ as solvent, which afforded **6** (16 mg), followed by **7** (16 mg). Finally, the dried *n*-hexane fraction (77 mg) was subjected to CPTLC, using a 2 mm $C_{18}$ RP rotor, and eluted with 65% MeCN:$H_2O$, which afforded 7-*O*-methylvestitol (**3**; 8 mg). A sub-fraction of DCM (15 mg) was subjected to prep-HPLC (Waters LC module I plus, using Phenomenex $C_{18}$, 2 mm), which afforded compound **1a+1b** (5 mg). The structures of (+)-vestitol (**2**), 7-*O*-methylvestitol (**3**), (+)-medicarpin (**4**) and 3,8-dihydroxy-9-methoxypterocarpan (**5**) were determined by physical and spectroscopic data ($^1$H and $^{13}$C NMR, see SI 1), and also by comparison with those reported [11,12]. The structures of the re-isolated compounds **6–8** and **10–12** were identified by NMR data [5,6] and by direct comparison (TLC, HPLC/LC–MS) with their respective authentic samples available in our laboratories. Finally, powdered leaves (560 g) and root bark (50 g) of *Machaerium* sp. were extracted using the method described previously [5,6], and compounds 6–8 and 10–12 were isolated from leaves as describe below.

The powdered leaf was percolated with *n*-hexane, followed by DCM and EtOH (each 3 × 2 L) to yield 5, 14, and 9 g of extracts, respectively. A portion of the DCM extract (10 g) was subjected to reversed phase (RP) cartridge (10 G, 60 mL Giga tube), and eluted with MeCN-$H_2O$ to afford 30 fractions. The combined fractions 20–21 (102 mg; eluted by 60–65% MeCN-H2O) were subjected to centrifugal preparative thin layer chromatograph (CPTLC, 1 mm Si gel P254 disc), eluting with 0.5–1% MeCN-DCM to yield **12** (11.4 mg). Fraction 22 (60 mg; eluted with 75% MeCN-$H_2O$) was further subjected to CPTLC (1 mm RP-C18 ChromatoRotor), eluted with 50–100% $H_2O$-$CH_3CN$ to afford 115 fractions, of which fractions 42–45 and 90–115 eluting with 80% and 90% MeCN-$H_2O$ yielded **10** (2.7 mg) and **11** (4.5 mg), respectively. Combined fractions 46–89 (28.4 mg) were enriched with **12** (+ traces of **10** + **11**). Similarly, RP cartridge purified fractions 23 and 24 (60 and 70 mg; eluted with 70 and 80% MeCN-$H_2O$, respectively) were further purified (1 mm RP-C18 ChromatoRotor) by eluting separately with 50–100% $H_2O$-MeCN to yield a mixture of **8** + **10** + **11** (32 mg and 31 mg respectively). The above enriched mixtures were further purified preparative RP-HPLC, using 90% MeCN-$H_2O$ as solvent to afford compounds **12**, **11** + **12**, **11**, and **8** (5, 32, 4 and 31.7 mg, respectively).

A portion of *n*-hexane extract (2.5 g) was fractionated with CPTLC (6 mm, Si gel P254 disc) eluting with 5% DCM in hexane to yield 10 fractions. The fractions 3–7 (840 mg) that enriched with compounds **6** and **7** were combined and further attempted to purify with an additional CPTLC (4 mm, Si gel P254 disc) eluting with 5% DCM in hexane to yield semi-pure **6** (40 mg), **6+7** (50 mg), and semi-pure **7** (6 mg). In addition, the presence of these compounds in leaves, stem bark, and root extracts were confirmed by HPLC and LC–MS (vide infra).

### 3.4. *Machaerifurogerol* (**1a**) *and 5-epi-Machaerifurogerol* (**1b**)

Amorphous solid; $[\alpha]^{26}{}_D$ +5.8 (*c* 0.05, MeOH); IR (KBr) $\upsilon$ max 3341 (OH), 2924, 1631, 1574, 1452, 1248, 961, 801, 613, 591 $cm^{-1}$; $^1$H and $^{13}$C NMR, see Table 1; HRESIMS m/z 379.1906 $[M + H]^+$ (calcd. for $C_{24}H_{27}O_4$, 379.1865).

### 3.5. *Identification of Compounds* **6–8** *and* **10–12** *by LC–MS*

LC–MS analysis was carried out on an Agilent system using Luna 5 μ C18 (2), 150 × 4.6 mm, λ 254, flow 1 mL/min, gradient 0–2 min [95% $H_2O$; 5% MeCN], 2–30 min, 5% MeCN→ 100% MeCN, 30–35 min 100% MeCN, 35–45 min [95% $H_2O$; 5% MeCN]. The retention times ($R_t$) of the compounds **6** (m/z 349.2 $[M + H]^+$; $C_{24}H_{29}O_2$), **8** (365.2 $[M + H]^+$; $C_{24}H_{29}O_3$), **7** (363.2 $[M + H]^+$; $C_{24}H_{27}O_3$), **10** (349.2 $[M + H]^+$; $C_{24}H_{29}O_2$); **11** (363.2 $[M + H]^+$; $C_{24}H_{27}O_3$), and **12** (365.2 $[M + H]^+$; $C_{24}H_{29}O_2$) were found to be 4.4, 4.5, 9.9, 10.0, 9.4, and 9.7 $min^{-1}$, respectively. Compounds **6–9** and **10–12** were identified from leaves, stem bark, and root extracts through HPLC and LC–MS.

### 3.6. *Antimicrobial Assays*

All organisms were obtained from the American Type Culture Collection (Manassas, VA, USA) or the National Collection of Type Cultures (Colindale, UK), unless specified otherwise.

These included the yeasts *Candida albicans* ATCC 90028, *C. glabrata* ATCC 90030, and *C. krusei* ATCC 6258; the fungi *Cryptococcus neoformans* ATCC 90113 and *Aspergillus fumigatus* ATCC 204305; and the bacteria *Escherichia coli* ATCC 35218, NCTC 12923, *Klebsiella pneumoniae* NCTC 13368, M6 (Colindale, UK), *Acinetobacter baumannii* AYE (ATCC BAA-1710), ATCC 17978, *Pseudomonas aeruginosa* ATCC 27853, PAO1 (Manoil collection, University of Washington, Washington, DC, USA), NCTC 13437, *Mycobacterium intracellulare* ATCC 23068, methicillin-resistant *Staphylococcus aureus* ATCC 33591 (MRSa), USA-300 MRSa (ATCC BAA-1717), USA-400 MRSa (ATCC BAA-1696), Mupirocin-resistant *S. aureus* (ATCC BAA-1708), *Enterococcus faecium* ATCC 700221 (VRE), *E. faecalis* ATCC 29212 (Vancomycin-sensitive) and *Enterococcus faecium* ATCC 51299 (Vancomycin-intermediate). Drug controls ciprofloxacin, methicillin and vancomycin (ICN Biomedicals, Aurora, OH, USA) for bacteria and amphotericin B (ICN Biomedicals) for yeasts and fungi were included in each assay. Susceptibility testing was performed using a modified version of the CLSI (formerly NCCLS) method [16–18]. *M. intracellulare* was tested using a modified Franzblau method [18]. Samples were serially diluted in 20% DMSO/saline and transferred in duplicates to 96-well flat-bottomed microplates. Microbial inocula were prepared by correcting the $OD_{630}$ of microbe suspensions in incubation broth to give final target inocula. All organisms were read at either 530 nm, using the Biotek Powerwave XS plate reader (Bio-Tek Instruments, Winooski, VT, USA) or 544ex/590em, (*M. intracellulare, A. fumigatus*) using the Polarstar Galaxy Plate Reader (BMG Lab Technologies, Ortenburg, Germany), prior to and after incubation. Minimum fungicidal or bactericidal concentrations were determined by removing 5 μL from each clear well, followed by transferring to agar, and incubating. The MFC/MBC was defined as the lowest test concentration that kills the organism (allows no growth on agar).

Gram-negative MICs were determined using the CLSI microbroth dilution method, modified as described previously [19]. Bacteria were added at a starting concentration of $5 \times 10^5$ cfu/mL and incubated for 20 h at 37 °C in the dark. Absorbance at $OD_{600}$ was then read using the CLARIOstar plate reader (BMG Lab Technologies, Germany). The MIC was defined as the lowest concentration where visible growth could not be detected, equivalent to an $OD_{600}$ of 0.1. MICs were also determined in the presence of the membrane permeabilizer, polymyxin-B-nonapeptide (PMBN) following the same method, with an additional step; after the 2-fold dilution of compound was prepared and before the bacteria were added, PMBN was added to all wells at a final concentration of 30 μg/mL. This concentration was shown to not significantly inhibit growth of the test panel.

### 3.7. Antimicrobial Combination Study by Checkerboard Method

The combination study of the compounds was carried out using a standard Checkerboard method [13,14]. Strains were grown on Eugon agar at 35 °C, prior to assays. Test samples were dissolved in DMSO (2 mg/mL) to the desired concentrations, and serially-diluted with 20% DMSO/saline. Samples were transferred to 96 well assay plates (10 μL) in a checkerboard layout. Inocula were prepared by suspending growth from agar in 0.9% saline, determining the $OD_{630}$, and correcting in incubation broth (cation-adjusted Mueller-Hinton, Difco) to afford $5 \times 10^5$ colony forming units per mL, after addition to samples (180 μL) using standard inocula calculations. Final sample test concentrations were 1/100th the DMSO stock concentrations. The assay plates were read at 530 nm prior to and after incubation at 35 °C for 18–20 h. $IC_{50}$s of each test compound were calculated using the XLfit 4.2 software (IDBS, Alameda, CA, USA) using the fit model 201. After incubation, all 96 wells were also pinned to Eugon Agar and incubated at 35 °C overnight to determine bactericidal activity. Fractional inhibitory concentrations (FICs) were calculated to evaluate possible synergy with FICS < 0.5 synergistic.

## 4. Conclusions

Based on our investigation carried out on the identity of the plant, and a re-examination of the voucher specimen (Rimachi # 12161) at the MOBOT, it can now be concluded that this species should be treated only as an unidentified species of *Machaerium* Pers., as determined by the collection information (Manuel Rimachi, Y. 12161). This appears to be the first report of macharifurogerol

(**1a**) and its epimer **1b** from a natural source. In addition, the isolation of isoflavons (**2** and **3**) and pterocarpans (**4** and **5**) from this *Machaerium* species (Rimachi 12161) illustrated that these isoflavonoids are typical chemotaxonomic markers of the genus *Machaerium* [11,12]. Machaeriols and its biogenetic precursor machaeridiols are only isolated from this species (#12161) from the genus *Machaerium*, which are analogous to hexahydrocannabinol (HHC) and dihydrocannabidiol base skeletons of *Cannabis* and its variants, in higher plants [6]. The only other bibenzyl analogue of $\Delta^9$-THC, perrottetinen, was previously reported from the liverwort *Radula perrottetii* [20]. It is intriguing to note that the strong MRSA and VRE inhibitory activities, together with their antiparasitic activities [5,6] of the isolated compounds of *Machaerium* (12161) is contributed by HHDBP machaeriols and their 5,6-*seco* analogs machaeridiols. The stereo-specific total synthesis of machaeriol A-D (**6–9**) and mechaeridiol B (**12**) was reported [21–24]. In addition, analogs of machaeriols and related HHC were recently synthesized, which showed anticancer activity [25]. It was anticipated that these phytocannabinoids could serve as potential template for anti-MRSA and anti-VRE lead candidates, because of their inherent inhibitory activities alone, as well as strong synergistic activity when tested in combination with machaeriol and machaeridiol. The observation of significant activity in permeabilized multidrug resistant Gram-negative pathogens, also offers the potential for optimization of the chemical scaffold to generate analogues with better cell permeability. These compounds might provide important new leads for WHO priority Gram-negative bacterial pathogens.

**Supplementary Materials:** The following are available online, NMR and HRMS spectra (Figures S1–S11) of compound **1**, and Table for NMR data (Table S1) of compounds **2–5** are provided in supporting information.

**Author Contributions:** I.M., J.M.S., and K.M.R., conceptualized the study; I.M., M.A.I., M.K., and J.Z., planned the experiments and spectral analysis for chemistry work; M.W., execute LCMS work; M.R.J., planned antimicrobial and Checkerboard assay at UM; J.M.S., C.H., M.C., B.M., and T.A.-A., executed antimicrobial work at KCL and PHE; V.R., and I.M., authenticated and provided critical information on plant authentication; M.A.I., M.R.J., M.A.I., V.R., M.W., J.Z., J.M.S., and K.M.R., prepared the original draft of the manuscript. All the authors contributed to the writing and editing of the manuscript. All authors have read and agreed to the published version of the manuscript.

**Funding:** This work was supported in part by the USDA Agricultural Research Service Specific Cooperative Agreement No. 58-6060-6-015, and the NIH, NIAID, Division of AIDS, grant no. AI 27094. Work at PHE was supported by Grant in aid funding through an Open Innovation programme (Project 111742).

**Acknowledgments:** The authors sincerely thank Manuel Rimachi Y. and (Late) Sydney T. McDaniel (Mississippi State University) for the collection of plant material; Andrew Townesmith, Missouri Botanical Garden, St. Louis, MO, USA; Andrew Sanders, UCR Herbarium, California, USA; and Fabiana Filardi, Instituto de Pesquisas Jardim Botânico do Rio de Janeiro, Brazil, for their expert opinion on the voucher specimen. Marwa Hasan, Y. Wang, F. Wiggers, and M. Wright, NCNPR, UM, for assisting and conducting chemistry work, HRMS, NMR experiments and biological assays, respectively.

**Conflicts of Interest:** The authors declare no conflict of interest.

## References

1. Mabberley, D.J. *Mabberley's Plant-Book: A Portable Dictionary of Plants, Their Classification and Uses*; Cambridge University Press: Cambridge, UK, 2017.
2. Amen, Y.M.; Marzouk, A.M.; Zaghloul, M.G.; Afifi, M.S. The genus *Machaerium* (Fabaceae): Taxonomy, phytochemistry, traditional uses and biological activities. *Natl. Prod. Res.* **2015**, *29*, 1388–1405. [CrossRef] [PubMed]
3. Joly, L.G.; Guerra, S.; Septimo, R.; Solis, P.N.; Correa, M.; Gupta, M.; Levy, S.; Sandberg, F. Ethnobotanical inventory of medicinal plants used by the Guaymi Indians in Western Panama. Part I. *J. Ethnopharmacol.* **1987**, *20*, 145–171. [CrossRef]
4. Polhill, R.M.; Raven, P.H. *Advances in Legume Systematics, Part 1*; Royal Botanic Gardens: Kew, UK, 1981.
5. Muhammad, I.; Li, X.-C.; Dunbar, D.C.; ElSohly, M.A.; Khan, I.A. Antimalarial (+)-trans-hexahydrodibenzopyran derivatives from *Machaerium multiflorum*. *J. Nat. Prod.* **2001**, *64*, 1322–1325. [CrossRef] [PubMed]
6. Muhammad, I.; Li, X.-C.; Jacob, M.R.; Tekwani, B.L.; Dunbar, D.C.; Ferreira, D. Antimicrobial and Antiparasitic (+)-trans-Hexahydrodibenzopyrans and Analogues from *Machaerium multiflorum*. *J. Nat. Prod.* **2003**, *66*, 804–809. [CrossRef] [PubMed]

7. Filardi, F.L.R.; Lima, H.C.D. The diversity of *Machaerium* (Leguminosae: Papilionoideae) in the Atlantic Forest: Three new species, nomenclatural updates, and a revised key. *Syst. Bot.* **2014**, *39*, 145–159. [CrossRef]
8. David, M.Z.; Daum, R.S. Community-associated methicillin-resistant *Staphylococcus aureus*: Epidemiology and clinical consequences of an emerging epidemic. *Clin. Microbiol. Rev.* **2010**, *23*, 616–687. [CrossRef]
9. Salgado, C.D. The risk of developing a vancomycin-resistant *Enterococcus* bloodstream infection for colonized patients. *Am. J. Infect. Control* **2008**, *36*, S175. [CrossRef]
10. Lozano, P.; Klitgaard, B.B. The genus *Machaerium* (Leguminosae: Papilionoideae: Dalbergieae) in Ecuador. *Brittonia* **2006**, *58*, 124–150. [CrossRef]
11. Piccinelli, A.L.; Campo Fernandez, M.; Cuesta-Rubio, O.; Márquez Hernández, I.; De Simone, F.; Rastrelli, L. Isoflavonoids isolated from Cuban propolis. *J. Agric. Food Chem.* **2005**, *53*, 9010–9016. [CrossRef] [PubMed]
12. Kurosawa, K.; Ollis, W.D.; Redman, B.T.; Sutherland, I.O.; Gottlieb, O.R. Vestitol and vesticarpan, isoflavonoids from *Machaerium vestitum*. *Phytochemistry* **1978**, *17*, 1413–1415. [CrossRef]
13. Norden, C.W.; Wentzel, H.; Keleti, E. Comparison of techniques for measurement of in vitro antibiotic synergism. *J. Infect. Dis.* **1979**, *140*, 629–633. [CrossRef] [PubMed]
14. Renneberg, J. Definitions of antibacterial interactions in animal infection models. *J. Antimicrob. Chemother.* **1993**, *31* (Suppl. D), 167–175. [CrossRef] [PubMed]
15. Ilias, M.; Samoylenko, V.; Gillium, V.D. Preparation of Pre-Coated Rp-Rotors and Universal Chromatorotors, Chromatographic Separation Devices and Methods for Centrifugal Preparative Chromatography. Google Patents US20140224740A1, 14 August 2014.
16. Wikler, M.A. Methods for dilution antimicrobial susceptibility tests for bacteria that grow aerobically: Approved standard. *CLSI (NCCLS)* **2006**, *26*, M7–A7.
17. Michael, A.; Vishnu, C.; Ana, E.; Mahmoud, A.; Linda, L.; Frank, C. *NCCLS, Reference Method for Broth Dilution Antifungal Susceptibility Testing of Yeasts*; Approved Standard, M27-A2; CLSI: Wayne, PA, USA, 2002.
18. NCCLS. *Susceptibility Testing of Mycobacteria, Nocardiae, and Other Aerobic Actinomycetes*; Approved Standard; NCCLS: Wayne, PA, USA, 2003.
19. Bock, L.; Hind, C.; Sutton, J.; Wand, M. Growth media and assay plate material can impact on the effectiveness of cationic biocides and antibiotics against different bacterial species. *Lett. Appl. Microbiol.* **2018**, *66*, 368–377. [CrossRef] [PubMed]
20. Toyota, M.; Shimamura, T.; Ishii, H.; Renner, M.; Braggins, J.; Asakawa, Y. New bibenzyl cannabinoid from the New Zealand liverwort *Radula marginata*. *Chem. Pharm. Bull.* **2002**, *50*, 1390–1392. [CrossRef] [PubMed]
21. Chittiboyina, A.G.; Reddy, C.R.; Watkins, E.B.; Avery, M.A. First synthesis of antimalarial Machaeriols A and B. *Tetrahedron Lett.* **2004**, *45*, 1689–1691. [CrossRef]
22. Huang, Q.; Wang, Q.; Zheng, J.; Zhang, J.; Pan, X.; She, X. A general route to 5, 6-seco-hexahydrodibenzopyrans and analogues: First total synthesis of (+)-Machaeridiol B and (+)-Machaeriol B. *Tetrahedron* **2007**, *63*, 1014–1021. [CrossRef]
23. Wang, Q.; Huang, Q.; Chen, B.; Lu, J.; Wang, H.; She, X.; Pan, X. Total Synthesis of (+)-Machaeriol D with a Key Regio-and Stereoselective SN2′ Reaction. *Angew. Chem. Int. Ed.* **2006**, *45*, 3651–3653. [CrossRef] [PubMed]
24. Xia, L.; Lee, Y.R. A short total synthesis for biologically interesting (+)-and (−)-machaeriol A. *Synlett* **2008**, *2008*, 1643–1646.
25. Thapa, D.; Lee, J.S.; Heo, S.W.; Lee, Y.R.; Kang, K.W.; Kwak, M.K.; Choi, H.G.; Kim, J.A. Novel hexahydrocannabinol analogs as potential anti-cancer agents inhibit cell proliferation and tumor angiogenesis. *Eur. J. Pharmacol.* **2011**, *650*, 64–71. [CrossRef] [PubMed]

**Sample Availability:** Samples of the compounds **1–8** and **10–12** are available from the authors.

**Publisher's Note:** MDPI stays neutral with regard to jurisdictional claims in published maps and institutional affiliations.

Article

# Antimalarials and Phytotoxins from *Botryosphaeria dothidea* Identified from a Seed of Diseased *Torreya taxifolia*

Mallika Kumarihamy [1,2,*], Luiz H. Rosa [3], Natascha Techen [1], Daneel Ferreira [2], Edward M. Croom, Jr. [2], Stephen O. Duke [4,†], Babu L. Tekwani [1,‡], Shabana Khan [1,2] and N. P. Dhammika Nanayakkara [1,*]

1 National Center for Natural Products Research, Research Institute of Pharmaceutical Sciences, School of Pharmacy, The University of Mississippi, University, MS 38677, USA; ntechen@olemiss.edu (N.T.); btekwani@southernresearch.org (B.L.T.); skhan@olemiss.edu (S.K.)
2 Division of Pharmacognosy, Department of BioMolecular Sciences, School of Pharmacy, The University of Mississippi, University, MS 38677, USA; dferreir@olemiss.edu (D.F.); emcroom@olemiss.edu (E.M.C.J.)
3 Departamento de Microbiologia, Universidade Federal de Minas Gerais, Belo Horizonte, MG, Brazil; lhrosa@icb.ufmg.br
4 Natural Products Utilization Research Unit, USDA-ARS, University, MS 38677, USA; sduke@olemiss.edu
* Correspondence: mkumarih@olemiss.edu (M.K.); dhammika@olemiss.edu (N.P.D.N.); Tel.: +1-662-915-1661 (M.K.); +1-662-915-1019 (N.P.D.N.)
† Current address: Affiliation 1.
‡ Current address: Department of Infectious Diseases, Division of Drug Discovery, Southern Research, Birmingham, AL 35205, USA.

**Abstract:** The metabolic pathways in the apicoplast organelle of *Plasmodium* parasites are similar to those in plastids in plant cells and are suitable targets for malaria drug discovery. Some phytotoxins released by plant pathogenic fungi have been known to target metabolic pathways of the plastid; thus, they may also serve as potential antimalarial drug leads. An EtOAc extract of the broth of the endophyte *Botryosphaeria dothidea* isolated from a seed collected from a *Torreya taxifolia* plant with disease symptoms, showed in vitro antimalarial and phytotoxic activities. Bioactivity-guided fractionation of the extract afforded a mixture of two known isomeric phytotoxins, FRT-A and flavipucine (or their enantiomers, sapinopyridione and (-)-flavipucine), and two new unstable $\gamma$-lactam alkaloids dothilactaenes A and B. The isomeric mixture of phytotoxins displayed strong phytotoxicity against both a dicot and a monocot and moderate cytotoxicity against a panel of cell lines. Dothilactaene A showed no activity. Dothilactaene B was isolated from the active fraction, which showed moderate in vitro antiplasmodial activity with high selectivity index. In spite of this activity, its instability and various other biological activities shown by related compounds would preclude it from being a viable antimalarial lead.

**Keywords:** *Torreya taxifolia*; plant pathogenic fungi; *Botryosphaeria dothidea*; malaria; phytotoxin; $\gamma$-lactam alkaloids

**Citation:** Kumarihamy, M.; Rosa, L.H.; Techen, N.; Ferreira, D.; Croom, E.M., Jr.; Duke, S.O.; Tekwani, B.L.; Khan, S.; Nanayakkara, N.P.D. Antimalarials and Phytotoxins from *Botryosphaeria dothidea* Identified from a Seed of Diseased *Torreya taxifolia*. *Molecules* **2021**, *26*, 59. https://dx.doi.org/10.3390/molecules26010059

Academic Editors: Valeria Patricia Sülsen and William Setzer

Received: 24 September 2020
Accepted: 22 December 2020
Published: 24 December 2020

**Publisher's Note:** MDPI stays neutral with regard to jurisdictional claims in published maps and institutional affiliations.

## 1. Introduction

Parasites of the genus *Plasmodium*, which cause malaria, contain an organelle called the apicoplast, and its functioning is essential for the survival in both the erythrocytic and the hepatic phases of development in mammalian hosts [1]. The apicoplast is similar to plastids of plants, as it is thought to be a vestigial plastid derived from endosymbiosis of a red alga by a heterotropic, unicellular eukaryote [2]. It retains plant-like metabolic pathways, which are absent in vertebrate hosts, making the enzymes of these pathways suitable targets for malaria drug discovery [3–6]. Some phytotoxins released by plant pathogenic fungi inhibit metabolic pathways of the plastid [7]. If the compounds produced by plant pathogenic fungi show phytotoxicity and malarial parasite death without causing cytotoxicity towards mammalian cells, it indicates that the mechanism of parasitic death may be due to the ability of the compounds to inhibit the plant-like metabolic pathways in the apicoplast. As part of our program to search for new antimalarials from plant pathogenic fungi [8–12], we investigated fungi from seeds of a diseased *Torreya taxifolia* Arnott. (Taxaceae). *T. taxifolia*, also known as Florida nutmeg, Florida torreya, stinking-cedar, or gopherwood, is a rare, critically endangered evergreen conifer endemic to three counties in Northern Florida [13–15]. The decline of the native population during the recent past has been attributed to both abiotic and biotic causes, including fungal diseases. Several fungi have been isolated from diseased *T. taxifolia* and some of them have been shown to cause leaf spots and canker disease in healthy plants [15–18]. For this study, seeds of *T. taxifolia* were collected from a tree with disease symptoms cultivated on the Biltmore Estate in Asheville, North Carolina.

From fragments of a surface-sterilized seed, several endophytic fungi were isolated. An EtOAc extract of the broth of one of these fungi grown in potato-dextrose liquid medium showed phytotoxic and antiplasmodial activities. This fungus (UM124) was identified as *Botryosphaeria dothidea* (*Botryosphaeriaceae*) by DNA analysis. Members of the family *Botryosphaeriaceae* (*Botryosphaeriales, Ascomycota*) cause leaf spots, fruit and root rots, and cankers in a variety of hosts [19], and *B. dothidea* has specifically been isolated from a large number of diseased and healthy woody plants, including many economically important crops [20]. A *Botryosphaeria* sp. strain has previously been isolated from *T. taxifolia* leaves infected with needle-spot disease [14]. A chemical investigation of endophytic *B. dothidea* has previously been carried out, and a variety of compounds, including a simple α-pyridone, were reported from a solid culture of this fungus [21].

Bioassay-guided fractionation of the active EtOAc extract resulted in the isolation and identification of a mixture of known, isomeric phytotoxins, FRT-A (**1**) [22] and flavipucine (**2**) [23] (or their enantiomers, sapinopyridione [24] and (-)-flavipucine [22]), as well as two new unstable α-alkyl-γ-lactam alkaloids, dothilactaenes A (**3**) and B (**4**), closely related to epolactaene [25] (Figure 1). While dothilactaene A showed no activity, dothilactaene B was isolated from the active fraction, which showed moderate, selective antiplasmodial activity. Two other components isolated from this fraction had the same molecular formula and similar NMR data indicating that they were diastereomers but due to the lack of sufficient materials their structural investigation was not completed. This is the first report of the isolation of a fungus producing phytotoxins from the seeds of diseased *T. taxifolia*. *B. dothidea* might play a significant role in decreasing the population of the endangered *T. taxifolia*.

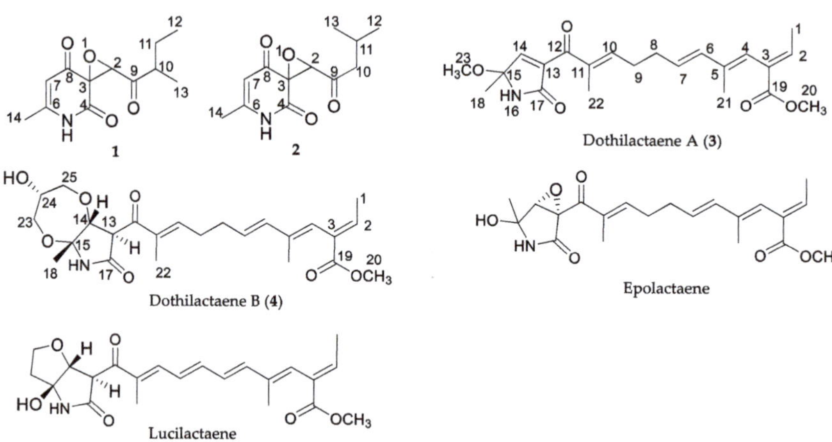


**Figure 1.** Structures of compounds.

## 2. Results and Discussion

The EtOAc extract of the fermentation broth of a fungus isolated from surface sterilized seed fragments of diseased *T. taxifolia* showed good phytotoxic activity against model plants, a dicot (lettuce, *Lactuca sativa* L.) and a monocot (bentgrass, *Agrostis stolonifera* L.). Furthermore, it showed good antiplasmodial activity against chloroquine-sensitive (D6) and -resistant (W2) strains of *Plasmodium falciparum* ($IC_{50}$ = 0.86 and 1.3 μg/mL, respectively), with low cytotoxicity (32 μg/mL) to mammalian kidney fibroblasts (Vero cells).

The analysis of the ITS genomic region of UM124 for a closest neighbor with published sequences showed that the highest identity was with various strains of the species *Botryosphaeria dothidea*. Analysis of 18S rDNA of the fungus gave 100% sequence identity to *B. dothidea*. The construction of the phylogenetic tree with different strains of *B. dothidea* involved 29 nucleotide sequences with a total of 488 positions in the final dataset. In addition, for the construction of the phylogenetic tree of UM124 with various taxa of the family *Botryosphaeriaceae*, 92 nucleotide sequences were analyzed with a total of 382 positions in the final dataset (Supplementary Materials: Tables S1 and S2 and Figures S1 and S2).

The EtOAc extract of the culture broth was fractionated by silica gel column chromatography and the active fractions were combined and separated by Sephadex LH-20 gel filtration using MeOH as the mobile phase. A fraction with high phytotoxic activity and no antiplasmodial activity afforded a white precipitate, and its NMR data (Table 1) indicated that it was a 3:1 mixture of two isomeric known 2,4-pyridione epoxides, fruit rot toxin A (FRT-A) (**1**) [22,23] and (+)-flavipucine [24] (or their enantiomers, sapinopyridione [25] and (-)-flavipucine [24,26,27]). FRT-A was previously reported from *B. berengeriana* [23].

**Table 1.** $^{1}H$- and $^{13}C$-NMR data for mixture of compounds **1** and **2** (3:1) in $CDCl_3$.

| Sapinopyridione [25] | | | Compound 1 | | (-)-Flavipucine [22,27] | | Compound 2 | |
|---|---|---|---|---|---|---|---|---|
| C | $\delta_C$ | $\delta_H$ | $\delta_C$ [a] | $\delta_H$ [b] (*J* in Hz) | $\delta_C$ | $\delta_H$ | $\delta_C$ [a] | $\delta_H$ [b] (*J* in Hz) |
| 2 | 68.6 | 3.94, s | 68.7 | 3.94, s | 68.8 | 3.86, s | 68.8 | 3.84, s |
| 3 | 59.9 | | 60.0 | | 60.0 | | 59.8 | |
| 4 | 168.2 | | 168.7 | | 168.8 | | 168.6 | |
| 6 | 154.9 | | 155.8 | | 156.2 | | 155.8 | |
| 7 | 107.1 | 5.64, s | 107.2 | 5.63, s | 107.3 | 5.64, m | 107.2 | 5.63, s |
| 8 | 186.2 | | 186.5 | | 186.9 | | 186.4 | |
| 9 | 206.5 | | 206.6 | | 203.6 | | 203.7 | |
| 10 | 44.3 | 3.28, ddq | 44.5 | 3.22, m | 49.6 | 2.68, m | 49.6 | 2.62, dd (16.7, 7.2) 2.68, dd (16.7, 7.2) |
| 11 | 25.3 | 1.79, ddq<br>1.52, ddq | 25.4 | 1.78, m<br>1.50, m | 24.1 | 2.1, m | 24.0 | 2.17 [c] |
| 12 | 11.1 | 0.97, t | 11.2 | 0.94, t (7.4) | 22.5 | 0.94, d (6.6) | 22.6 | 0.92, d ( 6.7) |
| 13 | 12.6 | 1.05, d | 12.7 | 1.02, d (6.6) | 22.8 | 0.98, d (6.6) | 22.9 | 0.96, d, (6.7) |
| 14 | 20.8 | 2.19, s | 20.8 | 2.16, s | 20.8 | 2.18, d (1.0) | 20.8 | 2.16, s |
| NH | | 8.38, br s | | 9.24, br s | | 9.14, br s | | 9.22, br s |

[a] Recorded at 100 MHz, [b] Recorded at 400 MHz, [c] Overlapped signals.

The subfraction with antiplasmodial activity ($IC_{50}$ = 0.68 and 0.78 μg/mL) was further separated by reverse-phase semi-preparative HPLC to afford an inactive and an active fraction ($IC_{50}$ < 0.523 μg/mL), with no cytotoxicity to Vero cells. The NMR data of the inactive fraction showed that it was more than 85% of a single compound with lipid impurities. Due to instability and limited material, further purification of this compound was not possible and was identified as dothilactaene A (**3**). The active fraction was separated by analytical HPLC to three peaks. The NMR data of the first peak showed it was a single compound with 90% purity with lipid contaminants and was identified as dothilactaene B (**4**).

The molecular weight of dothilactaene A (**3**) was determined as $C_{22}H_{29}NO_5$ by HRESIMS. The $^{1}H$- and $^{13}C$-NMR data of **3** (Table 2) indicated the presence of 22 carbon resonances that consisted of five quaternary, three carbonyl, six methine, two methylene, and six methyl carbons. Analysis of the $^1$H- and $^{13}$C-NMR data indicated that it was an $\alpha$-alkyl-$\gamma$-lactam derivative, related to epolactaene [28] (Figure 1). Comparison of the $^1$H and $^{13}$C-NMR data of these compounds (Table 2) showed that they had the same side chain but differed in the 2-pyrrolidone ring. These differences were attributable to the replacement of the epoxide and the hydroxy groups in the 2-pyrrolidone ring of epolactaene by a double bond and methoxy group, respectively, in **3**. In the HMBC spectrum of **3**, cross-peaks of $H_3$-1 ($\delta_H$ 1.72) with C-2 ($\delta_C$ 139.9) and C-3 ($\delta_C$ 130.3); H-2 ($\delta_H$ 6.96) with C-3 ($\delta_C$ 130.3) and C-19 ($\delta_C$ 167.8); H-4 ($\delta_H$ 5.97) with C-21 ($\delta_C$ 14.3), C-19 ($\delta_C$ 167.8), and C-2 ($\delta_C$ 139.9); $H_3$-21 ($\delta_H$ 1.62) with C-6 ($\delta_C$ 135.6) and C-4 ($\delta_C$ 123.1); $H_2$-8 ($\delta_H$ 2.32) with C-6 ($\delta_C$ 135.6), C-7 ($\delta_C$ 128.2), C-9 ($\delta_C$ 28.4), and C-10 ($\delta_C$ 148.6); and $H_3$-22 ($\delta_H$ 1.89) with C-10 ($\delta_C$ 148.6) and C-12 ($\delta_C$ 191.4) confirmed the structure of the side chain. Further, HMBC correlations of the -$OCH_3$ singlet ($\delta_H$ 3.18) with C-15 ($\delta_C$ 89.2); the 18-$CH_3$ singlet ($\delta_H$ 1.61) with C-15 ($\delta_C$ 89.2) and C-14 ($\delta_C$ 149.5); and the H-14 olefinic proton singlet ($\delta_H$ 6.83) with C-12 ($\delta_C$ 191.4), C-13 ($\delta_C$ 139.1), and C-17 ($\delta_C$ 168.3) confirmed the structure of the 2-pyrrolidone moiety and its link to the side chain. Other COSY and HMBC correlations supported this structure. The (*E*)-configuration of the $\Delta^{6(7)}$ double bond was determined by the coupling constant ($J$ = 15.2 Hz). ROESY data indicated that the $\Delta^{2(3)}$, $\Delta^{4(5)}$ and $\Delta^{10(11)}$ olefinic bonds were all (*E*) configured.

**Table 2.** $^1$H- and $^{13}$C-NMR spectroscopic data for compounds **3** and **4**.

| | 3 ($CDCl_3$) | | 4 ($CDCl_3$:methanol-$d_4$ 4:1) | |
|---|---|---|---|---|
| | $\delta_C$ [a] | $\delta_H$ [b] ($J$ in Hz) | $\delta_C$ [a] | $\delta_H$[b] ($J$ in Hz) |
| 1 | 15.9 | 1.72, d (7.1) | 15.8 | 1.55, d (7.1) |
| 2 | 139.9 | 6.96, q (7.1) | 140.1 | 6.76, q (7.1) |
| 3 | 130.3 | | 130.4 | |
| 4 | 123.1 | 5.97, br s | 122.7 | 5.76, s |
| 5 | 137.7 | | 138.0 | |
| 6 | 135.6 | 6.23, d (15.6) | 135.0 | 6.10, d (15.6) |
| 7 | 128.2 | 5.68, dt (15.2, 6.6) | 128.8 | 5.60, dt (15.4, 6.9) |
| 8 | 31.5 | 2.32, m | 31.6 | 2.20, m |
| 9 | 28.4 | 2.45, m | 29.4 | 2.30, m |
| 10 | 148.6 | 6.56, t (6.4) | 146.6 | 6.59, t (4.6) |
| 11 | 137.8 | | 138.4 | |
| 12 | 191.4 | | 196.2 | |
| 13 | 139.1 | | 50.0 | 3.21, s |
| 14 | 149.5 | 6.83, s | 76.5 | 4.29, s |
| 15 | 89.2 | | 83.9 | |
| 17 | 168.3 | | 168.3 | |
| 18 | 24.8 | 1.61, s | 20.3 | 1.46, s |
| 19 | 167.8 | | 168.3 | |
| 20 | 51.9 | 3.74, s | 51.9 | 3.55, s |
| 21 | 14.3 | 1.62, s | 14.3 | 1.45, s |
| 22 | 11.1 | 1.89, s | 11.6 | 1.67, s |
| 23 | 50.8 | 3.18, s | 60.7 | 3.45, 3.49, m |
| 24 | | | 70.4 | 3.65, m |
| 25 | | | 61.9 | 3.39, 3.58, m [c] |

[a] Recorded at 150 mHz, [b] Recorded at 600 mHz, [c] Overlapped signals.

Dothilactaene B (**4**) had the molecular formula, $C_{24}H_{33}NO_7$, by HRESIMS data. Its NMR spectra showed resonances due to the same side chain as **3**, and COSY and HMBC data provided confirmatory evidence (Figure 2). Comparison of the remaining resonances with those of **3** indicated the absence of the olefinic bond and the methoxy group in the 2-pyrrolidone ring and the presence of three oxygenated carbons due to a glycerol moiety. The COSY and HMBC correlations (Figure 2) of **4** showed, respectively, an oxygenated methine (H-24; $\delta_H$ 3.65) coupled with two oxygenated methylenes ($H_2$-23; $\delta_H$ 3.45, 3.49, and $H_2$-25; $\delta_H$ 3.39, 3.58), and the methylene resonance at $\delta_H$ 3.39 (H-25) correlating with

C-23 ($\delta_C$ 60.6) and C-24 ($\delta_C$ 70.4), confirming the presence of a disubstituted glycerol moiety. Even though it is not common, secondary metabolites with glycerol moieties have previously been isolated from endophytic fungi [29–32]. In the HMBC spectrum, H-14 ($\delta_H$ 4.29) had a cross peak with the oxygenated methylene at $\delta_C$ 60.7 (C-23) of glycerol, indicating one of its linkages. To satisfy the molecular formula $C_{24}H_{33}NO_7$ and the index of hydrogen deficiency, there should be another ether linkage to the 2-pyrrolidone moiety.

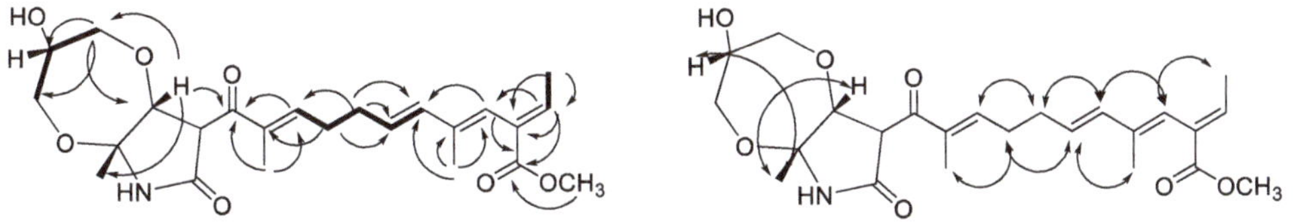

**Figure 2.** HMBC (⌒) and COSY (━) correlations of **4** ROESY (↙↘) correlations of **4**.

Even though no other HMBC cross peaks were visible between the glycerol moiety and the 2-pyrrolidone ring, this link should most probably be between the remaining primary hydroxy group of the former and OH-15 of the latter. An HMBC cross peak between H-14 ($\delta_H$ 4.29) and the C-12 carbonyl ($\delta_C$ 196.2) confirmed the link between the 2-pyrrolidone ring and the side chain. Large coupling ($J$ = 15.6 Hz) between H-6 and H-7 showed that the $\Delta^{6(7)}$ olefinic bond was in an *E* (*trans*) configuration. ROESY correlations (Figure 2) indicated the (*E*) configuration for the other double bonds in the side chain and the cofacial orientation of $CH_3$-18, H-14, and H-24. Comparison of $^3J_{13,14}$ value of **4** with those reported for fusarin A [33] and lucilactanen [34,35] indicates an *anti* configuration of H-13/H-14. These data permitted assignment of the (13*S**, 14*R**, 15*R**, 24*R**) relative configuration. However, the limited sample quantity and the instability precluded electronic circular dichroism (ECD) studies to assign its absolute configuration.

The active fraction afforded two additional components, which had the same molecular formula, $C_{24}H_{33}NO_7$, as that of **4**. NMR data analysis (Supplementary Materials) showed that they have the same gross structure but differ from **4** by the resonances of the 2-pyrrolidone ring and the glycerol moiety suggesting them to be diastereoisomers. Due to the lack of sufficient materials, their structural investigation was not completed.

These diastereomers probably formed from the precursor **6** by Michael addition (Scheme 1) [33]. All the fusarin [33], epolactaene [28], and lucilactaene [34] type compounds so far reported from natural sources have a free hydroxy group at C-15. It is very likely that compound **5** is the precursor of both **3** and **4**. The stereocenter at C-15 undergoes racemization (or epimerization) easily through ring opening, even under mild conditions [33,36]. The formation of methyl or glycerol ethers stabilizes this compound [36], but it is very likely that the product would be racemic in the case of **3** and diastereomeric in the case of **6**. In the chemical synthesis of lucilactaene, the Michael addition occurs spontaneously between the hydroxy group of the hydroxyethyl moiety and $\Delta^{13(14)}$ olefinic bond [36]. The glycerol moiety in **6** also has a stereogenic center, and a similar Michael addition with the primary hydroxyl group of the glycerol would create additional stereogenic centers, yielding diastereomers.

In **4**, the NMR resonance of H-13 was very weak and integrated to less than one proton, and the resonance due to C-13 appeared very small. The 2-pyrrolidone ring in this type of compound could undergo keto-enol tautomerism [36,37], and in deuterated protic solvents it is highly likely that H-13 was partially exchanged with deuterium in $CDCl_3$/methanol-$d_4$ in the presence of traces of water or HOD [38] (Scheme 2). The NMR spectra of fusarin A and lucilactaene have previously been recorded in $CDCl_3$ [33–36].

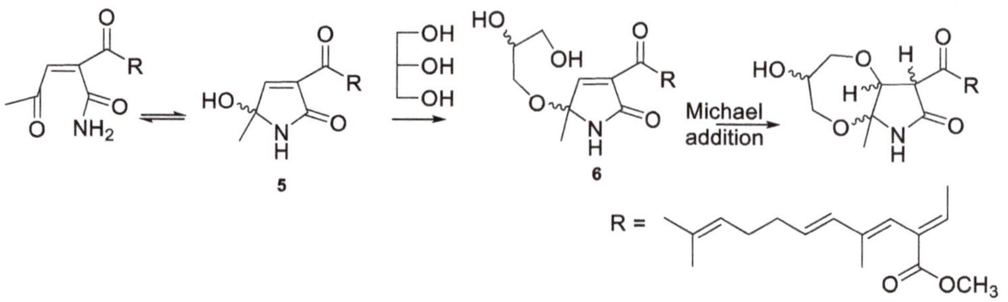

Scheme 1. Possible route of formation of the diastereomers of compound **4**.

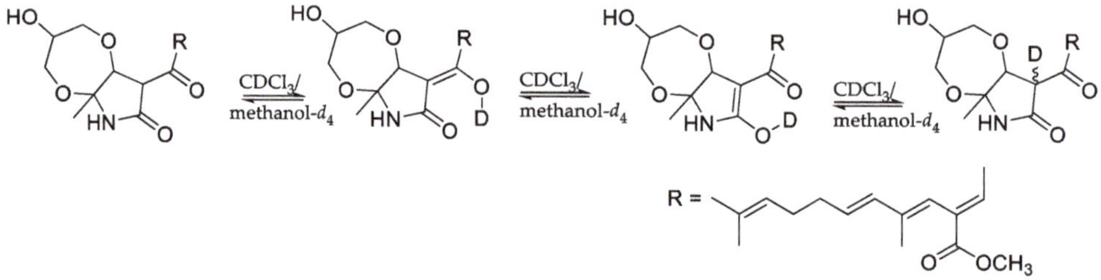

Scheme 2. Deuterium exchange of the 2-pyrrolidone ring of compound **4**.

Even though a number of fusarins [33], closely related compounds with fully unsaturated sidechains, have been isolated from different fungi, epolactaene [25] is the only other compound with the same sidechain as those of compounds **3** and **4** that has been reported so far. Unlike our previous studies [8–12] where active metabolites were found in fungi at the stationary phase of growth, which was typically after three weeks, antiplasmodial compound (**4**) in *B. dothidea* appeared in the log phase of growth after about two weeks and disappeared rapidly before the stationary phase of growth. These compounds may be biosynthetic intermediates rather than stable end products. Compounds **3** and **4** are the first natural fusarin type compounds with ether linkages at C-15.

Biological Activity

The 3:1 mixture of compounds **1** and **2** showed strong phytotoxic activity for both representative monocots and dicots (Table 3). Others have reported **1**, **2**, or both compounds to be phytotoxic [22,23,25], using bioassays that did not utilize whole plants. However, they were devoid of antiplasmodial activity (Table 4). The mixture of compound **1** & **2** was evaluated for toxicity against tumor cell lines SK-MEL, KB, BT-549, and SK-OV-3, as well as against the kidney epithelial cell line, LLC-PK11. The mixture showed cytotoxicity towards SK-MEL and SK -OV-3 cancer cell lines and kidney epithelial cells LLC-PK11, but no toxicity against KB and BT-549 cancer cell lines (Table 5). Previously, the enantiomers and racemate of compound **2** have shown moderate antibacterial activity against *Bacillus subtilis* and strong cytotoxic activity against human leukemia HL-60 cells that was comparable to the activity shown by the positive control, irinotecan [39].

**Table 3.** Phytotoxic activity of mixture of compounds 1 and 2 [a].

| Compound | Lettuce | Bentgrass |
|---|---|---|
| EtOAc extract | no growth | no growth |
| 3:1 mixture of **1** and **2** | no growth | no growth |

[a] Concentration = 1 mg/mL.

**Table 4.** Antiplasmodial activity of compounds **3** and **4**.

| Compound | Chloroquine-Sensitive (D6)-Clone | | Chloroquine-Resistant (W2)-Clone | | Cytotoxicity to Vero Cells |
|---|---|---|---|---|---|
| | $IC_{50}$ µg/mL (µM) | SI | $IC_{50}$ µg/mL (µM) | SI | $IC_{50}$ µg/mL (µM) |
| 3:1 mixture of **1** and **2** | NA | | NA | | NC |
| **3** | NA | | NA | | NC |
| Fraction containing **4** [a] | <0.523 (<1.17) | >9 | <0.523 (<1.17) | >9 | NC |
| Chloroquine [b] | 0.016 (0.03) | 496.6 | 0.16 (0.31) | 48.1 | NC |
| Artemisinin [b] | 0.0056 (0.02) | 845 | 0.003 (0.01) | 1690 | NC |

[a] this fraction showed 100% inhibition at the lowest concentration tested (0.523 µg/mL). [b] Positive controls, $IC_{50}$ = concentration causing 50% growth inhibition, NA = not active at the highest concentration tested (4.76 µg/mL), NC = no cytotoxic at the highest concentration tested (4.76 µg/mL), S. I. (selectivity index) = $IC_{50}$ for cytotoxicity/$IC_{50}$ for antiplasmodial activity.

**Table 5.** Cytotoxic activity [$IC_{50}$ (µM)] of **1** and **2** mixture.

| Compound | SK-MEL | KB | BT-549 | SK-OV-3 | LLC-$PK_{11}$ |
|---|---|---|---|---|---|
| 3:1 mixture of **1** and **2** | 5.05 | 21.9 | 20.3 | 6.7 | 7.6 |
| Doxorubicin [a] | 1.29 | 2.12 | 1.83 | 1.47 | 1.28 |

[a] Positive control, $IC_{50}$ = concentration causing 50% growth inhibition, SK-MEL = human malignant melanoma, KB = human epidermal carcinoma, BT-549 = human breast carcinoma (ductal), SK-OV-3 = ovarian carcinoma, LLC-$PK_{11}$ = kidney epithelial cells.

Compound **3** was found to be inactive in antiplasmodial assays. The fraction containing compound **4** showed moderate in vitro antiplasmodial activity against chloroquine-sensitive (D6) and -resistant (W2) strains of *P. falciparum* ($IC_{50}$ < 0.523 µg/mL) with no cytotoxicity to Vero cells (Table 4). Lack of sufficient material and instability prevented further biological studies on these compounds.

It is interesting that a related compound with a fully unsaturated sidechain, lucilactaene (Figure 1), has also shown potent antiplasmodial activity [35]. It has been found that the tetrahydropyran ring and methylation of the acid group of the sidechain are essential for the activity of this compound. Chemical instability and various other biological activities reported for this class of compounds [33–36] would preclude them from being potential antimalarial agents.

## 3. Materials and Methods

### 3.1. General Experimental Procedures

NMR spectra were recorded on a Varian 400 MHz and/or Varian 600 MHz spectrometer (Varian, Palo Alto, CA, USA ) using $CDCl_3$ or $CDCl_3$/methanol-$d_4$ (4:1) as the solvent, unless otherwise stated. MS analyses were performed on an Agilent Series 1100 SL equipped with an ESI source (Agilent Technologies, Palo Alto, CA, USA). Column chromatography was carried out on silica gel 60 (230–400 mesh) (Sigma-Aldrich, St. Louis, MO, USA) and Sephadex LH-20 (105 × 3 and 69 × 2 $cm^2$) (GE Healthcare Bio-Science, Marlborough, MA, USA). HPLC analysis was carried out on a Hewlett Packard 1100 series instrument with Luna C18 columns (10µ $C_{18}$ 250 × 4.6 $mm^2$, 10 micron; 10µ $C_{18}$ 250 × 2 $mm^2$, 10 micron; Phenomenex (Torrance, CA, USA) as the stationary phase and methanol-water (1:4) as the mobile phase. TLC spots were detected under UV light and by heating after spraying with anisaldehyde reagent.

### 3.2. Isolation of the Fungus from a Seed of Diseased T. taxifolia

Seeds were collected from a *T. taxifolia* tree with disease symptoms cultivated on the Biltmore Estate in Asheville, North Carolina. A voucher of the *T. taxifolia* Arn. plant was identified by E. M. Croom, Jr. and deposited in the University of Mississippi Pullen Herbarium. The voucher accession number is MISS 55406.

A seed of *T. taxifolia* was surface disinfected by immersing in 70% EtOH (1 min) and 2% NaOCl (3 min), followed by washing with sterile distilled water (2 min). The seed was subsequently fragmented and plated onto Petri dishes containing potato dextrose agar (PDA; BD Difco, Franklin Lakes, NJ, USA) supplemented with 200 mg/L chloramphenicol to avoid bacterial contamination. The plates were incubated at 25 °C for 60 days. Hyphal growth was monitored over an 8-week period. Using an aseptic technique, the endophyte was transferred to PDA contained in 60-mm Petri plates. The long-term preservation of filamentous fungal colonies was carried out in cryotubes containing 15% sterile glycerol at −80 °C.

### 3.3. Identification of the Fungus by DNA Analysis

DNA isolation, PCR amplification, cloning, and sequencing were performed as described in Kumarihamy et al. 2019 [12]. Homology searches were performed with the Basic Local Alignment Search Tool (BLAST) [40]. The UM124 sequence was submitted to GenBank (Accession MK679616).

A phylogenetic tree [12] was constructed to identify close relatives of UM124 with the best 19 hits from BLAST and already published sequences in Genebank of various *Botryosphaeria* species. In addition, a phylogenetic tree of UM124 and sequences from various taxa of the family *Botryosphaeriaceae* was constructed (Supplementary Materials: Tables S1 and S2 and Figures S1 and S2).

### 3.4. Fermentation, Extraction, and Purification

*B. dothidea* was cultured in 80 conical flasks (1 L) containing 500 mL of potato dextrose broth and incubated at 27 °C for 14 days on an orbital shaker at 100 rpm. The mycelium was separated by filtration, and the broth was extracted with an equal amount of EtOAc (×3). The EtOAc extract was evaporated to give a black residue (3.15 g).

A portion of the EtOAc extract (3 g) was chromatographed over silica gel and eluted with a gradient of hexanes, $CH_2Cl_2$ and MeOH to yield 15 fractions. Fractions which showed antimalarial activity were combined (550 mg), chromatographed over Sephadex LH-20 (105 × 3 cm), and eluted with MeOH to give 12 fractions. A white precipitate observed in subfraction 11 was separated and washed with $Et_2O$ to yield a 3:1 mixture (10 mg) of **1** and **2** as white amorphous powders. Their identity was confirmed by comparing $^1$H and $^{13}$C NMR, and HRESIMS data with the literature data [22–26].

Subfractions 6–10, which showed antimalarial activity, were combined (60 mg) and further separated using a $C_{18}$ reversed-phase preparative HPLC column (Luna 10μ $C_{18}$ 250 × 10 mm, 10 micron) and was eluted with MeOH-$H_2O$ (1:4) at a flow rate of 3.0 mL/min to give two major peaks. Peak one, which showed no antimalarial activity, was further purified by Sephadex LH-20 (60 × 2 cm) gel filtration with MeOH (100%) to give a new compound (**3**, 1.5 mg).

Peak 2, which showed moderate antimalarial activity, was purified by using $C_{18}$ reversed-phase analytical HPLC chromatography (Luna 10μ $C_{18}$ 250 × 4.6 mm, 10 micron) and was eluted with MeOH-$H_2O$ (1:4) at a flow rate of 1.5 mL/min to give **4** (1.0 mg) and two additional components.

Compound **3**: HRESIMS $[M + H]^+$ *m/z* 388.2184 (calcd for $[C_{22}H_{29}NO_5 + H]^+$ 388.2124), $^1$H- and $^{13}$C-NMR data: see Table 2.

Compound **4**: HRESIMS $[M + H]^+$ *m/z* 448.2268 (calcd for $[C_{24}H_{33}NO_7 + H]^+$ 448.2335), $^1$H- and $^{13}$C-NMR data: see Table 2.

### 3.5. In Vitro Antiplasmodial Assay

The antiplasmodial assay was performed against D6 (chloroquine sensitive) and W2 (chloroquine resistant) strains of *P. falciparum* using the in vitro assay as reported earlier [41].

Artemisinin and chloroquine were included as the drug controls, and $IC_{50}$ values were computed from the dose-response curves.

### 3.6. *In Vitro Phytotoxicity Assay*

Herbicidal or phytotoxic activity of the extract and compounds was performed according to a published procedure [42] using bentgrass (*Agrostis stolonifera*) and lettuce (*Lactuca sativa* cv. L., Iceberg), in 24-well plates. Phytotoxicity was ranked visually. The ranking of phytotoxic activity was based on a scale of 0 to 5 with 0 showing no effect and 5 showing no growth.

### 3.7. *In Vitro Cytotoxicity Assay*

In vitro cytotoxicity was determined against a panel of mammalian cells that included kidney fibroblast (Vero), kidney epithelial (LLC-$PK_{11}$), malignant melanoma (SK-MEL), oral epidermal carcinoma (KB), breast ductal carcinoma (BT-549), and ovarian carcinoma (SK-OV-3) cells [12]. Cells were seeded to the wells of a 96-well plate at a density of 25,000 cells/well and incubated for 24 h. Samples at different concentrations were added and plates were again incubated for 48 h. The number of viable cells was determined by using neutral red dye and $IC_{50}$ values were obtained from dose response curves. Doxorubicin was used as a positive control.

## 4. Conclusions

Bioactivity-guided fractionation of the EtOAc extract of the broth of *B. dothidea* isolated from a seed of a diseased *T. taxifolia* plant afforded a mixture of two known isomeric 2,4-pyridione epoxides, FRT-A (**1**) and flavipucine (**2**) (or their enantiomers, sapinopyridione and (−)-flavipucine) and two new $\alpha$-alkyl-$\gamma$-lactam alkaloids, dothilactaene A (**3**), and dothilactaene B (**4**). The mixture of 2,4-pyridione epoxides, displayed strong phytotoxicity against both a dicot and a monocot and moderate cytotoxicity against a panel of cell lines but no antiplasmodial activity. Dothilactaene A showed no activity. Dothilactaene B was isolated from the active fraction, which showed moderate in vitro antiplasmodial activity with high selectivity index. In spite of this activity, its instability and various other biological activities shown by related compounds would preclude it from being a viable antimalarial lead.

**Supplementary Materials:** The following are available online. Figure S1: Constructed tree using Neighbor-Joining method using MEGA X software to match UM124 to already published sequences of family *Botryosphaeriaceae* taxa to help identifying close relatives. Figure S2: Constructed tree using Neighbor-Joining method using MEGA X software to match UM124 to already published sequences to help identifying close relatives. Figures S3–S24: NMR spectra and HRMS data of compounds **1–4**, Figures S25–S38: NMR spectra and HRMS data of two additional components (compounds **7** and **8**) isolated from the active fraction. Table S1: Best nineteen hits 100% sequence identity, Table S2: ITS sequences of taxa of the *Botryosphaeriaceae* used for alignment. Table S3: $^{1}$H and $^{13}$C NMR data for mixture of compounds **1** & **2**.

**Author Contributions:** N.P.D.N., B.L.T., and S.O.D. conceived and designed experiments and reviewed the manuscript. M.K. carried out fungal culture, extraction, isolation and identification of compounds and wrote the original draft. L.H.R. isolated fungi from the infected *T. taxifolia* seeds and reviewed the manuscript. S.K. supervised in vitro antiplasmodial and cytotoxic assays, analyzed the data, and reviewed the manuscript. D.F. assisted with determination of structures of the compounds critically and reviewed the manuscript. E.M.C.J. identified and provided the infected plant material and reviewed the manuscript. N.T. identified the fungi using ITS DNA analysis and reviewed the manuscript. All authors have read and agreed to the published version of the manuscript.

**Funding:** This research was support in part, by the United State of Department of Agriculture, ARS, Specific Cooperative Agreement No. 58-6060-6-015. L.H.R. received funding from the National Council for Scientific and Technology Development (CNPq), Brazil.

**Data Availability Statement:** The data presented in this study are available in this article or supplementary material.

**Acknowledgments:** We thank Bharathi Avula and Frank Wiggers, NCNPR, University of Mississippi, for recording MS and NMR spectra, respectively, and Marsha Wright, John Trott, and Robert Johnson for biological testing.

**Conflicts of Interest:** The authors declare no conflict of interest. The funders had no role in the design of the study; in the collection, analyses, or interpretation of data; in the writing of the manuscript, or in the decision to publish the results.

**Sample Availability:** Samples of the compounds **1–4** are not available from the authors.

## References

1. Lim, L.; McFadden, G.I. The evolution, metabolism, and functions of the apicoplast. *Phil. Trans. R. Soc. B* **2010**, *365*, 749–763. [CrossRef] [PubMed]
2. Arisue, N.; Hashimoto, T. Phylogeny and evolution of apicoplasts and apicomplexan parasites. *Parasitol. Int.* **2015**, *64*, 254–259. [CrossRef] [PubMed]
3. Ralph, S.A.; van Dooren, G.G.; Waller, R.F.; Crawford, M.J.; Fraunholz, M.J.; Foth, B.J.; Tonkin, C.J.; Roos, D.S.; McFadden, G.I. Tropical infectious diseases: Metabolic maps and functions of the *Plasmodium falciparum* apicoplast. *Nat. Rev. Micro.* **2004**, *2*, 203–216. [CrossRef] [PubMed]
4. Ralph, S.A.; D'Ombrain, M.C.; McFadden, G.I. The apicoplast as an antimalarial drug target. *Drug Resist. Update.* **2001**, *4*, 145–151. [CrossRef] [PubMed]
5. Botté, C.Y.; Dubar, F.; McFadden, G.I.; Maréchal, E.; Biot, C. *Plasmodium falciparum* apicoplast drugs: Targets or off-targets? *Chem. Rev.* **2012**, *112*, 1269–1283. [CrossRef]
6. Uddin, T.; McFadden, G.I.; Goodman, C.D. Validation of putative apicoplast-targeting drugs using a chemical supplementation assay in cultured human malaria parasites. *Antimicrob. Agents Chemother.* **2018**, *62*, e01161-17. [CrossRef]
7. Franck, E.D.; Duke, S. Natural compounds as next-generation herbicides. *Plant Physiol.* **2014**, *166*, 1090–1105.
8. Bajsa, J.; Singh, K.; Nanayakkara, D.; Duke, S.O.; Rimando, A.M.; Evidente, A.; Tekwani, B.L. A Survey of synthetic and natural phytotoxic compounds and phytoalexins as potential antimalarial compounds. *Biol. Pharm. Bull.* **2007**, *30*, 1740–1744. [CrossRef]
9. Herath, H.M.B.T.; Herath, W.H.M.W.; Carvalho, P.; Khan, S.I.; Tekwani, B.L.; Duke, S.O.; Tomaso-Peterson, M.; Nanayakkara, N.P.D. Biologically active tetranorditerpenoids from the fungus *Sclerotinia homoeocarpa* causal agent of dollar spot in turfgrass. *J. Nat. Prod.* **2009**, *72*, 2091–2097. [CrossRef]
10. Kumarihamy, M.; Fronczek, F.R.; Ferreira, D.; Jacob, M.; Khan, S.I.; Nanayakkara, N.P.D. Bioactive 1,4-dihydroxy-5-phenyl-2-pyridinone alkaloids from *Septoria pistaciarum*. *J. Nat. Prod.* **2010**, *73*, 1250–1253. [CrossRef]
11. Kumarihamy, M.; Khan, S.I.; Jacob, M.; Tekwani, B.L.; Duke, S.O.; Ferreira, D.; Nanayakkara, N.P.D. Antiprotozoal and antimicrobial compounds from *Septoria pistaciarum*. *J. Nat. Prod.* **2012**, *75*, 883–889. [CrossRef] [PubMed]
12. Kumarihamy, M.; Ferreira, D.; Croom, E.M., Jr.; Sahu, R.; Tekwani, B.L.; Duke, S.O.; Khan, S.; Techen, N.; Nanayakkara, N.P.D. Antiplasmodial and cytotoxic cytochalasins from an endophytic fungus, *Nemania* sp. UM10M, isolated from a diseased *Torreya taxifolia* leaf. *Molecules* **2019**, *24*, 777. [CrossRef]
13. Schwartz, M.W.; Hermann, S.M. The continuing population decline of *Torreya taxifolia* Arn. *Bull. Torrey Bot. Club* **1993**, *120*, 275–328. [CrossRef]
14. Schwartz, M.W.; Hermann, S.M.; Vogel, C. The catastrophic loss of *Torreya taxifolia*: Assessing environmental induction of disease hypotheses. *Ecol. Appl.* **1995**, *5*, 501–516. [CrossRef]
15. Schwartz, M.W.; Hermann, S.M.; van Mantgem, P.J. Estimating the magnitude of decline of the Florida torreya (*Torreya taxifolia* Arn.). *Biol. Conserve.* **2000**, *95*, 77–84. [CrossRef]
16. Lee, J.C.; Yang, X.; Schwartz, M.; Strobel, G.; Clardy, J. The relationship between an endangered North American tree and an endophytic fungus. *Chem. Biol.* **1995**, *2*, 721–727. [CrossRef]
17. Smith, J.A.; O'Donnell, K.; Mount, L.L.; Shin, K.; Detemann, R. A novel *Fusarium* species causes a canker disease of the critically endangered conifer, *Torreya taxifolia*. *Plant Dis.* **2011**, *95*, 633–639. [CrossRef] [PubMed]
18. Aoki, T.; Smith, J.A.; Mount, L.L.; Geiser, D.M.; O'Donnell, K. *Fusarium torreyae* sp. nov., a pathogen causing canker disease of Florida torreya (*Torreya taxifolia*), a critically endangered conifer restricted to northern Florida and southwestern Georgia. *Mycologia* **2013**, *105*, 312–319. [CrossRef]
19. Yang, T.; Groenewald, J.Z.; Cheewangkoon, R.; Jami, F.; Abdollahzadeh, J.; Lombard, L.; Crous, P.W. Families, genera, and species of Botryosphaeriales. *Fungal Biol.* **2017**, *121*, 322–346. [CrossRef]
20. Marsberg, A.; Kemler, M.; Jami, F.; Nagel, J.H.; Postma-Smidt, A.; Naidoo, S.; Wingfield, M.J.; Crous, P.W.; Spatafora, J.W.; Hesse, C.N.; et al. *Botryosphaeria dothidea*: A latent pathogen of global importance to woody plant health. *Mol. Plant Pathol.* **2017**, *18*, 477–488. [CrossRef]
21. Xiao, J.; Zhang, Q.; Gao, Y.Q.; Tang, J.J.; Zhang, A.L.; Gao, J.M. Secondary metabolites from the endophytic *Botryosphaeria dothidea* of *Melia azedarach* and their antifungal, antibacterial, antioxidant, and cytotoxic activities. *J. Agric. Food Chem.* **2014**, *62*, 3584–3590. [CrossRef] [PubMed]
22. Sassa, T.; Onuma, Y. Isolation and identification of fruit rot toxins from the fungus caused *Macrphoma* fruit rot of apple. *Agric. Biol. Chem.* **1983**, *47*, 1155–1157.

23. Sassa, T.; Uchie, K.; Kato, H.; Onuma, Y. Decomposition of fruit rot toxin A, a host-selective phytotoxin from *Botryosphaeria berengeriana*. *Agric. Biol. Chem.* **1987**, *51*, 271–272. [CrossRef]
24. Loesgen, S.; Bruhn, T.; Meindl, K.; Dix, I.; Schulz, B.; Zeeck, A.; Bringmann, G. (+)-Flavipucine, the missing member of the pyridione epoxide family of fungal antibiotics. *Eur. J. Org. Chem.* **2011**, *2011*, 5156–5162. [CrossRef]
25. Evidente, A.; Fiore, M.; Bruno, G.; Sparapano, L.; Motta, A. Chemical and biological characterisation of sapinopyridione, a phytotoxic 3,3,6-trisubstituted-2,4-pyridione produced by *Sphaeropsis sapinea*, a toxigenic pathogen of native and exotic conifers, and its derivatives. *Phytochemistry* **2006**, *67*, 1019–1028. [CrossRef]
26. Grandolini, G.; Casinovi, C.G.; Radics, L. On the biosynthesis of flavipucine. *J. Antibiot.* **1987**, *40*, 1339–1340. [CrossRef]
27. Findlay, J.A.; Krepinsky, J.; Shum, A.; Casinovi, C.G.; Radics, L. The structure of isoflavipucine. *Can. J. Chem.* **1977**, *55*, 600–603. [CrossRef]
28. Kakeya, H.; Takahashi, I.; Okada, G.; Isono, K.; Osada, H. Epolactaene, a novel neuritogenic compound in human neuroblastoma cells, produced by a marine fungus. *J. Antibiot.* **1995**, *48*, 733–735. [CrossRef]
29. Weber, D.; Gorzalczany, S.; Martino, V.; Acevedo, C.; Sterner, O.; Anke, T. Metabolites from endophytes of the medicinal plant *Erythrina crista-galli*. *Z. Naturforsch. C J. Biosci.* **2005**, *60*, 467–477. [CrossRef] [PubMed]
30. Gao, S.S.; Li, X.M.; Du, F.Y.; Li, C.S.; Proksch, P.; Wang, B.G. Secondary metabolites from a marine-derived endophytic fungus *Penicillium chrysogenum* QEN-24S. *Mar. Drugs* **2010**, *9*, 59–70. [CrossRef]
31. Yanga, X.; Liu, Y.; Li, S.; Yang, F.; Zhao, L.; Peng, L.; Ding, Z. Antimicrobial metabolites from endophytic *Streptomyces sp.* YIM61470. *Nat. Prod. Commun.* **2014**, *9*, 1287–1288. [CrossRef] [PubMed]
32. Chagas, F.O.; Caraballo-Rodríguez, A.M.; Dorrestein, P.C.; Pupo, M.T. Expanding the chemical repertoire of the endophyte *Streptomyces albospinus* RLe7 reveals amphotericin B as an inducer of a fungal phenotype. *J. Nat. Prod.* **2017**, *80*, 1302–1309. [CrossRef] [PubMed]
33. Kleigrewe, K.; Aydin, F.; Hogrefe, K.; Piecuch, P.; Bergander, K.; Würthwein, E.U.; Humpf, H.U. Structure elucidation of new fusarins revealing insights in the rearrangement mechanisms of the *Fusarium* mycotoxin fusarin C. *J. Agric. Food Chem.* **2012**, *60*, 5497–5505. [CrossRef] [PubMed]
34. Kakeya, H.; Kageyama, S.; Nie, L.; Onose, R.; Okada, G.; Beppu, T.; Norbury, C.J.; Osada, H. Lucilactaene, a new cell cycle inhibitor in p53-transfected cancer cells, produced by a *Fusarium* sp. *J Antibiot.* **2001**, *54*, 850–854. [CrossRef] [PubMed]
35. Kato, S.; Motoyama, T.; Futamura, Y.; Uramoto, M.; Nogawa, T.; Hayashi, T.; Hirota, H.; Tanaka, A.; Takahashi-Ando, N.; Kamakura, T.; et al. Biosynthetic gene cluster identification and biological activity of lucilactaene from *Fusarium sp.* RK97-94. *Biosci. Biotechnol. Biochem.* **2020**, *84*, 1303–1307. [CrossRef] [PubMed]
36. Yamaguchi, J.; Kakeya, H.; Uno, T.; Shoji, M.; Osada, H.; Hayashi, Y. Determination by asymmetric total synthesis of the absolute configuration of lucilactaene, a cell-cycle inhibitor in p53-transfected cancer cells. *Angewandte Chemie Int. Ed. Engl.* **2005**, *44*, 3110–3115. [CrossRef] [PubMed]
37. Lin, Y.; Wang, L.; Wang, Y.; Wang, W.; Hao, J.; Zhu, W. Bioactive natural products of *Aspergillus sp.* OUCMDZ-1914, an aciduric fungus from the mangrove soils. *Chin. J. Org. Chem.* **2015**, *35*, 1955–1960. [CrossRef]
38. Nichols, M.A.; Waner, M.J. Kinetic and mechanistic studies of the deuterium exchange in classical keto-enol tautomeric equilibrium reactions. *J. Chem. Educ.* **2010**, *87*, 952–955. [CrossRef]
39. Kusakabe, Y.; Mizutani, S.; Kamo, S.; Yoshimoto, T.; Tomoshige, S.; Kawasaki, T.; Takasawa, R.; Tsubaki, K.; Kuramochi, K. Synthesis, antibacterial and cytotoxic evaluation of flavipucine and its derivatives. *Bioorg. Med. Chem. Lett.* **2019**, *29*, 1390–1394. [CrossRef]
40. Altschul, S.F.; Gish, W.; Myers, E.W.; Lipman, D.J. Basic local alignment search tool. *J. Mol. Biol.* **1990**, *215*, 403–410. [CrossRef]
41. Bharate, S.B.; Khan, S.I.; Yunus, N.A.; Chauthe, S.K.; Jacob, M.R.; Tekwani, B.L.; Khan, I.A.; Singh, I.P. Antiprotozoal and antimicrobial activities of *O*-alkylated and formylated acylphloroglucinols. *Bioorg. Med. Chem.* **2007**, *15*, 87–96. [CrossRef] [PubMed]
42. Dayan, F.E.; Romagni, J.G.; Duke, S.O. Investigating the mode of action of natural phytotoxins. *J. Chem. Ecol.* **2000**, *26*, 2079–2094. [CrossRef]

*Article*

# Anti-Inflammatory, Antidiabetic Properties and In Silico Modeling of Cucurbitane-Type Triterpene Glycosides from Fruits of an Indian Cultivar of *Momordica charantia* L.

**Wilmer H. Perera [1], Siddanagouda R. Shivanagoudra [1], Jose L. Pérez [1], Da Mi Kim [2], Yuxiang Sun [2], Guddadarangavvanahally K. Jayaprakasha [1] and Bhimanagouda S. Patil [1,*]**

[1] Vegetable and Fruit Improvement Center, Department of Horticultural Sciences, Texas A&M University, 1500 Research Parkway, Suite A120, College Station, TX 77843, USA; Wilmer.Perera@gmail.com (W.H.P.); siddu4191@gmail.com (S.R.S.); ppmd_4@tamu.edu (J.L.P.); gkjp@tamu.edu (G.K.J.)

[2] Department of Nutrition, Texas A&M University, College Station, TX 77843, USA; dmkim5322@tamu.edu (D.M.K.); yuxiangs@tamu.edu (Y.S.)

* Correspondence: b-patil@tamu.edu; Tel.: +1-979-458-890; Fax: +1-979-862-4522

**Abstract:** Diabetes mellitus is a chronic disease and one of the fastest-growing health challenges of the last decades. Studies have shown that chronic low-grade inflammation and activation of the innate immune system are intimately involved in type 2 diabetes pathogenesis. *Momordica charantia* L. fruits are used in traditional medicine to manage diabetes. Herein, we report the purification of a new 23-*O*-β-D-allopyranosyl-5β,19-epoxycucurbitane-6,24-diene triterpene (charantoside XV, **6**) along with 25ξ-isopropenylchole-5(6)-ene-3-*O*-β-D-glucopyranoside (**1**), karaviloside VI (**2**), karaviloside VIII (**3**), momordicoside L (**4**), momordicoside A (**5**) and kuguaglycoside C (**7**) from an Indian cultivar of *Momordica charantia*. At 50 μM compounds, **2–6** differentially affected the expression of pro-inflammatory markers *IL-6*, *TNF-α*, and *iNOS*, and mitochondrial marker *COX-2*. Compounds tested for the inhibition of α-amylase and α-glucosidase enzymes at 0.87 mM and 1.33 mM, respectively. Compounds showed similar α-amylase inhibitory activity than acarbose (0.13 mM) of control (68.0–76.6%). Karaviloside VIII (56.5%) was the most active compound in the α-glucosidase assay, followed by karaviloside VI (40.3%), while momordicoside L (23.7%), A (33.5%), and charantoside XV (23.9%) were the least active compounds. To better understand the mode of binding of cucurbitane-triterpenes to these enzymes, in silico docking of the isolated compounds was evaluated with α-amylase and α-glucosidase.

**Keywords:** *Momordica charantia*; cucurbitane-type triterpene glycosides; charantoside XV; α-amylase; α-glucosidase; anti-inflammatory activity; in silico study

**Citation:** Perera, W.H.; Shivanagoudra, S.R.; Pérez, J.L.; Kim, D.M.; Sun, Y.; K. Jayaprakasha, G.; S. Patil, B. Anti-Inflammatory, Antidiabetic Properties and In Silico Modeling of Cucurbitane-Type Triterpene Glycosides from Fruits of an Indian Cultivar of *Momordica charantia* L. *Molecules* **2021**, *26*, 1038. https://doi.org/10.3390/molecules26041038

Academic Editors: Ilias Muhammad and Nanayakkara Dhammika

Received: 18 January 2021
Accepted: 12 February 2021
Published: 16 February 2021



## 1. Introduction

The genus *Momordica* L., with around 59 species, is one of the most abundant genera in the Cucurbitaceae family. *Momordica charantia* L. is a taxon extensively studied for its antidiabetic properties [1]. Two varieties have been taxonomically identified: *M. charantia* L. (var. *muricata* (Willd.) Chakrav. and var. *charantia*) [1]. The *muricata* variety produces small size fruits, whereas *charantia* typically produces larger fruits [1]. Several phenotypes of *M. charantia* var. *charantia* appear in nature, displaying noteworthy differences in fruit size and shape. The most common fruit from the Indian cultivar has a greenish color, a narrower shape, and a serrated surface. *M. charantia* var. *charantia* L. is commonly consumed in traditional Asian cuisines and is well-known for its unique taste and flavor [2,3].

Cucurbitane-type triterpenes are the most representative sub-class of compounds in *M. charantia*, and some investigations suggest that they are responsible for their antidiabetic property [4,5]. More than 270 cucurbitane-type triterpenes have been isolated from different plant organs from 1980 to date. However, these compounds are found in very low concentration and their composition has been reported to vary between cultivars. Our comprehensive survey of the literature indicates that the Indian phenotype has received less attention than the Chinese or Sri Lanka phenotypes. Octonorcucurbitacins A, B and C (342 Da) are the smallest cucurbitane-type compounds reported from *M. charantia*, whereas momordicoside T (1110 Da) the largest one with four sugar units, one glucose linked at position C-25, two glucose units, and one xylose attached at C-3 as follows: -Glc[Glc(1-4)]Xyl(1–4) [4–6]. Additionally, several aglycones, mono, di, and tri-glycoside triterpene derivatives have also been isolated from bitter melon. The glycosidic moieties attached to these triterpenes typically consist of D series of allose, galactose, glucose and xylose monosaccharide units, mainly at positions C-23 and C-25 of the side chain, and C-3 and C-7 of the A and B rings, respectively [4,7–9].

Many purified compounds, as well as crude extracts, have been screened in response to various ethnomedicinal claims, concluding that bitter melon may play a potential role in the management of various chronic diseases by working as an antidiabetic, antioxidant, antiviral, antiobesity, and anticancer agent among others [8–14]. Moreover, some studies have been conducted in fruit, seeds, and leaves showing hypoglycemic activity in both diabetic animal and human models [15,16]. Thus, *M. charantia* var. *charantia* stands out as a promising natural alternative to reduce the risk and/or manage diabetes mellitus type II [17].

Diabetes is a chronic disease that affects millions of people worldwide. Therefore, it is vital to find new natural prevention and management strategies for this disease and related complications. Results from our previous studies indicated that cucurbitane-type triterpenoids isolated from Chinese cultivar might modulate biological activities involved in the pathogenicity of diabetes [18,19]. Due to these molecules structural complexity, much more work is warranted to test biological activities and better understand the structure-activity relationship. Therefore, herein we describe the isolation and structure elucidation of a new cucurbitane-type triterpene together with six known compounds from acetone extract of the Indian cultivar of *M. charantia*. The anti-inflammatory activity of purified compounds using lipopolysaccharide (LPS)-activated RAW264.7 macrophage cells, inhibition of the α-amylase, and α-glucosidase enzymes were presented. Furthermore, the molecular interaction of purified compounds with α-amylase and α-glucosidase enzymes was also studied using *in silico* molecular docking.

## 2. Results and Discussion

### 2.1. Isolation of Compounds

The isolation of the compounds presented was based on two main approaches, silica gel and reversed-phase chromatographic separations. Compound **6** was purified as a white amorphous powder, $[\alpha]^{25}{}_{D}$-71 (*c* 0.1, MeOH). HRESIMS of compound **6** showed a molecular ion at $m/z$ 687.3967 [M + Na]$^+$ (calculated 687.4047 ± 0.008 $m/z$ [M + Na]$^+$ for $C_{37}H_{60}O_{10}$). The molecular formula of compound **6** was compared with reported triterpenes isolated from *M. charantia* and our in-house database [20]. No match was observed, suggesting an unknown compound. The $^{13}$C-NMR spectrum of compound **6** showed resonances of 37 carbons. The 2D-gHMQC spectrum revealed the presence of sixteen methines, seven methylenes, seven methyl, one methoxy, and six quaternary carbons. Six tertiary groups appeared at $\delta_H(\delta_c)$ 0.84 (14.8); 1.74 (26.7); 1.77 (19.1); 1.18 (20.9); 0.87 (24.6); 0.87 (20.6) corresponding to $CH_3$-18, $CH_3$-26, $CH_3$-27, $CH_3$-28, $CH_3$-29 and $CH_3$-30, respectively and one secondary $\delta_H(\delta_c)$ 0.96 (14.7) (d, $J$ = 6.4 Hz, $CH_3$-21). The methoxy group was observed at $\delta_H(\delta c)$ 3.41 s, 3H, assigned as H-1" (58.2, C-1"), one acetal group at $\delta_H(\delta c)$ 4.70 s, 1H, assigned as H-19 (113.3, C-19), three olefinic protons at $\delta_H(\delta c)$ 5.98 [dd, $J$ = 9.8 Hz, 2.1 Hz, H-6] (132.2, C-6), $\delta_H(\delta c)$ 5.58 [dd, $J$ = 9.8 Hz, 3.4 Hz, H-7] (133.8, C-7) and $\delta_H(\delta c)$ 5.24 [d, $J$ = 10.2 Hz, H-24] (124.6, C-24), and one quaternary olefin carbon $\delta_C$ 137.4 (C-25). One anomeric proton at $\delta_H(\delta c)$ 4.73 [d ($J$ = 8 Hz, 1H] (102.6) with a beta linkage corresponding to one hexose appeared in the spectrum together with their respectively five oxygenated signals $\delta_H(\delta c)$ 3.32, H-2′ (73.1, C-2′), 4.05, H-3′ (73.0, C-3′),

3.48, H-4′ (68.5, C-4′), 3.58, H-5′ (75.7, C-5′) and 3.60, 3.76, H-6′ (63.1, C-6′). $^{1}$H and $^{13}$C NMR chemical shifts are shown in Table 1.

**Table 1.** $^{1}$H and $^{13}$C NMR chemical shifts of compound **6**.

| Position | Compound 6 (MeOH-$d_4$) | |
|---|---|---|
| | $\delta_H$ | $\delta_C$ |
| 1 | 1.15; 1.50 m, 2H | 18.5 |
| 2 | 0.82; 1.84 m, 2H | 28.3 |
| 3 | 3.36 m, 1H | 77.8 |
| 4 | – | 38.5 |
| 5 | – | 88.1 |
| 6 | 5.98 dd ($J$ = 9.8, 2.1 Hz, 1H) | 132.2 |
| 7 | 5.58 dd ($J$ = 9.8, 3.7 Hz, 1H) | 133.8 |
| 8 | 2.86 m, 1H | 43.2 |
| 9 | – | 49.7 |
| 10 | 2.44 m, 1H | 42.1 |
| 11 | 1.54; 1.75 m, 2H | 24.3 |
| 12 | 1.54; 1.54 m, 2H | 32.0 |
| 13 | – | 46.5 |
| 14 | – | 49.0 |
| 15 | 1.32; 1.32 m, 2H | 35.0 |
| 16 | 1.34; 1.34 m, 2H | 29.0 |
| 17 | 1.97 m, 1H | 47.7 |
| 18 | 0.84 s, 3H | 14.8 |
| 19 | 4.70 s, 1H | 113.3 |
| 20 | 1.77 dd ($J$ = 11.3, 1.0 Hz, 1H) | 42.1 |
| 21 | 0.96 d ($J$ = 6 Hz, 3H) | 14.7 |
| 22 | 3.63 m, 1H | 77.3 |
| 23 | 4.29 m, 1H | 81.5 |
| 24 | 5.24 d ($J$ = 10.2 Hz, 1H) | 124.6 |
| 25 | – | 137.4 |
| 26 | 1.74, 3H | 26.7 |
| 27 | 1.77, 3H | 19.1 |
| 28 | 1.18 s (3H) | 20.9 |
| 29 | 0.87 s (3H) | 24.6 |
| 30 | 0.87 s (3H) | 20.6 |
| 23-*O*-All | | |
| 1′ | 4.73 d ($J$ = 8 Hz, 1H) | 102.6 |
| 2′ | 3.32, 1H | 73.1 |
| 3′ | 4.05, 1H | 73.0 |
| 4′ | 3.48, 1H | 68.8 |
| 5′ | 3.58, 1H | 75.7 |
| 6′ | 3.60; 3.76, 2H | 63.1 |
| 1″ | 3.41, 3H | 58.2 |

Acid hydrolysis of **6** furnished D-allose, which was identified by comparison of the HPLC retention times of thiocarbamoyl thiazolidine derivative with sugar thiocarbamoyl thiazolidine derivative standards [21]. The NMR data of compound **6** suggested the presence of a 5β,19-epoxycucurbitane triterpenes with a methoxy group and a sugar moiety attached [22]. The position of the methoxy group was established through $^{3}J$ HMBC correlations between $\delta_H$ 3.41 ppm ($OCH_3$) and the acetal carbon at position C-19 ($\delta$c 113.3), indicating that the methoxy group was attached at C-19.

The relative position of the monosaccharide attachment was confirmed by the assignment of gHMBC spectrum. $^{3}J$ HMBC correlations between anomeric proton H-1′ (4.73 ppm) and C-23 (81.5 ppm) confirmed the attachment of the sugar at C-23. The position of proton H-2′ of the allose was confirmed through the $^{1}$H-$^{1}$H COSY correlation between H-2′ (4.05 ppm) and the anomeric proton H-1′ (4.73 ppm). TOCSY correlations were helpful to identify protons corresponding to the sugar moiety and HMBC, and $^{1}$H-$^{1}$H COSY to assign remaining proton and carbon chemical shifts. The HMBC correlations between C-24 and $CH_3$-26 and $CH_3$-27; C-25 and H-23, C-22 and H-23 together with C-20 and H-22, suggested that the side chain was -CH($CH_3$)CH(OH)CH(O-All)=C($CH_3$)$_2$. Crucial HMBC and $^1$H-$^1$H COSY correlations are shown in Figure 1A. This type of side-chain has been described previously for karavilosides IV, VIII, IX and X, 23-*O*-β-allopyranosyl-cucurbita-5,24-dien-7α,3β, 22(*R*), 23(*S*)-tetraol 3-O-*β*-allopyranoside; momordicosides M, N and O isolated from *M. charantia* cultivated in Sri Lanka and China [7,23–25]. The stereochemistry of the stereocenters C-22 and C-23 was established by NOESY experiment. A correlation of H-22 was observed with H-21, while no correlation was found for H-22 and H-23 (22S and 22R). C-19 configuration was deduced as *R* due to the correlation between proton H-19 with H-1 and H-2, only observed in *R* configuration (Figure 1B).

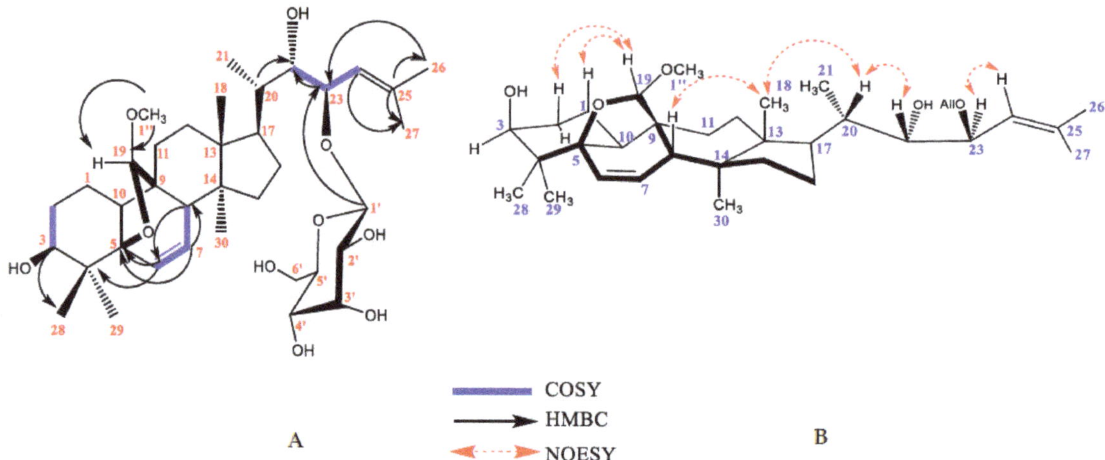


**Figure 1.** Crucial HMBC, $^1$H-$^1$H COSY (**A**) and NOESY correlations (**B**) for compound **6**.

This result was also supported by a comparison of the NMR chemical shifts of H-8, C-8, C-9, C-10, C-11, and C-19 previously reported for several 5β,19*R*- and *S*-epoxycucurbitane triterpenes with different substituents attached at C-19 (methoxy, ethoxy, and hydroxyl groups) [21,26]. However, the main difference in chemical shifts was observed for H-8 and C-8: *R* configuration of H-8 range from 2.82 to 3.37 ppm and from 41 to 43 ppm for C-8, whereas for *S* configuration H-8 (2.13–2.32 ppm) and C-8 (49.5–50.1 ppm). Chemical shifts for compound **6** appeared at H-8 (2.86 ppm) and C-8 (43.2 ppm), supporting the *R* configuration of C-19. Hence, the structure of compound **6** was established as 19(*R*)-methoxy-5β,19-epoxycucurbita-6,24-diene-3β,22*S*,23*R*-triol-23-*O*-β-D-allopyranoside (charantoside XV) and shown in Figure 2.

Compounds **1–5** and **7** were identified as known monoglycosides by spectroscopic ($^1$H- and $^{13}$C-NMR spectra) and spectrometric analyses (HRESIMS), and by comparison with the data reported in the literature as follows: 25ξ-isopropenylchole-5(6)-ene-3-*O*-β-D-glucopyranoside (**1**), karaviloside VI (**2**), karaviloside VIII (**3**), momordicoside L (**4**), momordicoside A (**5**) and kuguaglycoside C (**7**) [7,21,27–29]. Structures of the isolated compounds and $^1$H- and $^{13}$C-NMR data are shown in Figure 2 and the Supporting Information, respectively. This is the first report describing the stereochemistry of the karaviloside VIII side chain to the best of our knowledge. Herein, we described the configuration of the stereocenters C-22 and C-23 through NOESY experiments as 22 *R* and 23*S* since H-22 showed correlation with H-20 and no correlation was observed between H-22 and H-23. Moreover, momordicosides A and L have been isolated from fruits of an Indian cultivar while the other compounds are mainly from Sri Lanka, China, and Japan cultivars.

Therefore, this is the first finding of compounds **1**, **2**, **3** and **7** in fruits of the Indian cultivar. On the other hand, kuguaglycoside C was reported as an anticancer agent, showing significant cytotoxicity against human neuroblastoma IMR-32 cells [30]. In the same way, momordicoside A showed weak activity on glucose transport type-4 (GLUT4) on translocation cells [4] and momordicoside L was assayed as a hypoglycemic and antiproliferative compound, showing positive glucose uptake activity and no activity against human breast adenocarcinoma (MCF-7), human medulloblastoma (Doay), human colon adenocarcinoma (WiDr) and human laryngeal carcinoma (HEp-2).

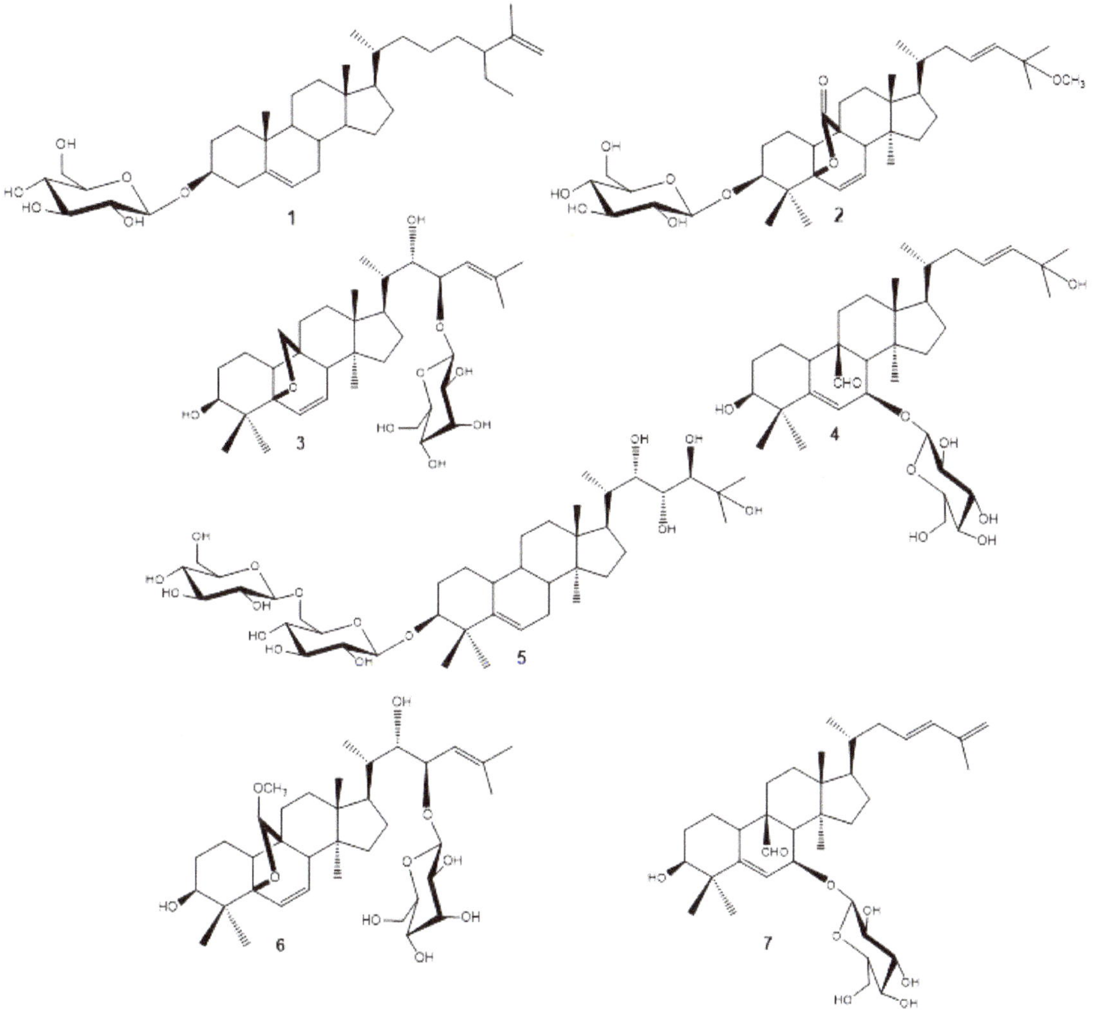

**Figure 2.** Chemical structures of compounds isolated and identified in the present study. (**1**) 25ξ-isopropenylchole-5(6)-ene-3-*O*-β-D-glucopyranoside, (**2**) karaviloside VI, (**3**) karaviloside VIII, (**4**) momordicoside L, (**5**) momordicoside A, (**6**) charantoside XV, and (**7**) kuguaglycoside C.

*2.2. Bioassays*

2.2.1. Anti-Inflammatory Activity

The anti-inflammatory activity of compounds **2–6** was analyzed. Compound **1** was assessed in our previous publication, and the low levels of compound **7** did not allow for further in-vitro assessments [18]. Interestingly, all the purified compounds exhibited anti-inflammatory properties by promoting the down-regulation of the pro-inflammatory gene markers *IL-6, TNF-α, COX-2,* and *iNOS* (Figure 3). All the compounds purified from the Indian bitter melon significantly decreased the expression of *IL-6* compared to LPS treated cells. The lowest expression of *IL-6* was observed in the cells treated with karaviloside VI, karaviloside VIII, momordicoside L, and momordicoside A. The downregulation of *iNOS* was also observed in all the compounds except momordicoside A. Similarly, momordicoside A, along with momordicoside L, significantly decreased *TNF-α* mRNA expression. Momordicoside L and karaviloside VI are also considerably reduced the expression of *COX-2* mRNA expression.

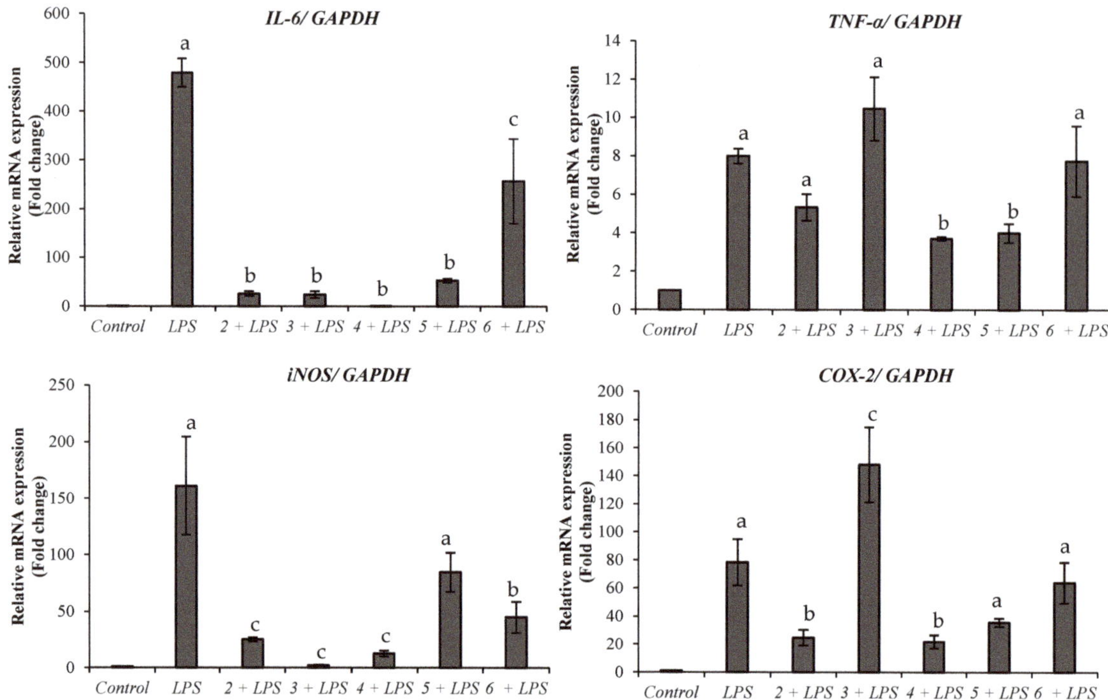


**Figure 3.** Effect of purified compounds on mRNA expression of *IL-6, TNF-α, iNOS,* and *COX-2* in LPS-induced murine macrophage RAW 264.7 cells. The cells were pretreated with 50 μM karaviloside VI (**2**), karaviloside VIII (**3**), momordicoside L (**4**), momordicoside A (**5**), and charantoside XV (**6**) followed by LPS (1 μg/mL) stimulation. Data are expressed as mean ± SD ($n$ = 9) and analyzed by one-way ANOVA with a Tukey post hoc test. Different letters within the same plot indicate the significant differences at $p < 0.05$.

Pro-inflammatory cytokines mediate the inflammatory response in humans. LPS is a potent pro-inflammatory agent causing an increase in the expression of pro-inflammatory genes such as *IL-6, TNF-α, COX-2,* and *iNOS*. High levels of *IL-6* and *TNF-α* have been reported in patients with type-2 diabetes and insulin resistance [15]. Additionally, *COX- 2* has been reported to be induced under hyperglycemic conditions [16]. Any agent that promotes the decrease in the expression of these genes has the potential to work as a viable anti-inflammatory agent. Our results indicate that the compounds isolated from bitter melon may have potential antidiabetic activities by modulating the inflammatory process. In past studies, various bitter melon extracts and compounds have been reported to have anti-inflammatory activities in different cell models [21,31–33]. Bitter melon ethyl acetate extracts have been reported to decrease the expression of *iNOS*, *COX-2*, *IL-6*, and *TNF-α* in RAW 264.6 macrophages, but purified compounds derived from those extracts were not evaluated [34]. As such, the anti-inflammatory activity presented in this study suggests that bitter melon compounds are potential agents against inflammation and possible diseases that arise from chronic inflammation, such as diabetes.

#### 2.2.2. α-Amylase and α-Glucosidase

Several strategies have been explored in the management of diabetes mellitus for either reducing glucose production by the liver or enhancing insulin sensitivity or secretion [35]. Among these approaches, the inhibition of the α-amylase and α-glucosidase enzymes are directed to manage the post-prandial hyperglycemia by delaying starch hydrolysis by cleaving 1,4-glucosidic linkages [36,37]. In this sense, the α-amylase inhibitory effect of compounds **2–6** was also evaluated in this paper. The inhibitory effect ranges from 68.0 to 76.6%, but no significant statistical difference was observed among the compounds (Figure 4A). Additionally, the α-glucosidase inhibitory effect ranges from 23.7 to 56.5%. Karaviloside VIII was the most active compound, followed by karaviloside VI (40.3%), while the less active compounds were momordicoside L (23.7%), momordicoside A (33.5%) and charantoside XV (23.9%). Assayed compounds showed a lower α-glucosidase inhibitory effect than acarbose (Figure 4B).

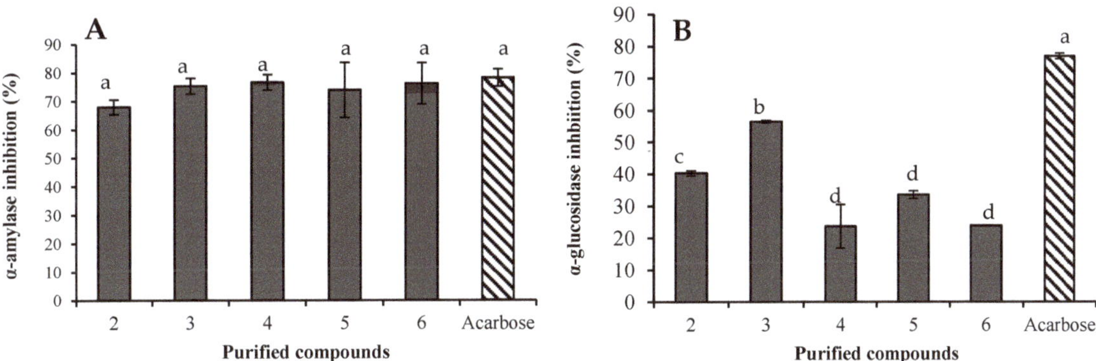


**Figure 4.** α-Amylase (**A**) and α-glucosidase (**B**) inhibitory effects of purified compounds: karaviloside VI (**2**), karaviloside VIII (**3**), momordicoside L (**4**), momordicoside A (**5**) and charantoside XV (**6**). Compounds were assayed at 0.67 mM and data are expressed as mean ± SD ($n = 3$), and analyzed by one-way ANOVA with a Tukey post hoc test. Different letters within the same plot indicate there were significant differences at $p < 0.05$.

### 2.3. *Molecular Docking*

Molecular docking studies were carried out to understand better the interactions of the isolated cucurbitane-triterpenes **2–7** with α-amylase and α-glucosidase as previously reported for a series of aglycone, monoglycoside cucurbitane-triterpenes, and 25ξ-isopropenylchole-5(6)-ene-3-*O*-β-D-glucopyranoside (**1**) isolated in this study [18].

#### 2.3.1. Molecular Docking Study with α-Amylase

Docking studies were carried out to understand the interaction of active compounds **2–6** inside the catalytic site of α-amylase. The crystalline structure of porcine pancreatic α-amylase (PDB ID: 1OSE) and active site amino residues are explained briefly in a previous publication [38]. The selection of the porcine pancreatic α-amylase crystal structure 1OSE in this study was based on three main reasons. First, we have used the porcine pancreatic α-amylase enzyme for our in vitro α-amylase inhibition study. Secondly, the selected protein forms a complex with the acarbose, which we used as a positive control or reference standard in our study. Finally, molecular models for the α-amylases from the human pancreas and human salivary are incredibly similar to the pig pancreatic molecular model. The homology modeling of the porcine and human pancreatic α-amylase is very similar (87.1%) compared with other amylases [39]. The binding energies of tested compounds with α-amylase ranged from −14.52 to −8.94 kcal/mol. The decreasing order of minimal binding energies in the case of α-amylase molecular docking studies are as follows: momordicoside A < kuguaglycoside C < karaviloside VIII < charantoside XV < karaviloside VI < momordicoside L (Table S1). The docking results of the isolated compounds showed the binding site as the same as the binding sites for acarbose. Indeed, the docking analysis predicted that acarbose, a competitive inhibitor of α-amylase, was surrounded by Glu233, Asp300, and Asp197, which are the part of the catalytic residues of α-amylase. [40]. The molecular docking study for 25ξ-isopropenylchole-5(6)-ene-3-*O*-β-D-glucopyranoside (**1**) was previously reported in our study [18]. The binding modes in the active site of α-amylase of all purified compounds, except **1**, are shown in Figure 5A–F. The binding energy and number of hydrogen bonds of the five triterpenes against porcine pancreatic α-amylase are shown in Table S1.

As shown in Figure 5A–F, all five triterpenes bound to the porcine pancreatic α-amylase through forming various hydrogen bonds. It was observed that the binding site for karaviloside VI (**2**) was close to the active site, allowing the interaction with Asp300, Ile235, and Trp59. Four hydrogen bonds were formed between compound **2** and the amino acid residues, including Glu240 (3 H-bonds), His201 (1 H-bond). The inhibition constant and binding energy for compound **2** is 18.79 nM and −10.54 kcal/mol (Table S1).

In case of karaviloside VIII (**3**), the docking results predicted that the compound was enfolded in the catalytic domain adopting the same conformation as the acarbose site of α-amylase, as shown in Figure 5B. It was surrounded by fourteen key amino acid residues. The likeliest docked interactions of karaviloside VIII (**3**) and α-amylase is shown in Figure 5B. The ligand was surrounded by amino acid residues located in domain A, making hydrogen bonds with Lys200, His201, Trp59, and Asp356 (Table S1). The principle interaction of karaviloside VIII (**3**) with α-amylase was surrounded by the key catalytic residue Trp59, and the interaction was crucial to inhibit the activity of α-amylase. Overall, the inhibition constant for compound **2** was 1.76 nM.

The refined docking of momordicoside L (**4**) showed weaker interactions than the other compounds because of the steric hindrance of the functional group at position 25, compared to compound **7**. The compound formed fewer H-bonds reflected in the lower binding energy of -8.94 kcal/mol. The 3D figure shows that momordicoside L interacted with sixteen key amino acid residues, including four conventional hydrogen bonds that were established between momordicoside L and Glu233, Ala307, His305 (Figure 5C). The inhibition constant was found to be 278.19 nM.

Figure 5D displays a 3D schematic interaction of momordicoside A (**5**) with α-amylase. Momordicoside A generated the best docking pose with a minimum binding energy of −14.52 kcal/mol, which indicates that momordicoside A showed the most stronger binding affinity with the α-amylase. Momordicoside A was found to anchor at the catalytic site of α-amylase by making ten conventional hydrogen bonds with Glu240, His101, Glu233, Asp197, Gly63 to residues resulting in potent inhibition of α-amylase. The strong inhibition activity of momordicoside A on α-amylase relies on the formation of multiple hydrogen bonds between hydroxyl groups and key residues of α-amylase. The molecular docking of momordicoside A provided supportive data for enzyme inhibition by predicting the binding site of α-amylase. We observed theoretical inhibition constant of 22.5 pM in the case of momordicoside A.

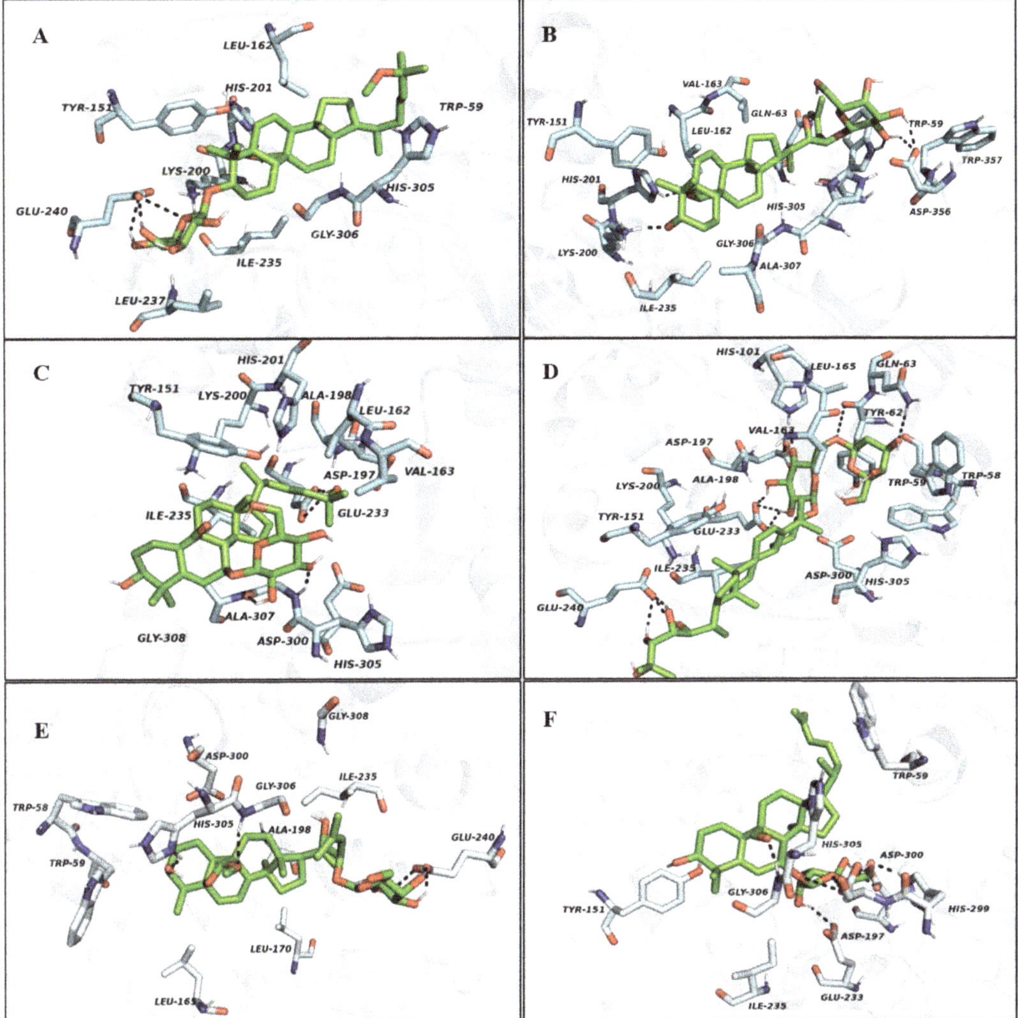


**Figure 5.** The 3D protein-ligand interactions for (**A**) karaviloside VI (**2**), (**B**) karaviloside VIII (**3**), (**C**) momordicoside L (**4**), (**D**) momordicoside A (**5**), (**E**) Charantoside XV (**6**), and (**F**) kuguaglycoside C (**7**) in the binding sites of α-amylase. Black dotted lines indicate hydrogen bonds between compounds and amino acid residues. Ligands in the active sites are denoted in green color. Active site residues are shown in cyans color.

Analysis of molecular docking for the optimized conformation for charantoside XV (**6**) is shown in Figure 5E. The compound was bound to the active site of α-amylase with binding energy −10.64 kcal/mol. charantoside XV interacted with sixteen crucial amino acid residues in domain A of α-amylase Table S1. The compound was able to interact with key amino acid residues, including Glu240, Gly306, by making three conventional hydrogen bonds. Besides, hydrophobic π-alkyl and alkyl interactions of charantoside XV with several amino acid residues, including Trr151, Lys200, Ala307, His201, Leu162, Val163, and Ile235. The inhibition constant of charantoside XV was 15.84 nM. The refined docking of the kuguaglycoside C (**7**) with α-amylase is shown in Figure 5F. The binding energy for compound **7** in the active site of α-amylase is −11.54 kcal/mol. We observed that the compound **7** was able to interact with Asp242, Ser240, Leu246, Asn247, Ser282, Ala281,

Asn302, Glu332, His280, Asp307, Thr310, Ser311, Lys156, Phe314, Leu313, Pro240 in the catalytic site of α-amylase (Figure 5F and Table S2). Also, the interactions calculated for the complex of kuguaglycoside C- α-amylase were effective because kuguaglycoside C (**7**) was buried entirely in the α-amylase binding pocket by forming ten hydrogen bonds with key amino acid residues, including Asp197, Glu233, His305, His299, and Asp300. The 2D figures for all compounds docked are compiled in supporting information (Figure S22).

Overall, the interactions between glucosides on the triterpenes and protein residues were suspected of playing an important role in determining the binding energy among triterpenes bearing the same backbone. Lower binding energy means that the ligand can more easily bind with the protein. Hence, regarding α-amylase molecular docking studies, we observed higher binding energy for momordicoside A and formed ten hydrogen bonds with the active site of α-amylase. Therefore, according to theoretical studies, momordicoside A was more likely to bind with α-amylase. However, in our in vitro studies, we did not observe a significant difference in the inhibitory activity, but we noticed momordicoside L showed comparatively higher inhibition among the six triterpenes. Therefore, our results suggested that in the inhibition process of triterpenes with α-amylase, lower binding energy does not necessarily lead to a higher inhibition activity, i.e., the inhibition activity of triterpene is affected by not only binding energy but also the chemical structure and glucoside type attached to a different position of triterpene. Besides, molecular docking studies are usually carried out under a theoretical vacuum condition, which deviates from real experimental conditions due to a low number of replications and prediction data [41].

#### 2.3.2. Molecular Docking Study with α-Glucosidase

The crystal structure of isomaltase from *Saccharomyces cerevisiae* (PDB ID: 3A4A) was used for the docking study. The 3D crystalline structures of α-glucosidase from *Saccharomyces cerevisiae* (maltase, EC 3.2.1.20) are unavailable in the PDB. However, there are crystalline X-ray structures of isomaltase or α-methylglucosidase have been deposited in the PDB. Previous studies have used isomaltase with a PDB ID 3AJ7 or 3A4A for molecular docking. This structure has a high-resolution X-ray structure, high sequence identity (72.51%) and sequence similarity score (0.54) with the *Saccharomyces cerevisiae* α-glucosidase MAL32 (UniProt entry P38158) [42,43]. Additionally, the authors identified several differences between residues lining the binding pockets of α-glucosidase and isomaltase, such as Phe157/Tyr158, Asp307/Glu204, Asp408/Glu411, Thr215/Val216, Ala278/Gln279, Val303/Thr306 or Ala178/Cys179, respectively. Most of these differences involve very similar residues. Hence we decided to use the crystal structure of isomaltase from *Saccharomyces cerevisiae* (PDB ID: 3A4A) for the molecular docking study with α-glucosidase.

The sequence alignment between α-glucosidase from bakers yeast (GI number 411229) and isomaltase (PDB ID: 3A4A) from *S. cerevisiae* have the structure identity and similarity of 73% and 85%, respectively. The key interactions of acarbose with *S. cerevisiae* α-glucosidase and *S. cerevisiae* isomaltase are quite similar.

The binding energy, interacting residues including H-bond interacting residues and Van der Waals interacting residues, along with the number of H-bonds with the crystal structure of isomaltase from *S. cerevisiae* for compounds **2–7**, are presented in Table S1. Karaviloside VI (**2**) had a binding energy of −10.54 kcal/mol and occupied the active region of isomaltase by interacting with sixteen amino acid residues (Table S1). In the conformation of isomaltase–karaviloside VI complex, the compound was able to establish four hydrogen bonds, including Glu332, Ala281, and Leu313. These hydrogen bonds overtly strengthened the interaction between karaviloside VI and isomaltase. The above interactions resulted in an inhibition constant of 12.48 nM. The 3D schematic interaction of karaviloside VI (**2**) is shown in Figure 6A. Indeed, the docking analysis of some cucurbitane triterpenes from Chinese bitter melon was carried out in our previous study [18,19] showed that compounds were surrounded by residues Glu277, His351, and Asp352, which are part of the catalytic residues of isomaltase.

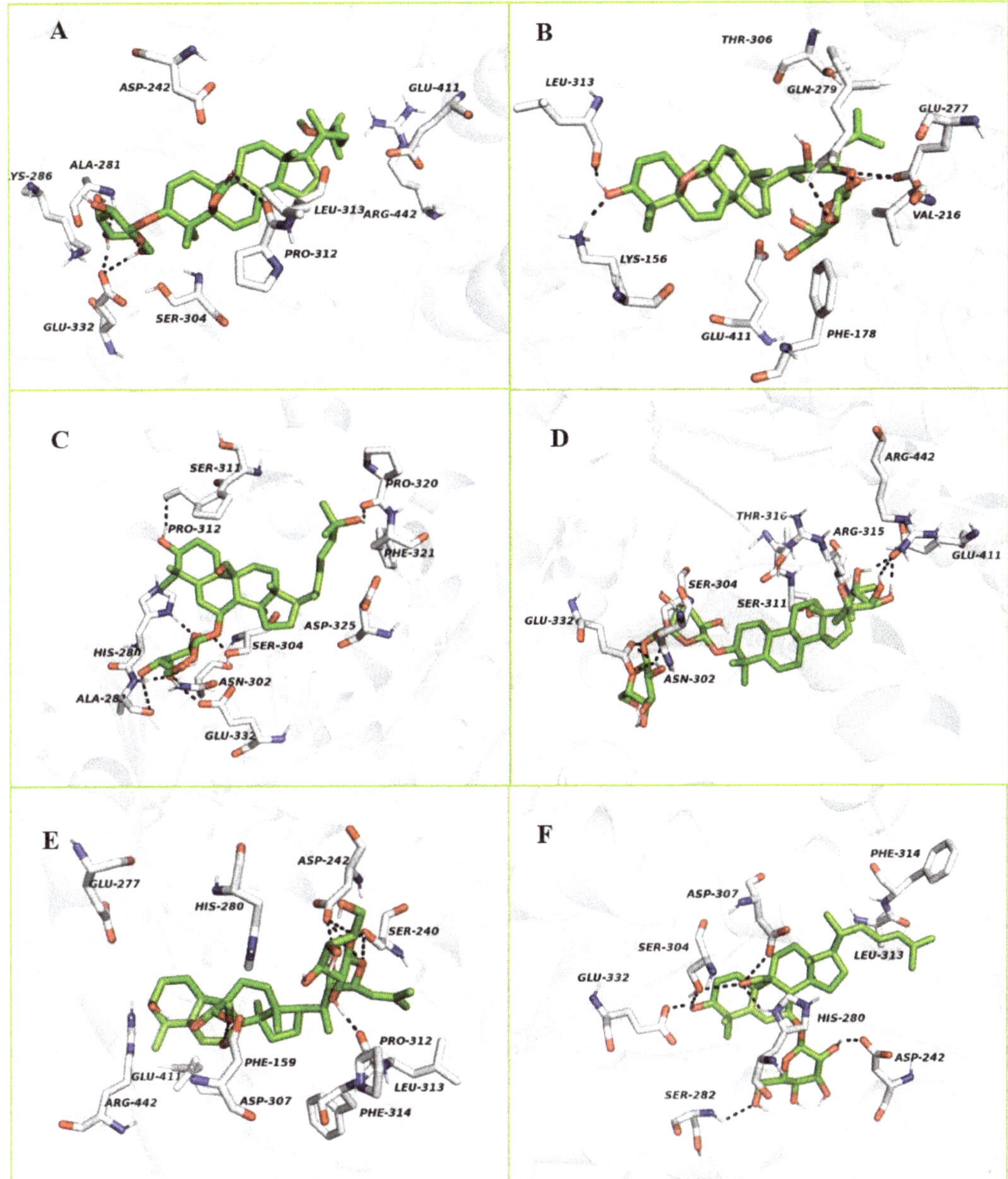


**Figure 6.** The 3D protein-ligand interactions for (**A**) karaviloside VI (**2**), (**B**) karaviloside VIII (**3**), (**C**) momordicoside L (**4**), (**D**) momordicoside A (**5**), (**E**) charantoside XV (**6**), and (**F**) kuguaglycoside C (**7**) in the binding sites of isomaltase. Black dotted lines indicate hydrogen bonds between compounds and amino acid residues. Ligands in the active sites are denoted in green color. Active site residues are shown in white color.

The 3D schematic interaction of karaviloside VIII (**3**) is shown in Figure 6B. Karaviloside VIII was oriented toward the core of the binding pocket and interacted closely with the important seven amino acid residues in the active site (Table S1). According to the Autodock 4.2 simulation results, the isomaltase–karaviloside VIII inhibitor complex showed −10.56 kcal/mol binding energy. Karaviloside VIII made six hydrogen bonds with Glu277 (two bonds), Gln279 (two bonds), Lys156, and Leu313. Moreover, two hydrogen bonds with Glu277 with a bond length of 1.4 and 1.5 Å, which is a key interaction to inhibit enzyme to a greater extent.

Momordicoside L (**4**) was surrounded by fourteen amino acid residues at the catalytic site of isomaltase. The 2D schematic and 3D interactions of momordicoside L and isomaltase are shown in Figure 6C. Momordicoside L generated the best docking pose with a minimum binding energy of −8.28 kcal/mol, and the inhibition constant of 852.19 nM. The ligand was surrounded by catalytic residue Asp307, and the interaction was likely crucial for inhibition of isomaltase. Eight conventional hydrogen bonds were observed between compound **4** and the isomaltase catalytic site residues, including Pro320, Pro312, His280, Ser304, Asn302, Ala281, and Glu232.

Similarly, momordicoside A (**5**) was bound to isomaltase's active site with minimum binding energy −12.48 kcal/mol. Figure 6D clearly shows how momordicoside A interacted with twenty-one crucial amino acid residues (Table S1). Compound **5** was able to establish six hydrogen bonds with Glu411, Ser314, and Glu332. The binding energy and inhibition constant for compound **5** are 706.7 pM and 12.48 kcal/mol, respectively.

The binding mode of charantoside XV (**6**) in the active site of isomaltase was shown in Figure 6E. The ligand was stabilized in an enzyme's active site by interacting with nine amino acid residues (Table S1). The compound was able to form seven hydrogen bonds with Pro312, Ser240 (two hydrogen bonds), Asp242 (three hydrogen bonds), and Asp307. Finally, the binding energy was −10.37 kcal/mol, and the inhibition constant was 25.02 nM.

The refined docking of kuguaglycoside C (**7**) with isomaltase generated the best pose with a minimum binding energy of −8.23 kcal/mol. The 3D schematic in Figure 6F shows that kuguaglycoside C established several hydrogen bonds within the isomaltase enzymatic pocket and interacted with sixteen amino acid residues (Table S1). Three conventional hydrogen bonds were established between kuguaglycoside C and the active pocket of isomaltase, Ser282, Glu332, and Asp242. The 2D docking images of all compounds are shown in the Supporting Information Figures S22 and S23.

## 3. Materials and Methods

### 3.1. Chemicals

D-(+)-Glucose, D-allose, D-galactose, $Ag_2CO_3$, L-cysteine methyl ester hydrochloride, phenyl isothiocyanate, pyridine, *i*PrOH, MTBE, porcine pancreatic α-amylase, acarbose, dinitrosalicylic acid, ACN and MeOH HPLC grade, and *n*-hexane, $CHCl_3$, MeOH, ACN, $(CH_3)_2CO$ and EtOAc technical grade were purchased from Sigma-Aldrich (St. Louis, MO, USA). Nano-pure water HPLC grade (18.2 MΩcm) was obtained from a NANO pure purification system (Barnstead/Thermolyne, Dubuque, IA, USA). Silica gel 60 F254 TLC plates, HCl and $H_3PO_4$ were purchased from EMD Millipore, Inc. (Darmstadt, Germany). Silica gel 60 Å 40-63 μm and AcOH glacial were purchased from VWR International LLC (West Chester, PA, Switzerland), while RP-$C_{18}$ Cosmosil 140 and formic acid were purchased from Nacalai Tesque, Inc., (Kyoto, Japan) and Alfa Aesar (Ward Hill, MA, USA), respectively. MeOH and pyridine NMR (perdeuterated) solvents were purchased from Cambridge Isotope Laboratories (Andover, MA, USA).

### 3.2. General Experimental Procedure

Optical rotations were measured in MeOH, using a Sac-i SACCHARIMETER instrument (ATAGO, Minato-ku, Tokyo, Japan. LC-HR-ESI-MS data was acquired using a maXis impact mass spectrometer (Bruker Daltonics, Billerica, MA, USA) coupled to a 1290 Agilent LC (Agilent, Santa Clara, CA, USA). The analysis of the fractions was performed using an Agilent HPLC 1200 Series (Agilent, Foster City, CA, USA) with a RP $C_{18}$ Gemini column (250 × 4.6 mm; 5 μm) (Phenomenex, Torrence, CA, USA) using an adequate gradient with acetonitrile: water acidified. Chromatographic separations were performed on RP-$C_{18}$

and silica cartridges and RP-$C_{18}$ columns using a CombiFlash Rf flash chromatography (Combiflash Rf, Teledyne ISCO, Lincoln, NE, USA) equipped with a quaternary pump. Waters DeltaPrep preparative HPLC (Waters Delta Prep 4000; Waters, Milford, MA, USA) was used for final purification of certain compounds using a Phenomenex Gemini column (250 × 21.2 mm, 5 μm).

1D ($^1$H; $^{13}$C and DEPT 135) and 2D NMR spectra [heteronuclear multiple-quantum correlation spectroscopy (gHMQC), double quantum filter correlation spectroscopy (DQ-COSY), Total correlation spectroscopy (TOCSY), heteronuclear multiple-bond correlation spectroscopy (gHMBC) and nuclear Overhauser effect spectroscopy (NOESY)] were recorded on a JEOL USA, Inc. spectrometer (Peabody, MA, USA) at 25 °C, operating at 400 MHz ($^1$H) and 100 MHz ($^{13}$C) using standard JEOL pulse programs. All samples were run in methanol-$d_4$, except for **2** and **3**, which, because of insufficient solubility in methanol-$d_4$, were run in pyridine-$d_5$. The chemical shifts are given in $\delta$ (ppm) and were referenced to residual solvent signals.

### 3.3. Plant Material

Fresh fruits of *Momordica charantia* (81 kg) were purchased from BCS Food Market, College Station, TX, USA. Bitter melons were cut into 0.5 inches size, seeds were removed manually, and the remaining pericarp was dried under shade, powdered using a home blender to get 60–80 mesh size, and stored at 25 °C until further use.

### 3.4. Soxhlet Extraction

Bitter melon powder (4.8 kg) was subjected to Soxhlet extraction using various solvents in a sequential manner with increasing polarities [18]. All the extracts were concentrated under vacuum separately and lyophilized to obtain dried powder as follows: *n*-hexane (57.6 g; 0.71%), ethyl acetate (91.8 g; 1.13%); acetone (161.9 g; 2.0%); methanol (131.5 g; 1.62%; methanol: water (1:1; *v*/*v*) (143.7 g; 1.77%) and methanol: water (1:3; *v*/*v*) (197.7 g; 2.44%). The extraction yields were expressed in the weight percentage of fresh fruit.

### 3.5. Purification Procedure

Flow chart of the isolation process of compounds **1**–**7** is presented in supporting information. Acetone extract (160 g) was impregnated with silica gel (~20 g) and loaded on an open column packed with 2 kg of silica gel. The extract was fractionated using step gradient elution using *n*-hexane, *n*-hexane: EtOAc (3:1, 1:1 and 1:3, *v*/*v*), EtOAc, EtOAc: MeOH (19:1, 9:1, 3:1, 1:1 and 1:3, *v*/*v*) to finish with MeOH. A total of 21 main fractions were pooled and dried under vacuum at 40 °C after analysis on silica gel TLC plates and sprayed with 10% $H_2SO_4$ in MeOH [18]. The solids from column fraction 1.11 (317.5 mg) were impregnated onto silica gel and fractionated by flash chromatography on a 40 g pre-packed cartridge and methyl *tert*-butyl ether (MTBE): acetone [$(CH_3)_2CO$] using flash chromatography. Fraction 2.4 (181 mg) was re-chromatographed using a 40 g packed RP-$C_{18}$ column and ran using MeOH and MeOH: isopropanol (*i*PrOH) (9:1, *v*/*v*) to give fraction 3.3 (61 mg), which was precipitated to afford compound **1** (39 mg).

The resulting supernatant from fractions 1.11 (2.17 g) and 1.12 (11.9 g) was combined and impregnated onto silica and submitted to flash chromatography using a 120 g cartridge and a mobile gradient phase consisting of MTBE: $(CH_3)_2CO$. Fraction 4.10 (1.22 g) was submitted to similar chromatography conditions using EtOAc: $(CH_3)_2CO$. Fraction 5.5 (624 mg) was re-chromatographed on RP-$C_{18}$ column (40 g) using MeOH: $H_2O$: AcOH (92:8:0.1; *v*/*v*/%) to MeOH to furnish compound **2** (28 mg) from fraction 6.2.

The supernatant and solids from fractions 1.15 (8.7 g) and 1.16 (9 g) respectively were impregnated with silica gel and subjected to flash chromatography using a 330 g cartridge and gradient eluted with EtOAc: $(CH_3)_2CO$ and $(CH_3)_2CO$: MeOH. Fraction 7.7 (8.8 g) was flash chromatographed using a RP-$C_{18}$ column (40 g) with a gradient consisting of MeOH:$H_2O$:AcOH (5:95:0.1 to 95:5:0.1 *v*/*v*/%). Fraction 8.7 (748 mg) was run in preparative HPLC using RP-$C_{18}$ column and MeOH: $H_2O$: formic acid gradient (5:95:0.1 to 95:5:0.1

*v*/*v*/%). Fraction 9.3 (55 mg) was re-chromatographed in RP-$C_{18}$ preparative HPLC using ACN: $H_2O$: formic acid (55:45: 0.1; *v*/*v*/%) to afford compound **3** (14 mg). Fractions 9.6 + 9.7 14–17 (136 mg) were subjected to RP-$C_{18}$ preparative HPLC using ACN: $H_2O$: formic acid (50:50: 0.1; *v*/*v*/%) to furnish compound **4** (13.9 mg) and compound **5** (31 mg). Fractions 8.5, 8.6, 7.4 and 7.6 were combined and ran using a 120 g RP-$C_{18}$ cartridge and a MeOH: $H_2O$: AcOH gradient. Fractions 11.8 (347 mg) and 11.10 (0.52 g) were individually re-chromatographed on RP-$C_{18}$ preparative HPLC using ACN: $H_2O$: formic acid (45:55:0.1; *v*/*v*/% and 60:40: 0.1) to give compound **6** (14.4 mg) and compound **7** (7.3 mg).

### 3.6. LC-HR-MS Analysis

LC-HR-ESI-MS were acquired using a maXis impact mass spectrometer (Bruker Daltonics, Billerica, MA, USA) coupled to a 1290 Agilent LC (Agilent) using a quadrupole time-of-flight mass detector equipped with an electrospray ionization interface controlled by Bruker software. Column fractions and purified compounds (2 µL) in MeOH were separated on Agilent Eclipse Plus RP-$C_{18}$ (2.1 × 50 mm; 1.8µm) (Agilent) at 40 °C with a flow rate of 0.2 mL/min. The column was eluted with 0.1% formic acid in water (A) and 0.1% formic acid in acetonitrile (B) as follows, 2–95% B for 0–9 min, 95–2%B for 9–11 min, the column was re-equilibrated for 2 min before the next injection. All acquisitions were performed under positive ionization mode with a capillary voltage of 4200 V. Nitrogen was used as nebulizer gas (4.0 bar) and, as well as drying gas at 12 L/min, source temperature was maintained at 250 °C. Full scan mass spectra were acquired from *m*/*z* 50–2000. Data processing was done using Data Analysis Version 4.3.

### 3.7. Determination of the Absolute Configuration of the Sugar Unit

Compounds **6** (1 mg) was hydrolyzed with HCl (1 N) at 80 °C (1 mL) over 2 h, followed by a liquid-liquid partition with EtOAc (2 × 1 mL). The aqueous layer was neutralized with $Ag_2CO_3$, and the supernatant was dried. L-cysteine methyl ester (2 mg) in pyridine (1 mL) was added and heated for 1 h at 60–70 °C. Phenylisothiocyanate (150 µL) was added and heated for an additional hour at 60–70 °C to form the thiocarbamoyl thiazoline derivative. The reaction mixture was analyzed by the HPLC method previously reported [44]. The absolute configuration of the sugar was determined by comparing the HPLC retention times of the prepared thiocarbamoyl thiazolidine derivatives to appropriate standards.

### 3.8. Physicochemical Parameters of Charantoside XV (6)

Amorphous white solid, $[\alpha]^{25}_{D}$—71 (*c* 0.1, MeOH). HRESIMS of compound **1** showed a molecular ion at *m*/*z* 687.3967 $[M + Na]^+$ (calculated 687.4047 *m*/*z* $[M + Na]^+$ for $C_{37}H_{60}O_{10}$). $^1$H- and $^{13}$C-NMR chemical shifts are presented in Table 1.

### 3.9. Bioassays

#### 3.9.1. Cell Culture and Quantitative Real-Time Polymerase Chain Reaction (qRT-PCR)

The anti-inflammatory activity of compounds purified from bitter melon was carried out using RAW 264.7 murine macrophage cells (ATCC, Rockville, MD, USA). The cells were cultured in RPMI 1640 medium supplemented with 10% (*v*/*v*) fetal bovine serum. Additionally, 100 U/mL of penicillin and 100 µg/mL streptomycin were added to the growth medium. Cultured cells were maintained at 37 °C in an incubator with 5% $CO_2$. After 80% cell confluency, spent media was replaced with fresh media. For the analysis of anti-inflammatory gene expression, RAW 264.7 cells were seeded into 6 well plates (5.0 × $10^5$ cells/well). After a 24 h incubation period, the cells were treated with a 50 µM solution of purified compounds for 1 h followed by the addition of LPS (1 µg/mL) to all cells except the control cells. The concentration of compounds used in this assay was determined according to previously published research articles [45]. The cells were incubated for an additional 18 h after which the total RNA was extracted from the cells. Total RNA extraction was carried out using the Aurum Total RNA Mini Kit. The quantity of RNA extracted from the cells was determined using a Nanodrop Spectrophotometer.

(Thermo-Fisher, Waltham, MA, USA) [19]. The purified RNA obtained from cell lysates were used as templates to synthesize cDNA. The synthesis procedure was carried out using the specified manufacture's protocols (iScript cDNA Synthesis Kit, Bio-Rad Inc, Hercules, CA, USA).

Further, real-time PCR was carried out using the manufacturer's specification for the Bio-Rad SYBR Green PCR Master Mix. The relative expression of *IL-6*, *TNF-α*, *COX-2*, and *iNOS* was compared and normalized to the expression of *GAPDH* from the respective treatment groups. Primer sequences for this study are available upon request. The negative control constituted of untreated cells, while the positive control consisted of LPS-stimulated cells.

#### 3.9.2. In Vitro α-Amylase Assay

The inhibition of α-amylase was measured using our previously published protocol using 96 well microplates [18]. Pure compounds (10 μL, 0.43 mM) and 140 μL of 1° saline were added to each well. Subsequently, 45 μL of 1% starch and α-amylase (10 mg/mL) was added. The reaction mixture was then incubated for 40 min at 25 °C followed by the addition of 50 μL of 3, 5-dinitrosalicylic acid (1%) in 20% Rochelle's salt. The reaction mixture was incubated for 1 h at 50 °C, and the absorbance of each well plate was recorded at 540 nm. The absorbance of each reaction mixture was plotted in the analytical curve previously obtained for dextrose (25, 50, 75, 100, 125, 150, 200 μg/mL). The results were expressed as a percentage of inhibition.

#### 3.9.3. In Vitro α-Glucosidase Assay

The α-glucosidase inhibitory activity was evaluated according to a previously published protocol [19]. The assay mixture consisted of 70 μL of 100 mM phosphate buffer (pH 6.8), 10 μL (0.67 mM) of compound dissolved in DMSO, and 20 μL of 1 U/mL α-glucosidase solution were added to 96 well plates in triplicates. The plate was incubated at 37 °C for 15 min, followed by the addition of 20 μL of the p-nitrophenyl α-D-glucopyranoside substrate. The reaction mixture was incubated at 40 °C for 30 min, and then 50 μL of 0.1 M $Na_2CO_3$ solution was added. The absorbance was recorded at 405 nm using a microplate reader, and results were expressed in percentage of inhibition. All experiments were conducted in triplicate.

### *3.10. In Silico Docking Study*

Purified compounds were docked with porcine pancreatic α-amylase (PDB ID: 1OSE) and α-glucosidase crystalline structure (PDB ID:3A4A) of isomaltase from *Saccharomyces cerevisiae*. The complex acarbose with porcine pancreatic α-amylase was removed using Biovia Discovery studio 4.5 software (Dassault Systems BIOVIA, Discovery Studio Modeling Environment, Release 4.5, San Diego, CA, USA, 2015). The methodology for the active site prediction, protein structure, and ligand preparations for molecular docking was provided in detail in our previous study [18]. Compounds were docked using the Lamarckian genetic algorithm in the AutoDock 4.2 Program (ADT, version: 1.5.6). The 2D visualization of ligand-protein interactions was analyzed using DS 4.5 (Dassault Systems), while PyMOL molecular graphics system (PyMOL Molecular Graphics System, San Carlos, CA, USA) was used to better visualize the 3D interactions between ligands and receptors.

### *3.11. Statistical Analysis*

Data from α-amylase, α-glucosidase and anti-inflammatory assays were compared by ANOVA using GraphPad InStat software (San Diego, CA, USA). Differences among means were determined by the least significant difference Tukey's post hoc test, with significance defined at $p < 0.05$ ($n = 3$ to 9).

## 4. Conclusions

The present study reports the purification and identification of a new 23-*O*-β-D-allopyranosyl-5β,19-epoxycucurbitane-6,24-diene triterpene (charantoside XV, **6**) along with five known cucurbitane-type triterpene glycosides and one sterol from crude acetone extract of an Indian cultivar of *Momordica charantia* L. var. *charantia*. Most of the isolated compounds downregulated the expression of pro-inflammatory *IL-6*, *TNF-α*, and *iNOS*, and mitochondrial marker *COX-2*. The most noteworthy activity was found in the downregulation of *IL-6* expression. With these findings, we evidence that the cucurbitane-type triterpene compounds could be responsible for the decreasing of the expression of defined markers. Further studies will incorporate western blot analysis to evaluate the induction of proteins related to inflammation in response to bitter melon triterpenes. Compounds also showed an antidiabetic potential by inhibition of the α-amylase and α-glucosidase carbolytic enzymes. Compounds docked as inhibitors of porcine pancreatic α-amylase complex showed binding energies from −14.53 to −8.94 kcal/mol and inhibition constants from 22.5 pM to 278.19 nM while when docked with α-glucosidase crystalline structure of isomaltase from *S. cerevisiae* binding energies ranged from −12.43 to −8.23 with inhibition constants from 706.7 pM to 929.04 nM.

**Supplementary Materials:** The following are available online.Scheme S1: Flow chart of the isolation process of compounds **1–7**, Figure S1. Chemical structure and LC-HRESIMS of compound **6**, Figure S2. 1H NMR spectrum of compound **6**, Figure S3. 13C NMR spectrum of compound **6**, Figure S4. DEPT-135 spectrum of compound **6**, Figure S5. DQCOSY spectrum of compound **6**, Figure S6. HMQC spectrum of compound **6**, Figure S7. HMBC spectrum of compound **6**, Figure S8. TOCSY spectrum of compound **6**, Figure S9. NOESY spectrum of compound **6**, Figure S10. 1H NMR spectrum of compound **1**, Figure S11. 13C NMR and DEPT 135 spectra of compound **1**, Figure S12. 1H NMR spectrum of compound **2**, Figure S13. 13C NMR and DEPT 135 spectra of compound **2**, Figure S14. 1H NMR spectrum of compound **3**, Figure S15. 13C NMR and DEPT 135 spectra of compound **3**, Figure S16. 1H NMR spectrum of compound **4**, Figure S17. 13C NMR and DEPT 135 spectra of compound **4**, Figure S18. 1H NMR spectrum of compound **5**, Figure S19. 13C NMR and DEPT 135 spectra of compound **5**, Figure S20. 1H NMR spectrum of compound **2**, Figure S21. 13C NMR and DEPT 135 spectra of compound **7**, Figure S22. 2D ligand-protein interactions of purified compounds with the α-amylase enzyme, Karaviloside VI (A), Karaviloside VIII (B), Momordicoside L (C), Momordicoside A (D), Charantoside XIV (E), and Kuguaglycoside C (F). Legends below the figure show the different types of bonding between compounds and the enzymatic pocket of α-amylase, Figure S23. 2D ligand-protein interactions of purified compounds with the isomaltase enzyme, Karaviloside VI (A), Karaviloside VIII (B), Momordicoside L (C), Momordicoside A (D), Charantoside XV (E), and Kuguaglycoside C (F). Legends below the figure show the different types of bonding between compounds and the enzymatic pocket of isomaltase, Table S1: Protein-ligand interactions and docking scores of isolated compounds (**2–7**) in α-amylase and isomaltase obtained using AutoDock 4.2.

**Author Contributions:** W.H.P., S.R.S., G.K.J., and B.S.P. conceived and designed the study. W.H.P. purified and elucidated compound structures; S.R.S., J.L.P., D.M.K., and Y.S. conducted the bioassays. All authors shared analyzing the data and writing the manuscript. All authors have read and agreed to the published version of the manuscript.

**Funding:** This study was supported by the United States Department of Agriculture-NIFA-SCRI-2017-51181-26834 through the National Center of Excellence for Melon at the Vegetable and Fruit Improvement Center of Texas A&M University.

**Institutional Review Board Statement:** Not applicable.

**Informed Consent Statement:** Not applicable.

**Data Availability Statement:** The data presented in this study are available in Supplementary Materials.

**Conflicts of Interest:** The authors declare no conflict of interest.

**Sample Availability:** Samples of compounds **1–7** are not available.

## References


1. Bharathi, L.K.; John, K.J. *Momordica Genus in Asia—An Overview*; Springer: New Delhi, India, 2013.
2. Snee, L.S.; Nerurkar, V.R.; Dooley, D.A.; Efird, J.T.; Shovic, A.C.; Nerurkar, P.V. Strategies to improve palatability and increase consumption intentions for *Momordica charantia* (bitter melon): A vegetable commonly used for diabetes management. *Nutr. J.* **2011**, *10*, 78–88. [CrossRef]
3. Deshaware, S.; Gupta, S.; Singhal, R.S.; Joshi, M.; Variyar, P.S. Debittering of bitter gourd juice using β-cyclodextrin: Mechanism and effect on antidiabetic potential. *Food Chem.* **2018**, *262*, 78–85. [CrossRef]
4. Tan, M.-J.; Ye, J.-M.; Turner, N.; Hohnen-Behrens, C.; Ke, C.-Q.; Tang, C.-P.; Chen, T.; Weiss, H.-C.; Gesing, E.-R.; Rowland, A. Antidiabetic activities of triterpenoids isolated from bitter melon associated with activation of the AMPK pathway. *Chem. Biol.* **2008**, *15*, 263–273. [CrossRef] [PubMed]
5. Nazaruk, J.; Borzym-Kluczyk, M. The role of triterpenes in the management of diabetes mellitus and its complications. *Phytochem. Rev.* **2015**, *14*, 675–690. [CrossRef]
6. Chang, C.-I.; Chen, C.-R.; Liao, Y.-W.; Shih, W.-L.; Cheng, H.-L.; Tzeng, C.-Y.; Li, J.-W.; Kung, M.-T. Octanorcucurbitane triterpenoids protect against tert-butyl hydroperoxide-induced hepatotoxicity from the stems of *Momordica charantia*. *Chem. Pharm. Bull.* **2010**, *58*, 225–229. [CrossRef] [PubMed]
7. Matsuda, H.; Nakamura, S.; Murakami, T. Structures of New Cucurbitane-Type Triterpenes and Glycosides, Karavilagenins D and E, and Karavilosides 6, 7, 8, 9, 10, and 11, from the Fruit of *Momordica charantia*. *Heterocycles* **2007**, *71*, 331–341.
8. Tuan, N.Q.; Lee, D.-H.; Oh, J.; Kim, C.S.; Heo, K.-S.; Myung, C.-S.; Na, M. Inhibition of proliferation of vascular smooth muscle cells by Cucurbitanes from *Momordica charantia*. *J. Nat. Prod.* **2017**, *80*, 2018–2025. [CrossRef]
9. Yue, J.; Xu, J.; Cao, J.; Zhang, X.; Zhao, Y. Cucurbitane triterpenoids from *Momordica charantia* L. and their inhibitory activity against α-glucosidase, α-amylase and protein tyrosine phosphatase 1B (PTP1B). *J. Funct. Foods* **2017**, *37*, 624–631. [CrossRef]
10. Akihisa, T.; Higo, N.; Tokuda, H.; Ukiya, M.; Akazawa, H.; Tochigi, Y.; Kimura, Y.; Suzuki, T.; Nishino, H. Cucurbitane-type triterpenoids from the fruits of Momordica charantia and their cancer chemopreventive effects. *J. Nat. Prod.* **2007**, *70*, 1233–1239. [PubMed]
11. Hasani-Ranjbar, S.; Nayebi, N.; Larijani, B.; Abdollahi, M. A systematic review of the efficacy and safety of herbal medicines used in the treatment of obesity. *World J. Gastroenterol.* **2009**, *15*, 3073–3085. [CrossRef] [PubMed]
12. Leung, L.; Birtwhistle, R.; Kotecha, J.; Hannah, S.; Cuthbertson, S. Anti-diabetic and hypoglycaemic effects of Momordica charantia (bitter melon): A mini review. *Br. J. Nutr.* **2009**, *102*, 1703–1708. [CrossRef]
13. Chen, J.-C.; Bik-San Lau, C.; Chan, J.Y.-W.; Fung, K.-P.; Leung, P.-C.; Liu, J.-Q.; Zhou, L.; Xie, M.-J.; Qiu, M.-H. The antigluconeogenic activity of cucurbitacins from *Momordica charantia*. *Planta Med.* **2015**, *81*, 327–332. [CrossRef]
14. Bao, B.; Chen, Y.-G.; Zhang, L.; Xu, Y.L.N.; Wang, X.; Liu, J.; Qu, W. *Momordica charantia* (Bitter Melon) reduces obesity-associated macrophage and mast cell infiltration as well as inflammatory cytokine expression in adipose tissues. *PLoS ONE* **2013**, *8*, e84075. [CrossRef]
15. Wang, X.; Bao, W.; Liu, J.; OuYang, Y.-Y.; Wang, D.; Rong, S.; Xiao, X.; Shan, Z.-L.; Zhang, Y.; Yao, P. Inflammatory markers and risk of type 2 diabetes: A systematic review and meta-analysis. *Diabetes Care* **2013**, *36*, 166–175.
16. Persaud, S.J.; Burns, C.J.; Belin, V.D.; Jones, P.M. Glucose-induced regulation of COX-2 expression in human islets of Langerhans. *Diabetes* **2004**, *53*, S190–S192. [CrossRef] [PubMed]
17. Abascal, K.; Yarnell, E. Using bitter melon to treat diabetes. *Altern. Complement. Ther.* **2005**, *11*, 179–184. [CrossRef]
18. Shivanagoudra, S.R.; Perera, W.H.; Perez, J.L.; Athrey, G.; Sun, Y.; Jayaprakasha, G.; Patil, B.S. Cucurbitane-type compounds from *Momordica charantia*: Isolation, in vitro antidiabetic, anti-inflammatory activities and in silico modeling approaches. *Bioorg. Chem.* **2019**, *87*, 31–42. [CrossRef] [PubMed]
19. Shivanagoudra, S.R.; Perera, W.H.; Perez, J.L.; Athrey, G.; Sun, Y.; Wu, C.S.; Jayaprakasha, G.; Patil, B.S. In vitro and in silico elucidation of antidiabetic and anti-inflammatory activities of bioactive compounds from *Momordica charantia* L. *Bioorg. Med. Chem.* **2019**, *27*, 3097–3109. [CrossRef]
20. Perez, J.L.; Jayaprakasha, G.; Patil, B.S. Separation and Identification of Cucurbitane-Type Triterpenoids from Bitter Melon. In *Instrumental Methods for the Analysis and Identification of Bioactive Molecules*; ACS Symposium Series 1185; American Chemical Society: Washington, DC, USA, 2014; pp. 51–78.
21. Liaw, C.-C.; Huang, H.-C.; Hsiao, P.-C.; Zhang, L.-J.; Lin, Z.-H.; Hwang, S.-Y.; Hsu, F.-L.; Kuo, Y.-H. 5β,19-epoxycucurbitane triterpenoids from *Momordica charantia* and their anti-inflammatory and cytotoxic activity. *Planta Med.* **2015**, *81*, 62–70. [CrossRef] [PubMed]
22. Nakamura, S.; Murakami, T.; Nakamura, J.; Kobayashi, H.; Matsuda, H.; Yoshikawa, M. Structures of new cucurbitane-type triterpenes and glycosides, karavilagenins and karavilosides, from the dried fruit of Momordica charantia L. in Sri Lanka. *Chem. Pharm. Bull.* **2006**, *54*, 1545–1550. [CrossRef] [PubMed]
23. Li, Q.Y.; Chen, H.B.; Liu, Z.M.; Wang, B.; Zhao, Y.Y. Cucurbitane triterpenoids from *Momordica charantia*. *Magn. Reson. Chem.* **2007**, *45*, 451–456. [CrossRef] [PubMed]
24. Liu, J.-Q.; Chen, J.-C.; Wang, C.-F.; Qiu, M.-H. New cucurbitane triterpenoids and steroidal glycoside from *Momordica charantia*. *Molecules* **2009**, *14*, 4804–4813. [CrossRef] [PubMed]
25. Peng, L.; Jian-Feng, L.; Li-Ping, K.; He-Shui, Y.; Zhang, L.-J.; Xin-Bo, S.; Bai-Ping, M. A new C30 sterol glycoside from the fresh fruits of *Momordica charantia*. *Chin. J. Nat. Med.* **2012**, *2*, 88–91.

26. Nhiem, N.X.; Kiem, P.V.; Minh, C.V.; Ban, N.K.; Cuong, N.X.; Ha, L.M.; Tai, B.H.; Quang, T.H.; Tung, N.H.; Kim, Y.H. Cucurbitane—type triterpene glycosides from the fruits of Momordica charantia. *Magn. Reson. Chem.* **2010**, *48*, 392–396.
27. Okabe, H.; Miyahara, Y.; Yamauchi, T.; Miyahara, K.; Kawasaki, T. Studies on the constituents of *Momordica charantia* LI Isolation and characterization of momordicosides A and B, glycosides of a pentahydroxy-cucurbitane triterpene. *Chem. Pharm. Bull.* **1980**, *28*, 2753–2762. [CrossRef]
28. Chen, J.C.; Lu, L.; Zhang, X.M.; Zhou, L.; Li, Z.R.; Qiu, M.H. Eight New Cucurbitane Glycosides, Kuguaglycosides A–H, from the Root of *Momordica charantia* L. *Helv. Chim. Acta* **2008**, *91*, 920–929. [CrossRef]
29. Tabata, K.; Hamano, A.; Akihisa, T.; Suzuki, T. Kuguaglycoside C, a constituent of *Momordica charantia*, induces caspase—independent cell death of neuroblastoma cells. *Cancer Sci.* **2012**, *103*, 2153–2158.
30. Hsiao, P.-C.; Liaw, C.-C.; Hwang, S.-Y.; Cheng, H.-L.; Zhang, L.-J.; Shen, C.-C.; Hsu, F.-L.; Kuo, Y.-H. Antiproliferative and hypoglycemic cucurbitane-type glycosides from the fruits of *Momordica charantia*. *J. Agric. Food Chem.* **2013**, *61*, 2979–2986. [CrossRef]
31. Kobori, M.; Nakayama, H.; Fukushima, K.; Ohnishi-Kameyama, M.; Ono, H.; Fukushima, T.; Akimoto, Y.; Masumoto, S.; Yukizaki, C.; Hoshi, Y. Bitter gourd suppresses lipopolysaccharide-induced inflammatory responses. *J. Agric. Food Chem.* **2008**, *56*, 4004–4011. [CrossRef]
32. Lii, C.-K.; Chen, H.-W.; Yun, W.-T.; Liu, K.-L. Suppressive effects of wild bitter gourd (Momordica charantia Linn. var. abbreviata ser.) fruit extracts on inflammatory responses in RAW 264.7 macrophages. *J. Ethnopharmacol.* **2009**, *122*, 227–233. [CrossRef]
33. Svobodova, B.; Barros, L.; Calhelha, R.C.; Heleno, S.; Alves, M.J.; Walcott, S.; Bittova, M.; Kuban, V.; Ferreira, I.C. Bioactive properties and phenolic profile of *Momordica charantia* L. medicinal plant growing wild in Trinidad and Tobago. *Ind. Crops Prod.* **2017**, *95*, 365–373. [CrossRef]
34. Hsu, C.-L.; Fang, S.-C.; Liu, C.-W.; Chen, Y.-F. Inhibitory effects of new varieties of bitter melon on lipopolysaccharide-stimulated inflammatory response in RAW 264.7 cells. *J. Funct. Foods* **2013**, *5*, 1829–1837.
35. Chaudhury, A.; Duvoor, C.; Reddy Dendi, V.S.; Kraleti, S.; Chada, A.; Ravilla, R.; Marco, A.; Shekhawat, N.S.; Montales, M.T.; Kuriakose, K. Clinical review of antidiabetic drugs: Implications for type 2 diabetes mellitus management. *Front. Endocrinol.* **2017**, *8*, 6. [CrossRef]
36. Alhadramy, M.S. Diabetes and oral therapies: A review of oral therapies for diabetes mellitus. *J. Taibah Univ. Med. Sci.* **2016**, *11*, 317–329.
37. Ceriello, A.; Davidson, J.; Hanefeld, M.; Leiter, L.; Monnier, L.; Owens, D.; Tajima, N.; Tuomilehto, J. Postprandial hyperglycaemia and cardiovascular complications of diabetes: An update. *Nutr. Metab. Cardiovasc. Dis.* **2006**, *16*, 453–456. [CrossRef] [PubMed]
38. Sui, X.; Zhang, Y.; Zhou, W. In vitro and in silico studies of the inhibition activity of anthocyanins against porcine pancreatic α-amylase. *J. Funct. Foods* **2016**, *21*, 50–57. [CrossRef]
39. Qian, M.; Spinelli, S.; Payan, F.; Driguez, H. Structure of a pancreatic α-amylase bound to a substrate analogue at 2.03 Å resolution. *Protein Sci.* **1997**, *6*, 2285–2296.
40. Lo Piparo, E.; Scheib, H.; Frei, N.; Williamson, G.; Grigorov, M.; Chou, C.J. Flavonoids for controlling starch digestion: Structural requirements for inhibiting human α-amylase. *J. Med. Chem.* **2008**, *51*, 3555–3561. [CrossRef]
41. Wang, Y.; Zhang, G.; Yan, J.; Gong, D. Inhibitory effect of morin on tyrosinase: Insights from spectroscopic and molecular docking studies. *Food Chem.* **2014**, *163*, 226–233. [CrossRef]
42. Yamamoto, K.; Miyake, H.; Kusunoki, M.; Osaki, S. Crystal structures of isomaltase from Saccharomyces cerevisiae and in complex with its competitive inhibitor maltose. *FEBS J.* **2010**, *277*, 4205–4214. [CrossRef]
43. Proença, C.; Freitas, M.; Ribeiro, D.; Oliveira, E.F.; Sousa, J.L.; Tomé, S.M.; Ramos, M.J.; Silva, A.M.; Fernandes, P.A.; Fernandes, E. α-Glucosidase inhibition by flavonoids: An in vitro and in silico structure–activity relationship study. *J. Enzyme Inhib. Med. Chem.* **2017**, *32*, 1216–1228. [CrossRef] [PubMed]
44. Perera, W.H.; Ghiviriga, I.; Rodenburg, D.L.; Alves, K.; Bowling, J.J.; Avula, B.; Khan, I.A.; McChesney, J.D. Rebaudiosides T and U, minor C-19 xylopyranosyl and arabinopyranosyl steviol glycoside derivatives from *Stevia rebaudiana* (Bertoni) Bertoni. *Phytochem.* **2017**, *135*, 106–114. [CrossRef] [PubMed]
45. Dirsch, V.M.; Kiemer, A.K.; Wagner, H.; Vollmar, A.M. Effect of allicin and ajoene, two compounds of garlic, on inducible nitric oxide synthase. *Atherosclerosis* **1998**, *139*, 333–339. [CrossRef]

*Article*

# Development of UHPLC/Q-TOF Analysis Method to Screen Glycerin for Direct Detection of Process Contaminants 3-Monochloropropane-1,2-diol Esters (3-MCPDEs) and Glycidyl Esters (GEs)

**Lauren Girard [1], Kithsiri Herath [1], Hernando Escobar [1], Renate Reimschuessel [2], Olgica Ceric [2] and Hiranthi Jayasuriya [1,*]**

[1] Center for Veterinary Medicine, Office of Research, Division of Residue Chemistry, Food and Drug Administration, 8401 Muirkirk Road, Laurel, MD 20708, USA; Lauren.Girard@fda.hhs.gov (L.G.); Kithsiri.Herath@fda.hhs.gov (K.H.); Hernando.EscobarLoaiza@fda.hhs.gov (H.E.)

[2] Center for Veterinary Medicine, Office of Research, Veterinary Laboratory Investigation & Response Network, Food and Drug Administration, 8401 Muirkirk Road, Laurel, MD 20708, USA; Renate.Reimschuessel@fda.hhs.gov (R.R.); Olgica.Ceric@fda.hhs.gov (O.C.)

* Correspondence: Hiranthi.Jayasuriya@fda.hhs.gov; Tel.: +1-240-402-6688

**Abstract:** The U.S. Food and Drug Administration's (FDA's) Center for Veterinary Medicine (CVM) has been investigating reports of pets becoming ill after consuming jerky pet treats since 2007. Renal failure accounted for 30% of reported cases. Jerky pet treats contain glycerin, which can be made from vegetable oil or as a byproduct of biodiesel production. Glycidyl esters (GEs) and 3-monochloropropanediol esters (3-MCPDEs) are food contaminants that can form in glycerin during the refining process. 3-MCPDEs and GEs pose food safety concerns, as they can release free 3-MCPD and glycidol in vivo. Evidence from studies in animals shows that 3-MCPDEs are potential toxins with kidneys as their main target. As renal failure accounted for 30% of reported pet illnesses after the consumption of jerky pet treats containing glycerin, there is a need to develop a screening method to detect 3-MCPDEs and GEs in glycerin. We describe the development of an ultra-high-pressure liquid chromatography/quadrupole time-of-flight (UHPLC/Q-TOF) method for screening glycerin for MCPDEs and GEs. Glycerin was extracted and directly analyzed without a solid-phase extraction procedure. An exact mass database, developed in-house, of MCPDEs and GEs formed with common fatty acids was used in the screening.

**Keywords:** UHPLC/Q-TOF-MS analysis; glycerin; process contaminants; 3-monochloropropane-1,2-diol esters; glycidyl esters

**Citation:** Girard, L.; Herath, K.; Escobar, H.; Reimschuessel, R.; Ceric, O.; Jayasuriya, H. Development of UHPLC/Q-TOF Analysis Method to Screen Glycerin for Direct Detection of Process Contaminants 3-Monochloropropane-1,2-diol Esters (3-MCPDEs) and Glycidyl Esters (GEs). *Molecules* **2021**, *26*, 2449. https://doi.org/10.3390/molecules26092449

Academic Editors: Muhammad Ilias and Dhammika Nanayakkara

Received: 22 February 2021
Accepted: 9 April 2021
Published: 22 April 2021



## 1. Introduction

In the late summer of 2007, the FDA became aware of reports of illness in dogs after consuming jerky pet treats (JPTs) [1,2]. For more than 10 years, the FDA dedicated extensive resources to the JPT investigation, including reviewing and investigating consumer complaint cases and collecting and testing JPT products (including products consumed by affected dogs) and animal diagnostic samples. Most of the cases (approximately 60%) involved gastrointestinal symptoms. Renal dysfunction or failure accounted for 30% of the reported cases, with a smaller subset reporting Fanconi syndrome, a dysfunction of the proximal renal tubules of the kidney in which glucose, amino acids, uric acid, phosphate, and bicarbonate are passed into the urine instead of being reabsorbed [3]. Fanconi syndrome is linked to exposure to certain toxins, medications, and infections in dogs as well as in people [4]. Exposure to toxins and drugs, including ethylene glycol (antifreeze), metals, chemotherapeutics, and expired tetracyclines and other antibiotics can damage the proximal renal tubule of the kidney [3,5]. Hooper commented that proximal renal tubular cells are exposed to the highest concentrations of renal-damaging substances, more so than distal tubular cells [6]. Damaged or dysfunctional proximal tubule cells can become "leaky", allowing glucose, bicarbonate, amino acids, and other substances to cross into the urine by hindering normal resorptive functions.

JPTs are often made from dried chicken (or duck) meat or sweet potatoes. As of 1 September 2018, FDA's Veterinary Laboratory Investigation and Response Network (Vet-LIRN) had collected and performed testing on more than 650 JPT samples related to more than 550 consumer complaints, as well as more than 400 retail samples of unopened product bags obtained from a store or shipment. Vet-LIRN did not subject every sample to the entire battery of testing due to limited resources and product availability. Sample testing was targeted based on the concerns with a particular product or brand and the symptoms displayed by the pet that consumed the product. The product-based testing plans targeted the main ingredients (chicken, duck, or sweet potatoes) and considered other information on the product labels, such as additional ingredients and/or their contaminants. For example, Vet-LIRN investigation revealed that some products contained glycerin (as high as 20%), although it was not listed on the label, and conducted additional testing for glycerin to follow up on this finding [2]. Glycerin is often added to JPTs to prevent excessive drying and improve the product's texture.

Glycerin is also used to produce many other products including drugs, cosmetics, foods, tobacco products, sweeteners, and toiletries. As glycerin may be produced during biodiesel production and is subjected to a refining process, it is important to ensure that there are methods to detect potential toxic contaminants that could arise during manufacture.

Recently we published a UHPLC/Q-TOF method to screen crude glycerin for toxic phorbol ester contaminants [7]. Another group of compounds that could be contaminants in glycerin comprises 3-monochloropropane-1,2-diol (MCPD), its esters (MCPDEs), and glycidyl esters (GEs), which carry a reactive epoxide moiety. These compounds are formed during the industrial processing of oils in the presence of chloride ions, glycerin, and high temperatures [8]. In 2015, Vet-LIRN tested 74 JPT samples for MCPD and found that 31 samples tested positive at concentrations ranging from 0.027 to 0.352 ppm. The significance of this finding was not clear, although the review of the available scientific literature did not suggest that these MCPD levels in JPTs would cause illness in pets.

Free MCPDs are present in the low mg/kg range in many foodstuffs such as acid-hydrolyzed vegetable protein, soy sauces, crackers, bread, toast and other bakery products, malt, meat products, and soups [9–13]. In most foodstuffs, only a small percentage of 3-MCPD is present as free 3-MCPD, while the majority is present as MCPDEs. Based on positive findings of MCPD in JPTs, Vet-LIRN investigation showed the need to evaluate JPTs and glycerin for their presence of MCPDEs; however, this type of testing is not available domestically.

An extensive review about 3-MCPDE toxicity was published by the Joint FAO/WHO Expert Committee on Food Additives (JEFCA) in 2016 [14]. In 2018, the same regulatory body established a group total daily intake (TDI) of 2 μg/kg bw/day for 3-MCPD and its esters [15]. 3-MCPDEs are potentially toxic compounds with kidneys as their main target, according to reports from studies in animals. In Wistar rats or Swiss mice dosed with different MCPDEs, nephrotoxic degenerative and inflammatory changes in kidneys were evidenced by organ weight increase, glomerular lesions, tubulotoxicity (necrosis, tubular epithelial hyperplasia, and multifocal hypertrophy), cellular infiltration in interstitial spaces, and fibrosis [16,17]. Rat bioavailability studies show that 3-MCPDEs are completely hydrolyzed by enzymatic reactions in the gastrointestinal tract with release of 3-MCPD to be distributed to blood, organs, and tissues [18,19]. From short-term studies in rats and mice, the most affected organ after 3-MCPD induced toxicity was the kidney, with similar organ degenerative changes found for parent 3-MCPD esters [20]. A more recent report describes induced acute renal failure with nephrotoxic structural changes after oral administration of 3-MCPD in male albino rats in a period as short as 7 days [21].

Potential severe toxicological effects induced in animals with the presence of 3-MCPDEs have raised concerns about the level of exposure to these food contaminants. Strong evidence shows that glycidol and GEs are potential genotoxic carcinogen agents and probably carcinogenic to humans [22]. Studies on bioavailability have shown that GEs are hydrolyzed (de-esterified) during digestion, and the free glycidol is almost completely released [14]. The GEs are therefore treated like glycidol from a toxicological point of view. Due to the genotoxic potential of glycidol, it is not possible to derive any safe intake quantities for GEs.

Due to the potential renal toxicity of MCPDEs, which are derived from MCPD, and the carcinogenicity of glycidol derived from GEs, we focused our method development efforts to screen glycerin for MCPDEs and GEs.

## 2. Results and Discussion

Glycerin is the major byproduct of biodiesel production. Centrifugation or gravitational settling is used to separate biodiesel from glycerin after transesterification [23,24]. The lower glycerin phase consists of glycerol and other impurities such as methyl esters; water; alcohol; salts; and unreacted mono-, di-, and triglycerides. The crude glycerin is subjected to an expensive refining process that includes treatments such as bleaching, deodorizing, and ion exchange. Contaminants such as free fatty acids and their salts are removed to meet the USP standards to be used in food production. Some steps such as the deodorization step are carried out at elevated temperatures of more than 200 °C.

The main factors for the formation of MCPDEs and GEs are the presence of chloride ions; glycerin; and tri-, di-, or monoacylglycerides and high temperatures. As glycerin is subjected to these conditions during production and refining, it is possible that it is tainted with such process contaminants.

Many edible oils also undergo a similar refining process to improve their quality. Evidence exists in the scientific literature to show the formation of MCPDEs and GEs during the industrial processing of oils [8,25–27]. These contaminants may be harmful to health and, therefore, undesirable in foods. There are many published studies to screen for these ester contaminants in edible oils [28–30], but to our knowledge, there are no methods published yet for glycerin.

The determination of MCPDEs is complicated due to structural diversity. Considering the possible positional isomers of MCPD, the formation of about 100 different ester compounds is possible. However, because of the relative abundance of the fatty acids, only seven esters (lauric, myristic, palmitic, stearic, oleic, linoleic, and linolenic acids) are considered in food analysis [13]. For GEs, the number of ester structures is smaller, caused by its single hydroxyl group and lack of positional isomers.

LC method development was challenging due to the polarity range of the compounds. We tried to use ACN/water gradients spanning 5–100% ACN in 15 min with formic acid in both solvents. The GEs and the mono-MCPDEs, which are polar, eluted from the column, but the nonpolar di-MCPDEs did not elute from the column with this solvent gradient. We also tried other solvent gradients reported in the literature such as ACN/ MeOH/water as solvent A and acetone as solvent B ramping up to 60% acetone [31]. Even though the nonpolar di-MCPDEs eluted from the column with this solvent gradient, the peak shapes were broad. The IPA/8% aqueous methanol with formic acid step gradient used in the LC is a modification of a method reported in the literature for the analysis of GEs and mono- and di-MCPDEs as contaminants in edible oils [28]. LC parameters including mobile phase, column type, column temperature, flow rate, and gradient conditions were varied to establish the optimal liquid chromatography. The optimized step gradient allowed us to retain all three classes of compounds of differing polarities on the column as well as elute the most nonpolar di-MCPDEs within the gradient.

Even though we used six representative compounds of GEs and MCPDEs in our mixed standard, we analyzed many others in our study. The most polar 1-lauroyl-3-chloropropanediol and the most nonpolar 1,2-distearoyl-3-chloropropanediol eluted within our gradient. The LC retention times ranged from 2.3 to 10.3 min for the six compounds in our mixed standard using this gradient (Figure 1). Average responses in Table 1 indicated that the sodium adducts were much more dominant than the protonated adducts for all six compounds. This is true and consistent for all GEs and MCPDEs we analyzed during this study. This was an important observation that allowed us to screen for lower levels of MCPDEs by detecting unique ion clusters containing chlorine with sodium adducts.

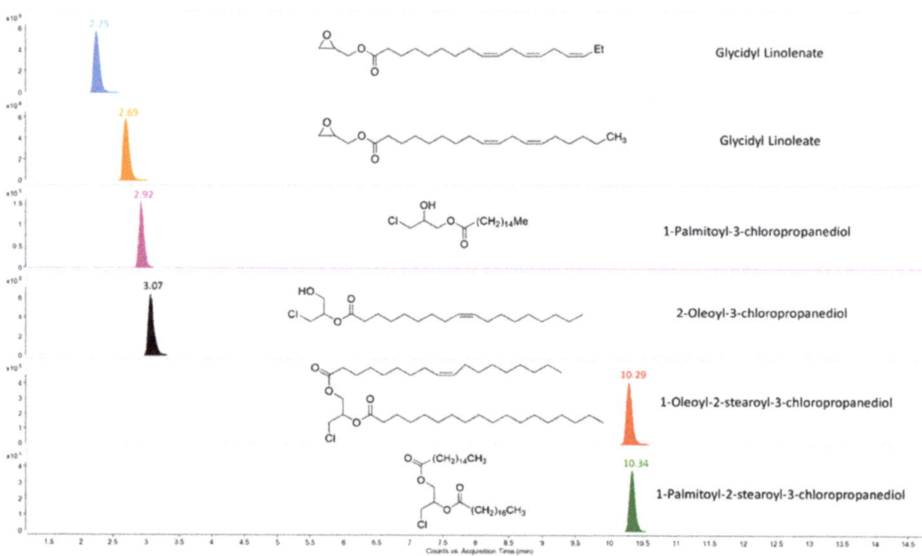


**Figure 1.** Extracted ion chromatograms for the mixed standard of glycidyl esters (GEs) and 3-monochloropropanediol esters (MCPDEs) in the glycerin matrix.

**Table 1.** Average response for protonated and sodium adducts of GEs and MCPDEs in the mixed standard.

| Compound | Exact Mass | (M+H) | (M+Na) |
|---|---|---|---|
| Glycidyl linolenate | 334.2508 | 2,209,671 | 34,331,876 |
| Glycidyl linoleate | 336.2664 | 3,308,840 | 45,741,444 |
| 1-Palmitoyl-3-chloropropanediol | 348.2431 | 0 | 1,517,938 |
| 2-Oleoyl-3-chloropropanediol | 374.2588 | 116,159 | 6,265,336 |
| 1-Oleoyl-2-stearoyl-3-chloropropanediol | 640.5197 | 0 | 7,484,321 |
| 1-Palmitoyl-2-stearoyl-3-chloropropanediol | 614.5041 | 0 | 8,217,162 |

### 2.1. Calibration Curves

Calibration curves were generated for selected GEs and mono- and di-MCPDEs in both solvents, methanol and glycerin-matrix (Figure 2). All calibration standards were analyzed in triplicate over the concentration range of 10 to 400 ppb for MCPDEs and 2 to 100 ppb for GEs. A quadratic regression algorithm with no weighting was used to compare the standard curves in solvent versus matrix. The average correlation coefficients for all calibration curves were >0.997. We observed matrix enhancement for di- and mono-MCPD esters; however, glycidyl esters showed a matrix suppression effect.

The calibration curves constructed for the mixed standard were used to determine the matrix effect. The calibration standards were used to validate the data analysis process.

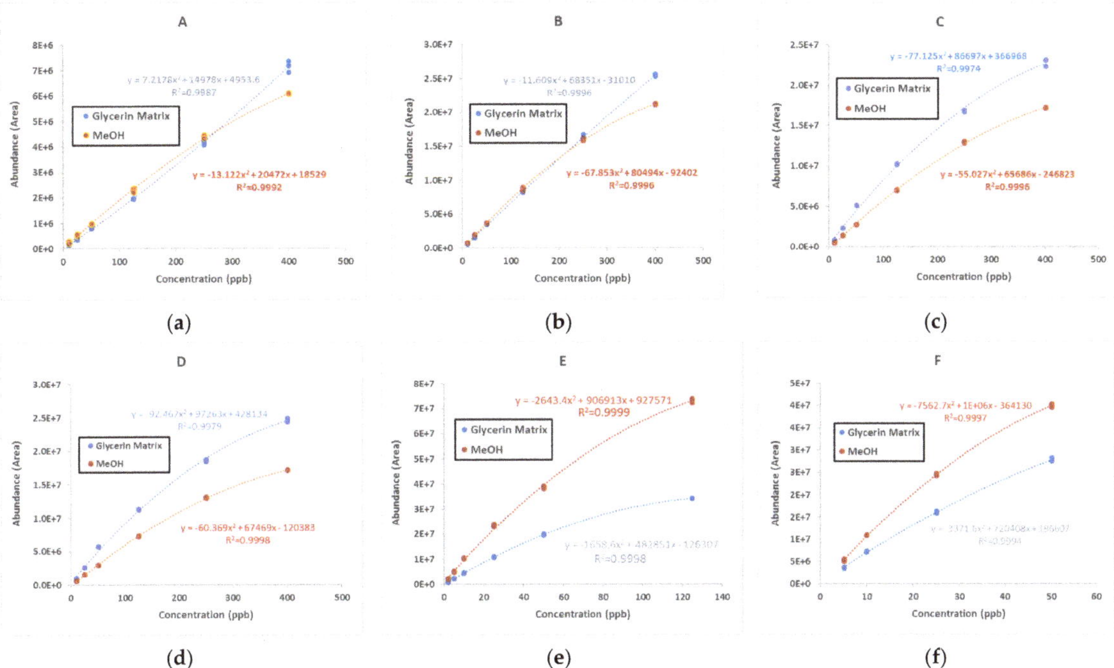


**Figure 2.** Calibration curves prepared in glycerin matrix and methanol for (**a**) palmitoyl-chloropropanediol, (**b**) oleyl-chloropropanediol, (**c**) 1-oleoyl-2-stearoyl-2-chloropropanediol, (**d**) 1-palmitoyl-2-stearoyl-3-chloropropanediol, (**e**) glycidyl linoleate, and (**f**) glycidyl linolenate.

Data analysis was performed with the Mass Hunter Qualitative Analysis software (B.06.00) using FbF (find by formula) algorithm to screen for compounds in the glycerin extracts using an in-house MCPDE and GE exact mass database as the formula source (Table 2). The in-house database was constructed by adding the exact masses of all possible esters of GEs and mono- and di-MCPDEs of lauric, myristic, palmitic, linolenic, linoleic, oleic, and stearic acids. The algorithm FbF was restricted to the *m/z* range of 50–1000 Da and completed within the timespan of the LC gradient. Singly charged ions and a minimum peak height of 10,000 ion counts were selected for FbF. The allowed adduct ions included $H^+$ and $Na^+$. FbF algorithm scores the database matches based on the similarity of each of the isotopic masses (Mass Match), isotope ratios (Abundance Match), isotope spacing (Spacing Match), and optionally, the retention time (RT Match).

**Table 2.** Database of MCPDEs and GEs.

| Formula | Exact Mass | Compound Name |
|---|---|---|
| **MCPD monoesters** | | |
| C17H33ClO3 | 320.2118 | Myristoyl-chloropropanediol |
| C15H29ClO3 | 292.1805 | Lauroyl-chloropropanediol |
| C19H37ClO3 | 348.2431 | Palmitoyl-chloropropanediol |
| C21H41ClO3 | 376.2744 | Stearoyl-chloropropanediol |
| C21H39ClO3 | 374.2588 | Oleyl-chloropropanediol |
| C21H37ClO3 | 372.2431 | Linoleoyl-chloropropanediol |
| C21H35ClO3 | 370.2275 | Linolenoyl-chloropropanediol |

**Table 2.** *Cont.*

| Formula | Exact Mass | Compound Name |
|---|---|---|
| **MCPD diesters** | | |
| C35H67ClO4 | 586.4728 | Di-palmitoyl-chloropropanediol |
| C27H51ClO4 | 474.3476 | Di-lauroyl-chloropropanediol |
| C29H55ClO4 | 502.3789 | Lauroyl-myristoyl-chloropropanediol |
| C33H57ClO4 | 552.3945 | Lauroyl-linolenoyl-chloropropanediol |
| C33H59ClO4 | 554.4102 | Lauroyl-linoleoyl-chloropropanediol |
| C39H67ClO4 | 634.4728 | Di-linoleoyl-chloropropanediol |
| C39H63ClO4 | 630.4415 | Di-linolenoyl-chloropropanediol |
| C35H61ClO4 | 580.4285 | Linolenoyl-myristoyl-chloropropanediol |
| C31H59ClO4 | 530.4102 | Di-myristoyl-chloropropanediol |
| C31H59ClO4 | 530.4102 | Lauroyl-Palmitoyl-chloropropanediol |
| C33H61ClO4 | 556.4258 | Lauroyl-Oleoyl-chloropropanediol |
| C39H65ClO4 | 632.4571 | Linoleoyl-Linolenoyl-chloropropanediol |
| C35H63ClO4 | 582.4415 | Myristoyl-Linoleoyl-chloropropanediol |
| C33H63ClO4 | 558.4415 | Myristoyl-palmitoyl-chloropropanediol |
| C33H63ClO4 | 558.4415 | Lauroyl-stearoyl-chloropropanediol |
| C35H65ClO4 | 584.4571 | Myristoyl-oleoyl-chloropropanediol |
| C39H71ClO4 | 638.5041 | Stearoyl-linoleoyl-chloropropanediol |
| C39H73ClO4 | 640.5197 | Oleoyl-stearoyl-chloropropanediol |
| C39H71ClO4 | 638.5041 | Linoleoyl-stearoyl-chloropropanediol |
| C37H71ClO4 | 614.5041 | Palmitoyl-stearoyl-chloropropanediol |
| C37H67ClO4 | 610.4728 | Palmitoyl-linoleoyl-chloropropanediol |
| C37H65ClO4 | 608.4571 | Palmitoyl-linolenoyl-chloropropanediol |
| C35H67ClO4 | 586.4728 | Myristoyl-stearoyl-chloropropanediol |
| C39H71ClO4 | 638.5041 | Di-oleoyl-chloropropanediol |
| C37H69ClO4 | 612.4884 | Oleoyl-palmitoyl-chloropropanediol |
| C39H75ClO4 | 642.5354 | Di-stearoyll-chloropropanediol |
| C39H67ClO4 | 634.4728 | Oleoyl-linolenoyl-chloropropanediol |
| C39H69ClO4 | 636.4884 | Oleoyl-linoleoyl-chloropropanediol |
| C37H69ClO4 | 612.4884 | Oleyl-palmitoyl-chloropropanediol |
| **Glycidyl esters (GEs)** | | |
| C15H28O3 | 256.2038 | Glycidyl laurate |
| C17H32O3 | 284.2351 | Glycidyl myristate |
| C21H40O3 | 340.2977 | Glycidyl stearate |
| C19H36O3 | 312.2664 | Glycidyl palmitate |
| C21H38O3 | 338.2821 | Glycidyl oleate |
| C21H36O3 | 336.2664 | Glycidyl linoleate |
| C21H34O3 | 334.2508 | Glycidyl linoleate |

FbF algorithm found all six GEs and MCPDEs in the calibration standards, thereby validating the data analysis process (see Figure 1). Our method was capable of detecting GEs and di-MCPDEs down to 2 ng/mL and mono-MCPDEs down to 5 ng/mL concentrations.

### 2.2. Analysis of Glycerin Samples

The refined and crude glycerin samples we obtained were diluted and subjected to the same screening procedure as the calibration standards. The screen against the database using FbF algorithm produced many false-positive hits for GEs that were eliminated by comparing the retention time and mass accuracy with standards. There were no hits for MCPDEs.

The data generated in the study were validated according to the HRMS guidance [32]. The mass accuracy was <5 ppm for all six compounds in the calibration standards. Each sample was analyzed in triplicate. The retention time shift was <0.2 min between the replicate injections. The detected levels were reproducible. The compound stability was monitored throughout the study by injecting the stock solutions of standards in methanol during the course of the study, and no instability was detected.

## 3. Materials and Methods

### 3.1. Materials

#### 3.1.1. Instrumentation

Samples were analyzed using an Agilent 1290 Infinity UHPLC coupled with 6550 Q-TOF mass spectrometer (Agilent Technologies, Palo Alto, CA, USA) in positive-ion mode.

#### 3.1.2. Chemicals

Glycidyl linolenate (CAS 51554-07-5), glycidyl linoleate (CAS 243085-63-3), 1-lauroyl-3-chloropropanediol (CAS 20542-96-5), 1-palmitoyl-3-chloropropanediol (CAS 30557-04-1), 2-oleoyl-3-chloropropanediol (CAS 915297-48-2), 1-palmitoyl-2-stearoyl-3-chloropropanediol (CAS 1185060-41-6), 1-oleoyl-2-stearoyl-3-chloropropanediol (CAS 1336935-05-7), and 1,2-distearoyl-3-chloropropanediol (CAS 72468-92-9) were purchased from Toronto Research Chemicals (Toronto, ON, Canada).

#### 3.1.3. Glycerin Samples

Sixteen food-grade glycerin samples and four crude-grade glycerin samples were obtained from various sources. A food-grade glycerin sample obtained from Procter & Gamble (Cincinnati, OH, USA) was used for constructing matrix calibration curves for the mixed standard.

### 3.2. Preparation of Stock Solutions and Working Standards

Primary stock solutions (1 mg/mL each) were prepared separately by dissolving the ester standard in methanol. There are two classes of MCPDEs, namely mono- and diesters. Two GEs (glycidyl linoleate and glycidyl linolenate), two MCPD monoesters (1-palmitoyl-3-chloropropanediol and 2-oleoyl-3-chloropropanediol), and two MCPD diesters (1-oleoyl-2-stearoyl-3-chloropropanediol and 1-palmitoyl-2-stearoyl-3-chloropropanediol) were selected to represent the two classes of MCPDEs and GEs. Working standard solutions were prepared at 50, 100, 500, and 1,000 µg/mL by diluting appropriate volumes of the primary stock solution with MeOH. All solutions were stored in glass vials at 4 °C or below.

### 3.3. Preparation of a Calibration Curve of the Mixed Standard in Glycerin and Methanol

Appropriate amounts of working standards of MCPDEs and GEs were spiked (added) into methanol and glycerin matrix to produce the following concentrations: 2, 5, 10, 25, 50, 125, 250, and 400 ng/glycerin. Mixed standard in glycerin was prepared using 1 g food grade glycerin (Procter and Gamble) in 2 mL methanol, spiked with appropriate volumes of working standards.

Calibration curves were constructed in both methanol and glycerin matrix.

### 3.4. Glycerin Sample Preparation for Screening for MCPDEs and GEs

Sixteen food-grade glycerin samples and four crude-grade glycerin samples (approximately 1 g each) were prepared separately in 4 mL glass amber vials. Each 1 g glycerin was weighed directly into the vial, and 500 μL of methanol was pipetted into each vial. Vials were capped and vortexed, and methanol was added dropwise to bring the total volume of each sample up to the 2 mL mark. Methanol was used as the solvent blank. Triplicate injections were performed for each sample.

### 3.5. UHPLC Analysis

We analyzed the diluted spiked glycerin without solid-phase extraction step using the QTOF6500 HRMS mass spectrometer coupled to the UHPLC system. A liquid chromatography system 1290 (Agilent Technologies, Santa Clara, CA, USA) with a BEHC18 column (2.0 mm × 100 mm) with 1.8 μm particles (Waters, Milford, MA, USA) at 30 °C was used for HPLC separation. Mobile phase A consisted of 92:8 methanol/water with 0.05% formic acid; mobile phase B consisted of 98:2 isopropanol/water with 0.05% formic acid. The flow rate was 300 μL/min, and the gradient used was as follows: from 0–3 min, mobile phase B 0–30%; from 3–10 min, mobile phase B 30–70%, followed by holding at 70% B for 5 min and returning to 0% B in 2 min. The first 1 min of the LC flow was diverted to the waste to prevent the glycerin matrix from entering the mass spectrometer.

### 3.6. HRMS Analysis

The Agilent 6550 QTOF instrument (Agilent Technologies, Santa Clara, CA, USA) is equipped with an Agilent Jet Stream Technology Dual Spray ESI source and an iFunnel. The instrument was calibrated in the extended dynamic range (2 GHz, High Res Mode) and lower mass range (*m/z* < 1700) in the positive ion mode. Data were collected in centroid format. Reference masses at *m/z* 121.0509 and *m/z* 922.0089 were continually introduced via a second sprayer for accurate mass calibration. The reference ions used were purine ($C_5H_4N_4$) at *m/z* 121.0509 and HP-921 (hexakis-(*1H,1H,3H*-tetrafluoropentoxy)phosphazene ($C_{18}H_{18}O_6N_3P_3F_{24}$)) at *m/z* 922.0089 for positive mode. The source conditions were as follows: sheath gas flow, 11 L min$^{-1}$; sheath gas temperature, 350 °C; nebulizer pressure, 40 psi; drying gas temperature, 150 °C; drying gas flow, 15 L min$^{-1}$; nozzle voltage, 380 V; fragmentor voltage, 360 V; capillary voltage, 3500 V. Nitrogen was used as source gas and as collision gas. Full scan MS were acquired over the mass range *m/z* 100–1000 at a scan speed of 2 scans/s. Agilent Mass Hunter workstation software version B.05.00 was used for data acquisition, and B.06.00 qualitative analysis was used for processing. Mass calibration was performed prior to analysis.

## 4. Conclusions

We accomplished our goal in this study by successfully developing a UHPLC/Q-TOF-based screening method to detect a variety of GEs and MCPDEs in glycerin samples (both crude and refined grades). It is a dilute-and-shoot method with minimal sample preparation that can be adapted to other matrices. The screening method by FbF against our in-house library produced many false positives for GEs, which were eliminated by comparing the retention times with standards. There were no hits for MCPDEs in the limited glycerin samples we analyzed.

**Author Contributions:** L.G., K.H., and H.J. conceptualized the study and planned the laboratory work; L.G. executed LCMS method development; L.G. and H.J. worked with the Mass Hunter software and FbF algorithm for data processing; L.G., H.J., and K.H. prepared the original draft. All the authors (L.G., K.H., H.E., R.R., O.C., and H.J.) reviewed and edited the manuscript. All authors have read and agreed to the published version of the manuscript.

**Funding:** This research received no external funding.

**Institutional Review Board Statement:** Not applicable.

**Informed Consent Statement:** Not applicable.

**Data Availability Statement:** Not applicable.

**Acknowledgments:** We thank Philip J. Kijak for administrative support and Matthew Bell for assistance in the graphical design of figures.

**Conflicts of Interest:** The authors declare no conflict of interest.

**Sample Availability:** Samples of the compounds are available from the authors.

## References


1. FDA. Food and Drug Administration Investigates Animal Illnesses Linked to Jerky Pet Treats. 2018. Available online: https://www.fda.gov/AnimalVeterinary/NewsEvents/ucm360951.htm (accessed on 12 March 2020).
2. FDA. Jerky Pet Treat Investigation Testing Rationale and Results for October 1, 2013–December 31, 2015. 2016. Available online: https://www.fda.gov/media/97687/download (accessed on 15 May 2020).
3. Carmichael, N.; Lee, J.; Giger, U. Fanconi syndrome in dog in the UK. *Vet. Rec.* **2014**, *174*, 357–358. [CrossRef]
4. Abraham, L.A.; Tyrrell, D.; Charles, J.A. Transient renal tubulopathy in a racing Greyhound. *Aust. Vet. J.* **2006**, *84*, 398–401. [CrossRef]
5. Thompson, M.F.; Fleeman, L.M.; Kessell, A.E.; Steenhard, L.A.; Foster, S.F. Acquired proximal renal tubulopathy in dogs exposed to a common dried chicken treat: Retrospective study of 108 cases (2007–2009). *Aust. Vet. J.* **2013**, *91*, 368–373. [CrossRef]
6. Hooper, A.N.; Roberts, B.K. Fanconi syndrome in four non-basenji dogs exposed to chicken jerky treats. *J. Am. Anim. Hosp. Assoc.* **2011**, *47*, e178–e187. [CrossRef]
7. Herath, K.; Girard, L.; Reimschuessel, R.; Jayasuriya, H. Application of time-of-flight mass spectrometry for screening of crude glycerins for toxic phorbol ester contaminants. *J. Chromatogr. B Analyt. Technol. Biomed. Life Sci.* **2017**, *1046*, 226–234. [CrossRef]
8. Hrncirik, K.; van Duijn, G. An initial study on the formation of 3-MCPD esters during oil refining. *Eur. J. Lipid Sci. Technol.* **2011**, *113*, 374–379. [CrossRef]
9. Arisseto, A.P.; Marcolino, P.F.; Vicente, E. 3-Monochloropropane-1,2-diol fatty acid esters in commercial deep-fat fried foods. *Food Addit. Contam. Part A* **2015**, *32*, 1431–1435. [CrossRef] [PubMed]
10. Mogol, B.A.; Pye, C.; Anderson, W.; Crews, C.; Gokmen, V. Formation of monochloropropane-1,2-diol and its esters in biscuits during baking. *J. Agric. Food Chem.* **2014**, *62*, 7297–7301. [CrossRef]
11. Hamlet, C.G.; Sadd, P.A.; Crews, C.; Velisek, J.; Baxter, D.E. Occurrence of 3-chloro-propane-1,2-diol (3-MCPD) and related compounds in foods: A review. *Food Addit. Contam.* **2002**, *19*, 619–631. [CrossRef] [PubMed]
12. Baer, I.B.; Calle, B.D.L.; Taylor, P. 3-MCPD in food other than soy sauce or hydrolysed vegetable protein (HVP). *Anal. Bioanal. Chem.* **2010**, *396*, 443–456. [CrossRef] [PubMed]
13. Dubois, M.; Tarres, A.; Goldmann, T.; Empl, A.M.; Donaubauer, A.; Seefelder, W. Comparison of indirect and direct quantification of esters of monochloropropanediol in vegetable oil. *J. Chromatogr. A* **2012**, *1236*, 189–201. [CrossRef] [PubMed]
14. Hoogenboom, L. EFSA Panel on Contaminants in the Food Chain. Risks for human health related to the presence of 3- and 2-monochloropropanediol (MCPD), and their fatty acid esters, and glycidyl fatty acid esters in food. *EFSA J.* **2016**, *14*, e04426. [CrossRef]
15. Knutsen, H.K.; Alexander, J.; Barregård, L.; Bignami, M.; Brüschweiler, B.; Ceccatelli, S.; Cottrill, B.; Dinovi, M.; Edler, L.; Grasl-Kraupp, B.; et al. Update of the risk assessment on 3-monochloropropane diol and its fatty acid esters. *EFSA J.* **2018**, *16*, e05083. [CrossRef] [PubMed]
16. Barocelli, E.; Corradi, A.; Mutti, A.; Petronini, P.G. Comparison between 3-MCPD and its palmitic esters in a 90-day toxicological study. *EFSA Support. Publ.* **2011**, *8*, 187E. [CrossRef]
17. Liu, M.; Gao, B.-Y.; Qin, F.; Wu, P.-P.; Shi, H.-M.; Luo, W.; Ma, A.-N.; Jiang, Y.-R.; Xu, X.-B.; Yu, L.-L. Acute oral toxicity of 3-MCPD mono- and di-palmitic esters in Swiss mice and their cytotoxicity in NRK-52E rat kidney cells. *Food Chem. Toxicol.* **2012**, *50*, 3785–3791. [CrossRef]
18. Abraham, K.; Appel, K.E.; Berger-Preiss, E.; Apel, E.; Gerling, S.; Mielke, H.; Creutzenberg, O.; Lampen, A. Relative oral bioavailability of 3-MCPD from 3-MCPD fatty acid esters in rats. *Arch. Toxicol.* **2013**, *87*, 649–659. [CrossRef]
19. Onami, S.; Cho, Y.M.; Toyoda, T.; Akagi, J.; Fujiwara, S.; Ochiai, R.; Tsujino, K.; Nishikawa, A.; Ogawa, K. Orally administered glycidol and its fatty acid esters as well as 3-MCPD fatty acid esters are metabolized to 3-MCPD in the F344 rat. *Regul. Toxicol. Pharmacol.* **2015**, *73*, 726–731. [CrossRef]
20. Bakhiya, N.; Abraham, K.; Gurtler, R.; Appel, K.E.; Lampen, A. Toxicological assessment of 3-chloropropane-1,2-diol and glycidol fatty acid esters in food. *Mol. Nutr. Food Res.* **2011**, *55*, 509–521. [CrossRef]
21. Mahmoud, Y.I.; Abo-Zied, F.S.; Salem, S.T. Effects of subacute 3-monochloropropane-1,2-diol treatment on the kidney of male albino rats. *Biotech. Histochem.* **2019**, *94*, 199–203. [CrossRef]
22. Irwin, R.D.; Eustis, S.L.; Stefanski, S.; Haseman, J.K. Carcinogenicity of glycidol in F344 rats and B6C3F1 mice. *J. Appl. Toxicol.* **1996**, *16*, 201–209. [CrossRef]
23. Zhao, Y.; Zhang, Y.; Zhang, Z.; Liu, J.; Wang, Y.L.; Gao, B.; Niu, Y.; Sun, X.; Yu, L. Formation of 3-MCPD Fatty Acid Esters from Monostearoyl Glycerol and the Thermal Stability of 3-MCPD Monoesters. *J. Agric. Food Chem.* **2016**, *64*, 8918–8926. [CrossRef]
24. Mbamalu, V.C. Glycerin and the Market. Master's Thesis, The University of Tennessee at Chattanooga, Chattanooga, TN, USA, May 2013.

25. Pudel, F.; Benecke, P.; Fehling, P.; Freudenstein, A.; Matthäus, B.; Schwaf, A. On the necessity of edible oil refining and possible sources of 3-MCPD and glycidyl esters. *Eur. J. Lipid Sci. Technol.* **2011**, *113*, 368–379. [CrossRef]
26. Destaillats, F.; Craft, B.D.; Sandoz, L.; Nagy, K. Formation mechanisms of monochloropropanediol (MCPD) fatty acid diesters in refined palm (Elaeis guineensis) oil and related fractions. *Food Addit. Contam. Part A* **2012**, *29*, 29–37. [CrossRef]
27. Destaillats, F.; Craft, B.D.; Dubois, M.l.; Nagy, K. Glycidyl esters in refined palm (Elaeis guineensis) oil and related fractions. *Part I: Formation mechanism. Food Chem.* **2012**, *131*, 1391–1398. [CrossRef]
28. MacMahon, S.; Mazzola, E.; Begley, T.H.; Diachenko, G.W. Analysis of processing contaminants in edible oils. Part 1. Liquid chromatography-tandem mass spectrometry method for the direct detection of 3-monochloropropanediol monoesters and glycidyl esters. *J. Agric. Food Chem.* **2013**, *61*, 4737–4747. [CrossRef] [PubMed]
29. MacMahon, S.; Begley, T.H.; Diachenko, G.W. Analysis of processing contaminants in edible oils. Part 2. Liquid chromatography-tandem mass spectrometry method for the direct detection of 3-monochloropropanediol and 2-monochloropropanediol diesters. *J. Agric. Food Chem.* **2013**, *61*, 4748–4757. [CrossRef] [PubMed]
30. MacMahon, S.; Ridge, C.D.; Begley, T.H. Liquid chromatography-tandem mass spectrometry (LC-MS/MS) method for the direct detection of 2-monochloropropanediol (2-MCPD) esters in edible oils. *J. Agric. Food Chem.* **2014**, *62*, 11647–11656. [CrossRef] [PubMed]
31. Hori, K.; Matsubara, A.; Uchikata, T.; Tsumura, K.; Fukusaki, E.; Bamba, T. High-throughput and sensitive analysis of 3-monochloropropane-1,2-diol fatty acid esters in edible oils by supercritical fluid chromatography/tandem mass spectrometry. *J. Chromatogr. A* **2012**, *1250*, 99–104. [CrossRef] [PubMed]
32. FDA. Acceptance Criteria for Confirmation of Identity of Chemical Residues using Exact Mass Data within the Office of Foods and Veterinary Medicine. 2015. Available online: https://www.fda.gov/media/96499/download (accessed on 20 August 2020).

Article

# Glycyrrhizic Acid Inhibits SARS-CoV-2 Infection by Blocking Spike Protein-Mediated Cell Attachment

Jingjing Li [1,2,3,†], Dongge Xu [1,2,4,†], Lingling Wang [1,2,5], Mengyu Zhang [1,2,4], Guohai Zhang [6], Erguang Li [1,2,7,*] and Susu He [1,2,4,*]

1 State Key Laboratory of Pharmaceutical Biotechnology, Medical School, Nanjing University, Nanjing 210093, China; lijingjing@topcelbio.com (J.L.); 13337718237@163.com (D.X.); wanglinling1010@163.com (L.W.); mg1935020@smail.nju.edu.cn (M.Z.)
2 Jiangsu Key Laboratory of Molecular Medicine, Medical School, Nanjing University, Nanjing 210093, China
3 Jiangsu Topcel Biological Technology Co., Ltd., Nanjing 210093, China
4 Yancheng Medical Research Centre, Medical School, Nanjing University, Yancheng 224000, China
5 Institute of Medical Virology, Nanjing Drum Tower Hospital, Medical School, Nanjing University, Nanjing 210093, China
6 State Key Laboratory for Chemistry and Molecular Engineering of Medicinal Resources, School of Chemistry and Pharmaceutical Sciences, Guangxi Normal University, Guilin 541006, China; zgh1207@gxnu.edu.cn
7 Shenzhen Institute of Nanjing University, Shenzhen 518000, China
* Correspondence: erguang@nju.edu.cn (E.L.); susuhetian@nju.edu.cn (S.H.)
† These authors contributed equally to this work.

**Citation:** Li, J.; Xu, D.; Wang, L.; Zhang, M.; Zhang, G.; Li, E.; He, S. Glycyrrhizic Acid Inhibits SARS-CoV-2 Infection by Blocking Spike Protein-Mediated Cell Attachment. *Molecules* **2021**, *26*, 6090. https://doi.org/10.3390/molecules26206090

Academic Editors: Muhammad Ilias and Raphaël E. Duval

Received: 10 August 2021
Accepted: 30 September 2021
Published: 9 October 2021

**Abstract:** Glycyrrhizic acid (GA), also known as glycyrrhizin, is a triterpene glycoside isolated from plants of *Glycyrrhiza* species (licorice). GA possesses a wide range of pharmacological and antiviral activities against enveloped viruses including severe acute respiratory syndrome (SARS) virus. Since the S protein (S) mediates SARS coronavirus 2 (SARS-CoV-2) cell attachment and cell entry, we assayed the GA effect on SARS-CoV-2 infection using an S protein-pseudotyped lentivirus (Lenti-S). GA treatment dose-dependently blocked Lenti-S infection. We showed that incubation of Lenti-S virus, but not the host cells with GA prior to the infection, reduced Lenti-S infection, indicating that GA targeted the virus for infection. Surface plasmon resonance measurement showed that GA interacted with a recombinant S protein and blocked S protein binding to host cells. Autodocking analysis revealed that the S protein has several GA-binding pockets including one at the interaction interface to the receptor angiotensin-converting enzyme 2 (ACE2) and another at the inner side of the receptor-binding domain (RBD) which might impact the close-to-open conformation change of the S protein required for ACE2 interaction. In addition to identifying GA antiviral activity against SARS-CoV-2, the study linked GA antiviral activity to its effect on virus cell binding.

**Keywords:** glycyrrhizin; SARS-CoV-2; surface plasmon resonance; autodocking

## 1. Introduction

The uncertainty of the coronavirus disease 2019 (COVID-19) pandemic situation remains elusive, although great success has been accomplished with the rapid development of protective vaccines against severe acute respiratory syndrome coronavirus 2 (SARS-CoV-2). There are currently no effective drugs for the treatment of COVID-19, although clinical trials and case reports involving antiviral and antiparasitic agents have yielded promising results that warrant further investigation [1,2]. Herbal medicines have a long history of vindicated clinical efficacy against infectious diseases. Clinical evidence shows that herbal medicines are effective against viral infections such as influenza, SARS, and SARS-CoV-2 by targeting virus cell entry, viral replication, and host antiviral immune response steps. During the pandemic of COVID-19, several drugs formulated according to principles of Chinese medicine showed therapeutic effects against mild and severe COVID-19. Among the drugs and formulae recommended by the Chinese authority for COVID-19 therapy, the dried root of *Glycyrrhiza* spp. (licorice) is among the most commonly used ingredients in the formulae [3]. The radices of *Glycyrrhiza* spp. have been used as an important ingredient of herbal medicines and also a flavoring agent in traditional formulae since antiquity [4,5]. The use of the dried roots of the plant can be traced back to ancient Assyrian, Egyptian, Chinese, and Indian cultures for symptoms that resemble those of viral respiratory tract infections such as dry cough or hoarse voice [5,6]. Recent reports also suggest that licorice extract may have a potential role in combating COVID-19 and associated conditions [7].

The main chemical component from *Glycyrrhiza* sp. is glycyrrhizic acid (GA), also known as glycyrrhizin, a triterpenoid saponin that can be extracted from the dried roots in high yields [8,9]. Numerous reports show that GA is effective against the infection of enveloped and nonenveloped viruses such as herpes viruses, respiratory syncytial virus, vaccinia virus, and SARS-CoV in cell culture studies [4,10,11]. Cinatl and colleagues found that glycyrrhizin was among the most active compounds tested in inhibiting replication of the SARS-CoV [12]. Thus, licorice and GA could be an old weapon against emerging diseases [13,14].

In this study, we investigated the antiviral effect of GA against SARS-CoV-2 using a pseudotyped lentivirus that has the SARS-CoV-2 S protein on its envelope (Lenti-S). We found that GA inhibited Lenti-S infection through inhibition of virus attachment to host cells. Preincubation of Lenti-S, rather than the host cells, with GA prior to the infection reduced Lenti-S infection, suggesting that GA primarily targets the virus rather than the host cells. GA interacted with the S protein with high affinity and blocked a recombinant S protein binding to the host cells. Thus, this study uncovered a mechanism by which GA blocks SARS-CoV-2 infection by impeding virus and host cell interaction.

## 2. Materials and Methods

### 2.1. Cells, Reagents, and Antibodies

Vero E6 and 293T cells were obtained from Cell Bank of Chinese Academy of Sciences and were cultured in Dulbecco's Modified Eagle Medium (DMEM) supplemented with 10% fetal bovine serum (FBS, $v/v$) at 37 °C in a humidified incubator. The medium also contained 10 mM HEPES (pH 7.4), 2 mM GlutaMAX, and penicillin–streptomycin–fungizone (100 units/mL of penicillin, 100 µg/mL of streptomycin, and 250 ng/mL of amphotericin B). The DMEM and all the supplemented ingredients were purchased from Invitrogen (Shanghai, China). An insect cell-expressed S protein (40589-V08B1, S1 + S2 ectodomain) and a polyclonal antibody cross-reactive to SARS-CoV-2 S (40150-T62) were purchased from Sino Biological (Beijing, China). The Steady-Lumi firefly luciferase reporter gene assay reagent was purchased from Beyotime (Nantong, China). An HIV-1 NL4-3 luciferase reporter vector that contains defective Nef, Env, and Vpr was purchased from MiaoLing (P20782, Wuhan, China). pCMV3-SARS-CoV-2-S (VG40589-UT, accession number MN908947.3) with codon-optimized *S* gene for mammalian cell expression and pCMV3-ACE2-HA (HG10108-CY) were purchased from Sino Biological. The pCMV3-SARS-CoV-2-S was modified in the lab by deletion of the last 19 amino acid residues to generate pCMV3-S since the endoplasmic reticulum (ER)-retention signal from the cytoplasmic tail was reported to interfere with virus preparation [15,16]. Glycyrrhizin ammonium salt (#50531, purity $\geq$ 95%) and HRP-conjugated anti-HA antibody (H3663) were purchased from Sigma-Aldrich. A stock solution of 50 mM GA was prepared by dissolving the compound in distilled water with pH adjusted to 7.4 using NaOH.

### 2.2. Spike Protein-Pseudotyped Virus (Lenti-S) Preparation

The spike protein-S-pseudotyped lentivirus (Lenti-S) was generated using a 2-plasmid system as previously reported [17]. Stocks of single-round infection of S protein-pseudotyped virus were produced by cotransfection of 293T cells ($1.0 \times 10^7$ cells per 10 cm dish) with 2 µg of pCMV-S plasmid and 8 µg of pNL4.3-Luc using PEI reagent. The supernatant was harvested 34 h following transfection. After clarification by centrifugation at 1500$\times$ $g$ followed by 0.45 µm filtration, pseudovirus-containing medium was collected and aliquots were stored at −80 °C. The virus was validated by testing for its ability to deliver the luciferase gene to ACE2-expressing HEK293T cells [18].

### 2.3. Infection Assay

Pseudotyped viruses provide an efficient way to determine an antiviral effect by measuring reporter gene expression. For gene-transducing assay, the permissible Vero E6 cells in 96-well plates (#3599, Costar, Corning, NY, USA) were infected with varying amounts of Lenti-S. Luciferase expression was determined at approximately 24 h PI using Steady-Lumi reagent on a GloMax 96 luminometer (Promega, Madison, WI, USA).

For blocking assay, Vero E6 cells were detached using 3 mM ethylenediaminetetraacetic acid (EDTA). After washing twice with serum-free medium (SFM), the cells were resuspended in ice-cold SFM-containing 2% ($v/v$) FBS ($10^7$ cells/mL). The cells were then aliquoted ($5 \times 10^5$ cells/sample), and duplicate samples were incubated on ice with Lenti-S in the absence or presence of GA or a blocking reagent as indicated. The mixtures were incubated on ice for 60 min with occasional mixing. At the end, the cells were then washed with ice-cold medium 3 times to remove non-bound virus then plated in 12-well plates (Corning™ Costar, #3512) without supplementation of GA or the blocking reagent. Luciferase expression in duplicate samples was determined at 24 h PI.

### 2.4. Protein Biotinylation and Binding Assay

An insect-cell-expressed S protein (2 μg), resuspended in 100 μL of 50 mM $NaHCO_3$, was labeled with freshly prepared sulfo-NHS-biotin (21217, Pierce, Carlsbad, CA, USA, 0.5 mg in 50 μL $NaHCO_3$). The reaction lasted for 10 min at room temperature. At the end, unreacted sulfo-NHS-biotin was quenched by reaction with 1 mg glycine dissolved in 20 μL $NaHCO_3$. The labeled protein was dialyzed against phosphate-buffered saline (MW cutoff 12 kD) and was used for protein-binding assay.

To measure the effect of GA on S protein binding, detached Vero E6 cells were incubated on ice with the biotinylated S protein in the absence or presence of varying amounts of GA or excess amounts of unlabeled S protein. After incubation on ice for 60 min, the cells were washed 3 times with ice-cold medium. After the cell lysates were separated by a 6% SDS-PAGE gel, cell-attached S protein was detected by immunoblotting analysis using an HRP-conjugated anti-biotin antibody (A0185, Sigma-Aldrich, St Louis, MO, USA). Anti-actin (sc-8432, Santa Cruz, Dallas, TX, USA) was used for loading controls.

### 2.5. Surface Plasmon Resonance (SPR) Studies

SPR analysis was performed using the Biacore T200 system and a CM5 sensor chip. The chip consists of two channels that were loaded with S protein. Briefly, the carboxymethylated dextran matrix was activated via the passage of EDC-HCl and NHS (0.2 M and 0.05 M in water, respectively). Then, the S protein at 20 μg/mL (in 1 mM sodium acetate) was passed over the surface. Any remaining un-reacted active ester groups were quenched by the passage of ethanolamine solution. The sensor chip surface was regenerated after each experiment by passing sodium dodecyl sulfate (SDS, 100 mM) over the surface for 2 min. This simple procedure reliably returned the original signal response seen before the binding experiment started. For testing, GA at concentrations was passed over the chip, and SPR angle changes were recorded and reported as response units (RUs). Data fitting was performed using the 1:1 Langmuir model in the BIAevaluation software package (GE Healthcare, Waukesha, WI, USA).

### 2.6. Autodocking

Docking analysis was performed using AutoDock Vina 1.1.2 (Windows version) with default scoring function [19]. The molecular structures of SARS-CoV-2 S protein (PDB id: 6vsb) and ACE2 (PDB id: 6m18) were downloaded from the Research Collaboratory for Structural Bioinformatics-Protein Databank (RCSB-PDB) [20]. The 3D structure of GA was from ZINC (https://zinc.docking.org/, id: 960251743495; last accessed on 20 September 2021).

The structures were prepared using AutoDock Tools (ADT) by addition of polar hydrogens and the Gasteiger charges. Structures were exported in the pdbqt format after assigning the AD4 (AutoDock 4) atom type [21]. In the AutoDock Vina configuration files, the parameter num_modes was set to 1000 and exhaustiveness to 100. We chose all the rotatable bonds in ligands to be flexible during the docking procedure, and we kept all the residues of S protein inside the binding pockets rigid. Multiple rounds of definition of the grid coordinate (x-, y-, and z-coordinates) with a defined grid box size were conducted to screen through the entire extracellular part of the S protein. The resulting docking structures were further analyzed in PyMOL.

### 2.7. Statistical Analysis

The assays were performed at least 2 times independently. Data were analyzed with Excel (Microsoft) for statistical significance using Student's *t* test. $p < 0.05$ was considered significant.

## 3. Results

### 3.1. Inhibition of GA against SARS-CoV-2 Infection against S Protein-Pseudotyped Virus

We used an S protein-pseudotyped lentivirus to determine whether GA had an antiviral effect against SARS-CoV-2. The lentivirus system is easy to construct and has been widely used for virus binding and infection assay. To this end, cells were infected with varying amounts of Lenti-S in the absence or presence of varying amounts of GA (Figure 1A). We tested GA at concentrations of 0.5–5 mM since previous studies showed that GA was previously reported active against a wide range of enveloped viruses at concentrations of 1–8 mM [10,12]. Vero E6 cells in 96-well plates were untreated or treated with GA at 0.5, 1, 2.5, and 5 mM 30 min prior to Lenti-S infection, and we found that GA treatment resulted in a reduction in luciferase activity (Figure 1B). At 2.5 and 5 mM, GA treatment reduced Lenti-S-mediated luciferase gene delivery by approximately 77% and 92%, respectively. The effect of GA on Lenti-S infection was specific since treatment of 293T cells transfected with pLenti-CMV-luc did not affect luciferase expression (Figure 1C).

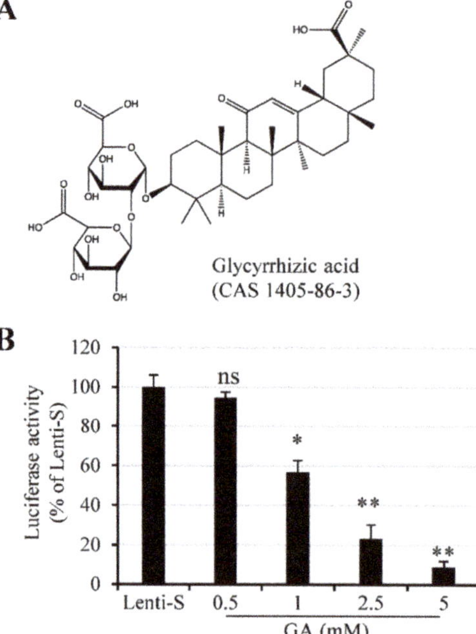


**Figure 1.** *Cont.*

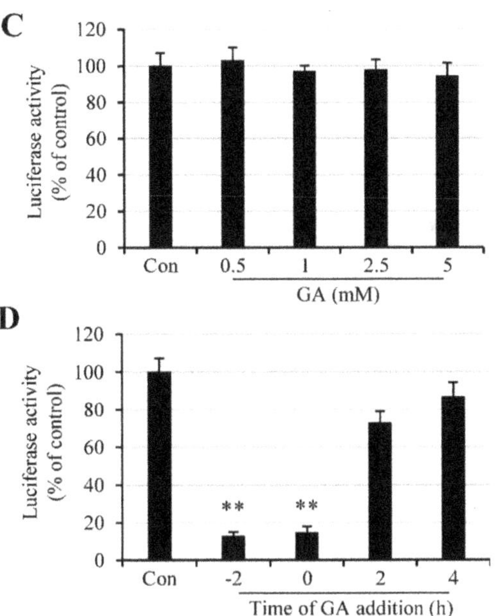


**Figure 1.** GA effect on Lenti-S infection. (**A**) Molecular structure of glycyrrhizic acid. (**B**) Effect of GA on Lenti-S infection. Vero E6 cells were infected with Lenti-S pseudovirus in the absence or presence of GA at indicated concentrations for 24 h. Luciferase activity was determined. Data are expressed as a percentage of untreated controls (Lenti-S). The experiment was performed twice, and data are mean ± SD of triplicate wells. Independent two-sample comparisons between non-GA treated sample and sample treated with GA at different concentrations, respectively, were determined by Student's t test. ns: no significance; *, $p < 0.05$; **, $p < 0.01$. (**C**) Effect of GA on luciferase expression in pCMV-luc-transfected cells. Monolayers of 293T cells were transfected with pCMV-luc for 16 h. The cells were then treated with GA at indicated concentrations. Luciferase activity was determined after 24 h incubation. GA treatment did not affect luciferase expression delivered by an encoding plasmid. The experiment was performed twice. The readings from untreated samples were used as a control for the calculation of relative luciferase activity. Data are mean ± SD of duplicate wells from 2 independent experiments. (**D**) Time of GA addition on Lenti-S-mediated luciferase gene delivery. Vero E6 cells were untreated or treated with 3 mM GA at 2 h prior to (−2 h), during (0), or at 2 and 4 h post Lenti-S infection. Luciferase activity was determined 24 h PI. The experiment was performed 2 times. Data from the untreated controls were used for the calculation of relative luciferase activity. **, $p < 0.01$.

The time effect of GA addition was tested by treating Vero E6 cells with 3 mM GA at 2 h prior to (−2 h), during (0 h), or at 2 h and 4 h post Lenti-S infection. Luciferase expression was determined at 24 h PI. As shown in Figure 1D, GA addition prior to or during Lenti-S inoculation significantly blocked Lenti-S infection. For comparison, addition of GA at 2 and 4 h post Lenti-S inoculation showed diminished effect against Lenti-S infection. This result suggests that GA likely targeted the early stages of Lenti-S infection.

### 3.2. GA Effect on S Protein Binding to Host Cells

Virus infection is initiated by receptor-mediated attachment, followed by cell entry via membrane fusion or endocytosis. Since Lenti-S is a pseudotyped virus that utilizes S protein for cell attachment and cell entry, we focused on whether GA targeted the S protein of SARS-CoV-2 for antiviral action. We performed a cell-binding assay using a biotin-labeled S protein to assess whether GA interfered with S protein attachment to host cells. In this regard, a biotinylated S protein was allowed to bind to Vero E6 cells in the presence or absence of GA (Figure 2A). After washing off non-bound S protein with an ice-cold medium, cell-bound S protein was detected by immunoblotting analysis. We used unlabeled S protein as a competitive reagent against biotin-labeled S protein binding to demonstrate the binding specificity of biotinylated S protein to host cells (Figure 2B). As shown in Figure 2C, GA treatment dose-dependently inhibited S protein binding to the cells. At 5 mM concentration, GA blocked S protein binding by more than 95%, suggesting that GA inhibited pseudovirus infection likely by blocking S protein-mediated cell attachment.

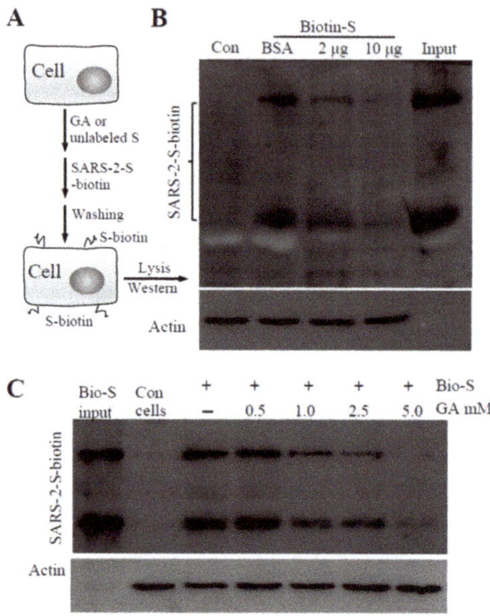


**Figure 2.** Effect of GA on recombinant S protein binding. (**A**) Schematic presentation of the experimental design. The cells were detached and incubated on ice for 60 min with GA or recombinant S protein (S unlabeled). Biotinylated S protein was then added for cell-attachment assay. After washing with ice-cold DMEM, cell-attached biotinylated S protein was detected by immunoblotting assay. (**B**) Biotinylated S protein binding to host cells in the presence of BSA or unlabeled S protein. Vero E6 cells resuspended in ice-cold DMEM containing 0.5% BSA were allowed to interact with Lenti-S in the presence or absence of unlabeled S protein (at 2 and 10 μg/mL). (**C**) Effect of GA on S protein binding to Vero E6 cells. A biotinylated S protein was allowed to bind to detached Vero E6 cells in the absence or presence of GA. Cell-bound biotinylated S protein was determined by immunoblotting assay. Actin was used as a loading control. Con, Vero E6 cells only. Input, biotinylated S protein for total binding.

### 3.3. *GA Treatment of Pseudovirus but Not the Cells Inhibits Pseudovirus Infection*

We performed an infection assay by pretreatment of the virus and the cells to preliminarily determine a primary target of the GA effect. In this regard, Lenti-S was pretreated with GA at a concentration of 3 mM on ice for 1 h. The treated virus was then 1:30 diluted with culture medium and used to infect the cells (final GA concentration in the culture medium was at approximately 0.1 mM; the concentration showed no antiviral activity). Virus infection was determined by measuring luciferase activity at 24 h post infection. Alternatively, we also treated the cells with 3 mM GA on ice for 1 h to determine whether GA targeted the host cells for its antiviral effect since the receptor ACE2 was a putative target of GA action [22]. After removal of GA by rinsing with fresh DMEM, the cells were then infected with Lenti-S for 24 h. Pretreatment of cells with GA showed marginal effect against Lenti-S infection, while pretreatment of pseudovirus with GA profoundly reduced pseudovirus infectivity (Figure 3), indicating that GA targeted virus particles for the antiviral effect.

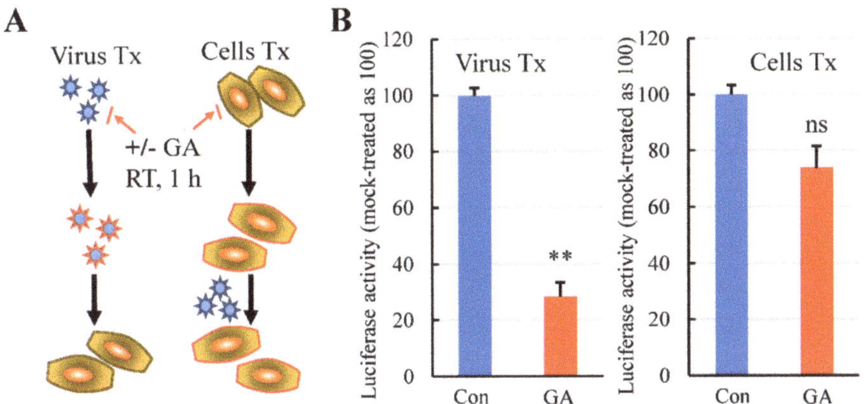


**Figure 3.** Pretreatment of Lenti-S or host cells with GA on Lenti-S mediated luciferase gene delivery. (**A**) Diagram showing Lenti-S or host cells were treated (Tx) with 3 mM GA prior to the infection assay. Both Lenti-S and host cells were untreated or treated with 3 mM GA for 1 h. At the end, the GA-treated Lenti-S was 1:30 diluted and used to infect untreated host cells (final concentration of GA in the medium was approximately 0.1 mM). In parallel, the medium of the GA-treated cells was replaced with fresh medium without GA followed by infection of untreated Lenti-S. Luciferase activity was determined 24 h later. (**B**) GA effect on Lenti-S (Virus Tx) and on host cells (Cells Tx). Luciferase activity was expressed as a percentage of untreated controls. Data are mean ± SD of duplicate wells from 2 experiments. ns, no significance; **, $p < 0.01$ by Student's $t$ test.

### 3.4. GA Interacts with S Protein

To more clearly demonstrate whether GA interacted with S protein, we performed an SPR assay to measure GA interaction with a recombinant S protein (Figure 4). The $k_a$ and $k_d$ were measured at approximately $7.6 \times 10^5$ $M^{-1}s^{-1}$ and $2.2 \times 10^{-4}$ $s^{-1}$, respectively, which translated to a calculated $K_D$ of 0.28 nM, if a 1:1 stoichiometry was used for GA and S protein interaction. The result shows clearly a direct interaction between GA and S protein, although this might have overestimated the affinity since the S protein is believed to be a homotrimeric protein which would have more than one GA binding site per protein. Regardless, the results together indicate that GA blocked Lenti-S infection through inhibition of S protein-mediated cell binding.

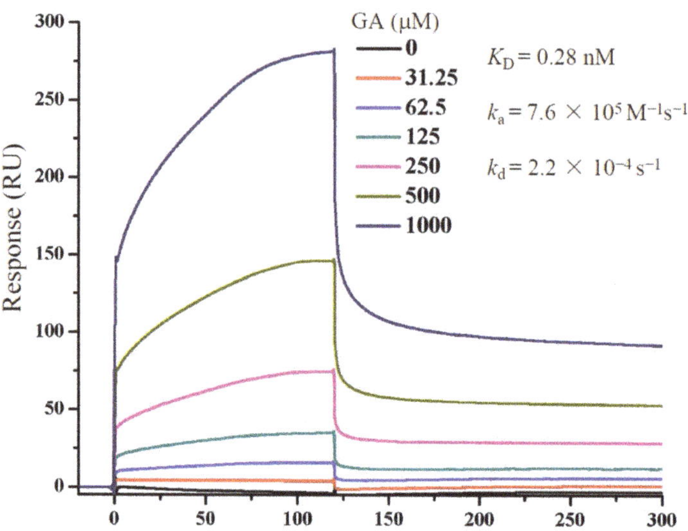


**Figure 4.** Determination of GA binding to S protein by SPR. S protein was immobilized on to a CM5 sensor chip. GA at concentrations as indicated was passed over the chip, and SPR angle changes were recorded using the Biacore T200 system and reported as response units (RUs). Data fitting was performed using the 1:1 Langmuir model in the BIAevaluation software package (GE Healthcare).

### *3.5. Autodocking Reveals GA-Binding Pockets on SARS-CoV-2 S Protein*

It was predicted that the RBD of the S protein contains binding pockets for natural products including GA [23]. We performed AutoDock Vina analysis by scanning through the entire extracellular domain of the S protein for the binding potentials. The first notion was GA might bind on the interaction interface of S protein–ACE2 since we found that GA treatment blocked Lenti-S infection and S protein attachment. Indeed, a binding pocket at the S-ACE2 interface was identified with a calculated binding energy of −8.0 kcal/mol (Figure 5A). The S protein presents in two different conformations, named open and closed states [24]. We also identified another binding pocket located at the inner side of the RBD with a binding energy of −7.0 kcal/mol (Figure 5B). SARS-CoV-2 opening is expected to be necessary for interacting with ACE2 at the host-cell surface and initiating the conformational changes leading to cleavage of the S2 site for efficient membrane fusion and viral entry [25]. It is possible that GA binding at this inner binding pocket impacted the close–open state transformation resulting in diminished open-state trimer to interact with ACE2. Thus, we also searched for the potential binding pocket at the ACE2 receptor. Similar to a previous report [22], a pocket at the interface binding to the S protein provided a binding energy of −4.1 kcal/mol (Figure 5C). Based on the predicted binding energy, we speculated that the primary binding site of GA should be on the S protein rather than the ACE2.

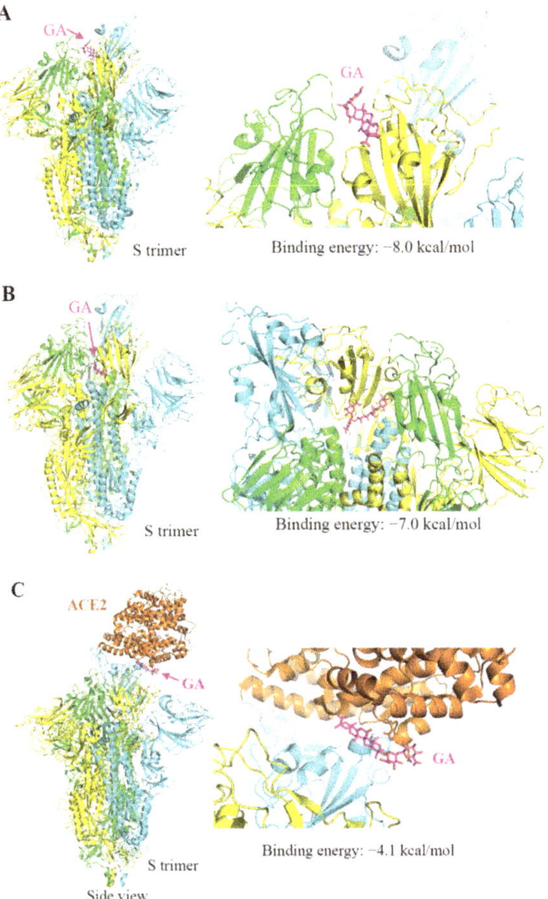


**Figure 5.** Autodocking analysis of GA interaction with SARS-CoV-2 S and ACE2 protein. Three protomers of SARS-Cov-2 S protein (PDB id: 6vsb) are shown in cyan, green, and yellow, respectively. ACE2 protein (PDB id: 6m18) is in orange. The structure of proteins is presented in ribbons. The structure of GA (ZINC id:960251743495) is shown in magenta as sticks. Arrows indicate the predicted binding site of GA. (**A**) Predicted binding of GA on the S protein at the S–ACE2 interface with a calculated binding energy of –8.0 kcal/mol. (**B**) Predicted binding of GA on a binding pocket located at the inner side of the RBD with a binding energy of −7.0 kcal/mol. (**C**) Predicted binding of GA on the ACE2 protein at the ACE2–S interface with a binding energy of −4.1 kcal/mol.

In summary, we showed here that GA interacted with SARS-CoV-2 S protein and blocked S protein-mediated cell binding for the antiviral activity of GA against SARS-CoV-2.

## 4. Discussion

Coronaviruses are a group of related viruses that cause diseases in humans and animals. In humans, coronaviruses cause respiratory tract infections, ranging from the common cold to the deadly diseases by SARS-CoV, MERS-CoV, and SARS-CoV-2. Due to the lack of medicines for COVID-19, repurposing currently existing and experimental drugs has been proposed as an alternative to uncover agents with therapeutic potentials. Traditional medicines have demonstrated records as anti-infectives throughout the history of mankind and have shown to be effective in China at alleviating COVID-19 symptoms or even reducing fatality. GA, a major component of *Glycyrrhiza* spp., possesses a wide range of pharmacological and biological activities, including antioxidant, antiviral, and anti-inflammatory effects [7,13,26,27]. Gowda and colleagues reported that GA could inhibit SARS-CoV-2-protein-induced high-mobility group box 1 (HMGB1) release and inhibits viral replication [28]. GA also targets SARS-CoV-2 main protease [29] and blocks proinflammatory response [30]. It is likely that GA can utilize multiple mechanisms against SARS-CoV-2 infection and disease [29,30]. To initiate a productive cycle of infection, a virus first attaches to a host cell, followed by a cell entry and replication process. In this study, we used a pseudotyped lentivirus and showed GA with antiviral activity. We focused on the early stages of virus infection using a pseudotyped virus system. This approach has been successfully used to construct pseudotyped lentiviruses for SARS-CoV, MERS-CoV, and recently SARS-CoV-2 and the corresponding mutants [31–34]. As a model for highly contagious pathogens, pseudotyped viruses are easy to construct and safe to use. It allows in particular the detailed studies involving virus attachment and virus-cell entry stages.

Several studies have predicted S protein or S–ACE2 interaction as potential targets for GA against SARS-CoV-2 [27,35]. Here we provided experimental evidence demonstrating that GA blocks S protein-mediated cell attachment for its antiviral effect. We found that GA interacted with the S protein with high affinity and blocked a recombinant S protein binding to the host cells. We also executed computational molecular docking to elucidate potential GA binding pockets on S protein. We screened through the entire extracellular domain of S protein by defining multiple grid boxes within this region. In addition to a previously revealed GA binding site at the interaction interface between the RBD and ACE2 protein [36], we also found a binding pocket at the inner side of the RBD. Based on the structural features, we predicted that the binding may have several impacts on the infectious activity of Lenti-S. First, S protein presents in two different conformations including a close state and an open state with one RBD of the trimer flipped out. The switch from close state to open state of S protein was necessary to establish an interaction with the ACE2 receptor. The binding site at the inner face of the RBD could impact this conformation transition. Secondly, it is also likely that GA binding at the inner side of the RBD might interfere with subsequent conformational change during the fusion stage.

In an assimilated SARS-CoV-2-infected mouse model, nanoparticles carrying GA demonstrated therapeutic effects through anti-inflammatory and antioxidant activities [37]. At the intracellular and circulating levels, GA binds to high-mobility group box 1 protein (HMGB1) to provide robust anti-inflammatory and neuroprotection [38,39]. GA attenuates pulmonary hypertension progression and pulmonary vascular remodeling in animal models [40–42]. As a hydrophilic compound, GA is not readily absorbed. After oral ingestion, glycyrrhizin is first hydrolyzed to 18 β-glycyrrhetinic acid by intestinal bacteria, which can be absorbed from the gut [43]. The metabolites in circulation, along with GA, can significantly reduce inflammatory cell infiltration and cytokine production during an infection [40,42].

Pompei and colleagues reported that GA was effective against a broad range of enveloped viruses [10]. Cinatl et al. showed that GA was effective against SARS-CoV [12,44]. The reported concentrations for GA antiviral effect in cell cultures generally vary between 1 and 5 mM concentrations, at which concentrations GA can form an emulsion or long-lasting foams in an aqueous solution [45–47]. It is well known that surfactants were able to inactivate enveloped viruses by causing protein aggregation, disruption of the envelope, or by distorting the shape of virions [48]. At the membrane level, GA induces cholesterol-dependent disorganization of lipid rafts which are important for the entry of coronavirus into cells [49,50]. GA was also reported to modulate the fluidity of the plasma membrane and HIV-1 envelope [45]. The fact that GA directly inactivated enveloped virus particles suggests that GA likely exerts its antiviral activity by destabilizing the envelope. Whether chemicals such as GA use the surfactant activity for their antiviral effect remains to be further studied.

Here we provided experimental and computational simulation data demonstrating that GA potentially targets S protein-mediated cell attachment for its antiviral activity.

GA interacted with the S protein with high affinity and blocked recombinant S protein binding to the host cells. Thus, this study uncovered a mechanism by which GA blocks SARS-CoV-2 infection, highlighting the potential of herbal medicine against emerging and reemerging infectious diseases.

**Author Contributions:** S.H., E.L., and J.L. conceived the concepts. J.L., D.X., L.W., S.H., G.Z., and M.Z. performed the experiments and analyzed the data. E.L., S.H., and J.L. curated the data. E.L., L.W., and S.H. wrote the draft. All authors have read and agreed to the published version of the manuscript.

**Funding:** This research was funded by grants from Ningxia Hui Autonomous Region (2017BN04, E.L.), Jiangsu Natural Science Foundation (BK20200316 to S.H.), NSFC (81871636 to E.L.), Central Universities Fundamental Research Funds (14380456 to E.L. and 14380470 to S.H.), Science, Technology and Innovation Commission of Shenzhen Municipality (JSGG 20200519160755008 to E.L.), and Nanjing Scientific and Technological Innovation Project for Returning Scholars from Abroad: (2021 to S.H.).

**Institutional Review Board Statement:** Not applicable.

**Informed Consent Statement:** Not applicable.

**Data Availability Statement:** Not applicable.

**Acknowledgments:** We thank Guang Yang for technical assistance and Jinhui Dou for comments.

**Conflicts of Interest:** The authors declare no conflict of interest.

**Sample Availability:** Not available.

## Abbreviations

| | |
|---|---|
| ACE2 | angiotensin-converting enzyme 2 |
| BSA | bovine serum albumin |
| COVID-19 | coronavirus disease 2019 |
| DMEM | Dulbecco's Modified Eagle Medium |
| EDTA | ethylenediaminetetraacetic acid |
| FBS | fetal bovine serum |
| GA | glycyrrhizic acid |
| RBD | receptor-binding domain |
| SARS-CoV-2 | severe acute respiratory syndrome coronavirus 2 |
| S protein | spike protein of SARS-CoV-2 |

## References

1. Li, T.; Lu, H.; Zhang, W. Clinical observation and management of COVID-19 patients. *Emerg. Microbes Infect.* **2020**, *9*, 687–690. [CrossRef]
2. Drayman, N.; DeMarco, J.K.; Jones, K.A.; Azizi, S.A.; Froggatt, H.M.; Tan, K.; Maltseva, N.I.; Chen, S.; Nicolaescu, V.; Dvorkin, S.; et al. Masitinib is a broad coronavirus 3CL inhibitor that blocks replication of SARS-CoV-2. *Science* **2021**, *373*, 931–936. [CrossRef] [PubMed]
3. National Health Commission of the People's Republic of China, Diagnosis and Treatment Protocol for COVID-19 Patients (Tentative 8th Edition). Available online: http://regional.chinadaily.com.cn/pdf/DiagnosisandTreatmentProtocolforCOVID-19Patients(Tentative8thEdition).pdf (accessed on 26 September 2021).
4. Fiore, C.; Eisenhut, M.; Krausse, R.; Ragazzi, E.; Pellati, D.; Armanini, D.; Bielenberg, J. Antiviral effects of Glycyrrhiza species. *Phytother. Res.* **2008**, *22*, 141–148. [CrossRef]
5. Shibata, S. A drug over the millennia: Pharmacognosy, chemistry, and pharmacology of licorice. *Yakugaku Zasshi* **2000**, *120*, 849–862. [CrossRef]
6. Fiore, C.; Eisenhut, M.; Ragazzi, E.; Zanchin, G.; Armanini, D. A history of the therapeutic use of liquorice in Europe. *J. Ethnopharmacol.* **2005**, *99*, 317–324. [CrossRef]
7. Gomaa, A.A.; Abdel-Wadood, Y.A. The potential of glycyrrhizin and licorice extract in combating COVID-19 and associated conditions. *Phytomed. Plus* **2021**, *1*, 100043. [CrossRef]
8. Rauchensteiner, F.; Matsumura, Y.; Yamamoto, Y.; Yamaji, S.; Tani, T. Analysis and comparison of Radix Glycyrrhizae (licorice) from Europe and China by capillary-zone electrophoresis (CZE). *J. Pharm. Biomed. Anal.* **2005**, *38*, 594–600. [CrossRef] [PubMed]

9. Isbrucker, R.A.; Burdock, G.A. Risk and safety assessment on the consumption of Licorice root (Glycyrrhiza sp.), its extract and powder as a food ingredient, with emphasis on the pharmacology and toxicology of glycyrrhizin. *Regul. Toxicol. Pharmacol.* **2006**, *46*, 167–192. [CrossRef]
10. Pompei, R.; Flore, O.; Marccialis, M.A.; Pani, A.; Loddo, B. Glycyrrhizic acid inhibits virus growth and inactivates virus particles. *Nature* **1979**, *281*, 689–690. [CrossRef]
11. Pompei, R.; Laconi, S.; Ingianni, A. Antiviral properties of glycyrrhizic acid and its semisynthetic derivatives. *Mini Rev. Med. Chem.* **2009**, *9*, 996–1001. [CrossRef] [PubMed]
12. Cinatl, J.; Morgenstern, B.; Bauer, G.; Chandra, P.; Rabenau, H.; Doerr, H.W. Glycyrrhizin, an active component of liquorice roots, and replication of SARS-associated coronavirus. *Lancet* **2003**, *361*, 2045–2046. [CrossRef]
13. Chrzanowski, J.; Chrzanowska, A.; Grabon, W. Glycyrrhizin: An old weapon against a novel coronavirus. *Phytother. Res.* **2021**, *35*, 629–636. [CrossRef]
14. Jezova, D.; Karailiev, P.; Karailievova, L.; Puhova, A.; Murck, H. Food Enrichment with Glycyrrhiza glabra Extract Suppresses ACE2 mRNA and Protein Expression in Rats-Possible Implications for COVID-19. *Nutrients* **2021**, *13*, 2321. [CrossRef]
15. Fukushi, S.; Mizutani, T.; Saijo, M.; Matsuyama, S.; Miyajima, N.; Taguchi, F.; Itamura, S.; Kurane, I.; Morikawa, S. Vesicular stomatitis virus pseudotyped with severe acute respiratory syndrome coronavirus spike protein. *J. Gen. Virol.* **2005**, *86*, 2269–2274. [CrossRef]
16. Lukassen, S.; Chua, R.L.; Trefzer, T.; Kahn, N.C.; Schneider, M.A.; Muley, T.; Winter, H.; Meister, M.; Veith, C.; Boots, A.W.; et al. SARS-CoV-2 receptor ACE2 and TMPRSS2 are primarily expressed in bronchial transient secretory cells. *EMBO J.* **2020**, *39*, e105114. [CrossRef] [PubMed]
17. Li, S.L.; Zhang, X.Y.; Ling, H.; Ikeda, J.; Shirato, K.; Hattori, T. A VSV-G pseudotyped HIV vector mediates efficient transduction of human pulmonary artery smooth muscle cells. *Microbiol. Immunol.* **2000**, *44*, 1019–1025. [CrossRef] [PubMed]
18. Li, W.; Moore, M.J.; Vasilieva, N.; Sui, J.; Wong, S.K.; Berne, M.A.; Somasundaran, M.; Sullivan, J.L.; Luzuriaga, K.; Greenough, T.C.; et al. Angiotensin-converting enzyme 2 is a functional receptor for the SARS coronavirus. *Nature* **2003**, *426*, 450–454. [CrossRef] [PubMed]
19. Trott, O.; Olson, A.J. AutoDock Vina: Improving the speed and accuracy of docking with a new scoring function, efficient optimization, and multithreading. *J. Comput. Chem.* **2010**, *31*, 455–461. [CrossRef]
20. Wrapp, D.; Wang, N.; Corbett, K.S.; Goldsmith, J.A.; Hsieh, C.L.; Abiona, O.; Graham, B.S.; McLellan, J.S. Cryo-EM structure of the 2019-nCoV spike in the prefusion conformation. *Science* **2020**, *367*, 1260–1263. [CrossRef]
21. Morris, G.M.; Huey, R.; Lindstrom, W.; Sanner, M.F.; Belew, R.K.; Goodsell, D.S.; Olson, A.J. AutoDock4 and AutoDockTools4: Automated docking with selective receptor flexibility. *J. Comput. Chem.* **2009**, *30*, 2785–2791. [CrossRef]
22. Ahmad, S.; Waheed, Y.; Abro, A.; Abbasi, S.W.; Ismail, S. Molecular screening of glycyrrhizin-based inhibitors against ACE2 host receptor of SARS-CoV-2. *J. Mol. Model.* **2021**, *27*, 206. [CrossRef]
23. Ye, C.; Gao, M.; Lin, W.; Yu, K.; Li, P.; Chen, G. Theoretical Study of the anti-NCP Molecular Mechanism of Traditional Chinese Medicine Lianhua-Qingwen Formula (LQF). *ChemRxiv* **2020**. Available online: https://chemrxiv.org/engage/chemrxiv/article-details/60c74908ee74301c74485bc74799cf (accessed on 26 September 2021).
24. Walls, A.C.; Park, Y.J.; Tortorici, M.A.; Wall, A.; McGuire, A.T.; Veesler, D. Structure, Function, and Antigenicity of the SARS-CoV-2 Spike Glycoprotein. *Cell* **2020**, *181*, 281–292. [CrossRef]
25. Xu, C.; Wang, Y.; Liu, C.; Zhang, C.; Han, W.; Hong, X.; Wang, Y.; Hong, Q.; Wang, S.; Zhao, Q.; et al. Conformational dynamics of SARS-CoV-2 trimeric spike glycoprotein in complex with receptor ACE2 revealed by cryo-EM. *Sci. Adv.* **2021**, *7*, eabe5575. [CrossRef]
26. Li, B.; Huang, H.; Wang, X.; Yan, W.; Huo, L. Possible Mechanisms of Glycyrrhizic Acid Preparation Against Coronavirus Disease 2019. *J. Pract. Card. Cereb. Pneumal Vasc. Dis.* **2020**, *28*, 13–18. (In Chinese)
27. Yu, S.; Zhu, Y.; Xu, J.; Yao, G.; Zhang, P.; Wang, M.; Zhao, Y.; Lin, G.; Chen, H.; Chen, L.; et al. Glycyrrhizic acid exerts inhibitory activity against the spike protein of SARS-CoV-2. *Phytomedicine* **2021**, *85*, 153364. [CrossRef] [PubMed]
28. Gowda, P.; Patrick, S.; Joshi, S.D.; Kumawat, R.K.; Sen, E. Glycyrrhizin prevents SARS-CoV-2 S1 and Orf3a induced high mobility group box 1 (HMGB1) release and inhibits viral replication. *Cytokine* **2021**, *142*, 155496. [CrossRef] [PubMed]
29. van de Sand, L.; Bormann, M.; Alt, M.; Schipper, L.; Heilingloh, C.S.; Steinmann, E.; Todt, D.; Dittmer, U.; Elsner, C.; Witzke, O.; et al. Glycyrrhizin Effectively Inhibits SARS-CoV-2 Replication by Inhibiting the Viral Main Protease. *Viruses* **2021**, *13*, 609. [CrossRef] [PubMed]
30. Zheng, W.; Huang, X.; Lai, Y.; Liu, X.; Jiang, Y.; Zhan, S. Glycyrrhizic Acid for COVID-19: Findings of Targeting Pivotal Inflammatory Pathways Triggered by SARS-CoV-2. *Front. Pharmacol.* **2021**, *12*, 631206. [CrossRef] [PubMed]
31. Nie, Y.; Wang, P.; Shi, X.; Wang, G.; Chen, J.; Zheng, A.; Wang, W.; Wang, Z.; Qu, X.; Luo, M.; et al. Highly infectious SARS-CoV pseudotyped virus reveals the cell tropism and its correlation with receptor expression. *Biochem. Biophys. Res. Commun.* **2004**, *321*, 994–1000. [CrossRef]
32. Wang, J.; Deng, F.; Ye, G.; Dong, W.; Zheng, A.; He, Q.; Peng, G. Comparison of lentiviruses pseudotyped with S proteins from coronaviruses and cell tropisms of porcine coronaviruses. *Virol. Sin.* **2016**, *31*, 49–56. [CrossRef] [PubMed]
33. Grehan, K.; Ferrara, F.; Temperton, N. An optimised method for the production of MERS-CoV spike expressing viral pseudotypes. *MethodsX* **2015**, *2*, 379–384. [CrossRef]

34. Ou, X.; Liu, Y.; Lei, X.; Li, P.; Mi, D.; Ren, L.; Guo, L.; Guo, R.; Chen, T.; Hu, J.; et al. Characterization of spike glycoprotein of SARS-CoV-2 on virus entry and its immune cross-reactivity with SARS-CoV. *Nat. Commun.* **2020**, *11*, 1620. [CrossRef] [PubMed]
35. Chen, H.; Du, Q. Potential natural compounds for preventing SARS-CoV-2 (2019-nCoV) infection. *Preprints* **2020**. Available online: https://www.preprints.org/manuscript/202001.200358/v202002 (accessed on 26 September 2021).
36. Sinha, S.K.; Prasad, S.K.; Islam, M.A.; Gurav, S.S.; Patil, R.B.; AlFaris, N.A.; Aldayel, T.S.; AlKehayez, N.M.; Wabaidur, S.M.; Shakya, A. Identification of bioactive compounds from Glycyrrhiza glabra as possible inhibitor of SARS-CoV-2 spike glycoprotein and non-structural protein-15: A pharmacoinformatics study. *J. Biomol. Struct. Dyn.* **2020**, *39*, 4686–4700. [CrossRef]
37. Zhao, Z.; Xiao, Y.; Xu, L.; Liu, Y.; Jiang, G.; Wang, W.; Li, B.; Zhu, T.; Tan, Q.; Tang, L.; et al. Glycyrrhizic Acid Nanoparticles as Antiviral and Anti-inflammatory Agents for COVID-19 Treatment. *ACS Appl. Mater. Interfaces* **2021**, *13*, 20995–21006. [CrossRef]
38. Mollica, L.; De Marchis, F.; Spitaleri, A.; Dallacosta, C.; Pennacchini, D.; Zamai, M.; Agresti, A.; Trisciuoglio, L.; Musco, G.; Bianchi, M.E. Glycyrrhizin binds to high-mobility group box 1 protein and inhibits its cytokine activities. *Chem. Biol.* **2007**, *14*, 431–441. [CrossRef] [PubMed]
39. Kim, S.W.; Jin, Y.; Shin, J.H.; Kim, I.D.; Lee, H.K.; Park, S.; Han, P.L.; Lee, J.K. Glycyrrhizic acid affords robust neuroprotection in the postischemic brain via anti-inflammatory effect by inhibiting HMGB1 phosphorylation and secretion. *Neurobiol. Dis.* **2012**, *46*, 147–156. [CrossRef]
40. Yang, P.S.; Kim, D.H.; Lee, Y.J.; Lee, S.E.; Kang, W.J.; Chang, H.J.; Shin, J.S. Glycyrrhizin, inhibitor of high mobility group box-1, attenuates monocrotaline-induced pulmonary hypertension and vascular remodeling in rats. *Respir. Res.* **2014**, *15*, 148. [CrossRef]
41. Marandici, A.; Monder, C. Inhibition by glycyrrhetinic acid of rat tissue 11 beta-hydroxysteroid dehydrogenase in vivo. *Steroids* **1993**, *58*, 153–156. [CrossRef]
42. Wang, C.Y.; Kao, T.C.; Lo, W.H.; Yen, G.C. Glycyrrhizic acid and 18beta-glycyrrhetinic acid modulate lipopolysaccharide-induced inflammatory response by suppression of NF-kappaB through PI3K p110delta and p110gamma inhibitions. *J. Agric. Food Chem.* **2011**, *59*, 7726–7733. [CrossRef] [PubMed]
43. Kočevar Glavač, N.; Kreft, S. Excretion profile of glycyrrhizin metabolite in human urine. *Food Chem.* **2012**, *131*, 305–308. [CrossRef]
44. Hoever, G.; Baltina, L.; Michaelis, M.; Kondratenko, R.; Baltina, L.; Tolstikov, G.A.; Doerr, H.W.; Cinatl, J., Jr. Antiviral activity of glycyrrhizic acid derivatives against SARS-coronavirus. *J. Med. Chem.* **2005**, *48*, 1256–1259. [CrossRef] [PubMed]
45. Harada, S. The broad anti-viral agent glycyrrhizin directly modulates the fluidity of plasma membrane and HIV-1 envelope. *Biochem. J.* **2005**, *392*, 191–199. [CrossRef]
46. Ralla, T.; Salminen, H.; Braun, K.; Edelmann, M.; Dawid, C.; Hofmann, T.; Weiss, J. Investigations into the Structure-Function Relationship of the Naturally-Derived Surfactant Glycyrrhizin: Emulsion Stability. *Food Biophys.* **2020**, *15*, 288–296. [CrossRef]
47. Selyutina, O.Y.; Apanasenko, I.E.; Kim, A.V.; Shelepova, E.A.; Khalikov, S.S.; Polyakov, N.E. Spectroscopic and molecular dynamics characterization of glycyrrhizin membrane-modifying activity. *Colloids Surf. B Biointerfaces* **2016**, *147*, 459–466. [CrossRef]
48. Predmore, A.; Li, J. Enhanced removal of a human norovirus surrogate from fresh vegetables and fruits by a combination of surfactants and sanitizers. *Appl. Environ. Microbiol.* **2011**, *77*, 4829–4838. [CrossRef]
49. Bailly, C.; Vergoten, G. Glycyrrhizin: An alternative drug for the treatment of COVID-19 infection and the associated respiratory syndrome? *Pharmacol. Ther.* **2020**, *214*, 107618. [CrossRef] [PubMed]
50. Selyutina, O.Y.; Shelepova, E.A.; Paramonova, E.D.; Kichigina, L.A.; Khalikov, S.S.; Polyakov, N.E. Glycyrrhizin-induced changes in phospholipid dynamics studied by (1)H NMR and MD simulation. *Arch. Biochem. Biophys.* **2020**, *686*, 108368. [CrossRef] [PubMed]

MDPI
St. Alban-Anlage 66
4052 Basel
Switzerland
Tel. +41 61 683 77 34
Fax +41 61 302 89 18
www.mdpi.com

*Molecules* Editorial Office
E-mail: molecules@mdpi.com
www.mdpi.com/journal/molecules

www.ingramcontent.com/pod-product-compliance
Lightning Source LLC
LaVergne TN
LVHW070134170726
843515LV00007B/1795

* 9 7 8 3 0 3 6 5 2 7 1 2 3 *